Experimental Methods in the Physical Sciences

Spectrophotometry: Accurate Measurement of Optical Properties of Materials

Experimental Methods in the Physical Sciences

Thomas Lucatorto, Albert C. Parr and Kenneth Baldwin
Editors in Chief

Experimental Methods in the Physical Sciences
Volume 46

Spectrophotometry: Accurate Measurement of Optical Properties of Materials

Edited by

Thomas A. Germer
National Institute of Standards and Technology, USA

Joanne C. Zwinkels
National Research Council, Canada

Benjamin K. Tsai
National Institute of Standards and Technology, USA

ELSEVIER

AMSTERDAM • BOSTON • HEIDELBERG • LONDON
NEW YORK • OXFORD • PARIS • SAN DIEGO
SAN FRANCISCO • SINGAPORE • SYDNEY • TOKYO
Academic Press is an imprint of Elsevier

Academic Press is an imprint of Elsevier
Radarweg 29, PO Box 211, 1000 AE Amsterdam, The Netherlands
225 Wyman Street, Waltham, MA 02451, USA
The Boulevard, Langford Lane, Kidlington, Oxford, OX5 1GB, UK

Library of Congress Cataloging-in-Publication Data
Spectrophotometry : accurate measurement of optical properties of materials / edited by Thomas A. Germer, Joanne C. Zwinkels, Benjamin K. Tsai.
pages cm — (Experimental methods in the physical sciences ; volume 46)
Includes bibliographical references and index.
ISBN 978-0-12-386022-4
1. Spectrophotometry. I. Germer, Thomas A., editor of compilation. II. Zwinkels, Joanne C., 1955- editor of compilation. III. Tsai, Benjamin K., editor of compilation. IV. Series: Experimental methods in the physical sciences ; v. 46.
QD117.S64S64 2014
543′.55–dc23

2014015452

British Library Cataloguing in Publication Data
A catalogue record for this book is available from the British Library

ISBN: 978-0-12-386022-4
ISSN: 1079-4042

For information on all Academic Press publications visit our website at store.elsevier.com

14 15 16 17 10 9 8 7 6 5 4 3 2 1

Contents

Contributors xv
Volumes in Series xvii
Preface xxi

1. Introduction 1

Thomas A. Germer, Joanne C. Zwinkels, and Benjamin K. Tsai

1.1 Opening Remarks 1
1.2 Uncertainties 4
1.3 Overview 8
References 9

2. Theoretical Concepts in Spectrophotometric Measurements 11

Thomas A. Germer, Joanne C. Zwinkels, and Benjamin K. Tsai

2.1 Introduction 12
2.2 Radiometric Quantities 13
2.2.1 Nonspectral Quantities 13
2.2.2 Spectral Quantities 18
2.2.3 Spectrally Weighted Quantities 19
2.3 Relationship Between Radiometric and Electromagnetic Quantities 20
2.3.1 Plane Waves and Irradiance 23
2.3.2 Spherical Waves and Intensity 24
2.3.3 Fourier Expansion and Radiance 25
2.4 The Spectrophotometric Quantities 26
2.4.1 Generalized Scattering Functions 27
2.4.2 Bidirectional Reflectance Distribution Function 28
2.4.3 Reflectance and Transmittance 30
2.4.4 Two Ideal Bidirectional Reflectance Distribution Functions 32
2.4.5 Absorptance 34
2.4.6 Fluorescence: Bispectral Luminescent Radiance Factor 36
2.4.7 Emittance and the Kirchhoff Relationship 38
2.5 Polarization 39
2.6 Reflection and Transmission from Flat Surfaces 43
2.6.1 Snell's Law of Refraction 43
2.6.2 Fresnel Reflection 44

2.6.3 Thin Films 48
2.6.4 Thick Films 49
2.7 Diffuse Scattering 52
2.7.1 Volume Scattering: Theory of Kubelka and Munk 52
2.7.2 Roughness: Facet Scattering Model 54
2.7.3 Roughness: First-Order Vector Perturbation Theory 59
References 65

3. Dispersive Methods 67

Arnold A. Gaertner, Howard W. Yoon, and Thomas A. Germer

3.1 Introduction 67
3.2 General Description 69
3.2.1 Bandpass Function 70
3.2.2 Gratings 74
3.2.3 Prisms 78
3.2.4 Resolving Power 81
3.3 Spectral Analyzer Design 82
3.3.1 Monochromator Design 82
3.3.2 Polychromator Design 84
3.4 Wavelength Calibration 84
3.5 Stray Light 87
3.6 Optical Radiation Sources 88
3.7 Optical Radiation Detectors 90
References 94

4. Fourier Transform Methods 97

Simon G. Kaplan and Manuel A. Quijada

4.1 Introduction: Ideal Michelson Interferometer 98
4.2 Real Fourier Transform Spectrometers 101
4.2.1 Finite Scan Length and Source Size: Spectral Resolution 101
4.2.2 Sampling the Interferogram: Spectral Bandwidth 106
4.2.3 Phase Error and Phase Correction 107
4.2.4 FTS Versus Dispersive Instruments 110
4.2.5 FTS Design Considerations 114
4.2.6 Instrument Design Examples 121
4.3 Sources of Uncertainty and Their Reduction 123
4.3.1 Noise 124
4.3.2 Interferometer Alignment Drift 124
4.3.3 Phase Error 125
4.3.4 Detector Nonlinearity 126
4.3.5 Interreflections 129

4.3.6 Nonsource Emission 130
4.3.7 Beam Geometry Errors 132
4.3.8 Polarization Effects 133
4.3.9 Stray Light 134
4.3.10 Atmospheric Absorption 134
4.3.11 Wavenumber Uncertainty 135
4.4 Measurement Applications 136
4.4.1 Measurement of Transmittance and Reflectance 136
4.4.2 Measurement of Refractive Index 137
4.4.3 Assessment of the Accuracy of FTS Measurements 137
4.5 Recommendations for Accurate FTS Measurements 138
4.5.1 Matching Instrument to Measurement Requirements 138
4.5.2 Instrument Software and Data Handling 139
4.5.3 Maintenance of Measurement Traceability 139
References 140

5. Regular Reflectance and Transmittance 143

Peter A. van Nijnatten

5.1 Introduction 144
5.2 Relevant Background Information 144
5.2.1 Measurement Geometry 144
5.2.2 Measurement Equation 145
5.2.3 Evaluation of Measurement Uncertainty 146
5.2.4 The Role of Integrating Spheres 147
5.2.5 Avoiding Spectral Artifacts 148
5.3 Measurements Near-Normal Incidence 150
5.3.1 Typical Configurations for Near-Normal Reflection Measurements 151
5.3.2 Measuring the Transmittance and Reflectance of Coated Substrates 153
5.3.3 Measuring Very Low Reflectance Materials 153
5.3.4 Measuring Specular Reflectance with an FTIR 154
5.3.5 Tools for Absolute Reflectance and Transmittance Measurements 156
5.4 Measurements at Oblique Incidence 158
5.4.1 Relevant Issues 159
5.4.2 Directional Transmittance Measurements 163
5.4.3 Relative Directional Reflection Measurements 166
5.4.4 Absolute Directional Reflection Measurements 169
5.5 Measuring the Reflectance of Highly Reflecting Materials 171
5.5.1 VW Method 172
5.5.2 IV Method 175
References 177

6. Diffuse Reflectance and Transmittance 179

Andreas Höpe

6.1 Introduction 180
6.2 Measurands 182
6.2.1 Reflectance 182
6.2.2 Reflectance Factor 182
6.2.3 Radiance Factor 183
6.2.4 Transmittance 183
6.2.5 Relationship Between Reflectance, Transmittance, and Absorptance 184
6.3 Notation of Diffuse Reflection Geometries 184
6.3.1 Recommended Integrating Sphere Geometries 186
6.4 Integrating Spheres 187
6.4.1 Historic Information 189
6.4.2 Setup 190
6.4.3 Construction Principles 190
6.4.4 Internal Coating Materials 191
6.4.5 Homogeneity 192
6.5 Absolute Sphere Methods for Measuring Diffuse Reflection 193
6.5.1 Basic Considerations 193
6.5.2 Method of Taylor (First Taylor) and Benford 195
6.5.3 Method of Taylor (Third Taylor) 197
6.5.4 Method of Sharp-Little 198
6.5.5 Method of van den Akker 199
6.5.6 Method of Korte–Schmidt 200
6.5.7 Coblentz Sphere 202
6.5.8 Goniometric Method 203
6.6 Diffuse Reflection Standards 203
6.6.1 Requirements 204
6.6.2 Common Diffuse Reflection Standards 205
6.6.3 Fluorescence of Diffuse Reflection Standards 210
6.6.4 Handling Recommendations 211
6.7 Relative Sphere Methods for Measuring Diffuse Reflection 212
6.8 Diffuse Transmittance Measurements 214
References 216

7. Spectral Fluorescence Measurements 221

Joanne C. Zwinkels, Paul C. DeRose, and James E. Leland

7.1 Introduction 222
7.1.1 Fluorescence Measurement Applications 223
7.2 Fundamental Concepts and Terminology 224
7.2.1 Quantities Common to Analytical and Colorimetric Applications 225
7.2.2 Specific to Analytical Applications 230

7.2.3 Specific to Colorimetric Applications 232
7.3 Measurement of Fluorescence Characteristics 238
7.3.1 Bispectral Measurements 238
7.3.2 Abridged Measurements 239
7.4 General Instrument Design and Measurement Considerations 240
7.4.1 Instrument Components 240
7.4.2 Spectral Measurement Methods 241
7.4.3 General Considerations and Good Laboratory Practices 243
7.5 Specialized Instrument Designs and Measurement Methods 244
7.5.1 Analytical Applications 244
7.5.2 Colorimetric Applications 247
7.5.3 Quantum Yield Measurements 256
7.6 Instrument Characteristics That Impact Spectral Fluorescence Measurements 260
7.6.1 Spectral Characteristics 260
7.6.2 Detector Nonlinearity 264
7.6.3 Polarization Effects 266
7.6.4 Stray Light 269
7.6.5 Instrument Lineshape 270
7.6.6 Sampling Effects 271
7.7 Sample Characteristics That Impact Fluorescence Measurements 274
7.7.1 Optical Density 274
7.7.2 Solvent/Matrix Properties 275
7.7.3 Refractive Index 275
7.7.4 Impurities 276
7.7.5 Luminescence Lifetime 277
7.7.6 Scattering 277
7.8 Standards for Spectral Fluorescence Measurements 278
7.8.1 Correcting Relative Excitation and Emission Spectra 278
7.8.2 Intensity Verification Standards 283
7.8.3 Fluorescent Color Standards 284
7.8.4 Quantum Yield Standards 286
References 287

8. Angle-Resolved Diffuse Reflectance and Transmittance 291

Thomas A. Germer, John C. Stover, and Sven Schröder

8.1 Introduction 292
8.2 Reference-Free Measurement Methods 294
8.2.1 The Over-Illumination Method 295
8.2.2 The Under-Illumination Method 297
8.3 Instrument Characterization 301
8.3.1 The Instrument Signature 301
8.3.2 Noise-Equivalent BRDF 303

8.3.3 Profile of Illumination Spot 303
8.3.4 The Field of View 304
8.3.5 Detector Location Sensitivity 304
8.3.6 Linearity 304
8.4 Goniometer Designs 304
8.4.1 In-Plane Scatterometer Designs 304
8.4.2 Out-of-Plane Scatterometer Designs 305
8.4.3 Multichannel Systems 309
8.5 Uncertainty Analysis 311
8.6 Normalization Schemes 315
8.6.1 Relative Normalization 315
8.6.2 Specular Normalization 316
8.7 Special Conditions or Considerations 316
8.7.1 Ultraviolet 316
8.7.2 Infrared 316
8.7.3 Polarization 316
8.7.4 Transmission Measurements 317
8.8 Applications 317
8.8.1 Pressed PTFE 318
8.8.2 Integration of BRDF 318
8.8.3 Diffuse Black Paint 321
8.8.4 Characterization of Surface Roughness 322
8.8.5 Scattering Characterization of Anisotropic Surface Structures 325
8.8.6 Predicting Out-of-Plane BRDF from In-Plane BRDF 326
References 328

9. Spectral Emissivity Measurements 333

Hiromichi Watanabe, Juntaro Ishii, Hidenobu Wakabayashi, Tomoyuki Kumano, and Leonard Hanssen

9.1 Introduction 334
9.2 Measurement Methods 336
9.3 Spectral Emissivity Measurements Near-Ambient Temperature 338
9.3.1 Measurement Apparatus Based on the Blackbody Comparison Method 338
9.3.2 Calibration of the System Response 340
9.3.3 Correction for Reflected Radiance of the Surroundings 342
9.3.4 Traceability and Uncertainty of Spectral Emissivity Measurements 342
9.4 Spectral Emissivity and Reflectance Measurements of Oxidized Metals 345
9.4.1 A Broad-Spectral-Range High-Speed Spectrophotometer System 345

9.5 Spectral Emissivity Measurements of Molten Metals at High Temperatures 349
9.5.1 Importance of Molten Metal Emissivity Data 349
9.5.2 Spectral Emissivity Measurements Based on Containerless Techniques 351
9.6 Spectral Emissivity Measurements of Ceramics 357
References 362

10. Color and Appearance 367

Maria E. Nadal, Dave Wyble, and Clarence J. Zarobila

10.1 Introduction 367
10.2 Spectral Attributes—Color 368
10.2.1 Definition 368
10.2.2 CIE System 370
10.2.3 Metamerism 376
10.2.4 Chromaticity Coordinates and Color Spaces 377
10.2.5 Color Difference Metrics 380
10.3 Color-Measuring Instruments 383
10.3.1 Geometric Considerations 385
10.3.2 Source Considerations 389
10.3.3 Detection and Signal Processing Considerations 390
10.3.4 Uncertainties 392
10.4 Gonioapparent Materials 394
10.5 Geometrical Attributes 396
10.5.1 Specular Gloss 397
10.5.2 Haze 400
10.5.3 Distinctness of Image 403
References 404

11. The Use of Spectrophotometry in the Pharmaceutical Industry 409

John P. Hammond

11.1 Introduction 410
11.2 Introduction to the Pharmaceutical Industry 410
11.2.1 Process Analytical Technology 411
11.2.2 Quality by Design 411
11.2.3 Product Development Cycle 412
11.2.4 Discovery Research 412
11.3 Quality System for the Analytical Laboratory 414
11.3.1 Consideration for Quality Systems in Data Quality Assurance/Control 414
11.4 UV and Visible Spectrophotometry 419
11.4.1 Qualification of UV and Visible Spectrophotometers 421
11.4.2 Validation and Verification 430
11.4.3 Pharmaceutical Applications (UV–vis) 430

11.5 **NIR Spectrometry** 433
11.5.1 Factors That Affect NIR Spectra 434
11.5.2 Instrumentation 435
11.5.3 Qualification of NIR Instruments 435
11.5.4 Method Validation 438
11.5.5 Pharmaceutical Applications (NIR) 439
11.6 **Mid-IR Spectrometry** 440
11.6.1 Qualification of IR Spectrophotometers 441
11.6.2 Validation and Verification 445
11.6.3 Pharmaceutical Applications (Mid-IR) 445
11.7 **Fluorescence Spectrometry** 446
11.7.1 Qualification of Fluorescence Instruments 447
11.7.2 Validation and Verification 453
11.7.3 Pharmaceutical Applications (Fluorescence) 454
11.8 **Where Next?** 454
References 455

12. Spectrophotometry Applications: Remote Sensing 457

Carol J. Bruegge, Roger Davies, Florian M. Schwandner, and Felix C. Seidel

12.1 **Introduction** 458
12.2 **Measurement of Atmospheric Carbon Dioxide** 458
12.3 **The Remote Sensing of Clouds in the Earth's Atmosphere** 464
12.3.1 Why Clouds? 465
12.3.2 The Spectrophotometry of Clouds: Forward Problem 466
12.3.3 The Spectrophotometry of Clouds: Inverse Problems 468
12.3.4 The Nakajima and King Approach for Plane-Parallel Clouds 468
12.3.5 The Optical Depth of Thin Clouds in General 469
12.3.6 The Optical Depth of Thick Clouds 469
12.3.7 Cloud Altitude 469
12.3.8 Cloud Motion Vectors 470
12.4 **The Retrieval of Snow Properties** 470
12.4.1 Snow-Covered Area 471
12.4.2 Snow-Grain Size 471
12.4.3 Snow Albedo 472
12.4.4 Radiative Forcing of Light-Absorbing Impurities in Snow 473
12.4.5 Benefits of Imaging Spectroscopy in Combination with Other Observations for Snow Remote Sensing 474
12.5 **Volcanic Unrest** 475
12.5.1 Early Remote Sensing Methods 475
12.5.2 Applications and Relevance 475
12.5.3 Imaging of Volcanic Landforms, Eruption Detection, and Degassing 477

12.6 Calibration 479
12.7 Summary 481
Acknowledgments 482
References 483

13. Microspectrophotometry 489

Paul C. Martin and Michael B. Eyring

13.1 Introduction 489
13.2 Microspectrophotometer Instrument Design and Construction 490
13.2.1 The Microscope 491
13.2.2 The Spectrophotometer 493
13.2.3 Computer Interface and Software Control 494
13.2.4 MSP System Capabilities 495
13.3 Using the MSP System 496
13.4 Current Applications of MSP in Industry and Research 498
13.4.1 Forensic Applications of MSP 498
13.4.2 Microspot Thin-Film Thickness 502
13.4.3 Graphene and Carbon Nanotubes 504
13.4.4 Heavy Element Chemistry 505
13.4.5 Lignins 506
13.4.6 Cellulosic Nanomaterials 507
13.4.7 The MSP Analysis of Gemstones and Glass Fragments 508
13.4.8 Identifying Protein Crystals 512
13.4.9 Gold and Silver Nanomaterials 514
13.5 Conclusion 515
References 516

Index 519

Contributors

Numbers in Parentheses indicate the pages on which the author's contributions begin.

Carol J. Bruegge (457), Jet Propulsion Laboratory, California Institute of Technology, Pasadena, California, USA

Roger Davies (457), Department of Physics, University of Auckland, Auckland, New Zealand

Paul C. DeRose (221), National Institute of Standards and Technology (NIST), NIST, Material Measurement Laboratory, Gaithersburg, Maryland, USA

Michael B. Eyring (489), Micro Forensics, Ltd., and Arizona Department of Public Safety Crime Laboratory, Phoenix, Arizona, USA

Arnold A. Gaertner (67), National Research Council Canada, Ottawa, Ontario, Canada

Thomas A. Germer (1, 11, 67, 291), National Institute of Standards and Technology, Gaithersburg, Maryland, USA

John P. Hammond (409), Starna Scientific Limited, Hainault, Essex, United Kingdom

Leonard Hanssen (333), National Institute of Standards and Technology, Gaithersburg, Maryland, USA

Andreas Höpe (179), Physikalisch-Technische Bundesanstalt, Braunschweig, Germany

Juntaro Ishii (333), National Metrology Institute of Japan, AIST, Tsukuba, Japan

Simon G. Kaplan (97), National Institute of Standards and Technology, Gaithersburg, Maryland, USA

Tomoyuki Kumano (333), Kobe City College of Technology, Kobe, Japan

James E. Leland (221), Copia LLC, Goshen, New Hampshire, USA

Paul C. Martin (489), CRAIC Technologies, San Dimas, California, USA

Maria E. Nadal (367), National Institute of Standards and Technology, Gaithersburg, Maryland, USA

Manuel A. Quijada (97), NASA Goddard Space Flight Center, Code 551, Greenbelt, Maryland, USA

Sven Schröder (291), Fraunhofer Institute for Applied Optics and Precision Engineering, Jena, Germany

Florian M. Schwandner (457), Jet Propulsion Laboratory, California Institute of Technology, Pasadena, California, USA

Felix C. Seidel (457), Jet Propulsion Laboratory, California Institute of Technology, Pasadena, California, USA

John C. Stover (291), The Scatter Works, Inc., Tucson, Arizona, USA

Benjamin K. Tsai (1, 11), National Institute of Standards and Technology, Gaithersburg, Maryland, USA

Peter A. van Nijnatten (143), OMT Solutions BV, Eindhoven, Netherlands

Hidenobu Wakabayashi (333), Kyoto University, Kyoto, Japan

Hiromichi Watanabe (333), National Metrology Institute of Japan, AIST, Tsukuba, Japan

Dave Wyble (367), Avian Rochester, LLC, Webster, New York, USA

Howard W. Yoon (67), National Institute of Standards and Technology, Gaithersburg, Maryland, USA

Clarence J. Zarobila (367), National Institute of Standards and Technology, Gaithersburg, Maryland, USA

Joanne C. Zwinkels (1, 11, 221), National Research Council Canada, NRC, Measurement Science and Standards, Ottawa, Ontario, Canada

Volumes in Series

Experimental Methods in the Physical Sciences

(Formerly Methods of Experimental Physics)

Volume 1. Classical Methods
Edited by Immanuel Estermann

Volume 2. Electronic Methods, Second Edition (in two parts)
Edited by E. Bleuler and R. O. Haxby

Volume 3. Molecular Physics, Second Edition (in two parts)
Edited by Dudley Williams

Volume 4. Atomic and Electron Physics - Part A: Atomic Sources and Detectors; Part B: Free Atoms
Edited by Vernon W. Hughes and Howard L. Schultz

Volume 5. Nuclear Physics (in two parts)
Edited by Luke C. L. Yuan and Chien-Shiung Wu

Volume 6. Solid State Physics - Part A: Preparation, Structure, Mechanical and Thermal Properties; Part B: Electrical, Magnetic and Optical Properties
Edited by K. Lark-Horovitz and Vivian A. Johnson

Volume 7. Atomic and Electron Physics - Atomic Interactions (in two parts)
Edited by Benjamin Bederson and Wade L. Fite

Volume 8. Problems and Solutions for Students
Edited by L. Marton and W. F. Hornyak

Volume 9. Plasma Physics (in two parts)
Edited by Hans R. Griem and Ralph H. Lovberg

Volume 10. Physical Principles of Far-Infrared Radiation
Edited by L. C. Robinson

Volume 11. Solid State Physics
Edited by R. V. Coleman

Volume 12. Astrophysics - Part A: Optical and Infrared Astronomy
Edited by N. Carleton
Part B: Radio Telescopes; Part C: Radio Observations
Edited by M. L. Meeks

Volume 13. Spectroscopy (in two parts)
Edited by Dudley Williams

Volume 14. Vacuum Physics and Technology
Edited by G. L. Weissler and R. W. Carlson

Volume 15. Quantum Electronics (in two parts)
Edited by C. L. Tang

Volume 16. Polymers - Part A: Molecular Structure and Dynamics; Part B: Crystal Structure and Morphology; Part C: Physical Properties
Edited by R. A. Fava

Volume 17. Accelerators in Atomic Physics
Edited by P. Richard

Volume 18. Fluid Dynamics (in two parts)
Edited by R. J. Emrich

Volume 19. Ultrasonics
Edited by Peter D. Edmonds

Volume 20. Biophysics
Edited by Gerald Ehrenstein and Harold Lecar

Volume 21. Solid State Physics: Nuclear Methods
Edited by J. N. Mundy, S. J. Rothman, M. J. Fluss, and L. C. Smedskjaer

Volume 22. Solid State Physics: Surfaces
Edited by Robert L. Park and Max G. Lagally

Volume 23. Neutron Scattering (in three parts)
Edited by K. Skold and D. L. Price

Volume 24. Geophysics - Part A: Laboratory Measurements;
Part B: Field Measurements
Edited by C. G. Sammis and T. L. Henyey

Volume 25. Geometrical and Instrumental Optics
Edited by Daniel Malacara

Volume 26. Physical Optics and Light Measurements
Edited by Daniel Malacara

Volume 27. Scanning Tunneling Microscopy
Edited by Joseph Stroscio and William Kaiser

Volume 28. Statistical Methods for Physical Science
Edited by John L. Stanford and Stephen B. Vardaman

Volume 29. Atomic, Molecular, and Optical Physics - Part A: Charged Particles; Part B: Atoms and Molecules; Part C: Electromagnetic Radiation
Edited by F. B. Dunning and Randall G. Hulet

Volume 30. Laser Ablation and Desorption
Edited by John C. Miller and Richard F. Haglund, Jr.

Volume 31. Vacuum Ultraviolet Spectroscopy I
Edited by J. A. R. Samson and D. L. Ederer

Volume 32. Vacuum Ultraviolet Spectroscopy II
Edited by J. A. R. Samson and D. L. Ederer

Volume 33. Cumulative Author Index and Tables of Contents, Volumes 1-32

Volume 34. Cumulative Subject Index

Volume 35. Methods in the Physics of Porous Media
Edited by Po-zen Wong

Volume 36. Magnetic Imaging and its Applications to Materials
Edited by Marc De Graef and Yimei Zhu

Volume 37. Characterization of Amorphous and Crystalline Rough Surface: Principles and Applications
Edited by Yi Ping Zhao, Gwo-Ching Wang, and Toh-Ming Lu

Volume 38. Advances in Surface Science
Edited by Hari Singh Nalwa

Volume 39. Modern Acoustical Techniques for the Measurement of Mechanical Properties
Edited by Moises Levy, Henry E. Bass, and Richard Stern

Volume 40. Cavity-Enhanced Spectroscopies
Edited by Roger D. van Zee and J. Patrick Looney

Volume 41. Optical Radiometry
Edited by A. C. Parr, R. U. Datla, and J. L. Gardner

Volume 42. Radiometric Temperature Measurements. I. Fundamentals
Edited by Z. M. Zhang, B. K. Tsai, and G. Machin

Volume 43. Radiometric Temperature Measurements. II. Applications
Edited by Z. M. Zhang, B. K. Tsai, and G. Machin

Volume 44. Neutron Scattering – Fundamentals
Edited by Felix Fernandez-Alonso, and David L. Price

Volume 45. Single-Photon Generation and Detection
Edited by Alan Migdall, Sergey Polyakov, Jingyun Fan, and Joshua Bienfang

Volume 46. Spectrophotometry: Accurate Measurement of Optical Properties of Materials
Edited by Thomas A. Germer, Joanne C. Zwinkels, and Benjamin K. Tsai

Preface

Spectrophotometry is the quantitative measurement of the optical properties of materials over a wide wavelength range encompassing the ultraviolet, visible, and infrared spectral regions. These spectral measurements include reflectance, transmittance, absorptance/emittance, scattering, and fluorescence, and their accurate measurement has an impact on a wide field of science and technology. The design and performance of optical instruments, ranging from low cost cell-phone cameras to high cost microlithography projection tools and satellite telescopes, requires knowledge of the optical properties of the components, such as their refractive index, roughness, subsurface scatter, and contamination. The pharmaceutical and chemical industries use optical absorption and fluorescence measurements to quantify concentration which is required for accurate dosing and elimination of contaminants. Global climate change simulations require accurate knowledge of the optical properties of materials, gases, and aerosols to calculate the net energy balance of our planet. The properties of thin films, even when they are not intended for optical applications, are often related to their optical reflection, transmission, and scattering properties. Commercial products are often selected by consumers based upon their appearance, a complex attribute that encompasses more specific attributes, such as color, gloss, and texture. Renewed interest in solar energy has driven the need to maximize the light capture efficiency of solar collectors.

This book is intended to be a hands-on text for those seeking to perform precise and accurate spectrophotometry of the optical properties of materials. Based on teaching experiences at our respective institutes, it is our aim to present material that helps the practitioner to set up and optimize the spectrophotometer to perform these various measurements, validate the instrument performance, and be aware of the various sources of errors that can impact the results. A number of our students have been from institutes interested in developing an independent capability for realizing spectrophotometric scales, and it is our intention to also provide the researcher the necessary framework for designing and characterizing reference instruments for traceable measurements of these optical quantities.

Chapter 1 introduces the topic of spectrophotometry and a short history of its development and present-day challenges. It includes a section outlining the standard methods and terminology used for evaluating and expressing uncertainty in any measurement. This will provide the reader the background needed to follow the examples of uncertainty analysis that are given for specific spectrophotometric measurements in some of the following chapters.

Chapter 2 describes some theoretical concepts underlying spectrophotometry and the optical properties of materials. It begins by defining the different radiometric quantities of radiance, irradiance, and intensity. These quantities are then related to the corresponding electromagnetic field quantities. This theoretical foundation allows the definition of the spectrophotometric quantities: reflectance, transmittance, emittance, diffuse reflectance, diffuse transmittance, and the bidirectional reflectance distribution function (BRDF). Relationships between these quantities are derived and explained. A review of the Fresnel relations is given, including expressions appropriate for thin films and thick films. The Kirchhoff relationship between the reflectance and emittance is derived. While the topic of BRDF modeling is beyond the scope of this text, Chapter 2 describes the theory of Kubelka and Munk, which is applicable to highly diffuse materials and used widely in color formulation and for determining molecular optical properties (absorption and scattering coefficients). Chapter 2 also describes two theories for scattering from rough surfaces, one appropriate for very rough surfaces and one appropriate for very smooth surfaces.

Chapters 3 and 4 discuss the means of obtaining wavelength separation and spectral resolution in spectrophotometry. Chapter 3 provides an overview of the use of grating and prism spectrometers in spectrophotometry, while Chapter 4 discusses the use of Fourier transform interferometric methods. Each of these methods has its advantages and disadvantages. Included in these chapters are discussions of how measurement noise is propagated, how nonlinearities affect results, how stray light or interreflections produce artifacts, and what determines the spectral resolution or bandpass of the measured spectrophotometric quantities. Methods to identify and alleviate these problems are described.

Chapter 5 covers the topic of specular reflectance of nonscattering samples such as mirrors and other samples with mirror-like reflection, and of regular transmittance of transparent samples such as glass filters. The instrumentation and procedures for measuring these properties via both absolute and relative methods are discussed and practical methods are given for improving the measurement accuracy. Sources of error are treated and representative uncertainty budgets are given for various examples. As a special topic, methods for accurate reflectance measurement at oblique incidence of very high reflectance materials like reference mirrors and laser mirrors are discussed. Examples are given for both Fourier transform infrared (FTIR) and ultraviolet, visible, and near-infrared (UV/Vis/NIR) spectrophotometry.

Chapter 6 describes the basic principles of diffuse reflectance and transmission measurements. Unlike the measurement of specular reflection, where the angle of incidence equals the angle of reflection, in the case of diffuse reflection, the incoming radiation is spread over the half-space above the surface, with a certain distribution specific to the surface or material under test. The integrating sphere with its ability to collect all or the diffuse-only component of the radiation reflected and transmitted by the sphere wall becomes important as a sampling device which can be configured to precisely measure these various diffuse quantities. This chapter presents the theory of integrating spheres and discusses both absolute and relative methods.

Chapter 7 covers the topic of spectral fluorescence measurements. These fluorescence measurements offer significant advantages in terms of sensitivity and selectivity, finding wide use in a range of applications in analytical and color technologies. The accurate measurement of fluorescent optical quantities, such as spectral excitation and emission curves and quantum yields, has become increasingly important because of the increasing use of fluorescent materials in manufacturing for enhancing appearance, for example, whiteness, brightness, colorfulness, conspicuity, and for bioanalytical applications, for example, medical diagnostics. Conventional spectrophotometric instrumentation and procedures may not give meaningful results because the measured spectral fluorescence data will depend not only on the intrinsic optical properties of the fluorescent sample but are strongly influenced by the instrument characteristics and its interaction with the sample and its environment. The extent of this distortion depends on the details of the instrument design, including its spectral, geometric, polarization, and temporal characteristics, and on the characteristics of the sample itself. This chapter discusses these basic principles, specialized terminology and instrumentation, and experimental calibration and measurement procedures that are used for reliable and accurate measurements of fluorescent materials. This chapter includes a description of both one-monochromator and two-monochromator methods.

The topic of BRDF measurements is covered in Chapter 8. These measurements, like the diffuse reflectance and transmittance measurements described in Chapter 6, cover the gamut from highly reflecting, diffuse materials to highly specular, low scatter, materials. If performed incorrectly, BRDF measurements can often be plagued with artifacts that make them notoriously inaccurate. For example, detector linearity, stray light, and poor sample treatment will contribute to errors. Chapter 8 discusses the various methods for performing these measurements, including both laser-based and nonlaser-based. A comparison is also made between methods that use a sphere source and measure radiance and those that use a source that under-fills the sample and measure power scattered per solid angle. A number of practical measurement applications and examples are given.

Chapter 9 covers the measurement of emittance. This is perhaps the least intuitive of all the spectrophotometric quantities. Anyone who has sensed the surprisingly high temperature of a shiny plated object in the hot summer sun may wonder how such a reflective surface gets so hot. The very low thermal emittance these objects exhibit is not detectable by the human eye, but its effects are detected with high sensitivity by infrared spectral reflectometers, which will be described in detail in this chapter, along with related techniques. The factors that can complicate the proper evaluation of emittance, notably scattering effects discussed in the previous chapter, are also presented here.

Chapter 10 covers the measurement of color and appearance. These measurements essentially constitute an example of the spectrophotometric measurements described in previous chapters, where the spectral measurements are limited to the wavelengths in the visible range which produce a sensation of color in the human eye. Since this color stimulus also depends upon the spectral properties of the illuminating source and the spectral sensitivity of the human visual system, unambiguous color specification requires that these influencing parameters be standardized. This standard system of color measurement or CIE colorimetry will be described. Oddly, while most measurements with physical detectors far outperform human senses, the human eye is amazingly sensitive to color differences and, coupled with color's commercial impact, makes this class of spectrophotometric measurement extremely important. Many modern commercial products are emblazoned with materials that change their appearance depending upon illumination and viewing conditions, making measurements of these materials particularly challenging. This chapter presents an overview of the approaches that are taken to obtain accurate color measurements. Other aspects of appearance, quantified by gloss, distinctness of image, and orange peel, are also discussed and their measurements described.

Chapters 11–13 cover three specific applications where spectrophotometry has, and will continue to have, significant impact. Chapter 11 covers the use of spectrophotometry in the pharmaceutical industry. Much of the driving force for accurate optical property measurements stems from the need to control the purity and dose of the drugs. Modern products are taking advantage of new delivery methods so that many of the products are no longer simple solutions or powders, but complex colloids. In many cases, the particle size and distribution is critical to the functionality of the drugs, and a high degree of quality control is necessary. Regulatory demands also place requirements on the purity and stability of the pharmaceutical products and spectrophotometric methods play an important role in ensuring compliance.

Chapter 12 covers the application of remote sensing, which is gaining significantly more importance as we learn about the impact that we have on our environment and seek to monitor and control that impact. This field spans applications from satellite imaging (including the calibration of those systems) to ground-based open-path monitoring. This chapter discusses three applications: the measurement of greenhouse gases, cloud radiative effects, and volcano monitoring. It concludes with a brief discussion on the calibration of a field spectrometer.

The final chapter, Chapter 13, deals with microspectrophotometry and its applications. Measurements of the diffuse and spectral reflectance, transmittance, and photoluminescence spectra of very small samples or samples with spatial resolution in the micrometer range are required in a variety of applications, ranging from process control of thin film thickness, pattern fidelity in the semiconductor industry, quality control in the ink, toner, and pigments industries, forensic analysis of transferred or trace materials in patent, tort, civil, or criminal cases, to more novel evaluations of protein crystals, silver and gold nanodispersions, graphene and carbon nanotubes to cellulosic nanomaterials, and lignin chemistry.

This spectrophotometry book project has been an incredible collaborative effort involving authors from around the world with both research and practical experience who are all experts in their field. We would like to thank all of these authors for their participation and excellent contributions to this text. We also would like to acknowledge the assistance and editorial support of Michael Jacobson of Optical Data Associates in the early stages of this book project and thank our families, friends, and colleagues for their encouragement and enduring patience.

Thomas A. Germer
National Institute of Standards and Technology, USA

Joanne C. Zwinkels
National Research Council, Canada

Benjamin K. Tsai
National Institute of Standards and Technology, USA
March 2014

Chapter 1

Introduction

Thomas A. Germer*, Joanne C. Zwinkels† and Benjamin K. Tsai*
**National Institute of Standards and Technology, Gaithersburg, Maryland, USA*
†National Research Council Canada, NRC, Measurement Science and Standards, Ottawa, Ontario, Canada

Chapter Outline

1.1 Opening Remarks	1	1.3 Overview	8
1.2 Uncertainties	4	References	9

Exper. *6. And as these Colours were not changeable by Refractions, so neither were they by Reflexions. For all white, grey, red, yellow, green, blue, violet Bodies, as Paper, Ashes, red Lead, Orpiment, Indico Bise, Gold, Silver, Copper, Grass, blue Flowers, Violets, Bubbles of Water tinged with various Colours, Peacock's Feathers, the Tincture of* Lignum Nephriticum, *and such-like, in red homogeneal Light appeared totally red, in blue Light totally blue, in green Light totally green, and so of other Colours. In the homogeneal Light of any Colour they all appeared totally of that same Colour, with this only Difference, that some of them reflected that Light more strongly, others more faintly. I never yet found any Body, which by reflecting homogeneal Light could sensibly change its Colour.*

Sir Isaac Newton [1]

1.1 OPENING REMARKS

Spectrophotometry is the quantitative measurement of the interaction of ultraviolet (UV), visible, and infrared (IR) radiation with a material and has an impact on a wide field of science and technology. The nature of this interaction depends upon the physical properties of the material, for example, transparent or opaque, smooth or rough, pure or contaminated, and thin or thick. Thus, spectrophotometric measurements can be used to quantify, in turn, these important physical properties of the material. The choices of spectrophotometric measurements include spectral reflectance, transmittance, absorptance, emittance, scattering, and fluorescence and can be classified as phenomenological optical properties of the material. Spectrophotometric measurements can also

Experimental Methods in the Physical Sciences, Vol. 46. http://dx.doi.org/10.1016/B978-0-12-386022-4.00001-7

be used to probe the intrinsic or internal physical nature of the material, such as its refractive index and extinction coefficient.

The design and performance of optical instruments, ranging from low-cost cell-phone cameras to high-cost microlithography projection tools and satellite telescopes, require knowledge of the optical properties of the components, such as their refractive index, roughness, subsurface scatter, and contamination. The pharmaceutical and chemical industries use optical absorption and fluorescence measurements to quantify concentration, required for accurate dosing and elimination of contaminants. Global climate change simulations require accurate knowledge of the optical properties of materials, gases, and aerosols to calculate the net energy balance of our planet. The properties of thin films, even when they are not intended for optical applications, are often related to their optical reflection, transmission, and scattering properties. Commercial products are often selected by consumers based upon appearance, a complex attribute that encompasses more specific terms, such as color, gloss, and texture. Renewed interest in solar energy has driven the need to maximize the light capture efficiency of solar collectors.

When we are asked to inspect a piece of material, it is our natural inclination to view it by holding it up to a light. The interaction of the light with the material gives us an overall impression of its quality. Our vision is also inherently multispectral, by providing color discrimination on a relatively high spatial resolution. Binocular vision, by allowing us to view the object from multiple directions simultaneously, gives us an ability to perform rudimentary tomography. The spectral, spatial, and directional properties permit us to identify materials, characterize topography, and observe defects, without ever coming into contact with the object. It is not surprising, then, that we seek to make measurements of optical properties of materials in order to better quantify what our own eyes sense qualitatively.

While certain aspects of optics, such as the laws governing refraction of light and the ray nature of light, were well established by the mid-1600s, it was Isaac Newton who discovered that white light was a mixture of colors that could be separated into its components using a prism. It could be argued that Newton performed the first spectrophotometric measurements of this light interaction with a prismatic material. This chapter's epigraph [1] gives an account of his discovery that, in the absence of fluorescence, rays of one color cannot be changed into rays of another, but that different materials simply reflect the colors in different amounts. Newton noted that the color purple was not in the rainbow, but could be created by mixing violet and red rays. He then proposed the basic structure of the color circle and noted that mixtures of any two opposing colors yield a neutral gray.

Newton, of course, used his eyes as the detector. While he could be quite quantitative in measuring angles of refraction, he had more difficulty in estimating intensity or quantifying color. Furthermore, because of his reluctance to accept the wave nature of light, he would never correlate the colors that he observed after dispersion through a prism with the corresponding wavelengths of light. Through the years, however, acceptance of the wave properties took hold, first through the double slit experiment of Young [2] and then through the progressive works of Augustin Fresnel, Michael Faraday, James Clerk Maxwell, and others. By the 1800s, the world was ready for precision measurements of wavelength, and the birth of quantitative spectroscopy occurred.

In the early 1920s, it was recognized that it was important that the results from spectrophotometric measurements not only provide qualitative information but are also reliable and meaningful quantitatively. The Optical Society of America convened a progress committee on spectrophotometry, which issued a report in 1925 [3]. This report gives an amazing account of the status of spectrophotometry at that time, and except for the obvious lack of automation, many of the issues that were covered in this report are still relevant today to obtaining meaningful spectrophotometric results. These include establishing a common terminology, spectral calibration, stray light exclusion, polarization, differentiation between diffuse and specular components, and precision and accuracy.

The first automated, recording spectrophotometer was developed between 1926 and 1928 by Hardy and his colleagues at the Massachusetts Institute of Technology [4]. Before this time, spectrophotometers were extremely tedious to use. In a retrospective written a decade later [5], Hardy pointed out that the first months of operation of this instrument were very exciting, that they measured everything within sight, and that it took less time to make a measurement than it took to decide whether the measurement would be significant. Their results brought hundreds of visitors to their laboratory, and they soon realized that virtually every industry was in need of such measurements. Soon, Hardy made arrangements with the General Electric Company [6] to commercialize the instrument. In the intervening years, there have been significant advances in the design of spectrophotometers including the emergence of faster multi-wavelength designs in the 1970s and the introduction of a commercially available diode array spectrophotometer in 1979. The variety of different types of spectrophotometers has also increased dramatically over the years, including many specialized features for measuring any type of sample and every type of optical property.

In its simplest form, a spectrophotometer contains three parts: a source, a sample holder, and a detector. The source usually contains some sort of spectrometer so that the optical radiation is monochromatic, covering a range of interest. The wavelength range for spectrophotometric measurements depends upon the application and can cover the ultraviolet, visible, or various ranges in the infrared. For example, for characterizing materials for use in solar energy applications, spectrophotometric measurements extending from about 200 to 2500 nm (i.e., the region of the solar spectrum) are important. The detector is designed to be sensitive over the range of interest, and an instrument might employ multiple detectors so that it can cover a broader wavelength range than that covered by a single detector. Operated in this fashion, a spectrophotometric curve for the sample can be obtained, by comparing at each wavelength the signal collected after interaction of the monochromatic source with the sample to the signal recorded without the sample in the measurement beam.

More complex measurements can be achieved on commercial instruments through the use of specialized accessories that modify the beam path, move or substitute the detector, or manipulate the sample orientation. In this manner, specular, diffuse, or angle-resolved reflectance or transmittance measurements can be performed over a wide range of wavelengths, making the commercial spectrophotometer a very versatile tool. Accurate measurements of optical properties of materials using spectrophotometric instrumentation remain a challenge, and the wide variety of modern instruments and applications has heightened this need for improved standardization and traceability. This book aims to address this need by providing both the novice and the experienced user of spectrophotometry an authoritative reference document with comprehensive terminology, guiding principles, and best measurement practices, including examples of important applications. In this text, we do not limit ourselves to measurements made on commercial instruments. In many cases, the commercial instrument is designed to rely upon a reference standard, with which a relative measurement is made. The reference standard, on the other hand, often has its reference values certified using an absolute method that does not rely upon a physical standard, but rather, upon methods used to realize the definition of the quantity.

1.2 UNCERTAINTIES

The accuracy of spectrophotometric measurements, and the derived optical properties of the material under test, depends upon the design and calibration of the spectrophotometer, the choice of reference standard, and the interaction of the sample with the measuring instrument. A discussion of high-accuracy measurements is not complete without a discussion of measurement uncertainties. That is, one cannot speak of "high accuracy" without asking the question, "Just how accurate is the measurement?" Many of the chapters of this text illustrate how such an estimate can be made for each measurement. In this introduction, we present the basic principles of how such an uncertainty analysis is performed in accordance with methods described in the ISO Guide to the Expression of Uncertainty in Measurements [7]. For a complete description, consult this reference or a suitably abridged version [8]. In the following, we attempt to reduce a few hundred pages into just a few.

Accuracy and precision are two concepts that are important to understand and differentiate in any field of metrology. When we perform any measurement, we never know what is the actual value of the measurand. Precision is a term that is used to loosely describe how close different measurements on the same instrument are to one another and is relatively simple to assess. Accuracy, on the other hand, is the closeness that a measurement result is to its true value. Unfortunately, accuracy is much harder to assess because it requires having a much higher degree of understanding of all the possible sources of error and their impact on the measurement. For absolute measurements, that is, those that do not rely on a reference material for their scale, the determination of the uncertainty can be quite complicated. For relative measurements, where the reference material comes with a stated uncertainty, the determination of the uncertainty is usually quite a bit simpler.

In short, each measurement has a measurement equation, which can be written as a functional relationship. For the general case, this is expressed as

$$y = f(x_1, x_2, \ldots, x_N), \tag{1.1}$$

where y is the *output* quantity of interest, and the x_i are all the *input* quantities that are combined through a functional relationship f to yield y. The function f is referred to as the measurement function. The quantities x_i include everything that affects the measurement result in any way. Often, we assume that the measurement function is of the form

$$f(x_1, x_2, \ldots, x_N) = \tilde{f}(x_1, x_2, \ldots, x_M) C_1(x_{M+1}) C_2(x_{M+2}) \ldots C_{N-M}(x_N), \tag{1.2}$$

where $\tilde{f}(x_1, x_2, \ldots, x_M)$ is the essential measurement function, which is that part used to calculate the output value, and $C_j(x_{M+j})$ are correction factors, which we assume are supposed to be unity, but which account for most of the nonidealities in the measurement, including nonlinearity and stray light. These correction factors are often not explicitly shown in the measurement equation, but nonetheless, their contributions to the combined uncertainty will appear in a measurement's uncertainty budget. It is recommended, however, that the measurement equation be written out in full, whenever possible.

For each of the identified input quantities x_i, we need an estimate of its standard uncertainty $u(x_i)$, which represents an estimate of the standard deviation of x_i. Furthermore, we need an estimate of the covariance, $u(x_i,x_j)$, between each x_i and x_j. The combined standard uncertainty, $u_c(y)$, is then determined, using the law of propagation of uncertainty from

$$u_c^2(y) = \sum_{i=1}^{N} \left(\frac{\partial f}{\partial x_i}\right)^2 u^2(x_i) + 2\sum_{i=1}^{N-1} \sum_{j=i+1}^{N} \frac{\partial f}{\partial x_i} \frac{\partial f}{\partial x_j} u(x_i, x_j). \tag{1.3}$$

The combined standard uncertainty represents the estimated standard deviation of the final measurement result y. It is often assumed that there is little or no correlation between the input quantities, that is, the covariance terms are negligible and the combined standard uncertainty can be calculated as the simple quadrature sum of the individual uncertainty components. However, for designing a measurement for optimum accuracy or for careful uncertainty analysis of any measurement result, it is important to have an eye out for covariance between terms, since, left unnoticed, they can significantly impact the uncertainty. While the factors in the first term in Eq. (1.3) always increase the uncertainty, those in the second term do not necessarily do so and in some cases can reduce the combined standard uncertainty.

It is common practice to identify estimates of uncertainty, $u(x_i)$, as either Type A or Type B. In a Type A evaluation, the input quantity is obtained using a statistical analysis. In a Type B evaluation, the uncertainty is estimated from other nonstatistical sources. Examples of Type A uncertainty estimates are those obtained from the standard deviation of the mean of a series of independent measurements, the estimated standard deviations of parameters obtained using the method of least squares to fit a curve of data, or those obtained from an analysis of variance (ANOVA). Examples of Type B uncertainty estimates are those based on experience, previous measurement results, uncertainties provided in calibration reports, and manufacturers' specifications.

The partial derivative in the first term on the right of Eq. (1.3) is often referred to as the sensitivity coefficient for the ith input quantity. This quantity is determined analytically, numerically, or experimentally (i.e., by measuring the change in y for a small intentional change in the value of x_i while keeping the other input quantities constant). In the absence of covariance terms, we often express the uncertainty budget with a table, listing the various contributions, their sensitivity coefficients $c_i = \partial f / \partial x_i$, their respective uncertainties $u(x_i)$, and their contributions $c_i u(x_i)$ to the combined standard uncertainty. The combined standard uncertainty is then determined by adding all the contributions in quadrature. As a result, one does not necessarily see Eq. (1.3) written out explicitly.

When the measurement equation contains a moderate to large number of independent inputs that each contribute to the uncertainty about equally and the degrees of freedom in the measurement are large, the result y will be normally distributed, and we will have 68% confidence that the true value will lie between $y - u_c(y)$ and $y + u_c(y)$. Since this is a fairly low level of confidence, it is standard practice to multiply the combined standard uncertainty by a coverage factor k to obtain an expanded uncertainty $U(y) = ku_c(y)$ so that in an interval between $y - U(y)$ and $y + U(y)$, we have a significantly higher level of confidence. For example, under the conditions given above, $k = 2$ corresponds to a 95% confidence, and $k = 3$ corresponds to a 99.7% confidence. For a small to moderate number of degrees of freedom, the interval of the Student's t-distribution that encompasses the fraction p of the distribution should be considered. Table 1.1 gives appropriate coverage factors for some representative degrees of freedom and some common confidence levels.

To estimate the effective degrees of freedom in the measurement, when the covariance terms can be ignored, one should use the Welch–Satterthwaite formula

TABLE 1.1 Appropriate Values of Coverage Factor k as a Function of the Number of Degrees of Freedom, ν, and the Confidence Level p

	p		
ν	*68.27%*	*95.45%*	*99.73%*
1	1.84	14	236
2	1.32	4.5	19
3	1.20	3.31	9.22
4	1.14	2.87	6.62
5	1.11	2.65	5.51
10	1.05	2.28	3.96
25	1.02	2.11	3.33
100	1.01	2.03	3.08
∞	1	2	3

$$\nu_{\text{eff}} = \frac{u_c{}^4(y)}{\sum_{i=1}^{N} \frac{c_i{}^4 u^4(x_i)}{\nu_i}}, \tag{1.4}$$

where ν_i is the number of degrees of freedom of $u(x_i)$. For Type A uncertainties, the number of degrees of freedom ν_i should be chosen by the appropriate statistical method. For example, if the value is taken from the average of n samples, and the standard uncertainty is taken as the standard deviation of the mean, then $\nu_i = n - 1$. For Type B uncertainties, and in the absence of any other guidance (e.g., that given on a calibration certificate), it is common practice that one takes $\nu_i \to \infty$, which eliminates its contribution to the denominator of Eq. (1.4).

In many cases, the values are expected to be normally distributed and statistical analysis can estimate the standard deviation. For many Type B estimates, we may only have a tolerance such that we know that x_i lies in an interval between $x_i - a$ and $x_i + a$. In this case, we assume a probability distribution that quantifies the different possible values of the input value. Then, the standard deviation of that probability distribution is taken to be the standard uncertainty of that input. For example, for a rectangular probability distribution, we find that the appropriate standard uncertainty is

$$u(x_i) = \frac{a}{\sqrt{3}}. \tag{1.5}$$

Key contributions to the Type A uncertainty are the measurement repeatability and reproducibility. Repeatability indicates the variation in the measurement result if the measurement is performed multiple times under the same conditions. This usually entails performing the measurement on the same instrument and by the same operator. Generally, an assessment of the repeatability involves removal of the sample under test and replacement, as if the measurement were being done each time independently. Sometimes, one distinguishes between short-term repeatability and long-term repeatability. These two cases are differentiated by the time interval between the successive measurements. Short-term repeatability tests may involve the operator removing the sample from the instrument and then immediately repeating the procedure from the start, while long-term repeatability tests may involve occasionally performing the same measurement over days, weeks, or months. However, strictly speaking, this requires that no other conditions that impact this sample measurement change over the course of the time interval of the repeat run, for example, the environmental conditions, or this would constitute an assessment of reproducibility due to this influence parameter.

Reproducibility indicates the variation in the measurement result when the conditions of the measurement have changed significantly. This usually entails different instruments, possibly even using different measurement methods or principles, or in different laboratories. When reporting reproducibility, it is important to describe the conditions that changed between the measurements.

1.3 OVERVIEW

This textbook is organized as follows. Chapter 2 describes a number of theoretical concepts important for the field of spectrophotometry, including defining radiometric quantities, expressing their relationships to the electromagnetic fields, defining the spectrophotometric quantities, and providing a mathematical foundation for describing reflection, transmission, absorption, emission, and scattering from materials. Chapters 3 and 4 are dedicated to describing the two primary methods, dispersive and interferometric, by which spectroscopic measurements are performed. Chapters 5–8 then discuss the measurements of reflection, transmission, fluorescence, and scattering, which are common to spectrophotometry over its wide range of application. Chapters 9 and 10 discuss two spectrophotometric measurements that are specific to the IR and visible ranges. These are emissivity and color, respectively, which, while being related to those discussed in Chapters 5–8, have their own specific measurement issues. Each of these core chapters provides detailed information on the considerations and approaches needed for precise and accurate measurements of optical properties using their particular spectrophotometric method. Finally, we devote three chapters, Chapters 11–13, to overviews of three specific industries or important users that illustrate the diverse applications of spectrophotometric measurements. We hope that this book will become the handbook of choice for those intending to make accurate spectrophotometric measurements.

REFERENCES

[1] I. Newton, Opticks: or, a Treatise of the Reflexions, Refractions, Inflexions and Colours of Light, Royal Society, London, 1704, p. 89.

[2] T. Young, Bakerian Lecture: experiments and calculations relative to physical optics, Phil. Trans. R. Soc. Lond. 94 (1804) 1–16.

[3] K.S. Gibson, et al., Spectrophotometry: report of O.S.A. Progress Committee for 1922–3, J. Opt. Soc. Am. 10 (1925) 169–241.

[4] A.C. Hardy, A recording photoelectric color analyser, J. Opt. Soc. Am. 18 (1929) 96–117.

[5] A.C. Hardy, History of the design of the recording spectrophotometer, J. Opt. Soc. Am. 28 (1938) 360–364.

[6] Identification of commercial names is not intended to imply recommendation or endorsement by the National Institute of Standards and Technology or the National Research Council of Canada.

[7] ISO Guide to the Expression of Uncertainty in Measurement, ISO, Geneva, Switzerland, 1995.

[8] B.N. Taylor, C.E. Kuyatt, Guidelines for Evaluating and Expressing the Uncertainty of NIST Measurement Results, National Institute of Standards and Technology, Gaithersburg, MD, 1994.

Chapter 2

Theoretical Concepts in Spectrophotometric Measurements

Thomas A. Germer*, Joanne C. Zwinkels[†] and Benjamin K. Tsai*
*National Institute of Standards and Technology, Gaithersburg, Maryland, USA
[†]National Research Council Canada, NRC, Measurement Science and Standards, Ottawa, Ontario, Canada

Chapter Outline

2.1 Introduction 12
2.2 Radiometric Quantities 13
2.2.1 Nonspectral Quantities 13
2.2.2 Spectral Quantities 18
2.2.3 Spectrally Weighted Quantities 19
2.3 Relationship Between Radiometric and Electromagnetic Quantities 20
2.3.1 Plane Waves and Irradiance 23
2.3.2 Spherical Waves and Intensity 24
2.3.3 Fourier Expansion and Radiance 25
2.4 The Spectrophotometric Quantities 26
2.4.1 Generalized Scattering Functions 27
2.4.2 Bidirectional Reflectance Distribution Function 28
2.4.3 Reflectance and Transmittance 30
2.4.4 Two Ideal Bidirectional Reflectance Distribution Functions 32
2.4.5 Absorptance 34
2.4.6 Fluorescence: Bispectral Luminescent Radiance Factor 36
2.4.7 Emittance and the Kirchhoff Relationship 38
2.5 Polarization 39
2.6 Reflection and Transmission from Flat Surfaces 43
2.6.1 Snell's Law of Refraction 43
2.6.2 Fresnel Reflection 44
2.6.3 Thin Films 48
2.6.4 Thick Films 49
2.7 Diffuse Scattering 52

Experimental Methods in the Physical Sciences, Vol. 46. http://dx.doi.org/10.1016/B978-0-12-386022-4.00002-9

2.7.1 Volume Scattering: Theory of Kubelka and Munk 52
2.7.2 Roughness: Facet Scattering Model 54
2.7.3 Roughness: First-Order Vector Perturbation Theory 59
References **65**

2.1 INTRODUCTION

Spectrophotometry deals with the measurement of the interaction of light with materials. Light can be reflected, transmitted, scattered, or absorbed, and a material can emit light, either because it has absorbed some light and reemits it, because it has gained energy in some other way (e.g., electroluminescence), or because it emits light due to its temperature (incandescence). The measurement of these spectrophotometric properties will be covered in various chapters in this book. To begin, however, we need to lay some groundwork. Precision measurements rely heavily on precise definitions of the quantities involved. In many cases, the lack of well-defined quantities has caused confusion and discrepancies between measurements. So, one important bit of knowledge you should take away from this book is that, when you make a measurement, you should know precisely what you are measuring, that what you are measuring is what you intended to measure, and that, if you are providing that measurement result to others, you communicate that information to them unambiguously. Reflectance, for example, is a relatively vague term. Do you mean the specular reflectance? Do you mean the total fraction of light that reflects into a backward hemisphere? Is the light to be incident unidirectionally along the surface normal, or are you diffusely illuminating the sample? If you are measuring the reflected, scattered radiation, and excluding the specular reflection, how close to the specular direction do you include? There are a myriad of answers to each of these questions, and nuances in between. There are many standard measurement configurations, but those standard configurations are often chosen more out of convenience than optimization for a specific application. Each chapter in this book will discuss these issues.

In this chapter, we will outline the framework by which we can precisely define our measurement by defining the terms and geometries used in spectrophotometry. Depending on the specific application and the particular measurement, many common geometries are employed; however, this chapter will attempt to discuss these geometric configurations in the most general sense, and subsequent chapters will provide more specific information.

We will describe the theoretical background needed to understand a large variety of spectrophotometric phenomena. This chapter is intended to be a reference to provide the reader the theory needed to interpret spectrophotometric measurements or to predict basic optical property quantities, and not a replacement for a full electromagnetics or optics textbook. In particular, we will outline the theory of reflectance and transmittance, and include the effects of thin and thick films. We will present the Kirchhoff relationship between reflectance and thermal emittance. Finally, we will discuss a number of basic scattering models that can be used to interpret and understand scatter measurements.

2.2 RADIOMETRIC QUANTITIES

In this section, we describe the different terms that are used for quantifying the radiation incident on or exiting a material per unit area, time, solid angle, bandwidth, or combination of the above. Most of the spectrophotometric quantities, described later in Section 2.4, are coefficients relating radiometric quantities described in this section. Each of these terms is defined because each has an associated ideal measurement or measurement condition that makes it useful in some application.

Figure 2.1 provides a basis for defining the measurement geometry. The symbols used for each of these quantities will be kept relatively consistent throughout this text. We are attempting to maintain consistency with the International Lighting Vocabulary (ILV), published by the Commission Internationale de l'Eclairage (CIE, International Commission on Illumination) [1]. However, because of the broad range of applications discussed in this book and the need to further distinguish quantities within those applications and between them, we have slightly modified these symbols, usually by altering or adding additional subscripts. Furthermore, we denote polar and azimuthal angles for directions with the symbols θ and ϕ, respectively, since the ILV symbols are not conducive to more general, nonplanar directions.

2.2.1 Nonspectral Quantities

Radiant power Φ is the amount of energy passing through, emitted from, or received by a surface per unit time, expressed in watts [W]. It is often referred to as radiant flux, although this term can be confusing, since many fields use the term *flux* to indicate the amount of something (energy, electrons, particles, water, etc.) passing through a unit area per unit time. The measurement of radiant power is generally performed using some sort of detector, which is under-filled by the light, so that it captures all of the light in the beam and is relatively insensitive to alignment and positioning. Highly accurate measurements of radiant power can be performed with electrical substitution radiometers [2], which capture all of the light incident upon them, and compare the heat load to that achieved by electrical, ohmic heating.

Radiant energy Q is the amount of energy in a light beam during a specified period of time, Δt, or within some sort of pulse. Expressed in joules [J], it is given by the time integral of the radiant power over a time interval Δt,

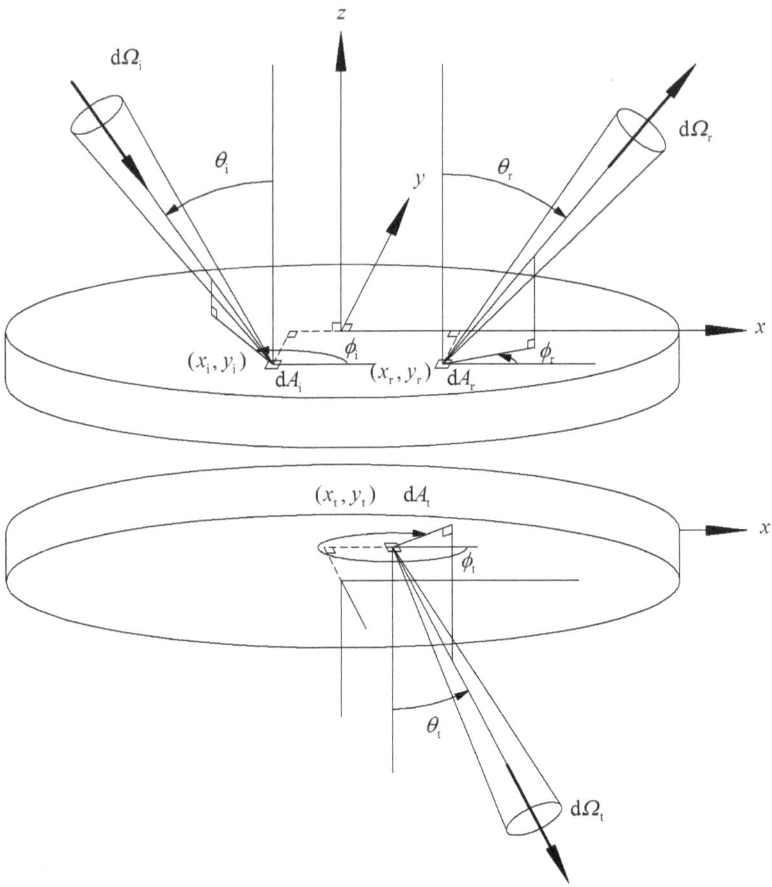

FIGURE 2.1 The geometry for a generalized description of reflection and transmission. The lower, transmission part of the figure shows the sample from below.

$$Q = \int_{\Delta t} \Phi \mathrm{d}t. \tag{2.1}$$

Measurements of radiant energy are generally restricted to pulsed light sources. In that case, Δt is longer than the pulse duration, so that the detector integration time captures the entire energy in the pulse. One example of such a detector is an absorbing material, well insulated from the environment, which is under-filled by the light, and whose peak temperature rise is recorded. Another example is an under-filled photovoltaic detector, which is connected to a charge accumulator.

Irradiance E is the amount of radiant power passing through a surface per unit area on the surface:

$$E = \mathrm{d}\Phi/\mathrm{d}A, \tag{2.2}$$

expressed in W m^{-2}. It is common to express the various radiometric quantities in terms of what appear to be derivatives. They are, however, not derivatives in the formal sense, but borrow the notation of an infinitesimal amount of some quantity per an infinitesimal amount of some other quantity. Thus, the notation used in Eq. (2.2) is really shorthand for

$$E = \lim_{A \to 0} \frac{\Phi}{A}. \tag{2.3}$$

A measurement of irradiance is generally performed by a detector of known area (possibly with an aperture defining that area) exposed at the location of interest and in a manner in which the light overfills that area.

It is important for us to indicate, either directly or by context, the element of surface that we are considering when we use the term irradiance, because light that is diverging will have an irradiance which falls off as a function of distance from the source, and the surface may not be perpendicular to the direction of propagation. The term can be used either to quantify light propagating in a well-defined direction or in many directions at once.

We will often be concerned with optical power that is converging onto or diverging from a surface. Thus, we need to consider the amount of power contained in a solid angle. Just as a planar angle θ is defined by the length of a section of a unit circle subtended by an angle, a solid angle Ω is defined by the area of a unit sphere subtended by a spread of directions. While there are 2π rad (radians) in a circle (the unit circle having circumference 2π), there are 4π sr (steradians) in a sphere (the unit sphere having area 4π). The differential solid angle can be expressed in spherical coordinates as $\mathrm{d}\Omega = \sin\theta\, \mathrm{d}\theta\, \mathrm{d}\phi$ (see Fig. 2.2). Thus, a right circular cone with a half angle α subtends the solid angle

$$\Omega = \int_0^{2\pi} \int_0^{\alpha} \sin\theta\, \mathrm{d}\theta\, \mathrm{d}\phi = 2\pi(1 - \cos\alpha). \tag{2.4}$$

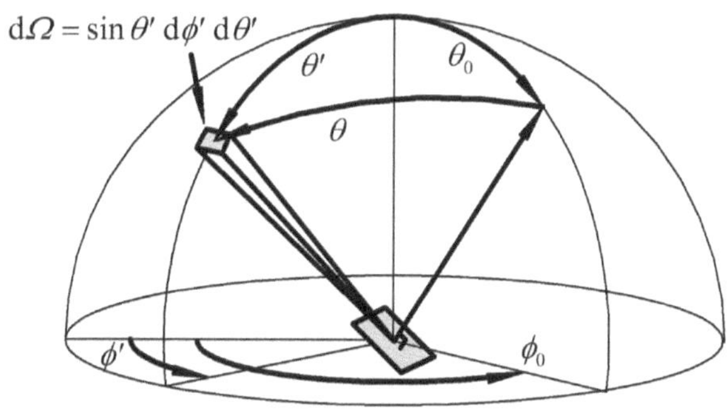

FIGURE 2.2 The projected differential solid angle. The surface element shown is oriented with a polar angle θ_0 and azimuthal angle ϕ_0. The projected solid angle onto the surface element is $\mathrm{d}\Omega_{proj} = \cos\theta\, \mathrm{d}\Omega$.

The total solid angle for a hemisphere ($\alpha = \pi/2$) is thus 2π sr. Note that sr, like rad, is a special unit with value unity. Being the ratio of two areas (like two lengths, as in a radian), it is dimensionless. Despite its lack of dimension, though, it should always be included with other units of a measurement quantity to avoid confusion. Carrying it as a unit reminds one that it may need to be canceled out by another corresponding solid angle.

A quantity related to the solid angle is the *projected* solid angle, given by

$$d\Omega_{\text{proj}} = \cos\theta \, d\Omega, \tag{2.5}$$

where θ is the angle the surface makes with respect to the direction of the solid angle. From Fig. 2.2 and some spherical trigonometry, we have

$$\cos\theta = \cos\theta_0 \cos\theta' + \sin\theta_0 \sin\theta' \cos(\phi' - \phi_0), \tag{2.6}$$

where θ_0 and ϕ_0 are the polar and azimuthal angles of the surface orientation. We can then find the projected solid angle for a right circular cone centered on the vertical direction in Fig. 2.2 by integration,

$$\Omega_{\text{proj}} = \int_0^{2\pi} \int_0^{\alpha} \sin\theta' \cos\theta \, d\theta' \, d\phi' = \pi \sin^2\alpha \cos\theta_0, \tag{2.7}$$

where α is the maximum half-cone angle collected. The total projected solid angle for a hemisphere ($\alpha = \pi/2$) onto the surface perpendicular to it is thus π. An aperture of radius a whose center is a distance r from a given point subtends the angle $\alpha = \tan^{-1}(a/r)$. Therefore, the solid angle subtended is

$$\Omega = 2\pi \frac{\sqrt{r^2 + a^2} - r}{\sqrt{r^2 + a^2}}, \tag{2.8}$$

while the projected solid angle is

$$\Omega_{\text{proj}} = \frac{\pi a^2 \cos\theta_0}{r^2 + a^2}. \tag{2.9}$$

We will find $d\Omega$ or the combination $\cos\theta \, d\Omega$ in a number of the following radiometric definitions. In practice, we must measure those quantities using finite-sized apertures. If the quantities are uniform over the apertures, Eqs. (2.8) and (2.9) can be used in the respective measurement equations.

Radiance L quantifies the amount of energy passing through a unit area, considered perpendicular to the direction of propagation, per unit solid angle. That is, if one considers a surface that is emitting or collecting the radiant power Φ, the radiance is given by

$$L = \frac{d\Phi}{\cos\theta \, dA \, d\Omega}, \tag{2.10}$$

where dA is the unit area and $d\Omega$ is a solid angle. Radiance, expressed in W m^{-2} sr^{-1}, can also be expressed in terms of the irradiance,

$$L = \frac{dE}{\cos\theta \, d\Omega}. \tag{2.11}$$

The most common instrument that senses radiance is a camera. The focal plane senses irradiance E and the solid angle $d\Omega$ is determined by the aperture. Consider Fig. 2.3, which shows a simple optical system with focal lengths f_1 and f_2 and aperture area dA_{lens}. From the magnification of the system, we see that

$$dA_1/f_1^2 = dA_2/f_2^2 \tag{2.12}$$

and

$$f_1^2 d\Omega_1 = f_2^2 d\Omega_2. \tag{2.13}$$

Therefore, if radiant power $d\Phi$ leaving dA_1 is incident upon dA_2, the radiance L_1 leaving dA_1 is the same as the radiance L_2 incident upon dA_2. Thus, in the absence of reflection or absorption losses, radiance is conserved through an optical system.

When the extent of a radiation source is limited, the irradiance decays according to the inverse square law at large distances. Radiant intensity is defined as the radiant power in a given direction per unit solid angle containing the given direction

$$I = d\Phi/d\Omega, \tag{2.14}$$

expressed in W sr^{-1}. Since $d\Omega = dA/r^2$, an irradiance E measured a distance r from the source will indicate a radiant intensity of

$$I = r^2 E. \tag{2.15}$$

Consider a circular source of radius a, with area $A = \pi a^2$, radiating with a constant radiance L. At a distance r perpendicular to the source, the irradiance is given by combining Eqs. (2.11) and (2.9) (letting $\theta_0 = 0$),

$$E = L\Omega_{proj} = L\frac{\pi a^2}{r^2 + a^2} = L\frac{A}{r^2}\left(1 - \frac{a^2}{r^2} + \cdots\right). \tag{2.16}$$

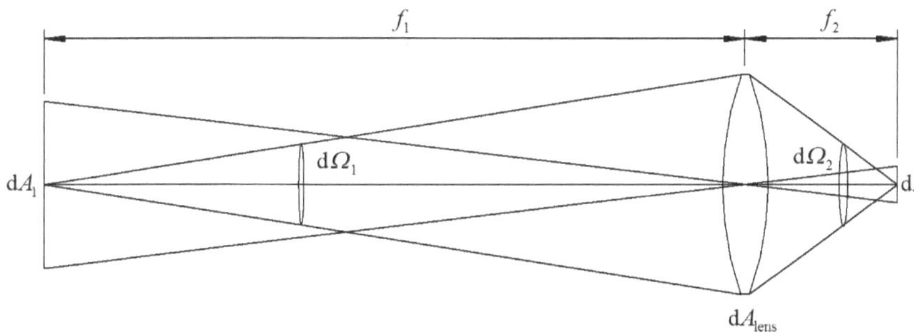

FIGURE 2.3 Illustration of the geometric extent.

Thus, the fractional deviation from inverse square behavior is approximately given by a^2/r^2. From Eqs. (2.15) and (2.16), we see that the intensity associated with the source is

$$I = AL \tag{2.17}$$

at large distances.

2.2.2 Spectral Quantities

There are two sets of quantities that are derived from the above quantities (radiant energy, radiant power, irradiance, radiance, and radiant intensity), but which add the spectral dimension to the mix. The first of these are the spectral quantities, which express the former quantities per unit wavelength range. The second are spectrally integrated quantities that incorporate a specific spectral weighting, for example, the photometric response of the human eye, solar spectral irradiance, erythemal action spectrum, etc. We will begin by discussing the spectral quantities. Each of the radiometric quantities described above can be expressed with its spectral counterpart. The spectral power is the amount of radiant power per unit spectral bandwidth

$$\Phi_\lambda = \mathrm{d}\Phi/\mathrm{d}\lambda, \tag{2.18}$$

and while it can be expressed in $\mathrm{W\,m^{-1}}$, it is usually expressed in $\mathrm{W\,nm^{-1}}$ or $\mathrm{W\,\mu m^{-1}}$. The spectral irradiance is the irradiance per unit wavelength

$$E_\lambda = \frac{\mathrm{d}E}{\mathrm{d}\lambda} = \frac{\mathrm{d}\Phi}{\mathrm{d}\lambda\,\mathrm{d}A}, \tag{2.19}$$

which is best expressed in $\mathrm{W\,m^{-2}\,nm^{-1}}$ or $\mathrm{W\,m^{-2}\,\mu m^{-1}}$. Similarly, the spectral radiance is

$$L_\lambda = \frac{\mathrm{d}L}{\mathrm{d}\lambda} = \frac{\mathrm{d}\Phi}{\cos\theta\,\mathrm{d}\lambda\,\mathrm{d}A\,\mathrm{d}\Omega}, \tag{2.20}$$

expressed in $\mathrm{W\,m^{-2}\,sr^{-1}\,nm^{-1}}$ or $\mathrm{W\,m^{-2}\,sr^{-1}\,\mu m^{-1}}$. Lastly, the spectral intensity is

$$I_\lambda = \frac{\mathrm{d}I}{\mathrm{d}\lambda} = \frac{\mathrm{d}\Phi}{\mathrm{d}\lambda\,\mathrm{d}\Omega}, \tag{2.21}$$

expressed in $\mathrm{W\,sr^{-1}\,nm^{-1}}$ or $\mathrm{W\,sr^{-1}\,\mu m^{-1}}$.

A blackbody, at an absolute temperature T radiates according to Planck's law, with a spectral radiance given by

$$L_{\mathrm{e},\lambda}(\lambda, T) = \frac{2hc_0^2}{\lambda^5}\frac{1}{\exp(hc_0/\lambda kT) - 1}, \tag{2.22}$$

where h is the Planck constant, c_0 is the speed of light in vacuum, and k is the Boltzmann constant. All of the above spectral quantities can also be expressed in terms of frequency bands dν instead of wavelength bands. So, for example, we can define spectral radiant power as

$$\Phi_\nu = \frac{\mathrm{d}\Phi}{\mathrm{d}\nu}. \tag{2.23}$$

Since $\mathrm{d}\nu = -c_0\lambda^{-2}\,\mathrm{d}\lambda = -(\nu^2/c_0)\,\mathrm{d}\lambda$,

$$\Phi_\nu = \frac{c_0}{\nu^2}\Phi_\lambda = \frac{\lambda^2}{c_0}\Phi_\lambda. \tag{2.24}$$

Thus, for example, Planck's law can be written

$$L_{e,\nu}(\nu, T) = \frac{2h\nu^3}{c_0^2}\frac{1}{\exp(h\nu/kT) - 1}. \tag{2.25}$$

It is worth noting that the spectral quantities are often not measured for monochromatic sources, such as lasers, because the bandwidth, dλ or dν, is very small and for many purposes is zero. Thus, one should not simply substitute the spectral quantities for the nonspectral quantities, in the belief that adding the spectral information will expand the information contained in the measurement. For example, consider regular reflectance, discussed later as being the ratio of reflected radiant power to incident radiant power. If one were to define the reflectance in terms of the spectral powers, one would be required to also measure the bandwidth of the light before and after reflection. Thus, the measurement quantity would depend on the coherence properties of the laser, in a manner, which may be completely unexpected or undesired.

2.2.3 Spectrally Weighted Quantities

Another set of quantities are defined with spectral weightings that account for spectral responses in specific applications. The most common one is the spectral response of the human eye, which has been standardized by the International Commission on Illumination (CIE). In this case, the luminous flux is the amount of visually perceived "power" and is the integral over wavelength of the measured spectral power weighted by the spectral response of the human eye:

$$\Phi_\mathrm{v} = K_\mathrm{m}\int V(\lambda)\,\Phi_\lambda\,\mathrm{d}\lambda. \tag{2.26}$$

[*Note*: the subscript here is the letter vee for "visual", rather than the previous section, which used the Greek letter nu for "frequency".] The function $V(\lambda)$, shown in Fig. 2.4, is called the spectral luminous efficiency function for photopic (light-adapted) vision, and the normalization factor $K_\mathrm{m} = 683.002\ \mathrm{lm\,W^{-1}}$. The normalization factor is chosen so that the spectral luminous efficiency $V(\lambda)$ has a maximum value of one. Likewise, there is the illuminance,

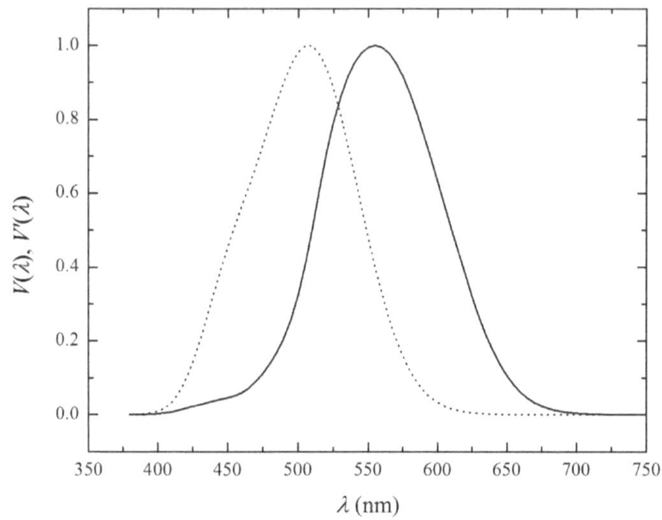

FIGURE 2.4 The CIE 1924 photopic spectral luminous efficiency function $V(\lambda)$ for a 2° observer (solid curve) and the CIE 1951 scotopic spectral luminous efficiency function $V'(\lambda)$ (dashed curve).

$$E_v = K_m \int V(\lambda) E_\lambda \, d\lambda, \tag{2.27}$$

the luminance,

$$L_v = K_m \int V(\lambda) L_\lambda \, d\lambda, \tag{2.28}$$

and the luminous intensity

$$I_v = K_m \int V(\lambda) I_\lambda \, d\lambda. \tag{2.29}$$

The candela (cd) is the SI base unit for luminous intensity and is defined as luminous intensity, in a given direction, of a source that emits monochromatic radiation of frequency 540×10^{12} Hz (that is, a wavelength of about 555 nm) and that has a radiant intensity in that direction of 1/683 W sr^{-1}. Thus, the unit of luminous flux is cd sr, which is defined as the lumen (lm). Likewise, the units for illuminance and luminance are lm m^{-2} and cd m^{-2}, respectively. Figure 2.4 shows the photopic spectral luminous efficiency function $V(\lambda)$, as well as the scotopic (dark-adapted) function $V'(\lambda)$.

2.3 RELATIONSHIP BETWEEN RADIOMETRIC AND ELECTROMAGNETIC QUANTITIES

The previous section described the radiometric quantities in terms of energy transport. It should be borne in mind, however, that the fundamental properties of optical radiation are governed by the laws of electromagnetics, and that beams of light are in fact waves in the electromagnetic field. In many applications, it is sufficient to think of optical radiation only in terms of energy transport, but doing so ignores such phenomena as diffraction and interference. Optical polarization can be treated with appropriate extensions of the energy transport theory (see Section 2.5, below), but it is ultimately an electromagnetic phenomenon caused by the vector nature of the electromagnetic field. Furthermore, the optical behavior of a material is determined by the spectral dependence of two parameters, which describe the speed and attenuation, respectively, of a plane electromagnetic wave propagating through the material. These two parameters are the optical or constitutive constants (e.g., real and imaginary parts of the refractive index) and can be derived from the spectral dependence of the measured spectrophotometric quantities (reflectance, transmittance, etc.). Interpretation of experimental spectrophotometric measurements in terms of the optical constants of the material requires a knowledge of Maxwell's equations, the nature of the interaction between electromagnetic fields and matter, and an understanding of the dependence of the optical constants on the wavelength or frequency and, in the case of anisotropic materials, the vibration directions of the propagating wave. Accurate theoretical predictive treatments of the optical properties of materials, such as scattering, diffraction, reflection, or transmittance, are ultimately based upon solving the associated electromagnetic problems. Relating the results of the electromagnetic simulations to the measured spectrophotometric quantities requires an understanding of how the electromagnetic quantities relate to the radiometric ones. This section will provide this necessary theoretical background. In particular, we will point out the relationships between irradiance and the plane-wave amplitude (Section 2.3.1), intensity and the spherical wave amplitude (Section 2.3.2), and radiance and the Fourier expansion of plane waves (Section 2.3.3). Later in this chapter, we will use these findings to calculate the reflectance, transmittance, and scattering properties in a number of examples (see Sections 2.6 and 2.7).

The properties of light in a medium are described by the time-dependent Maxwell equations [3–5]:

$$\begin{aligned} \nabla\times\mathbf{E} &= -\frac{\partial\mathbf{B}}{\partial t}, \\ \nabla\times\mathbf{H} &= \mathbf{J}+\frac{\partial\mathbf{D}}{\partial t}, \\ \nabla\cdot\mathbf{D} &= \rho, \\ \nabla\cdot\mathbf{B} &= 0, \end{aligned} \tag{2.30}$$

where a bold-faced character denotes a vector quantity and $\mathbf{E}$ is the electric field, $\mathbf{H}$ is the magnetic field, $\mathbf{D}$ is the electric displacement field, $\mathbf{B}$ is the magnetic induction field, $\mathbf{J}$ is the current density, and ρ is the charge density. These quantities are a function of location $\mathbf{r}$ and time t. If we assume that we have monochromatic light and the material responses are linear, each of these fields is given by $\mathbf{E}=\hat{\mathbf{E}}\exp(-i\omega t)$, $\mathbf{H}=\hat{\mathbf{H}}\exp(-i\omega t)$, etc., where ω is the angular frequency, related to the natural frequency $\nu=\omega/(2\pi)$. We use the complex number notation because it often makes the mathematics easier. If a complex field satisfies Maxwell's equations, so does its complex conjugate. The sum of any two fields that satisfy Eqs. (2.30) is also a solution. Thus, it is implicit that, when we use the complex notation, we are referring to the real part of the expression. Note that our choice of the $\exp(-i\omega t)$ time dependence with the real part of a quantity being the physical value [as opposed to assigning the actual field to the sum of the quantity and its complex conjugate, and as opposed to using $\exp(i\omega t)$] will have consequences on the behavior of other quantities, which will be mentioned as they are introduced. In the absence of time-dependent currents and charges, the frequency dependent Maxwell equations are thus

$$\begin{aligned} \nabla\times\hat{\mathbf{E}} &= i\omega\hat{\mathbf{B}}, \\ \nabla\times\hat{\mathbf{H}} &= -i\omega\hat{\mathbf{D}}, \\ \nabla\cdot\hat{\mathbf{D}} &= 0, \\ \nabla\cdot\hat{\mathbf{B}} &= 0. \end{aligned} \tag{2.31}$$

The relationships between $\hat{\mathbf{E}}$ and $\hat{\mathbf{D}}$ and between $\hat{\mathbf{H}}$ and $\hat{\mathbf{B}}$ can be quite complicated in the most general case. However, for most media and for fields within the linear regime, we have the constitutive relations:

$$\begin{aligned} \hat{\mathbf{D}} &= \epsilon_0\,\overleftrightarrow{\boldsymbol{\epsilon}}\cdot\hat{\mathbf{E}}, \\ \hat{\mathbf{B}} &= \mu_0\,\overleftrightarrow{\boldsymbol{\mu}}\cdot\hat{\mathbf{H}}, \end{aligned} \tag{2.32}$$

where $\overleftrightarrow{\boldsymbol{\epsilon}}$ is the relative electric permittivity tensor, $\overleftrightarrow{\boldsymbol{\mu}}$ is the relative magnetic permeability tensor, and ϵ_0 and μ_0 are the electric and magnetic constants, respectively. In the case of an anisotropic material, the magnitude of the induced conduction and polarization varies with the direction of the applied electric field, and, consequently, $\overleftrightarrow{\boldsymbol{\epsilon}}$ and $\overleftrightarrow{\boldsymbol{\mu}}$ must be expressed as tensor quantities. That is, the electric field $\hat{\mathbf{E}}$ and the displacement field $\hat{\mathbf{D}}$ are not necessarily in the same direction. In cases where the materials are isotropic, though, we can treat these tensors as scalars, ϵ and μ, respectively. In a homogeneous and isotropic material, Eqs. (2.31) and (2.32) reduce to the Helmholtz equation

$$\nabla\times\left(\nabla\times\hat{\mathbf{E}}\right)=\omega^2\epsilon\epsilon_0\mu\mu_0\hat{\mathbf{E}}. \tag{2.33}$$

The speed of light in vacuum is $c_0=(\epsilon_0\mu_0)^{-1/2}$, while the phase velocity in an isotropic material is $c=(\epsilon\mu)^{-1/2}c_0$. The complex index of refraction is $\tilde{n}=(\epsilon\mu)^{1/2}$. Note, because of our $\exp(-i\omega t)$ convention, ϵ, μ, and $\tilde{n}$ are chosen so that their imaginary parts are nonnegative. Lastly, ϵ is often referred to as the dielectric constant (which is actually not a constant, but nearly always a function of frequency), and for nonmagnetic materials, $\mu=1$.

2.3.1 Plane Waves and Irradiance

Plane-wave, monochromatic radiation is defined by a field of infinite extent in space, traveling in a specific direction, having a time dependence that is sinusoidal for all of time, and having a phase front that is planar. Real optical fields are never this ideal. At best, the field coming out of many lasers has a finite extent and only has sinusoidal time dependence for a relatively short period of time. Natural light, such as that coming from the sun, has light coming from many different directions (albeit centered about a small cone from the direction of the sun) and has many different wavelengths. While the monochromatic plane-wave field is a relatively abstract quantity, it nevertheless is a very useful concept from which much of electromagnetism, and thus optics, is based.

Consider light propagating in a direction given by a unit vector $\hat{\mathbf{k}}$. It is convenient to define a propagation vector $\mathbf{k}=k\hat{\mathbf{k}}$, where $k=2\pi/\lambda=\omega/c$. The wavelength associated with this wave is $\lambda=c/v$.The wave consists of an electric and magnetic field, perpendicular to each other and perpendicular to the direction of propagation. In this case, the electric field and magnetic fields are given by the real parts of the complex functions

$$\begin{aligned}\hat{\mathbf{E}}&=\hat{\mathbf{E}}_0\exp(\mathbf{ik}\cdot\mathbf{r}),\\ \hat{\mathbf{H}}&=\hat{\mathbf{H}}_0\exp(\mathbf{ik}\cdot\mathbf{r})\end{aligned} \tag{2.34}$$

Here, $\hat{\mathbf{E}}_0$ is a constant vector, perpendicular to $\mathbf{k}$, and is often referred to as the amplitude of the wave. From the first of Eq. (2.31) and the second of Eq. (2.32), $\mathbf{H}_0$ is also a constant vector, given by

$$\hat{\mathbf{H}}_0=\frac{1}{\mu\mu_0\omega}\mathbf{k}\times\hat{\mathbf{E}}_0=\sqrt{\frac{\epsilon\epsilon_0}{\mu\mu_0}}\hat{\mathbf{k}}\times\hat{\mathbf{E}}_0. \tag{2.35}$$

By defining $\hat{n}=(\epsilon/\mu)^{1/2}=\tilde{n}/\mu$ and the impedance of free space $Z_0=(\epsilon_0/\mu_0)^{1/2}$, we can simplify Eq. (2.35) and save a little typographical effort in the future,

$$\hat{\mathbf{H}}_0=\hat{n}Z_0\,\hat{\mathbf{k}}\times\hat{\mathbf{E}}_0. \tag{2.36}$$

The Poynting vector,

$$\mathbf{S}=\text{Re}\mathbf{E}\times\text{Re}\mathbf{H} \tag{2.37}$$

is useful for calculating the instantaneous energy flow across a surface per unit time per unit area. The radiant power crossing a surface element $\text{d}\mathbf{A}$ is $\text{d}\Phi=\mathbf{S}\cdot\text{d}\mathbf{A}$. Thus, the instantaneous irradiance is $E=|\mathbf{S}|\cos\theta$, where θ is the angle between the direction of propagation and the surface element $\text{d}\mathbf{A}$. When we calculate the intensities, it must be borne in mind that we are interested in the time average of the Poynting vector, not necessarily the instantaneous Poynting vector. For the conventions that we chose above, the time average of the Poynting vector is

$$\langle \mathbf{S} \rangle = \frac{1}{2}\mathrm{Re}\left(\hat{\mathbf{E}}_0 \times \hat{\mathbf{H}}_0^*\right). \tag{2.38}$$

The irradiance onto a surface tilted at an angle θ with respect to the direction of propagation is thus

$$E = \langle \mathbf{S} \rangle \cos\theta = \frac{\cos\theta}{2}\hat{n}Z_0\left\langle |\hat{\mathbf{E}}_0|^2 \right\rangle. \tag{2.39}$$

Since, for most materials in the optical regime, $\mu = 1$, the quantities $\hat{n}$ and $\tilde{n}$ are essentially the same and treated as the same. The latter is the complex index of refraction, which determines the refractive properties of the medium, while the quantity $\hat{n}$ is important for determining other optical properties that are introduced later, such as the reflection coefficients.

2.3.2 Spherical Waves and Intensity

A second form of wave that is useful to discuss is a spherical wave diverging from or converging onto a point in space. While plane waves form a complete set of functions that span all the propagating solutions to Maxwell's equations in free space, they form only one particular set and, as a set, are not unique. They are convenient for applications where the solutions to Maxwell's equations are best described in Cartesian coordinates, such as the reflection from a planar surface, or when describing the local properties of a wave, far the source. However, in some applications, such as the scattering by small particles, it is more convenient to express the solutions in spherical coordinates. A spherical, monochromatic wave radiating from the origin has a form far from the origin

$$\begin{aligned} \hat{\mathbf{E}} &\approx \frac{1}{kr}\hat{\mathbf{E}}_1(\hat{\mathbf{r}})\exp(ikr), \\ \hat{\mathbf{H}} &\approx \frac{1}{kr}\hat{\mathbf{H}}_1(\hat{\mathbf{r}})\exp(ikr), \end{aligned} \tag{2.40}$$

where $r = |\mathbf{r}|$ is the distance from the origin. That is, the fields have the oscillatory radial dependence, but the amplitudes decay inversely with distance from the origin. The functions $\hat{\mathbf{E}}_1(\hat{\mathbf{r}})$ and $\hat{\mathbf{H}}_1(\hat{\mathbf{r}})$ are not constants, but depend only upon the direction $\hat{\mathbf{r}} = \mathbf{r}/r$. Nonetheless, it is still informative to determine the radiometric properties of these waves. The irradiance associated with this wave is, from Eq. (2.39),

$$E = \frac{1}{2k^2r^2}\hat{n}Z_0|\hat{\mathbf{E}}_1(\hat{\mathbf{r}})|^2. \tag{2.41}$$

Thus, the radiant intensity for such a field, from Eq. (2.15), is

$$I = \frac{\mathrm{d}\Phi}{\mathrm{d}\Omega} = r^2 \frac{\mathrm{d}\Phi}{\mathrm{d}A} = r^2 E = \frac{\hat{n} Z_0}{2k^2} |\hat{\mathbf{E}}_1(\hat{\mathbf{r}})|^2. \tag{2.42}$$

For theoretical calculations of the interaction of electromagnetic radiation with small (at maximum, smaller than several wavelengths) particles, the electric and magnetic fields are often expanded in each material region in terms of vector spherical harmonics. A very nice discussion of vector spherical harmonics, their numerous properties, and how they are used to calculate the scattering by a homogeneous sphere (Mie scattering) is given in [6]. For large r, the vector spherical harmonics behave like Eqs. (2.40).

2.3.3 Fourier Expansion and Radiance

In this section, we will describe the connection between the radiance and its corresponding electromagnetic description for light radiating from a surface. This relationship is useful when one wishes to calculate the diffuse scattering properties of materials where the quantities of fundamental importance in the calculation (which involves solving Maxwell's equations) are given in terms of the field quantities. In a homogeneous half space, for $z > 0$, with no sources ($\mathbf{J} = 0$ and $\rho = 0$), the electric and magnetic fields can be expanded as integrals of plane waves, in what is often called the Fourier expansion:

$$\begin{aligned}\hat{\mathbf{E}} &= \frac{1}{4\pi^2} \int \mathrm{d}^2 k \mathbf{e}(\mathbf{k}) \exp(\mathrm{i}\mathbf{k}\cdot\mathbf{r}),\\ \hat{\mathbf{H}} &= \frac{1}{4\pi^2} \int \mathrm{d}^2 k \mathbf{h}(\mathbf{k}) \exp(\mathrm{i}\mathbf{k}\cdot\mathbf{r}),\end{aligned} \tag{2.43}$$

where the vector $\mathbf{k}$ is constrained to have $k_z = \left(\epsilon\mu\omega^2/c_0{}^2 - k_x^2 - k_y^2\right)^{1/2}$, and

$$\mathbf{h}(\mathbf{k}) = \hat{n} Z_0 \hat{\mathbf{k}} \times \mathbf{e}(\mathbf{k}). \tag{2.44}$$

In order that the fields propagate or decay evanescently in the positive z-direction, the sign of k_z is chosen so that $\mathrm{Re}k_z + \mathrm{Im}k_z > 0$. The $\mathbf{e}(\mathbf{k})$ can be obtained by integration over the $z = 0$ plane,

$$\mathbf{e}(\mathbf{k}) = \int \hat{\mathbf{E}}(\mathbf{r}) \exp(-\mathrm{i}\mathbf{k}\cdot\mathbf{r}) \mathrm{d}^2 r. \tag{2.45}$$

The irradiance onto the $z = 0$ plane is the component of the Poynting vector in the z-direction and is

$$\begin{aligned}E &= \langle \mathbf{S}\cdot\hat{\mathbf{z}} \rangle\\ &= \frac{\hat{n} Z_0}{8\pi^2} \int \left\langle \mathrm{Re}\left\{\mathbf{e}^*(\mathbf{k}) \times \left[\hat{\mathbf{k}} \times \mathbf{e}(\mathbf{k})\right]\right\}\cdot\hat{\mathbf{z}} \right\rangle \mathrm{d}^2 k\\ &= \frac{\hat{n} Z_0}{8\pi^2} \int_{k_x^2 + k_y^2 < \epsilon\mu\omega^2/c_0^2} \cos\theta \left\langle |\mathbf{e}(\mathbf{k})|^2 \right\rangle \mathrm{d}^2 k,\end{aligned} \tag{2.46}$$

where we use the vector identity $\mathbf{a}\times(\mathbf{b}\times\mathbf{c})=(\mathbf{a}\cdot\mathbf{c})\mathbf{b}-(\mathbf{a}\cdot\mathbf{b})\mathbf{c}$ and the fact $\mathbf{e}(\mathbf{k})\cdot\hat{\mathbf{k}}=0$. The bounds on the final integral in Eq. (2.46) simply limit the integral to propagating waves; evanescent waves do not transport energy, since $\mathrm{Re}\,\mathbf{k}\cdot\hat{\mathbf{z}}=0$. The angle $\theta=\cos^{-1}\left(\hat{\mathbf{k}}\cdot\hat{\mathbf{z}}\right)$ is the angle of propagation from z-direction. With the change of variables

$$\mathrm{d}^2\mathbf{k}=k^2\cos\theta\sin\theta\,\mathrm{d}\theta\,\mathrm{d}\phi=k^2\cos\theta\,\mathrm{d}\Omega, \tag{2.47}$$

the irradiance can be written

$$E=\frac{k^2\hat{n}Z_0}{8\pi^2}\int\cos^2\theta\left\langle|\mathbf{e}(\mathbf{k})|^2\right\rangle\mathrm{d}\Omega. \tag{2.48}$$

It is straightforward then to identify the radiance as

$$L=\frac{k^2\hat{n}Z_0}{8\pi^2}\cos\theta\left\langle|\mathbf{e}(\mathbf{k})|^2\right\rangle. \tag{2.49}$$

The integral in Eq. (2.45) can be performed at any value of vertical height z. Say we evaluate the spectrum at $z=z_1$ and obtain the spectrum of plane waves $\mathbf{e}_1(\mathbf{k})$ and then evaluate the spectrum at $z=z_2$ and obtain the spectrum of plane waves $\mathbf{e}_2(\mathbf{k})$. Then, it is not too difficult to show that

$$\mathbf{e}_1(\mathbf{k})=\mathbf{e}_2(\mathbf{k})\exp[\mathrm{i}k_z(z_2-z_1)]. \tag{2.50}$$

That is, the two spectra only differ in phase. Thus, when we evaluate the radiance using Eq. (2.49), we obtain the same result. This is another approach towards demonstrating that the radiance is a conserved quantity along a beam path.

2.4 THE SPECTROPHOTOMETRIC QUANTITIES

In spectrophotometry, we are concerned with the linear optical properties of materials, how light incident upon a material is reflected, transmitted, absorbed, and scattered by a medium. There are a number of other optical properties that usually do not fall under the purview of spectrophotometry: inelastic scattering, such as Raman or Brillouin scattering, and nonlinear processes, such as harmonic generation, four-wave mixing, self-focusing, and Kerr effects. These effects tend to be weak, as in the case of inelastic scattering, or require extremely high fields, in the case of nonlinear processes. Thus, we are primarily concerned with processes that are shown in Fig. 2.1. We choose a coordinate system in which the origin is at the middle of the face of the material, with the x and y axes being in the plane of the material and the z axis being along the center surface normal. Light of wavelength λ_i is shown incident with a polar angle θ_i, measured from the surface normal, and an azimuthal angle ϕ_i, measured from the projection of the incident direction onto the xy plane and measured from the x direction. A portion of the incident light may be incident upon a surface element of area dA_i at location with coordinates (x_i, y_i) and come from a solid angle $d\Omega_i$.

2.4.1 Generalized Scattering Functions

We begin by describing the most general case [7]. We define a scattering function $S_{\lambda t}$ that relates the spectral energy $dQ_{\lambda r}$ emitted into a solid angle $d\Omega_r$ about direction (θ_r, ϕ_r), by a surface element dA_r located at position (x_r, y_r), in a wavelength band $d\lambda$ about wavelength λ, and in some time interval dt_r about time t_r, after some energy dQ_i is incident upon it from direction (θ_i, ϕ_i), onto position (x_i, y_i), with wavelength μ, at time t_i:

$$dQ_{\lambda r} = S_{\lambda t}(\theta_i, \phi_i, x_i, y_i, \mu, t_i; \theta_r, \phi_r, x_r, y_r, \lambda, t_r)dQ_i \cos\theta_r dA_r d\Omega_r d\lambda dt_r. \quad (2.51)$$

A similar function can be defined in transmission, and we will use the subscript t instead of r for all the quantities to distinguish them from reflective quantities. The only assumption made in Eq. (2.51) is that the material behaves in a linear fashion to the amount of incident light. There are materials for which the absorption saturates under sufficient illumination power or whose emittance properties depend upon the illumination power. These materials have a nonlinear dependence upon the illumination power and do not follow Eq. (2.51). Furthermore, the response of these materials often depend upon the coherence aspects of the light, which make it difficult to assume a simple intensity relationship like that found in Eq. (2.51). Fortunately, a vast majority of materials, under relatively low levels of light, follow Eq. (2.51) to a high degree.

Optical phenomena occur over various time scales. For visible light, the period of oscillation of the electromagnetic field is roughly 2 fs. The time it takes for light to transmit through a material of, for example 1 cm, is tens of picoseconds. When light diffuses, it follows a longer path than when it would transmit or reflect directly, and the light emitted by a highly diffuse and nonabsorbing material can be in the material several nanoseconds. Fluorescence occurs usually nanoseconds to microseconds after initial excitation, and phosphorescence can continue to emit light for seconds, minutes, or even hours after excitation. In most cases of interest in spectrophotometry, the scattering function only depends upon on the time difference $t_r - t_i$. This may not be the case if measurements are being performed while the sample is being degraded somehow or otherwise dynamically modified. Furthermore, in many cases, we may either continuously illuminate the sample or have a sensor that integrates over a relatively long period of time. Then, we can make the substitution $dQ_i = d\Phi_i dt_i$, integrate over dt_i, and divide by dt_r to yield a time-independent scattering function, defined by the relationship between the incident power $d\Phi_i = d\Phi_i(\theta_i, \phi_i, x_i, y_i, \mu)$ and a scattered spectral power $d\Phi_{\lambda r} = d\Phi_\lambda(\theta_r, \phi_r, x_r, y_r, \lambda)$:

$$d\Phi_{\lambda r} = S_\lambda(\theta_i, \phi_i, x_i, y_i, \mu; \theta_r, \phi_r, x_r, y_r, \lambda)\cos\theta_r d\Phi_i dA_r d\Omega_r d\lambda, \tag{2.52}$$

or, by dividing by cos $\theta_r dA_r d\Omega_r d\lambda$, the spectral radiance $dL_{\lambda r} = dL_{\lambda r}(\theta_r, \phi_r, x_r, y_r, \lambda)$

$$dL_\lambda = S_\lambda(\theta_i, \phi_i, x_i, y_i, \mu; \theta_r, \phi_r, x_r, y_r, \lambda)\, d\Phi_i, \tag{2.53}$$

where

$$S_\lambda(\theta_i, \phi_i, x_i, y_i, \mu; \theta_r, \phi_r, x_r, y_r, \lambda) = \int S_{\lambda t}(\theta_i, \phi_i, x_i, y_i, \mu, t_i; \theta_r, \phi_r, x_r, y_r, \lambda, t_r)\, dt_r. \tag{2.54}$$

In many cases, our scattering function is elastic, so that the scattered wavelength is the same as the incident wavelength. (This is not the case for fluorescent or phosphorescent materials.) For elastic scattering, we can further integrate over reflected wavelength to obtain a relationship between the reflected radiance $dL_r = dL_r(\theta_r, \phi_r, x_r, y_r)$ and the incident radiant power $d\Phi_i = d\Phi_i(\theta_i, \phi_i, x_i, y_i, \mu)$,

$$dL_r = S(\theta_i, \phi_i, x_i, y_i, \mu; \theta_r, \phi_r, x_r, y_r)\, d\Phi_i, \tag{2.55}$$

where

$$S(\theta_i, \phi_i, x_i, y_i, \mu; \theta_r, \phi_r, x_r, y_r) = \int S_\lambda(\theta_i, \phi_i, x_i, y_i, \mu; \theta_r, \phi_r, x_r, y_r, \lambda)\, d\lambda. \tag{2.56}$$

The function S has been called the bidirectional scattering-surface reflectance distribution function (BSSRDF) [7, 8]. For a homogeneous and isotropic material, the function only depends on the distance between (x_i, y_i) and (x_r, y_r),

$$\left[(x_r - x_i)^2 + (y_r - y_i)^2\right]^{1/2}. \tag{2.57}$$

It is very rare that one can make a similar assumption about the angular dependence.

2.4.2 Bidirectional Reflectance Distribution Function

Consider the situation where we illuminate a static, nonfluorescent material uniformly, from the direction (θ_i, ϕ_i) with irradiance E_i [8, 9]. Then, $d\Phi_i = E_i dA_i$ is also uniform, and $\mu = \lambda$. The reflected radiance L_r is then given by integrating Eq. (2.55),

$$L_r = \int S(\theta_i, \phi_i, x_i, y_i, \lambda; \theta_r, \phi_r, x_r, y_r) E_i\, dA_i. \tag{2.58}$$

Since E_i is uniform, that is, not a function of x_i or y_i, it can be taken out of the integral, and we find that

$$L_r = f_r(\theta_i, \phi_i, \lambda; \theta_r, \phi_r, x_r, y_r) E_i, \tag{2.59}$$

where we introduce the bidirectional reflectance distribution function (BRDF),

$$f_r(\theta_i, \phi_i, \lambda; \theta_r, \phi_r, x_r, y_r) = \int S(\theta_i, \phi_i, x_i, y_i, \lambda; \theta_r, \phi_r, x_r, y_r)\, dA_i. \tag{2.60}$$

Equation (2.59) is usually used as the definition of the BRDF. Bear in mind, however, that this definition assumes that the illumination is uniform. Many measurements of the BRDF use an illumination that is not uniform and has finite extent. In this case, one tends to not measure the radiance, but rather the radiant intensity. Eq. (2.55) can be written as

$$d\Phi_r = S(\theta_i, \phi_i, x_i, y_i, \lambda; \theta_r, \phi_r, x_r, y_r)\cos\theta_r\, d\Phi_i\, d\Omega_r\, dA_r. \tag{2.61}$$

Integrating over dA_r, we have

$$d\Phi_r = f_r(\theta_i, \phi_i, x_i, y_i, \lambda; \theta_r, \phi_r)\cos\theta_r\, d\Phi_i\, d\Omega_r, \tag{2.62}$$

where

$$f_r(\theta_i, \phi_i, x_i, y_i, \lambda; \theta_r, \phi_r) = \int S(\theta_i, \phi_i, x_i, y_i, \lambda; \theta_r, \phi_r, x_r, y_r)\, dA_r. \tag{2.63}$$

Equations (2.60) and (2.63) appear similar, and indeed are equivalent, provided the sample is uniform. Eq. (2.60) is integrated over incident position and is a function of reflected position, while Eq. (2.63) is integrated over reflected position and is a function of incident position; this is a result of the former determining the radiance as a function of position, while the latter obtains its spatial resolution using the spatial profile of the incident beam. If the illumination spot is small, then we have

$$f_r = \frac{d\Phi_r}{\Phi_i \cos\theta_r\, d\Omega_r}. \tag{2.64}$$

From Eq. (2.64), it can be seen that the BRDF is the power scattered per unit projected solid angle. The BRDF can be thought of as a function of the incident direction and a distribution in the reflected direction. The value of the BRDF is always positive, has units of sr^{-1}, and can have arbitrarily large values.

In the above, we discussed everything in terms of the light being reflected back into the same hemisphere as the incident light. The various definitions given also apply when discussing transmitted light, and we use the subscript t to indicate that we are measuring in a transmitted geometry. For example, we can define the bidirectional transmittance distribution function (BTDF) in a manner analogous to the BRDF, Eq. (2.59):

$$L_t = f_t(\theta_i, \phi_i, \lambda; \theta_t, \phi_t, x_t, y_t)E_i, \tag{2.65}$$

where L_t is the diffusely scattered transmitted radiance. The BTDF and the BRDF, taken together, is sometimes referred to as the bidirectional scattering distribution function (BSDF).

Helmholtz reciprocity states that, if a uniform material is illuminated with a irradiance from a direction (θ_i, ϕ_i) with a collimated source and the radiance is measured in a direction (θ_r, ϕ_r), one would obtain the same result if the incident and viewing directions are interchanged. This is equivalent to saying that

$$f_r(\theta_i, \phi_i, \theta_r, \phi_r) = f_r(\theta_r, \phi_r, \theta_i, \phi_i). \tag{2.66}$$

This rule applies if the material obeys time-reversal symmetry. The only materials that do not follow time-reversal symmetry are those that exhibit a magneto-optical response or those that are photorefractive (that is, the light is modifying the material) [10]. In the transmitted geometry (provided you could measure radiance *inside* a material), the relationship must account for the different indices of refraction in the incident and transmitting region. Consider the case where the light is scattered by the first surface of a transparent optic. If one considers the external BTDF, we have

$$L_t(\theta_t, \phi_t) = f_t(\theta_i, \phi_i, \theta_t, \phi_t) E_i(\theta_i, \phi_i), \tag{2.67}$$

where f_t follows Helmholtz reciprocity, and the radiance L_t and irradiance E_i are evaluated outside the material. We can define the internal BTDF by

$$L_t'(\theta_t, \phi_t) = f_t'(\theta_i, \phi_i, \theta_t, \phi_t) E_i(\theta_i, \phi_i), \tag{2.68}$$

where L_t' is the radiance inside the material. Using Snell's law (especially Eq. (2.142), discussed later in Section 2.6.1), we can show that

$$\tilde{n}_i^2 f_t'(\theta_i, \phi_i, \theta_r, \phi_r) = \tilde{n}_r^2 f_t'(\theta_r, \phi_r, \theta_i, \phi_i). \tag{2.69}$$

Reciprocity is particularly useful for checking the validity of measurement and models. If a measurement or a model does not follow the reciprocity relationships, then it must be incorrect, unless the material is not time-reversal symmetric or the model is attempting to account for that. Care must be taken when the material modifies the polarization state of the light so that the appropriate reciprocal measurement is performed. The polarimetric generalization to Eq. (2.66) is given in Section 2.5, below.

2.4.3 Reflectance and Transmittance

Reflectance is defined as the ratio of collected reflected radiant power to incident radiant power. Due to conservation of energy, reflectance can have any value in the interval 0 to 1. The incident radiant power may be from a single direction, a range of directions, or from the entire hemisphere. Furthermore, if there are multiple directions, the incident radiant power may or may not be uniform in direction. Similarly, the collected light may be restricted to a small range of directions, a wide range of directions, or the full hemisphere, and the collection efficiency may or may not be uniform. Thus, the geometry of one reflectance measurement may differ significantly from another, and reflectance measurements can only be compared if they have identical conditions.

Suppose we have a radiance incident upon a surface that is a function of position (x_i, y_i) and direction (θ_i, ϕ_i),

$$L_i = L_i(\theta_i, \phi_i, x_i, y_i). \tag{2.70}$$

The radiant power coming from the solid angle $d\Omega_i$ and incident on the area element dA_i is

$$d\Phi_i(\theta_i, \phi_i, x, y) = L_i(\theta_i, \phi_i, x, y)\cos\theta_i \, d\Omega_i \, dA_i. \tag{2.71}$$

From Eqs. (2.55) and (2.71), the radiance emitted by the surface from $d\Phi_i$ is

$$dL_r(\theta_r, \phi_r, x_r, y_r) = S(\theta_i, \phi_i, x_i, y_i; \theta_r, \phi_r, x_r, y_r)L_i(\theta_i, \phi_i, x_i, y_i)\cos\theta_i \, d\Omega_i \, dA_i. \tag{2.72}$$

The radiant power of reflected light emitted into the solid angle $d\Omega_r$ from the area element dA_r is then

$$\begin{aligned} d\Phi_r &= \cos\theta_r \, dL_r(\theta_r, \phi_r, x_r, y_r) dA_r \, d\Omega_r \\ &= S(\theta_i, \phi_i, x_i, y_i; \theta_r, \phi_r, x_r, y_r)L_i(\theta_i, \phi_i, x_i, y_i)\cos\theta_r \cos\theta_i \, d\Omega_r \, dA_r \, d\Omega_i \, dA_i. \end{aligned} \tag{2.73}$$

The reflectance is the ratio of the radiant power collected from the surface to the radiant power incident onto the surface. The incident directions span a total solid angle, represented by Ω_i, and an area, represented by A_i. The collected directions span a total solid angle, represented by Ω_r, and an area A_r. The total incident radiant power is thus given by

$$\Phi_i(\Omega_i, A_i; L_i) = \int_{\Omega_i}\int_{A_i} \cos\theta_i L_i(\theta_i, \phi_i, x, y) dA_i \, d\Omega_i, \tag{2.74}$$

while the collected radiant power is

$$\begin{aligned} &\Phi_r(\Omega_i, A_i, \Omega_r, A_r; L_i) \\ &= \int_{A_r}\int_{\Omega_r}\int_{A_i}\int_{\Omega_i} \cos\theta_r \cos\theta_i \, S(\theta_i, \phi_i, x_i, y_i; \theta_r, \phi_r, x_r, y_r)L_i(\theta_i, \phi_i, x_i, y_i) \\ &\quad d\Omega_i \, dA_i \, d\Omega_r \, dA_r. \end{aligned} \tag{2.75}$$

Thus, the reflectance for this given geometry and illumination condition is given by

$$\rho(\Omega_i, A_i, \Omega_r, A_r; L_i) = \frac{\Phi_r(\Omega_i, A_i, \Omega_r, A_r; L_i)}{\Phi_i(\Omega_i, A_i; L_i)}. \tag{2.76}$$

Equation (2.76), with Eqs. (2.74) and (2.75), is a very general statement, probably too general for any reasonable application. However, it stresses the importance of carefully defining the geometric conditions under which the reflectance measurement is made. We will make a number of approximations that simplify this expression significantly. In the first step, we will assume that the sample is uniform. In that case, integration of Eq. (2.75) over dA_i, together with Eq. (2.60), yields

$$\Phi_r(\Omega_i, \Omega_r; L_i, A_i) = \int_{\Omega_r}\int_{A_i}\int_{\Omega_i} \cos\theta_r \cos\theta_i f_r(\theta_i, \phi_i; \theta_r, \phi_r) L_i(\theta_i, \phi_i, x_i, y_i) d\Omega_i dA_i d\Omega_r. \quad (2.77)$$

Thus, the reflectance is

$$\rho(\Omega_i, A_i, \Omega_r; L_i) = \frac{\int_{\Omega_r}\int_{A_i}\int_{\Omega_i} \cos\theta_r \cos\theta_i f_r(\theta_i, \phi_i; \theta_r, \phi_r) L_i(\theta_i, \phi_i, x_i, y_i) d\Omega_i dA_i d\Omega_r}{\int_{\Omega_i}\int_{A_i} \cos\theta_i L_i(\theta_i, \phi_i, x_i, y_i) dA_i d\Omega_i}. \quad (2.78)$$

If the radiance is uniform in space, but not necessarily isotropic, so that $L_i(\theta_i, \phi_i, x_i, y_i) = L_i(\theta_i, \phi_i)$, then integration of dA_i in the numerator and denominator of Eq. (2.78) will cancel, so that

$$\rho(\Omega_i, \Omega_r; L_i) = \frac{\int_{\Omega_r}\int_{\Omega_i} \cos\theta_r \cos\theta_i f_r(\theta_i, \phi_i; \theta_r, \phi_r) L_i(\theta_i, \phi_i) d\Omega_i d\Omega_r}{\int_{\Omega_i} \cos\theta_i L_i(\theta_i, \phi_i) d\Omega_i}. \quad (2.79)$$

Finally, if the radiance is uniform and isotropic, so that $L_i(\theta_i, \phi_i) = L_i$, a constant within $d\Omega_i$, then integration over $d\Omega_i$ will cancel, leading to

$$\rho(\Omega_i, \Omega_r) = \frac{\int_{\Omega_r}\int_{\Omega_i} \cos\theta_r \cos\theta_i f_r(\theta_i, \phi_i; \theta_r, \phi_r) d\Omega_i d\Omega_r}{\int_{\Omega_i} \cos\theta_i d\Omega_i}. \quad (2.80)$$

Furthermore, if the incident light is collimated, integration over $d\Omega_i$ in the numerator will cancel the denominator, so that the directional–hemispherical reflectance is

$$\rho(\theta_i, \phi_i; \Omega_r) = \int_{\Omega_r} \cos\theta_r f_r(\theta_i, \phi_i, \theta_r, \phi_r) d\Omega_r. \quad (2.81)$$

Again, we have discussed everything with respect to reflectance towards the backward hemisphere. The transmission equivalent of the reflectance ρ is the transmittance τ. All of the above expressions for reflectance can be expressed as transmittance τ, with the substitution $r \rightarrow t$. Thus, the directional–hemispherical transmittance is $\tau(\theta_i, \phi_i; 2\pi)$, given by

$$\tau(\theta_i, \phi_i; \Omega_t) = \int_{\Omega_t} \cos\theta_t f_t(\theta_i, \phi_i, \theta_t, \phi_t) d\Omega_t. \quad (2.82)$$

2.4.4 Two Ideal Bidirectional Reflectance Distribution Functions

There are two ideal cases for the BRDF. The first is the specular reflector, with reflectance ρ_{sp}. For such a surface, the radiance reflected from the surface, $L_r(\theta_r, \phi_r)$, is equal to the product of ρ_{sp} and the radiance onto the surface in the specular direction, $L_i(\theta_i, \phi_i \pm \pi)$; that is,

$$L_r(\theta_r, \phi_r) = \rho_{sp} L_i(\theta_i, \phi_i \pm \pi). \tag{2.83}$$

Applying the definition of the BRDF, Eq. (2.59), with Eq. (2.11), and integrating, we find ourselves looking for a function that satisfies

$$\rho_{sp} L_i(\theta_i, \phi_i \pm \pi) = \int \cos\theta_i L_i(\theta_i, \phi_i) f_r(\theta_i, \phi_i, \theta_r, \phi_r) d\Omega_i \tag{2.84}$$

for any incident radiance $L_i(\theta_i, \phi_i)$. This need is precisely what is defined by the Dirac delta-function, $\delta(x)$, which is defined, in one dimension, as the function (technically, a distribution) that satisfies

$$\int \delta(x - x_0) f(x) dx = f(x_0) \tag{2.85}$$

for all $f(x)$. The function $\delta(x)$ can be thought of as being zero everywhere, except at $x = 0$, where it has a large enough value that its integral satisfies Eq. (2.85). It can be approximated as the limit of a Gaussian or sinc function normalized to unity as the width of that function approaches zero. For a purely specular surface, we find that the BRDF is

$$f_{r,sp}(\theta_i, \phi_i, \theta_r, \phi_r) = \rho_{sp} \frac{\delta(\theta_i - \theta_r)\delta(\phi_i - \phi_r \pm \pi)}{\sin\theta_i \cos\theta_i}. \tag{2.86}$$

Such a specular reflecting surface is never perfectly realized. Real surfaces, owing to their imperfect flatness, always exhibit some width to their BRDF. Furthermore, because there is always some residual roughness, if only at the atomic level, BRDF at directions far from the specular direction is always nonzero.

A perfectly diffuse, or Lambertian, surface is one in which the radiance is uniform for all directions, for any illumination. In that case, the BRDF is also uniform. Considering Eq. (2.81), we have, for a unit-reflectance Lambertian surface,

$$\rho(\theta_i, \phi_i; 2\pi) = f_r \int_{2\pi} \cos\theta_r d\Omega_r = f_r \pi, \tag{2.87}$$

where $\rho(\theta_i, \phi_i; 2\pi)$ is the diffuse hemispherical reflectance for incident direction (θ_i, ϕ_i). Because of Helmholtz reciprocity, the BRDF cannot depend upon incident direction, if it is uniform in the scattering direction. Thus, the BRDF for this surface is

$$f_{r,\text{Lambert}} = \frac{\rho(2\pi)}{\pi}. \tag{2.88}$$

where $\rho(2\pi)$ is the hemispherical reflectance. Thus, a perfectly reflecting Lambertian surface will have a BRDF of $f_{r,\text{Lambert}} = 1/\pi \approx 0.3183\ \text{sr}^{-1}$. No real surfaces exhibit perfect Lambertian character. Often, for example, the BRDF is somewhat higher in directions normal to the surface or in the specular direction and decay below the mean value at directions close to the surface horizon.

2.4.5 Absorptance

When light is incident upon a medium, it is either reflected [characterized by its directional–hemispherical reflectance $\rho(\theta_i, \phi_i; 2\pi)$], transmitted [characterized by its directional–hemispherical transmittance $\tau(\theta_i, \phi_i; 2\pi)$], or absorbed. Thus, we define the absorptance as the fraction of the incident light that is absorbed by

$$\alpha(\theta_i, \phi_i) = 1 - \rho(\theta_i, \phi_i; 2\pi) - \tau(\theta_i, \phi_i; 2\pi), \tag{2.89}$$

which must, like the reflectance and transmittance, be in the range $0 \leq \alpha(\theta_i, \phi_i) \leq 1$.

When light is transmitted through a material, some of it is lost to absorption. For a plane-wave incident and simple homogeneous materials, we see that there is a spatial dependence

$$\hat{\mathbf{E}} = \hat{\mathbf{E}}_0 \exp(\mathrm{i}\mathbf{k}\cdot\mathbf{r}) = \hat{\mathbf{E}}_0 \exp\left[\mathrm{i}\left(k_x x + k_y y + k_z z\right)\right]. \tag{2.90}$$

Since $k_z = (\omega/c_0)(\epsilon\mu - \sin^2\theta_i)^{1/2}$, for normal incidence,

$$k_z = \frac{\omega\tilde{n}}{c_0} = \frac{2\pi\tilde{n}}{\lambda_0}, \tag{2.91}$$

(λ_0 is the vacuum wavelength) the imaginary part contributes to a z dependence to the amplitude of $\exp(-2\pi \mathrm{Im}(\tilde{n})z/\lambda_0)$, so that the irradiance decays as $\exp(-a_n z)$. Thus, we have a Naperian, natural absorption coefficient a_n given by

$$a_n = \frac{4\pi \mathrm{Im}\,\tilde{n}}{\lambda_0}. \tag{2.92}$$

The imaginary part of the refractive index, also known as the extinction coefficient, defines the absorptive capacity of the material. In practical applications, we often use expressions based upon exponential decay in the common basis. That is, we express the irradiance decay behavior as 10^{-az} rather than $\exp(-a_n z)$. In this case, the common absorption coefficient a is related to the Naperian absorption coefficient a_n by

$$a = a_n / \log_e(10) \approx 0.4343 a_n. \tag{2.93}$$

It is important that one not confuse these two absorptions coefficients, and the best practice is to always specify which basis we are working in, rather than rely on nomenclature alone. The net absorption can also be expressed by the Naperian absorbance, $A_n = a_n z$ or by the common absorbance, $A = az$. The internal transmittance, not accounting for reflection losses, is then

$$\tau = \exp(-A_n) = 10^{-A}. \tag{2.94}$$

The optical theorem describes the connection between a scattering function and the extinction coefficient. Consider an incident plane wave on an object, which then scatters with a spherical wave, as in Eq. (2.40), where the spherical wave amplitude is proportional to the amplitude of the incident plane wave,

$$\hat{\mathbf{E}}_s(\hat{\mathbf{r}}) = \overleftrightarrow{\mathbf{X}}(\hat{\mathbf{r}}) \cdot \hat{\mathbf{E}}_i. \tag{2.95}$$

The matrix $\overleftrightarrow{\mathbf{X}}(\hat{\mathbf{r}})$ is the amplitude scattering matrix. If we let the incident plane wave be propagating in the z-direction, the total field is

$$\hat{\mathbf{E}} = \hat{\mathbf{E}}_i \exp(ikz) + \frac{1}{kr} \overleftrightarrow{\mathbf{X}}(\hat{\mathbf{r}}) \cdot \hat{\mathbf{E}}_i \exp(ikr). \tag{2.96}$$

Near the forward scattering direction, x and y are much smaller than z, so that

$$r \cong z + \left(x^2 + y^2\right)/2z. \tag{2.97}$$

The irradiance in these forward directions is then

$$E = \frac{\hat{n}Z_0}{2} \left| \hat{\mathbf{E}}_i \exp(ikz) + \frac{1}{kr} \overleftrightarrow{\mathbf{X}}(\hat{\mathbf{r}}) \cdot \hat{\mathbf{E}}_i \exp(ikz) \exp\left[ik\left(x^2 + y^2\right)/2z\right] \right|^2. \tag{2.98}$$

Performing the magnitude-square and keeping the lowest order terms, we can show that the irradiance is

$$E = \frac{\hat{n}Z_0}{2} \left|\hat{\mathbf{E}}_i\right|^2 \left(1 + \frac{2}{kz} \mathrm{Re}\left[\overleftrightarrow{\mathbf{X}}(\hat{\mathbf{z}}) \cdot \hat{\mathbf{e}}_i \exp\left[ik\left(x^2 + y^2\right)/2z\right]\right]\right), \tag{2.99}$$

where the $\hat{\mathbf{e}}_i = \hat{\mathbf{E}}_i / |\hat{\mathbf{E}}_i|$ is the direction of the incident electric field. If we integrate over an area A, which is large enough that the integral converges,

$$\int_A \exp\left[ik\left(x^2 + y^2\right)/2z\right] \mathrm{d}x\mathrm{d}y \to \frac{2\pi z i}{k}, \tag{2.100}$$

we obtain the radiant power in the forward direction,

$$\Phi = \frac{\hat{n}Z_0}{2} \left|\hat{\mathbf{E}}_i\right|^2 \left[A - \frac{2\pi}{k} \mathrm{Im}\left(\overleftrightarrow{\mathbf{X}}(\hat{\mathbf{z}}) \cdot \hat{\mathbf{e}}_i\right)\right]. \tag{2.101}$$

Thus, the scatterer reduces the radiant power in the beam as if it had a cross sectional area given by

$$\sigma_{\mathrm{ext}} = \frac{2\pi}{k} \mathrm{Im}\left(\overleftrightarrow{\mathbf{X}}(\hat{\mathbf{z}}) \cdot \hat{\mathbf{e}}_i\right), \tag{2.102}$$

which we call the extinction cross section. Eq. (2.102) is an expression of the optical theorem, which relates the extinction cross section of a scatterer to the imaginary part of the scattering function, evaluated in the forward scattering direction. Suppose we have N/V such scatterers per unit volume. Upon passing through a length dz through the medium, the total number of scatterers encountered per unit area will be N dz/V, and the effective area lost to the beam per unit area will be $N\ \sigma_{\mathrm{ext}}$ dz/V. Thus, the irradiance will follow

$$\mathrm{d}E = -\frac{NE\sigma_{\mathrm{ext}}}{V}\,\mathrm{d}z. \tag{2.103}$$

The solution to Eq. (2.103) is $E(z) = E(0)\exp(-a_{\mathrm{n,ext}}z)$, where the Naperian extinction coefficient is

$$a_{\mathrm{n,ext}} = \frac{N\sigma_{\mathrm{ext}}}{V}. \tag{2.104}$$

We see that the extinction coefficient is directly proportional to the concentration of absorbing or scattering entities. This is often referred to as the Beer–Lambert law of absorption. In the absence of scattering, the extinction coefficient is entirely due to absorption. The Beer–Lambert law can also be expressed in terms of the intrinsic molecular property, the specific absorbance, ε, which is the absorbance per unit path length and unit concentration. That is, the absorbance is given by

$$A = \varepsilon Cl, \tag{2.105}$$

where l is the path length, and $C = N/V$ is the concentration. It should also be noted that for many chemical applications of spectrophotometry, the specific absorbance ε is also known as the molar absorption coefficient or the molar extinction coefficient.

The optical density D_τ, after correction for reflection losses, is identical to the absorbance A and is related to the internal or bulk transmittance τ by

$$D_\tau = -\log_{10}\tau, \tag{2.106}$$

where τ needs to be specified with a geometry. The absorbance A, not to be confused with the absorptance α, defined in Eq. (2.89), is often used in chemical absorption applications or in colored or neutral-density filters.

2.4.6 Fluorescence: Bispectral Luminescent Radiance Factor

Consider the situation where the light emitted (luminesced) by the material has a different wavelength λ from the incident wavelength μ. Eq. (2.53), above, describes the spectral radiance as a function of the radiant power incident upon the surface at a specific location, and from a specific direction. In a manner nearly identical to how we developed the BRDF, we can develop a bispectral BRDF (BBRDF), which for uniform illumination is the ratio of the spectral luminescent radiance to the incident irradiance,

$$f_\lambda = \frac{L_\lambda}{E_\mathrm{i}} \tag{2.107}$$

or, for a small spot illumination, is

$$f_\lambda = \frac{\mathrm{d}\Phi_{\lambda\mathrm{r}}}{\Phi_\mathrm{i}\cos\theta_\mathrm{r}\,\mathrm{d}\Omega_\mathrm{r}\,\mathrm{d}\lambda}, \tag{2.108}$$

where $\mathrm{d}\Phi_{\lambda\mathrm{r}}$ is the reflected spectral radiant power in solid angle $\mathrm{d}\Omega_\mathrm{r}$. Eqs. (2.107) or (2.108) can be used for describing the angular and spectral distribution of light emitted by a material due to fluorescence. If the scattering properties are entirely elastic, that is, with no fluorescence, the BBRDF is related to the BRDF as

$$f_\lambda = f_\mathrm{r}\delta(\lambda - \mu). \tag{2.109}$$

The term BBRDF has not found its way into the standard nomenclature and there may be other descriptive names for it. For a more complicated illumination $L_{\lambda,\mathrm{i}}(\theta_\mathrm{i}, \phi_\mathrm{i}, \mu)$, the spectral radiance is given by integration of the BBRDF over the incident illumination,

$$L_{\lambda,\mathrm{r}}(\Omega_\mathrm{i}, L_\mathrm{i}, M; \theta_\mathrm{r}, \phi_\mathrm{r}, \lambda) = \int_M \int_{\Omega_\mathrm{i}} \cos\theta_\mathrm{r}\cos\theta_\mathrm{i} f_\lambda(\theta_\mathrm{i}, \phi_\mathrm{i}, \mu; \theta_\mathrm{r}, \phi_\mathrm{r}, \lambda) L_{\lambda,\mathrm{i}}(\theta_\mathrm{i}, \phi_\mathrm{i}, \mu)\,\mathrm{d}\Omega_\mathrm{i}\,\mathrm{d}\mu, \tag{2.110}$$

where M represents the spectral bandwidth of the incident light. The term defined in the ILV, which is the common term in the standards literature to describe luminescence, is the bispectral luminescent radiance factor, $\beta_{\mathrm{L},\lambda}$, defined as the spectral radiance, normalized by the radiance of a perfectly reflecting Lambertian diffuser identically irradiated and viewed. For unidirectional illumination,

$$\beta_{\mathrm{L},\lambda}(\theta_\mathrm{i}, \phi_\mathrm{i}, \mu; \theta_\mathrm{r}, \phi_\mathrm{r}, \lambda) = \frac{f_\lambda(\theta_\mathrm{i}, \phi_\mathrm{i}, \mu; \theta_\mathrm{r}, \phi_\mathrm{r}, \lambda)}{f_{\mathrm{r,Lambert}}} = \pi f_\lambda(\theta_\mathrm{i}, \phi_\mathrm{i}, \mu; \theta_\mathrm{r}, \phi_\mathrm{r}, \lambda). \tag{2.111}$$

In the fluorescence community, there is often a distinction between radiant properties and quantum properties. The radiant energy associated with a single photon of frequency v or wavelength λ is given by

$$Q_\mathrm{p} = hv = \frac{hc_0}{\lambda}. \tag{2.112}$$

Thus, the quantum equivalent to the radiant power is given by

$$\Phi_\mathrm{q} = \frac{\Phi}{Q_\mathrm{q}} = \frac{\Phi\lambda}{hc_0}, \tag{2.113}$$

which represents the number of photons per unit time (photon flux). For elastic properties of light, such as reflectance, BRDF, or absorptance, there need not be any distinction between the radiant quantities and the quantum quantities, since, being ratios of quantities with identical wavelength, the quantity in Eq. (2.111) and (2.112) is a common constant. However, for fluorescence, the quantum properties will differ from the radiant properties. For example, a quantum BBRDF $f_{p\lambda}$ can be defined as

$$f_{p\lambda} = \frac{d\Phi_{pr}}{\Phi_{pi} \cos\theta_r \, d\Omega_r \, d\lambda}, \tag{2.114}$$

where $d\Phi_{p\lambda r}$ is the emitted spectral photon flux at wavelength λ and Φ_{pi} is the incident photon flux at excitation wavelength, μ. Combining Eqs. (2.113) and (2.114), we see that the quantum BBRDF is related to the BBRDF by

$$f_{p\lambda} = f_{\lambda} \frac{\lambda}{\mu}. \tag{2.115}$$

Similarly, a quantum bispectral luminescent radiance factor, $\beta_{Lp,\lambda}$, can be defined, which is related to $\beta_{L,\lambda}$ by

$$\beta_{Lp,\lambda} = \beta_{L,\lambda} \frac{\lambda}{\mu}. \tag{2.116}$$

2.4.7 Emittance and the Kirchhoff Relationship

Figure 2.5 shows an object inside a cavity. The material surrounding the cavity is held at a temperature T_1. It radiates according to Planck's law with a spectral radiance given by $\varepsilon_1(\lambda, \theta, \phi) L_{e,\lambda}(\lambda, T_1)$ where $\varepsilon_1(\lambda, \theta, \phi)$ is its emissivity and $L_{e,\lambda}(\lambda, T_1)$ is given by Eq. (2.22). Similarly, the object inside the cavity is at temperature T_2 and emits a spectral radiance $\varepsilon_2(\lambda, \theta, \phi) L_{e,\lambda}(\lambda, T_2)$. The inward and outward spectral radiances are each the sum of that thermally radiated by the respective material and that reflected and transmitted by it. Thus, we have, for each wavelength λ,

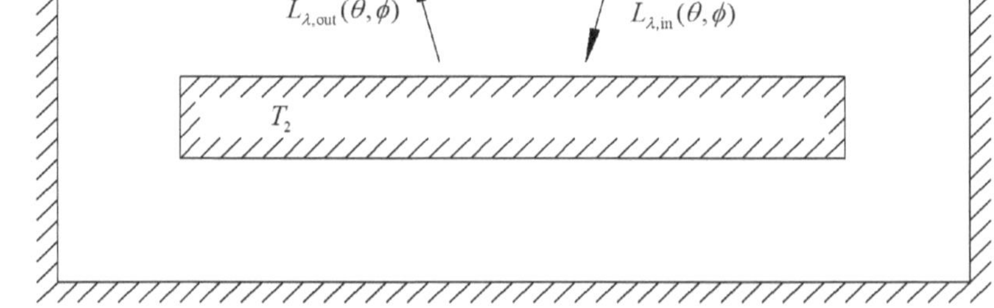

FIGURE 2.5 An object at temperature T_2 inside a cavity at temperature T_1.

$$
\begin{aligned}
L_{\lambda,\mathrm{out}}(\theta,\phi) &= \varepsilon_2(\lambda,\theta,\phi)L_{\mathrm{e},\lambda}(\lambda,T_2) \\
&\quad + \int f_{\mathrm{r}2}(\theta,\phi,\theta',\phi')L_{\lambda,\mathrm{in}}(\theta',\phi')\cos\theta'\,\mathrm{d}\Omega' \\
&\quad + \int f_{\mathrm{t}2}(\theta,\phi,\theta',\phi')L_{\lambda,\mathrm{in}}(\theta',\phi')\cos\theta'\,\mathrm{d}\Omega', \\
L_{\lambda,\mathrm{in}}(\theta,\phi) &= \varepsilon_1(\lambda,\theta,\phi)L_{\mathrm{e},\lambda}(\lambda,T_1) \\
&\quad + \int f_{\mathrm{r}1}(\theta,\phi,\theta',\phi')L_{\lambda,\mathrm{out}}(\theta',\phi')\cos\theta'\,\mathrm{d}\Omega' \\
&\quad + \int f_{\mathrm{t}1}(\theta,\phi,\theta',\phi')L_{\lambda,\mathrm{out}}(\theta',\phi')\cos\theta'\,\mathrm{d}\Omega'.
\end{aligned}
\tag{2.117}
$$

In equilibrium, there is no net transfer of energy, so that $L_{\lambda,\mathrm{in}} = L_{\lambda,\mathrm{out}}$. In this condition, it follows that

$$
\begin{aligned}
\varepsilon_1(\lambda,\theta,\phi) &= 1 - \int f_{\mathrm{r}1}(\theta,\phi,\theta',\phi')\cos\theta'\,\mathrm{d}\Omega' - \int f_{\mathrm{t}1}(\theta,\phi,\theta',\phi')\cos\theta'\,\mathrm{d}\Omega', \\
\varepsilon_2(\lambda,\theta,\phi) &= 1 - \int f_{\mathrm{r}2}(\theta,\phi,\theta',\phi')\cos\theta'\,\mathrm{d}\Omega' - \int f_{\mathrm{t}2}(\theta,\phi,\theta',\phi')\cos\theta'\,\mathrm{d}\Omega', \\
T_1 &= T_2.
\end{aligned}
\tag{2.118}
$$

The first two conditions must be met for all directions, wavelengths, and polarizations. From Eq. (2.81), it can be seen that the integrals in Eq. (2.118) are simply the reflectances and transmittances of the two materials. Furthermore, from Eq. (2.89), it can be seen that

$$
\begin{aligned}
\varepsilon_1(\lambda,\theta,\phi) &= \alpha_1(\theta,\phi), \\
\varepsilon_2(\lambda,\theta,\phi) &= \alpha_2(\theta,\phi).
\end{aligned}
\tag{2.119}
$$

The Eq. (2.119) is often referred to as the Kirchhoff law of radiation, that the emissivity of a material is given by its absorptance at each wavelength. It also holds that the inward and outward spectral radiances are given by

$$
L_{\lambda,\mathrm{in}}(\theta,\phi) = L_{\lambda,\mathrm{out}}(\theta,\phi) = L_{\mathrm{e},\lambda}(\lambda,T). \tag{2.120}
$$

Thus, the spectral radiance inside a closed cavity is given by the Planck spectral radiance, regardless of the emissivity of the walls. This statement, in effect, says that the radiation field is in thermal equilibrium with the walls.

2.5 POLARIZATION

The radiometric quantities, defined in Section 2.2, and the spectrophotometric quantities, given in Section 2.4, were all given without regard to the polarization state of the light. However, as we know from Section 2.3, the electric and magnetic fields, **E** and **H**, are vector quantities. Both of these amplitudes, being transverse to the direction of propagation (at least in nonabsorbing media), can be characterized by the projections of one of the fields onto two orthogonal directions. The field chosen to represent the polarization is usually the electric field, but the reader should be aware that history has not been consistent in this choice. Thus, statements like, "The polarization is chosen to be in the direction of the electric field," are not superfluous. This section is a very brief introduction to polarization. Further reading can be found in [11, 12].

Next, we must choose a basis set for defining the polarization. Here, again, there can be numerous choices. In some applications, a basis set that is specific to the source and detector is most convenient. For example, if the source, sample, and detector define a plane, then the basis set for both the incident and exiting light can be chosen such that one vector is perpendicular to this plane and the other vectors, specific to each direction, are in the plane.

For spectrophotometric measurements, however, this is not the most convenient basis. Instead, we will use a basis set defined by the plane of incidence, which contains the incident direction and the surface normal, and the plane of exitance, which contains the exiting direction and the surface normal. Light having an electric field perpendicular to the respective plane will be considered s-polarized (s being from the German *senkrecht*), while that parallel to this plane will be p-polarized (p being from the German *parallel*). These polarization states are also alternatively referred to as transverse electric (TE) and transverse magnetic (TM), respectively.

Then comes the question of the handedness between these vectors. We choose, somewhat arbitrarily, for **s**, **p**, and **k** to form a right-handed basis set, where **s** and **p** are unit vectors in the directions of the electric field for s- and p-polarized light, respectively. If the electric field amplitude is $\hat{\mathbf{E}}$, the components along the **s** and **p** directions are $E_s = \mathbf{s}\cdot\hat{\mathbf{E}}$ and $E_p = \mathbf{p}\cdot\hat{\mathbf{E}}$, respectively.

The Stokes formalism is often used to characterize the intensity and polarization of a light beam. The Stokes vector is defined as

$$\mathbf{S} = \begin{pmatrix} \langle |E_s|^2 \rangle + \langle |E_p|^2 \rangle \\ \langle |E_s|^2 \rangle - \langle |E_p|^2 \rangle \\ \mathrm{Re}\left\langle E_s E_p^* \right\rangle \\ \mathrm{Im}\left\langle E_s E_p^* \right\rangle \end{pmatrix}. \tag{2.121}$$

These four elements have a fundamental connection to the complex Hermitian coherency matrix,

$$\begin{pmatrix} \langle E_s E_s^* \rangle & \left\langle E_s E_p^* \right\rangle \\ \langle E_p E_s^* \rangle & \left\langle E_p E_p^* \right\rangle \end{pmatrix}, \tag{2.122}$$

which contains all of the information in the electric field. That is, the sum and difference between the diagonal coherency matrix elements make up the first two elements of the Stokes vector, while the sum and difference of the off-diagonal elements make up the second two elements. The Stokes vector has the property that

$$[\mathbf{S}]_1^{\ 2} \geq [\mathbf{S}]_2^{\ 2} + [\mathbf{S}]_3^{\ 2} + [\mathbf{S}]_4^{\ 2}. \tag{2.123}$$

The averages in Eq. (2.122) can be either temporal or spatial averages, depending upon the application. For totally polarized light, the left and right sides of Eq. (2.123) are equal.

The Stokes vector given in Eq. (2.121) can be shown to be equivalent to

$$\mathbf{S} = \begin{pmatrix} \langle |E_{\mathrm{s}}|^2 \rangle + \langle |E_{\mathrm{p}}|^2 \rangle \\ \langle |E_{\mathrm{s}}|^2 \rangle - \langle |E_{\mathrm{p}}|^2 \rangle \\ \langle |E_{45^\circ}|^2 \rangle - \langle |E_{-45^\circ}|^2 \rangle \\ \langle |E_{\mathrm{rcp}}|^2 \rangle - \langle |E_{\mathrm{lcp}}|^2 \rangle \end{pmatrix}, \tag{2.124}$$

where light polarized at 45° is along the $(\mathbf{s}+\mathbf{p})/2^{1/2}$ direction and light polarized at −45° is along the $(\mathbf{s}-\mathbf{p})/2^{1/2}$ direction. Light that is right circularly polarized (rcp) is polarized along $(\mathbf{s}-i\mathbf{p})/2^{1/2}$, while that of left circularly polarized (lcp) is polarized along $(\mathbf{s}+i\mathbf{p})/2^{1/2}$. Light that is right circularly polarized has its fields at a given point in space rotating in a clockwise fashion when viewing into the beam. Equivalently, the fields a right-hand circularly polarized light trace out a right-handed screw in space.

The conventions mentioned above depend upon the time dependence used in the complex notation. Furthermore, one can consider the handedness from the perspective of the source (looking down the beam) or from the perspective of the detector (looking up the beam). Different fields have been known to use different conventions.

Any radiant quantity can be generalized as a Stokes vector. Thus, we have

$$\begin{aligned} Q &\rightarrow \mathbf{Q}, \\ \Phi &\rightarrow \mathbf{\Phi}, \\ L &\rightarrow \mathbf{L}, \\ E &\rightarrow \mathbf{E}, \end{aligned} \tag{2.125}$$

for the radiant energy, power, radiance, and irradiance, respectively. The first element of a Stokes vector is the unpolarized quantity.

The relationship between ingoing and outgoing Stokes vectors [assuming it is a linear function, such as Eq. (2.51)] can be characterized by a linear transformation, referred to as a Mueller Matrix. The 4×4 Mueller matrix can be used to generalize any of the intensity-transforming quantities. For example, the scattering function S, given in Eq. (2.51), is a relationship between the radiant energy and the spectral radiance. Thus, Eq. (2.51) can be rewritten

$$\mathrm{d}\mathbf{Q}_{\lambda\mathrm{r}} = \mathbf{S}_{\lambda t} \cos\theta_{\mathrm{r}} \mathrm{d}\mathbf{Q}_{\mathrm{i}}\, \mathrm{d}A_{\mathrm{r}}\, \mathrm{d}\Omega_{\mathrm{r}}\, \mathrm{d}\lambda\, \mathrm{d}t_{\mathrm{r}}, \tag{2.126}$$

where $\mathrm{d}\mathbf{Q}_{\lambda\mathrm{r}}$ is the reflected Stokes vector spectral radiant energy, $\mathbf{S}_{\lambda\mathrm{r}}$ is the Mueller matrix scattering function, and $\mathrm{d}\mathbf{Q}_{\mathrm{i}}$ is the incident Stokes vector spectral radiant energy. Similarly, the BRDF is also a Mueller matrix,

$$f_r \rightarrow \mathbf{f}_r. \tag{2.127}$$

In general, the reflectances and transmittances are slightly more complicated to describe, because they require specifying the geometry of the incident and collected light. For example, the specular reflectance and regular transmittance are both, like the BRDF, Mueller matrices, because they are relationships between two Stokes vectors:

$$\begin{aligned} \rho_{sp} &= \boldsymbol{\rho}_{sp}, \\ \tau_{sp} &= \boldsymbol{\tau}_{sp}. \end{aligned} \tag{2.128}$$

The diffuse reflectance and the absorptance, however, are more complicated. Consider trying to generalize Eq. (2.79). Because there is an integral over the incident radiance and the BRDF, there is not necessarily a unique way to generalize Eq. (2.79). However, one can characterize the diffuse-directional reflectance for unpolarized diffuse illumination, in which case it would be appropriate to characterize it simply by a Stokes vector. Because of the relationship between emittance and absorptance, emittance would also be a Stokes vector.

Let us now consider how the Stokes parameters change upon time reversal. In the conventions we are using, where we have defined **s**, **p**, and **k** to form a right-handed basis set, the first and second Stokes parameters clearly do not change sign upon time reversal. The directions of 45° and −45° polarizations, however, switch in space. Thus, the third Stokes parameter changes sign upon time reversal. For the circularly polarized states, the direction of rotation of the polarization reverses, as well as the direction of propagation. Thus, circularly polarized light does not change its handedness upon time reversal, and the fourth Stokes parameter does not change sign. The forward and backwards Stokes vectors, **S** and $\mathbf{S}^{TR}$, respectively, are related by

$$\mathbf{S}^{TR} = \begin{pmatrix} 1 & & & \\ & 1 & & \\ & & -1 & \\ & & & 1 \end{pmatrix} \mathbf{S}. \tag{2.129}$$

Since a Mueller matrix relates two Stokes vectors, a Mueller matrix upon time reversal will obey

$$\mathbf{M}^{TR} = \begin{pmatrix} 1 & & & \\ & 1 & & \\ & & -1 & \\ & & & 1 \end{pmatrix} \mathbf{M}^{T} \begin{pmatrix} 1 & & & \\ & 1 & & \\ & & -1 & \\ & & & 1 \end{pmatrix}. \tag{2.130}$$

Thus, Helmholtz reciprocity for the Mueller matrix BRDF, using the polarization convention described here, can be expressed as:

$$[\mathbf{f}_r(\theta_i, \phi_i, \theta_r, \phi_r)]_{jj'} = \begin{cases} -[\mathbf{f}_r(\theta_r, \phi_r, \theta_i, \phi_i)]_{j'j} & \text{for } jj' = 31, 32, 34, 13, 23, 43 \\ [\mathbf{f}_r(\theta_r, \phi_r, \theta_i, \phi_i)]_{j'j} & \text{otherwise} \end{cases} \tag{2.131}$$

2.6 REFLECTION AND TRANSMISSION FROM FLAT SURFACES

2.6.1 Snell's Law of Refraction

A plane wave that is incident upon the flat interface between two materials, 1 and 2, will refract so that its direction of propagation changes. The propagation vector in medium 1 is

$$\mathbf{k}_1 = k_1 \sin\theta_1 \cos\phi_1 \hat{\mathbf{x}} + k_1 \sin\theta_1 \sin\phi_1 \hat{\mathbf{y}} - k_1 \cos\theta_1 \hat{\mathbf{z}}, \tag{2.132}$$

while that in medium 2 is

$$\mathbf{k}_2 = k_2 \sin\theta_2 \cos\phi_2 \hat{\mathbf{x}} + k_2 \sin\theta_2 \sin\phi_2 \hat{\mathbf{y}} - k_2 \cos\theta_2 \hat{\mathbf{z}}. \tag{2.133}$$

The magnitudes of the vectors k_1 and k_2 are constrained to be $(\omega/c_0)\widetilde{n}_1$ and $(\omega/c_0)\widetilde{n}_2$ respectively. If the interface is along the xy plane, then in order for the phase to be consistent across the interface at every point on it, the x and y components of the propagation vectors must match. Thus

$$\begin{aligned} \widetilde{n}_1 \sin\theta_1 \cos\phi_1 &= \widetilde{n}_2 \sin\theta_2 \cos\phi_2, \\ \widetilde{n}_1 \sin\theta_1 \sin\phi_1 &= \widetilde{n}_2 \sin\theta_2 \sin\phi_2. \end{aligned} \tag{2.134}$$

As a result,

$$\widetilde{n}_1 \sin\theta_1 = \widetilde{n}_2 \sin\theta_2 \tag{2.135}$$

and $\phi_1 = \phi_2$. Eq. (2.135) is usually referred to as Snell's law. The z components of the propagation vectors then follow

$$\begin{aligned} k_{1,z} &= [(\omega\widetilde{n}_1/c_0)^2 - k_{1,x}^2 - k_{1,y}^2]^{1/2}, \\ k_{2,z} &= [(\omega\widetilde{n}_2/c_0)^2 - k_{2,x}^2 - k_{2,y}^2]^{1/2}. \end{aligned} \tag{2.136}$$

Suppose that we have a bundle of rays, contained in a solid angle $\mathrm{d}\Omega_1$ incident upon that interface. If we differentiate Eq. (2.135), we have

$$\widetilde{n}_1 \cos\theta_1 \, \mathrm{d}\theta_1 = \widetilde{n}_2 \cos\theta_2 \, \mathrm{d}\theta_2. \tag{2.137}$$

The solid angle is

$$\mathrm{d}\Omega_1 = \sin\theta_1 \, \mathrm{d}\theta_1 \, \mathrm{d}\phi_1 = \frac{\widetilde{n}_2}{\widetilde{n}_1} \sin\theta_2 \frac{\widetilde{n}_2 \cos\theta_2}{\widetilde{n}_1 \cos\theta_1} \mathrm{d}\theta_2 \, \mathrm{d}\phi_2 = \frac{\widetilde{n}_2^2 \cos\theta_2}{\widetilde{n}_1^2 \cos\theta_1} \mathrm{d}\Omega_2. \tag{2.138}$$

Thus, Snell's law for solid angles is

$$\widetilde{n}_1^2 \cos\theta_1 \, \mathrm{d}\Omega_1 = \widetilde{n}_2^2 \cos\theta_2 \, \mathrm{d}\Omega_2. \tag{2.139}$$

The quantity

$$G = \int_A \int_\Omega \cos\theta \, \mathrm{d}\Omega \, \mathrm{d}A \tag{2.140}$$

is often called the *geometric extent* of a beam of rays. Using Eq. (2.139), we see that the quantity $\widetilde{n}^2 G$ is an invariant in a beam. Consider the definition of radiance, given in Eq. (2.10). If we consider a radiance incident on a boundary between two materials, having indices $\tilde{n}_1$ on one side and $\tilde{n}_2$ on the other, and assume there is no reflectance or absorption (such as when the interface contains an antireflection coating), then the radiant powers on the two sides are equal:

$$L_1 \cos\theta_1 \, dA \, d\Omega_1 = L_2 \cos\theta_2 \, dA \, d\Omega_2. \tag{2.141}$$

Combining Eqs. (2.139) and (2.141), we have

$$\frac{L_1}{\tilde{n}_1^2} = \frac{L_2}{\tilde{n}_2^2}. \tag{2.142}$$

Thus, the quantity $L/\tilde{n}^2$ is also an invariant, provided there are no losses.

2.6.2 Fresnel Reflection

Boundary conditions require that the components of both the electric field **E** and magnetic field **H** transverse to the surface be continuous across the boundary [3–5]. In addition, the components of the electric displacement **D** and magnetic induction **B** normal to the surface must be continuous. Consider a plane-wave incident from a medium of complex refractive index $\tilde{n}_1$ to a medium of complex refractive index $\tilde{n}_2$ (see Fig. 2.6). If we consider s-polarization, the in-surface boundary conditions can be written

$$\begin{aligned} E_i + E_r &= E_t, \\ -H_i \cos\theta_1 + H_r \cos\theta_1 &= -H_t \cos\theta_2, \end{aligned} \tag{2.143}$$

where we use the nonbold-faced symbols, E and H, to represent the magnitudes of **E** and **H**, respectively, and not the irradiance. From Eq. (2.36), the relationship between the electric and magnetic fields is

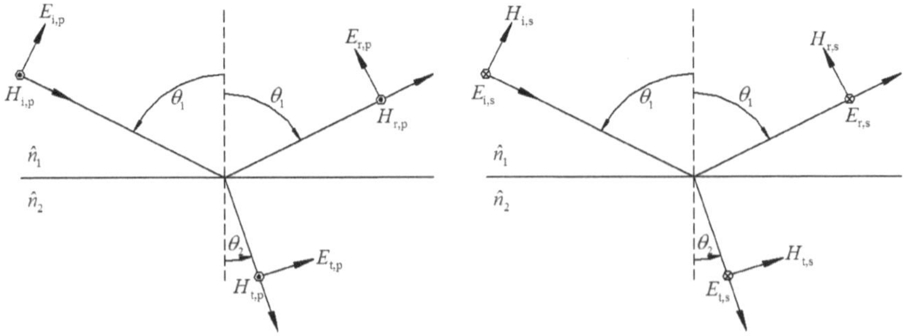

FIGURE 2.6 Electric and magnetic fields for reflection of plane waves from the interface between two media for (left) p-polarization and (right) for s-polarization.

$$\begin{aligned} H_{\mathrm{i}} &= \hat{n}_1 Z_0 E_{\mathrm{i}}, \\ H_{\mathrm{r}} &= \hat{n}_1 Z_0 E_{\mathrm{r}}, \\ H_{\mathrm{t}} &= \hat{n}_2 Z_0 E_{\mathrm{t}}. \end{aligned} \tag{2.144}$$

If we make the substitutions

$$\begin{aligned} E_{\mathrm{r}} &= r_{\mathrm{s}} E_{\mathrm{i}}, \\ E_{\mathrm{t}} &= t_{\mathrm{s}} E_{\mathrm{i}}, \end{aligned} \tag{2.145}$$

we can solve Eqs. (2.143), (2.144), and (2.145) for the coefficients r_{s} and t_{s},

$$\begin{aligned} r_{\mathrm{s}} &= \frac{\hat{n}_1 \cos\theta_1 - \hat{n}_2 \cos\theta_2}{\hat{n}_1 \cos\theta_1 + \hat{n}_2 \cos\theta_2}, \\ t_{\mathrm{s}} &= \frac{2\hat{n}_1 \cos\theta_1}{\hat{n}_1 \cos\theta_1 + \hat{n}_2 \cos\theta_2}. \end{aligned} \tag{2.146}$$

[Note that $\mu = 1$ and thus $\hat{n} = \tilde{n}$ for all but the most exotic optical materials (magnetic materials being the exception). Furthermore, $\tilde{n}$ is used to calculate the cosine of the angle θ_1 and θ_2 from Eq. (2.135), while $\hat{n}$ is then used to match the boundary conditions.] In a similar fashion, we consider p-polarization. The in-surface boundary conditions can be written

$$\begin{aligned} E_{\mathrm{i}} \cos\theta_1 - E_{\mathrm{r}} \cos\theta_1 &= E_{\mathrm{t}} \cos\theta_2, \\ H_{\mathrm{i}} + H_{\mathrm{r}} &= H_{\mathrm{t}}. \end{aligned} \tag{2.147}$$

If we make the substitutions

$$\begin{aligned} E_{\mathrm{r}} &= r_{\mathrm{p}} E_{\mathrm{i}}, \\ E_{\mathrm{t}} &= t_{\mathrm{p}} E_{\mathrm{i}}, \end{aligned} \tag{2.148}$$

we can solve Eqs. (2.144), (2.147), and (2.148) for the coefficients r_{p} and t_{p},

$$\begin{aligned} r_{\mathrm{p}} &= \frac{\hat{n}_2 \cos\theta_1 - \hat{n}_1 \cos\theta_2}{\hat{n}_2 \cos\theta_1 + \hat{n}_1 \cos\theta_2}, \\ t_{\mathrm{p}} &= \frac{2\hat{n}_1 \cos\theta_1}{\hat{n}_2 \cos\theta_1 + \hat{n}_1 \cos\theta_2}. \end{aligned} \tag{2.149}$$

The coefficients in Eqs. (2.146) and (2.149) are often referred to as the Fresnel reflection coefficients. Using Eq. (2.39), the ratio of the irradiances in reflection is the regular, or specular, reflectance

$$\rho = \frac{\langle |E_r|^2 \rangle}{\langle |E_i|^2 \rangle} = |r_{\mathrm{s,p}}|^2, \tag{2.150}$$

while that in transmission is the regular transmittance

$$\tau = \left| \frac{\cos\theta_2}{\cos\theta_1} \right| \frac{\hat{n}_2}{\hat{n}_1} |t_{\mathrm{s,p}}|^2. \tag{2.151}$$

It can be shown that $\tau + \rho = 1$. At normal incidence, the reflectance is given by letting $\theta_1 = \theta_2 = 0$ in either Eqs. (2.146) or (2.149)

$$\rho(0°) = \left| \frac{\hat{n}_2 - \hat{n}_1}{\hat{n}_2 + \hat{n}_1} \right|^2, \quad (2.152)$$

and the transmittance is given by

$$\tau(0°) = \frac{4\hat{n}_2 \hat{n}_1}{(\hat{n}_2 + \hat{n}_1)^2}. \quad (2.153)$$

Figure 2.7 shows the reflectance as a function of incident angle for four different materials at a wavelength of 633 nm: glass ($\tilde{n} = 1.5$), silicon ($\tilde{n} = 3.874 + 0.015i$), tungsten ($\tilde{n} = 3.640 + 2.915i$), and gold ($\tilde{n} = 0.181 + 3.074i$). Near normal incidence, s- and p-polarization yield the same values and are relatively flat. At grazing incidence, the reflectance approaches 1. For s-polarization, the transition from its value at normal incidence to that at grazing incidence is gradual and monotonic, increasing most quickly at large angles. For p-polarization, the reflectance always decreases until some angle and then increases to 1. When the materials are nonabsorbing and nonmagnetic ($\tilde{n}_1 = \hat{n}_1$ and $\tilde{n}_2 = \hat{n}_2$ are real), the reflectance for p-polarization vanishes at the Brewster angle given by

$$\theta_B = \tan^{-1}(\tilde{n}_2/\tilde{n}_1). \quad (2.154)$$

When the index of the transmitting medium is below that of the incident medium, the behavior is different. Figure 2.8 shows the reflectance as a function of angle for a number of different internal reflection configurations, with

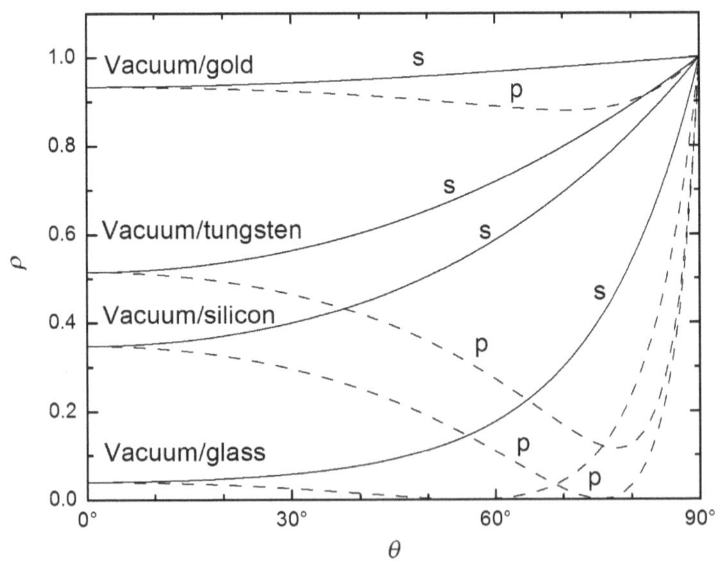

FIGURE 2.7 The reflectance as a function of angle and polarization for four different materials at $\lambda = 633$ nm.

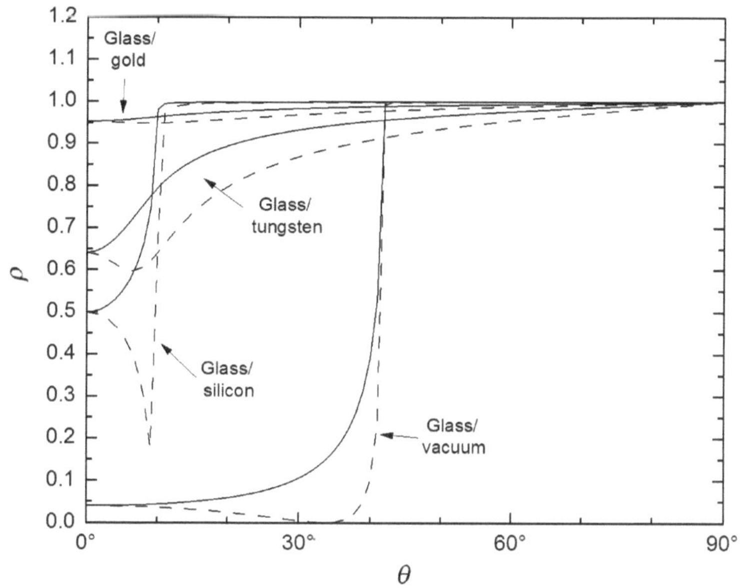

FIGURE 2.8 The internal reflection from four different interfaces at $\lambda = 633\,\text{nm}$.

glass being the incident medium. In the case of the glass to vacuum interface, the reflectances for the two polarizations are the same at normal incidence. The s-polarized reflectance gradually increased to 1, reaching that value at what is referred to as the critical angle θ_c, after which the interface is totally reflecting. This condition exists when the waves do not propagate in the transmitting medium. When the indices are real, the critical angle is given from Snell's law, Eq. (2.135) with $\theta_2 = 90°$ by

$$\theta_c = \sin^{-1}(\tilde{n}_2/\tilde{n}_1). \tag{2.155}$$

Like the previous case, the two polarizations give different results, with the reflectance of p-polarization always being lower than that of s-polarization and exhibiting a Brewster angle, where the reflectance is zero.

Past the critical angle, there exists a field on the other side of the boundary that is evanescent. That is, it decays exponentially with a decay constant given by the imaginary part of the z-component of the **k** vector. For weakly absorbing materials, this exponential tail extends into the material and results in some absorption. The curves for the glass/silicon reflectance in Fig. 2.8 appear similar to those for the glass/vacuum reflectance, in that there exists a quasi-Brewster angle and a critical angle. However, the transition at the critical angle is not as abrupt and the reflectance only gradually reaches 1. For the glass/tungsten reflectance, the features are washed out even more, and for the glass/gold reflectance, structure in the internal reflectance is completely washed out.

2.6.3 Thin Films

Figure 2.9 shows the electric and magnetic fields in the presence of a thin film. In each film, there is an upward propagating wave and a downward propagating wave. At each boundary, we have to match the conditions that the tangential components of the electric and magnetic fields must be continuous.

Consider a thin film shown in s-polarization shown in Fig. 2.9 (right). At the top interface, we have the boundary conditions

$$\begin{gathered} E_i + E_r = E_u + E_d, \\ -H_i\cos\theta_1 + H_r\cos\theta_1 = H_u\cos\theta_2 - H_d\cos\theta_2. \end{gathered} \tag{2.156}$$

At the lower interface, the boundary conditions are

$$\begin{gathered} E_u\exp(-ikl\cos\theta_2) + E_d\exp(ikl\cos\theta_2) = E_t, \\ H_u\exp(-ikl\cos\theta_2)\cos\theta_2 - H_d\exp(ikl\cos\theta_2)\cos\theta_2 = -H_t\cos\theta_3. \end{gathered} \tag{2.157}$$

By eliminating the fields inside the layer (E_u, H_u, E_d, and H_d), and using Eq. (2.145), we can write the boundary condition expressions in a matrix form as

$$\begin{aligned} &\begin{pmatrix} 1 \\ n_1\cos\theta_1 \end{pmatrix} + \begin{pmatrix} 1 \\ -n_1\cos\theta_1 \end{pmatrix} r_s \\ &= \begin{pmatrix} \cos(kl\cos\theta_2) & -\dfrac{i}{p_s}\sin(kl\cos\theta_2) \\ -ip_s\sin(kl\cos\theta_2) & \cos(kl\cos\theta_2) \end{pmatrix} \begin{pmatrix} 1 \\ n_3\cos\theta_3 \end{pmatrix} t_s, \end{aligned} \tag{2.158}$$

where $p_s = n_2\cos\theta_2$. Similarly, for p-polarization, shown in Fig. 2.9 (left), the top interface boundary conditions are

$$\begin{gathered} E_i\cos\theta_1 - E_r\cos\theta_1 = -E_u\cos\theta_2 + E_d\cos\theta_2, \\ H_i + H_r = H_u + H_d, \end{gathered} \tag{2.159}$$

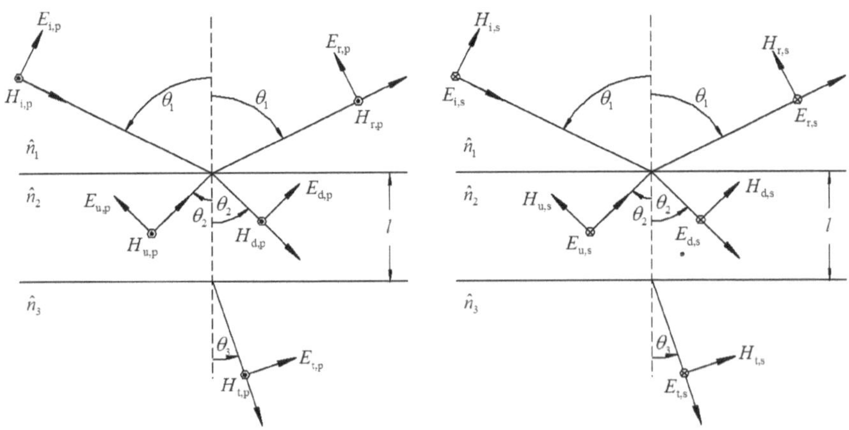

FIGURE 2.9 Electric and magnetic fields for reflection of plane waves from a dielectric film for (left) p-polarization and (right) for s-polarization.

while those at the lower interface are

$$\begin{aligned} -E_u \exp(-ikl\cos\theta_2)\cos\theta_2 + E_d \exp(ikl\cos\theta_2)\cos\theta_2 = E_t \cos\theta_3, \\ H_u \exp(-ikl\cos\theta_2) + H_d \exp(ikl\cos\theta_2) = H_t. \end{aligned} \tag{2.160}$$

Again, by eliminating the fields inside the layer and using Eq. (2.148), we can write the boundary condition expressions in a matrix form as

$$\begin{pmatrix} \cos\theta_1 \\ n_1 \end{pmatrix} + \begin{pmatrix} -\cos\theta_1 \\ n_1 \end{pmatrix} r_p = \begin{pmatrix} \cos(kl\cos\theta_2) & -\frac{i}{p_p}\sin(kl\cos\theta_2) \\ -ip_p \sin(kl\cos\theta_2) & \cos(kl\cos\theta_2) \end{pmatrix} \begin{pmatrix} \cos\theta_3 \\ n_3 \end{pmatrix} t_p, \tag{2.161}$$

where $p_p = n_2/\cos\theta_2$. The reason we wrote Eqs. (2.158) and (2.161) as matrix expressions is to facilitate the calculation of the reflectance and transmission properties of multilayer films. The matrices in Eqs. (2.158) and (2.161) are referred to as the transfer matrices. If we write the transfer matrix as

$$M_{s,p} = \begin{pmatrix} A_{s,p} & B_{s,p} \\ C_{s,p} & D_{s,p} \end{pmatrix} = \begin{pmatrix} \cos(kl\cos\theta_2) & -\frac{i}{p_{s,p}}\sin(kl\cos\theta_2) \\ -ip_{s,p} \sin(kl\cos\theta_2) & \cos(kl\cos\theta_2) \end{pmatrix}, \tag{2.162}$$

the reflection and transmission coefficients can be determined from

$$r_s = \frac{A_s n_1 \cos\theta_1 + B_s n_1 n_3 \cos\theta_1 \cos\theta_3 - C_s - D_s n_3 \cos\theta_3}{A_s n_1 \cos\theta_1 + B_s n_1 n_3 \cos\theta_1 \cos\theta_3 + C_s + D_s n_3 \cos\theta_3}, \tag{2.163}$$

$$t_s = \frac{2n_1\cos\theta_1}{A_s n_1 \cos\theta_1 + B_s n_1 n_3 \cos\theta_1 \cos\theta_3 + C_s + D_s n_3 \cos\theta_3}, \tag{2.164}$$

$$r_p = \frac{A_p n_1 \cos\theta_3 + B_p n_1 n_3 - C_p \cos\theta_1 \cos\theta_3 - D_p n_3 \cos\theta_1}{A_p n_1 \cos\theta_3 + B_p n_1 n_3 + C_p \cos\theta_1 \cos\theta_3 + D_p n_3 \cos\theta_1}, \tag{2.165}$$

$$t_p = \frac{2n_1\cos\theta_1}{A_p n_1 \cos\theta_3 + B_p n_1 n_3 + C_p \cos\theta_1 \cos\theta_3 + D_p n_3 \cos\theta_1}. \tag{2.166}$$

The advantage of taking the extra steps to create the transfer matrix is that the reflection and transmission from multiple layers can be solved by multiplying respective matrices, for example,

$$M = M_1 \cdot M_2 \cdot \ldots \cdot M_{n-1} \cdot M_n, \tag{2.167}$$

where M_1 is the transfer matrix for the topmost layer, etc.

2.6.4 Thick Films

The description of the reflectance and transmittance, given in the previous section assumed that the films are thin enough that the light is coherent between the interfaces. When the films are sufficiently thick and the light is suitably polychromatic, then the fields do not interfere, and our arguments need to be based upon intensities. The characteristic length at which this occurs is the temporal coherence length of the light source. If the frequency bandwidth is $\Delta\nu$, the coherence time Δt is approximately given by $1/\Delta\nu$, being affected to some degree by the band shape. The distance over which the light is coherent in a material having an index of refraction of $\widetilde{n}$ is thus

$$l_{\mathrm{coh}} = c_0/(\widetilde{n}\Delta\nu), \tag{2.168}$$

or, in terms of spectral bandwidth $\Delta\lambda$,

$$l_{\mathrm{coh}} = \lambda^2/(\widetilde{n}\Delta\lambda). \tag{2.169}$$

For $\lambda = 550\,\mathrm{nm}$, $\Delta\lambda = 5\,\mathrm{nm}$, and $\widetilde{n} = 1.5$, the coherence length is about 40 μm. For films thicker than this distance, the fields directly impinging on an interface will not be coherent with the reflected field. In that case, the discussion in the preceding section does not apply. Instead, we need to think in terms of irradiances, rather than fields.

Suppose we have a layer, which is significantly thicker than the coherence length of the light, as shown in Fig. 2.10. Each interface j has a reflectance for upward propagating incidence, $\rho_{j,\mathrm{u}}$, a reflectance for downward propagating incidence, $\rho_{j,\mathrm{d}}$, and a transmittance τ_j that are calculated from Eqs. (2.150) and (2.151), respectively, for the appropriate polarization. At the top interface, we have conservation of irradiance, which can be written as

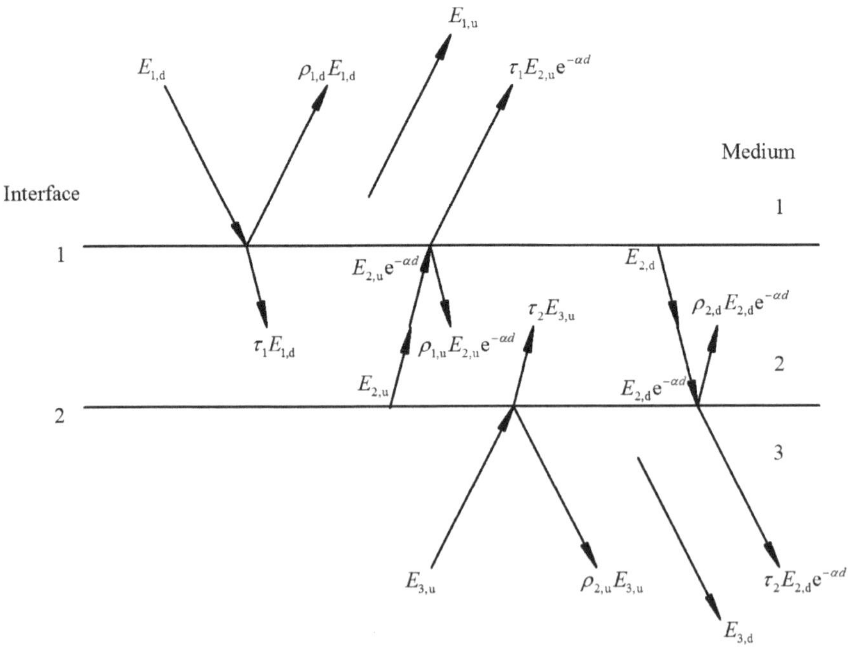

FIGURE 2.10 Reflection and transmission from a thick film.

$$E_{1,\mathrm{d}} - E_{1,\mathrm{u}} = E_{2,\mathrm{u}} \exp(-ad) - E_{2,\mathrm{d}}, \tag{2.170}$$

where a is the absorption coefficient for medium 2. The irradiances in medium 2 are evaluated at their interface of origin, the lower interface for the upward propagating irradiance and the upper interface for the downward propagating irradiance. Letting $E_{i,q}$ be the irradiance in medium i propagating in the upward (u) or downward (d) direction,

$$\begin{aligned} E_{1,\mathrm{u}} &= \rho_{1,\mathrm{d}} E_{1,\mathrm{d}} + \tau_1 E_{2,\mathrm{u}} \exp(-ad), \\ E_{2,\mathrm{d}} &= \tau_1 E_{1,\mathrm{d}} + \rho_{1,\mathrm{u}} E_{2,\mathrm{u}} \exp(-ad), \\ E_{2,\mathrm{u}} &= \rho_{2,\mathrm{d}} E_{2,\mathrm{d}} \exp(-ad) + \tau_2 E_{3,\mathrm{u}}, \\ E_{3,\mathrm{d}} &= \tau_2 E_{2,\mathrm{d}} \exp(-ad) + \rho_{2,\mathrm{u}} E_{3,\mathrm{u}}. \end{aligned} \tag{2.171}$$

We can solve the system of equations Eq. (2.171) for $E_{1,\mathrm{u}}$ and $E_{3,\mathrm{d}}$, eliminating $E_{2,\mathrm{u}}$ and $E_{2,\mathrm{d}}$,

$$\begin{aligned} E_{1,\mathrm{u}} &= \rho_{12,\mathrm{d}} E_{1,\mathrm{d}} + \tau_{12} E_{3,\mathrm{u}}, \\ E_{3,\mathrm{d}} &= \tau_{12} E_{1,\mathrm{d}} + \rho_{12,\mathrm{u}} E_{3,\mathrm{u}}, \end{aligned} \tag{2.172}$$

where

$$\begin{aligned} \rho_{12,\mathrm{d}} &= \frac{\rho_{1,\mathrm{d}} + \rho_{2,\mathrm{d}}\left(\tau_1^2 - \rho_{1,\mathrm{u}}\rho_{1,\mathrm{d}}\right)\exp(-2ad)}{1 - \rho_{1,\mathrm{u}}\rho_{2,\mathrm{d}}\exp(-2ad)}, \\ \rho_{12,\mathrm{u}} &= \frac{\rho_{2,\mathrm{u}} + \rho_{1,\mathrm{u}}\left(\tau_2^2 - \rho_{2,\mathrm{d}}\rho_{2,\mathrm{u}}\right)\exp(-2ad)}{1 - \rho_{1,\mathrm{u}}\rho_{2,\mathrm{d}}\exp(-2ad)}, \\ \tau_{12} &= \frac{\tau_1\tau_2\exp(-ad)}{1 - \rho_{1,\mathrm{u}}\rho_{2,\mathrm{d}}\exp(-2ad)}. \end{aligned} \tag{2.173}$$

Thus, $\rho_{12,\mathrm{d}}$ is the combined reflectance for downward propagating light, $\rho_{12,\mathrm{u}}$ is the combined reflectance for upward propagating light, and τ_{12} is the combined transmittance (which is reciprocal). For a thick transparent plate of index 1.5, the normal incidence reflectance and reflectance at each interface, from Eqs. (2.152) and (2.153), are 0.04 and 0.96, respectively. The net reflectance and transmittance, from Eq. (2.173), is about 0.077 and 0.923, and differ from the single-pass values (0.08 and 0.92) by about 0.003. For a plate of index 4.0, the reflectance and transmittance from each interface is 0.36 and 0.64, respectively. In this case, use of Eq. (2.173) is significantly more important, and yields 0.529 and 0.470, respectively.

Multiple layers can be treated with successive application of Eq. (2.173). For example, with three interfaces, we can first calculate $\rho_{12,\mathrm{d}}$, $\rho_{12,\mathrm{u}}$, $\tau_{12,\mathrm{d}}$, and $\tau_{12,\mathrm{u}}$, then substitute them for $\rho_{1,\mathrm{d}}$, $\rho_{1,\mathrm{u}}$, $\tau_{1,\mathrm{d}}$, and $\tau_{1,\mathrm{u}}$, respectively, and substitute $\rho_{3,\mathrm{d}}$, $\rho_{3,\mathrm{u}}$, $\tau_{3,\mathrm{d}}$, and $\tau_{3,\mathrm{u}}$ for $\rho_{2,\mathrm{d}}$, $\rho_{2,\mathrm{u}}$, $\tau_{2,\mathrm{d}}$, and $\tau_{2,\mathrm{u}}$, respectively. Note that for the first application of Eq. (2.173), the upward and downward reflectances are the same. However, for subsequent layers, this does not hold true.

2.7 DIFFUSE SCATTERING

We will discuss three models for light scattering. The first model, discussed in Section 2.7.1, applies to volume scattering and is based on a diffuse scattering model, that of Kubelka and Munk [13], which describes a solution to the one-dimensional radiative transfer equation and was originally developed to predict the color of paints, given the pigment scattering and absorption properties. The second model, described in Section 2.7.2, is applicable to very rough surfaces and assumes that the light is specularly reflected from a distribution of microfacets on the surface. Finally, the third model, discussed in Section 7.3, applies to rough surfaces in the smooth surface limit.

2.7.1 Volume Scattering: Theory of Kubelka and Munk

Volume diffuse scatterers occur everywhere from paint layers to atmospheric clouds. The scattering by pigments in a paint layer was studied by Kubelka and Munk in 1931 [13]. Although their approach is in many ways oversimplified, in that it does not account for the angular dependence to the scattering function, it nonetheless is an important step in understanding how pigments behave in diffuse layers. In some respects, it represents the first attempt at solving what is referred to as the radiative transfer equation. The approach is to only consider two irradiances, one propagating upward and one propagating downward. A more general approach would include the angular dependence of the scattering, often encapsulated in what is called a phase function, together with the angular distribution of the light, for example, the radiance function [14]. That problem can become significantly complicated, especially when one considers anisotropic, aligned scatterers, or includes polarization. Thus, we forego a discussion of more complex, accurate approaches, which is better suited for more specialized texts.

Consider a downward propagating irradiance $E_{\mathrm{d}}(z)$ and an upward propagating irradiance $E_{\mathrm{u}}(z)$ (see Fig. 2.11). Within each infinitesimal layer of thickness $\mathrm{d}z$ at depth z, there is an absorbed irradiance given by $aE_j(z)\,\mathrm{d}z$ and an irradiance $sE_j(z)\,\mathrm{d}z$ scattering into the opposing direction ($j=\mathrm{u},\mathrm{d}$). The total loss of irradiance to $E_j(z)$ in that infinitesimal layer is the sum of the absorption and scattering losses, $(a+s)E_j(z)\,\mathrm{d}z$. Thus, considering only these two directions, we can write down a pair of coupled differential equations

$$\frac{\mathrm{d}}{\mathrm{d}z}\begin{pmatrix} E_{\mathrm{d}}(z) \\ E_{\mathrm{u}}(z) \end{pmatrix} = \begin{pmatrix} a+s & -s \\ s & -(a+s) \end{pmatrix} \begin{pmatrix} E_{\mathrm{d}}(z) \\ E_{\mathrm{u}}(z) \end{pmatrix}. \tag{2.174}$$

There are standard methods for solving first-order coupled linear differential equations like this. If we let $e=a+s$ be the extinction coefficient and $\eta=(e^2-s^2)^{1/2}$, we can write the solution to Eq. (2.174) as

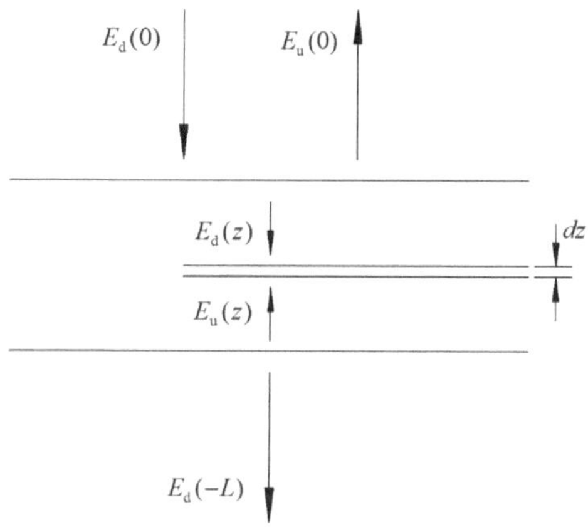

FIGURE 2.11 Volume scattering within a layer of paint.

$$\begin{pmatrix} E_d(z) \\ E_u(z) \end{pmatrix} = \begin{pmatrix} \cosh\eta z + (e/\eta)\sinh\eta z & -(s/\eta)\sinh\eta z \\ (s/\eta)\sinh\eta z & \cosh\eta z - (e/\eta)\sinh\eta z \end{pmatrix} \begin{pmatrix} E_d(0) \\ E_u(0) \end{pmatrix}. \tag{2.175}$$

We then solve Eq. (2.175) with the boundary condition that the substrate reflectance is ρ_0, so that the upward irradiance at that interface is related to the downward irradiance at that interface by $E_u(-L) = \rho_0 E_d(-L)$. We also express the total reflectance of the combination of the layer with the substrate as $\rho = E_u(0)/E_d(0)$, while the fraction of incidence irradiance that reaches the substrate is $\tau = E_d(-L)/E_d(0)$. We then solve for ρ and τ and arrive at a general expression for the reflectance and transmittance of the layer with the substrate:

$$\begin{aligned} \rho &= \frac{\rho_0 \eta \cosh\eta L + (-e\rho_0 + s)\sinh\eta L}{\eta\cosh\eta L + (e - \rho_0 s)\sinh\eta L}, \\ \tau &= \frac{\eta\cosh^2\eta L - \eta\sinh^2\eta L}{\eta\cosh\eta L + (e - \rho_0 s)\sinh\eta L}. \end{aligned} \tag{2.176}$$

There are a number of limiting cases to Eq. (2.176) that are of interest. If there is no absorption, $\alpha = 0$, the expressions simplify significantly

$$\begin{aligned} \rho_{\alpha=0} &= \frac{\rho_0 + sL(1-\rho_0)}{1 + sL(1-\rho_0)}, \\ \tau_{\alpha=0} &= \frac{1}{1 + sL(1-\rho_0)}. \end{aligned} \tag{2.177}$$

Regardless of the substrate reflectance, the reflectance of the combined system approaches unit reflectance as the thickness increases, while its transmittance goes to zero. That is, thick clouds are white viewed from above and black when viewed from below. Furthermore, if the substrate reflectance

$\rho_0 = 0$ (i.e., it is black), we have particularly simple expressions for the reflectance and transmittance:

$$\rho_{\alpha=0,\rho_0=0} = \frac{sL}{1+sL},$$
$$\tau_{\alpha=0,\rho_0=0} = \frac{1}{1+sL}. \tag{2.178}$$

If there is no scattering, $s=0$, then we find that Eq. (2.176) reduces to

$$\rho_{\sigma=0} = \rho_0 \exp(-2aL),$$
$$\tau_{\sigma=0} = \exp(-aL). \tag{2.179}$$

That is, we get the Beer–Lambert law for the transmittance, and the reflectance shows the Beer–Lambert law applied twice with the additional reflectance of the substrate. For an infinitely thick layer, $L \to \infty$, and provided there is at least some scattering, that is, $s \neq 0$, we have

$$\rho_\infty = 1 + \frac{a}{s} - \sqrt{\frac{a^2}{s^2} + 2\frac{a}{s}},$$
$$\tau_\infty = 0, \tag{2.180}$$

where ρ_∞ is often referred to as the intrinsic reflectance. In this case, the substrate reflectance does not appear in the result, and the reflectance of the system only depends upon the ratio of the absorption to the scattering coefficients, a/s. Eq. (2.180) can be solved for a/s, in what is called the remittance function,

$$\frac{a}{s} = \frac{(1-\rho_\infty)^2}{2\rho_\infty}. \tag{2.181}$$

That is, the ratio of the absorption to scattering coefficients, a/s, can be determined from the reflectance of a thick sample, using Eq. (2.181). Note that for many color applications of spectrophotometry for textiles and paints, the light absorption and scattering coefficients in Kubelka–Munk analysis are also often denoted by the symbols $K=a$ and $S=s$, respectively.

2.7.2 Roughness: Facet Scattering Model

Scattering from a rough interface can be challenging to predict in the most general case. Electromagnetic models, based upon the Maxwell equations, can become extremely complicated, and there is no single solution that works for all conditions. The most general approaches invariably become numerically intensive, and their solutions neither provide much physical insight nor allow the roughness to be determined from scattering measurements. In two regimes, however, theoretical approaches can be taken that have relatively simple interpretations and allow for roughness information to be extracted from data. In this section, we will discuss a ray tracing approach that is valid for large scale roughness, where the radius of curvature of the surface is always much greater than the wavelength of the light and the correlation lengths are small. In the next section, we will describe a first-order perturbation approach that is valid for small levels of roughness.

In the approach taken for large scale roughness, we are going to assume that a ray tracing, or facet, approach can be used. Rays of light are incident upon the surface, and the directions they scatter are dependent upon the orientation of facets that they strike [15, 16]. Figure 2.12 illustrates the specular reflection from a surface whose surface is slowly modulating in space. Rays are incident onto the surface with an incident angle θ_i onto a surface whose local surface normal is given by the unit vector $\hat{\mathbf{n}}$ (having a polar angle θ_n and azimuthal angle ϕ_n). The local incident angle is given by the angle ι. The direction of propagation of the incident light is given by the unit vector

$$\hat{\mathbf{k}}_i = \hat{\mathbf{x}} \sin\theta_i - \hat{\mathbf{z}} \cos\theta_i, \tag{2.182}$$

where $\hat{\mathbf{x}}$ and $\hat{\mathbf{z}}$ are unit vectors along the x and z directions, respectively. The direction reflected by the local surface is given by the unit vector

$$\hat{\mathbf{k}}_r = \hat{\mathbf{x}} \sin\theta_r \cos\phi_r + \hat{\mathbf{y}} \sin\theta_r \sin\phi_r + \hat{\mathbf{z}} \cos\theta_r, \tag{2.183}$$

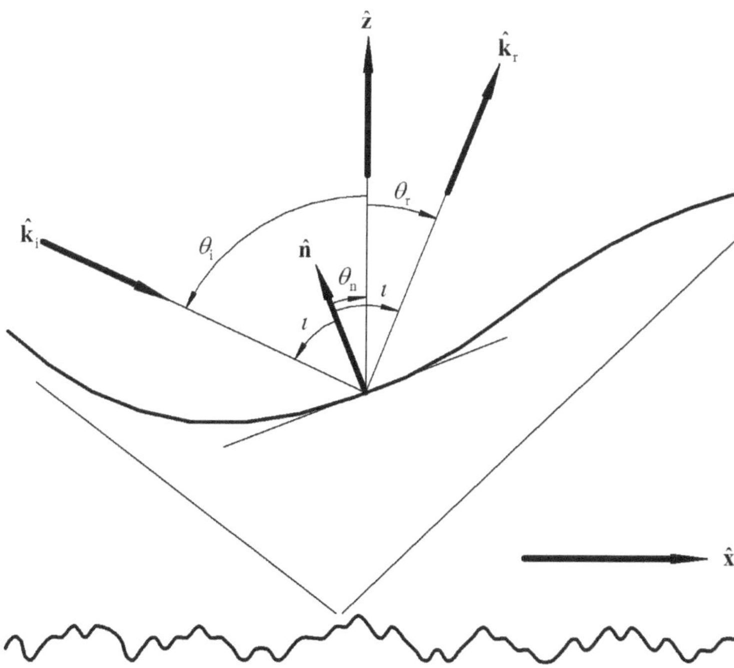

FIGURE 2.12 The facet scattering model.

where $\hat{\mathbf{y}}$ is a unit vector along the y direction. A facet that reflects from $\hat{\mathbf{k}}_i$ toward $\hat{\mathbf{k}}_r$ is aligned with its surface normal along the unit vector

$$\hat{\mathbf{n}} = (\mathbf{k}_r - \mathbf{k}_i)/\|\mathbf{k}_r - \mathbf{k}_i\| \\ = \frac{\hat{\mathbf{x}}(\cos\phi_r \sin\theta_r - \sin\theta_i) - \hat{\mathbf{y}}\sin\phi_r \sin\theta_r + \hat{\mathbf{z}}(\cos\theta_i + \cos\theta_r)}{\sqrt{2 + 2\cos\theta_i \cos\theta_r - 2\cos\phi_r \sin\theta_i \sin\theta_r}}. \tag{2.184}$$

The cosine of the local incident angle is determined by the scalar product $\hat{\mathbf{k}}_r \cdot \hat{\mathbf{n}}$, which is found to be

$$\cos\iota = \sqrt{(1 + \cos\theta_i \cos\theta_r - \cos\phi_r \sin\theta_i \sin\theta_r)/2}. \tag{2.185}$$

The cosine of the polar orientation of this facet is the z-component of $\hat{\mathbf{n}}$,

$$\cos\theta_n = \hat{\mathbf{n}} \cdot \hat{\mathbf{z}} = \frac{\cos\theta_i + \cos\theta_r}{2\cos\iota}, \tag{2.186}$$

while the tangent of the azimuthal orientation can be determined from the ratio of the y to the x components of $\hat{\mathbf{n}}$:

$$\tan\phi_n = \frac{-\sin\phi_r \sin\theta_r}{\cos\phi_r \sin\theta_r - \sin\theta_i}. \tag{2.187}$$

The projected area of the surface covered with facets having this orientation within a solid angle $d\Omega_n = \sin\theta_n d\theta_n d\phi_n$ is related to an orientation probability distribution function $P(\theta_n, \phi_n)$ by

$$dA_n = P(\theta_n, \phi_n) d\Omega_n dA, \tag{2.188}$$

where dA is an element of area on the mean, flat surface. We can make a change of variables from facet orientation (θ_n, ϕ_n) to scattering angle (θ_r, ϕ_r) by evaluating the Jacobian, which, with a bit of algebra, is

$$\frac{\partial(\theta_n, \phi_n)}{\partial(\theta_r, \phi_r)} = \frac{\sin\theta_r}{4\cos\iota \sin\theta_n}. \tag{2.189}$$

Hence,

$$d\Omega_n = \frac{d\Omega_r}{4\cos\iota}. \tag{2.190}$$

From Fig. 2.13, the radiant power incident onto the facets is given by

$$d\Phi_i = \frac{E}{\cos\theta_i} \frac{\cos\iota}{\cos\theta_n} dA_n, \tag{2.191}$$

where E is the incident irradiance. The radiant power scattered into $d\Omega_r$ is the product of $d\Phi_i$ and the reflectance of the facet, ρ, which, by combining Eqs. (2.188) and (2.191), is given by

$$d\Phi_i = \frac{E\rho P(\theta_n, \phi_n)}{4\cos\theta_i \cos\theta_n} d\Omega_r \, dA. \tag{2.192}$$

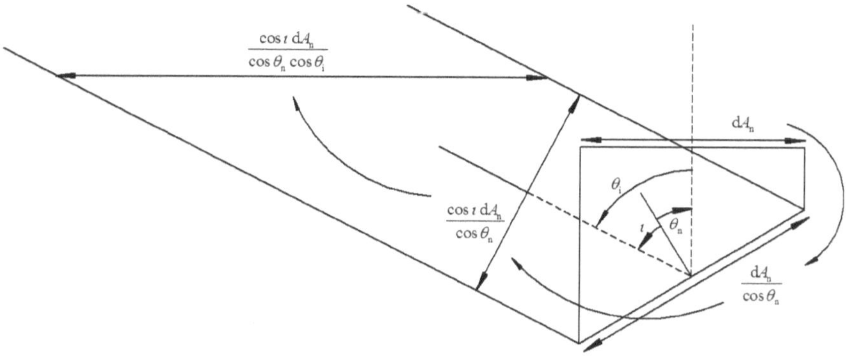

FIGURE 2.13 The projection of the incident beam striking a facet onto the surface.

Thus, from Eq. (2.10), the radiance is

$$L = \frac{E\rho P(\theta_n, \phi_n)}{4\cos\theta_i \cos\theta_r \cos\theta_n}, \tag{2.193}$$

and, from its definition Eq. (2.59), the BRDF is

$$f_r = \frac{\rho P(\theta_n, \phi_n)}{4\cos\theta_i \cos\theta_r \cos\theta_n}. \tag{2.194}$$

The distribution of facet orientations can be defined as above in Eq. (2.188) as a distribution in angles, or it can be defined in terms of slopes

$$\begin{aligned} \zeta_x &= \tan\theta_n \cos\phi_n, \\ \zeta_y &= \tan\theta_n \sin\phi_n. \end{aligned} \tag{2.195}$$

In that case, we could have written Eq. (2.188) as

$$dA_n = P(\zeta_x, \zeta_y)\, d\zeta_x\, d\zeta_y\, dA. \tag{2.196}$$

Since, from Eqs. (2.195),

$$\frac{d\zeta_x d\zeta_y}{d\Omega_n} = \frac{1}{\sin\theta_n} \frac{\partial(\tan\theta_n, \tan\phi_n)}{\partial(\theta_n, \phi_n)} = \sec^3\theta_n, \tag{2.197}$$

we have

$$dA_n = P(\zeta_x, \zeta_y) \sec^3\theta_n\, d\Omega_n\, dA. \tag{2.198}$$

The probability distribution functions are thus related by

$$P(\zeta_x, \zeta_y) = P(\theta_n, \phi_n) \cos^3\theta_n. \tag{2.199}$$

Therefore, combining Eqs. (2.194) and (2.199), the BRDF can also be written as

$$f_\mathrm{r} = \frac{\rho P(\zeta_x, \zeta_y)}{4\cos\theta_\mathrm{i}\cos\theta_\mathrm{r}\cos^4\theta_\mathrm{n}}. \tag{2.200}$$

The slope (ζ_x, ζ_y) is given by the facet orientation tangents, which in turn is found from $\zeta_x = [\hat{\mathbf{n}}]_x/[\hat{\mathbf{n}}]_z$ and $\zeta_y = [\hat{\mathbf{n}}]_y/[\hat{\mathbf{n}}]_z$. Thus, from Eq. (2.184), the facet slopes reflecting from $\hat{\mathbf{k}}_\mathrm{i}$ to $\hat{\mathbf{k}}_\mathrm{r}$ are related to the scattering geometry by

$$\begin{aligned}\zeta_x &= \frac{\sin\theta_\mathrm{r}\cos\phi_\mathrm{r} - \sin\theta_\mathrm{i}}{\cos\theta_\mathrm{r} + \cos\theta_\mathrm{i}},\\ \zeta_y &= \frac{\sin\theta_\mathrm{r}\sin\phi_\mathrm{r}}{\cos\theta_\mathrm{r} + \cos\theta_\mathrm{i}}.\end{aligned} \tag{2.201}$$

At this point, we have not discussed the reflectance ρ of the facets, which will be dependent upon the local incident angle on the facet and the polarization. Given a specific scattering geometry, the local angle of incidence is given by ι, determined from Eq. (2.185). In a coordinate system in which the scattering plane, containing the incident and scattering directions, defines our polarizations, the reflectance coefficients are given by $r_\mathrm{s}(\iota)$ and $r_\mathrm{p}(\iota)$, which for a single interface are given by Eqs. (2.146) and (2.149), respectively, evaluated with $\theta_1 = \iota$. In that local coordinate system, our reflectance coefficients would form a reflection matrix

$$\begin{pmatrix} r_\mathrm{s}(\iota) & 0 \\ 0 & r_\mathrm{p}(\iota)\end{pmatrix}. \tag{2.202}$$

However, such a coordinate system is not always convenient, and it is not the coordinate system that we have chosen to use in other parts of this chapter. We need then, to appropriately rotate the matrix given in Eq. (2.202) to the global coordinate system. To accomplish this task, we first write down unit vectors describing the polarization bases for the local and global coordinate systems. For the local coordinates, we have

$$\begin{aligned}\hat{\mathbf{s}}_\mathrm{i,local} &= \hat{\mathbf{s}}_\mathrm{r,local} = \mathbf{k}_\mathrm{i}\times\mathbf{k}_\mathrm{r}/\|\mathbf{k}_\mathrm{i}\times\mathbf{k}_\mathrm{r}\|,\\ \hat{\mathbf{p}}_\mathrm{i,local} &= \mathbf{k}_\mathrm{i}\times\hat{\mathbf{s}}_\mathrm{i,local}/\|\mathbf{k}_\mathrm{i}\times\hat{\mathbf{s}}_\mathrm{i,local}\|,\\ \hat{\mathbf{p}}_\mathrm{r,local} &= \mathbf{k}_\mathrm{r}\times\hat{\mathbf{s}}_\mathrm{r,local}/\|\mathbf{k}_\mathrm{r}\times\hat{\mathbf{s}}_\mathrm{r,local}\|.\end{aligned} \tag{2.203}$$

For the global coordinates, we have

$$\begin{aligned}\hat{\mathbf{s}}_\mathrm{i,global} &= \hat{\mathbf{y}},\\ \hat{\mathbf{p}}_\mathrm{i,global} &= \cos\theta_\mathrm{i}\hat{\mathbf{x}} + \sin\theta_\mathrm{i}\hat{\mathbf{z}},\\ \hat{\mathbf{s}}_\mathrm{r,global} &= -\sin\phi_\mathrm{r}\hat{\mathbf{x}} + \cos\phi_\mathrm{r}\hat{\mathbf{y}},\\ \hat{\mathbf{p}}_\mathrm{r,global} &= -\cos\theta_\mathrm{i}\cos\phi_\mathrm{r}\hat{\mathbf{x}} - \cos\theta_\mathrm{i}\sin\phi_\mathrm{r}\hat{\mathbf{y}} + \sin\theta_\mathrm{r}\hat{\mathbf{z}}.\end{aligned} \tag{2.204}$$

The polarization matrix can be transformed according to

$$\begin{pmatrix}\hat{\mathbf{s}}_\mathrm{r,global}\cdot\hat{\mathbf{s}}_\mathrm{r,local} & \hat{\mathbf{s}}_\mathrm{r,global}\cdot\hat{\mathbf{p}}_\mathrm{r,local}\\ \hat{\mathbf{p}}_\mathrm{r,global}\cdot\hat{\mathbf{s}}_\mathrm{r,local} & \hat{\mathbf{p}}_\mathrm{r,global}\cdot\hat{\mathbf{p}}_\mathrm{r,local}\end{pmatrix}\begin{pmatrix} r_\mathrm{s}(\iota) & 0\\ 0 & r_\mathrm{p}(\iota)\end{pmatrix}\begin{pmatrix}\hat{\mathbf{s}}_\mathrm{i,local}\cdot\hat{\mathbf{s}}_\mathrm{i,global} & \hat{\mathbf{s}}_\mathrm{i,local}\cdot\hat{\mathbf{p}}_\mathrm{i,global}\\ \hat{\mathbf{p}}_\mathrm{i,local}\cdot\hat{\mathbf{s}}_\mathrm{i,global} & \hat{\mathbf{p}}_\mathrm{i,local}\cdot\hat{\mathbf{p}}_\mathrm{i,global}\end{pmatrix}. \tag{2.205}$$

After a bit of algebra, we obtain the reflection matrix in the global coordinates

$$\begin{pmatrix} r_{ss} & r_{sp} \\ r_{ps} & r_{pp} \end{pmatrix}, \tag{2.206}$$

where

$$\begin{aligned} r_{ss} &= [r_p(\imath)\sin\theta_i\sin\theta_r\sin^2\phi_r + a_2a_3r_s(\imath)]/a_1, \\ r_{ps} &= \sin\phi_r[a_3r_p(\imath)\sin\theta_i - a_2r_s(\imath)\sin\theta_r]/a_1, \\ r_{sp} &= \sin\phi_r[a_2r_p(\imath)\sin\theta_r - a_3r_s(\imath)\sin\theta_i]/a_1, \\ r_{pp} &= [a_2a_3r_p(\imath) + r_s(\imath)\sin\theta_i\sin\theta_r\sin^2\phi_r]/a_1, \end{aligned} \tag{2.207}$$

and

$$\begin{aligned} a_1 &= \sin^2 2\imath, \\ a_2 &= \cos\theta_i\sin\theta_r + \sin\theta_i\cos\theta_r\cos\phi_r, \\ a_3 &= \sin\theta_i\cos\theta_r + \cos\theta_i\sin\theta_r\cos\phi_r. \end{aligned} \tag{2.208}$$

The average, unpolarized reflectance of the surface facet, ρ, is given by

$$\rho = \frac{1}{2}\left(|r_{ss}|^2 + |r_{sp}|^2 + |r_{sp}|^2 + |r_{pp}|^2\right), \tag{2.209}$$

which, after significant simplification, is given by

$$\rho = \frac{1}{2}\left(|r_s|^2 + |r_p|^2\right). \tag{2.210}$$

The reflectance is sometimes treated as being further reduced by a shadow and obscuration function [16, 17].

2.7.3 Roughness: First-Order Vector Perturbation Theory

The theory of scattering by a slightly rough surface was originally considered by Rayleigh [18] in 1907 and extended by Rice [19] in 1951 and Barrick and Peake [20] in 1968. The approach takes a first-order vector perturbation theory approach to the vector fields and is often referred to as Rayleigh–Rice theory. The results apply when the surface is essentially mirror-like. In this section, we will derive expressions that relate the BRDF to the power spectral density (PSD) function of the surface profile.

Suppose we have an interface between two media, that rather than being planar, as was done previously in Section 2.6.2, deviates from planar by a height function $\Delta z(x,y)$. In the following, we assume that $\Delta z(x,y)$ is much smaller than the wavelength of the light. We will use this assumption and the Maxwell equations (in particular, the boundary conditions imposed by them) to derive the BRDF for this surface. We begin by writing down an approximation to the unit surface normal vector:

$$\hat{\mathbf{n}} = -\left(\frac{\partial \Delta z(x,y)}{\partial x}\right)\hat{\mathbf{x}} - \left(\frac{\partial \Delta z(x,y)}{\partial y}\right)\hat{\mathbf{y}} + \hat{\mathbf{z}}, \tag{2.211}$$

where $\hat{\mathbf{x}}$ and $\hat{\mathbf{y}}$ are perpendicular unit vectors in the mean plane of the sample, and $\hat{\mathbf{z}}$ is a unit vector in the mean surface normal. Eq. (2.211) is an approximation, in that the component along $\hat{\mathbf{z}}$ needs to be reduced to keep the vector normalized. Thus, in addition to $\Delta z(x,y)$ being much smaller than the wavelength, we will assume that the surface slopes are small.

We will consider the electric and magnetic fields to be the sum of zero-order electric and magnetic fields ($\mathbf{E}_1^{(0)}$,$\mathbf{H}_1^{(0)}$,$\mathbf{E}_2^{(0)}$, and $\mathbf{H}_2^{(0)}$ in media 1 and 2, respectively) , given by the solution to the electromagnetic problem for the flat surface (see Fresnel theory in Section 2.6.2, above), and first-order electric and magnetic fields ($\mathbf{E}_1^{(1)}$,$\mathbf{H}_1^{(1)}$,$\mathbf{E}_2^{(1)}$, and $\mathbf{H}_2^{(1)}$ in media 1 and 2, respectively). Figure 2.14 shows the basis vectors and symbols used for the waves and angles in this section. Subscripts include the region the field or angle applies (1 or 2), the polarization (s or p), and for the zero-order incident field, an additional subscript (i). Superscripts indicate whether a symbol applies to the zero-order (0) or first-order (1) fields. Medium 1 will be treated as vacuum ($\hat{n}_1 = \tilde{n}_1 = 1$), and medium 2 will be treated as nonmagnetic ($\mu = 1$ and $\hat{n}_2 = \tilde{n}_2$). For the first-order fields, there are additional out-of-plane angles $\phi_j^{(1)}$ that are not shown in Fig. 2.14. We let the zero-order fields be given by

$$\begin{aligned} \mathbf{E}_j^{(0)}(x,y,z) &= \mathbf{e}_j^{(0)}(k_0,z)\exp(ik_0x), \\ \mathbf{H}_j^{(0)}(x,y,z) &= \mathbf{h}_j^{(0)}(k_0,z)\exp(ik_0x), \end{aligned} \tag{2.212}$$

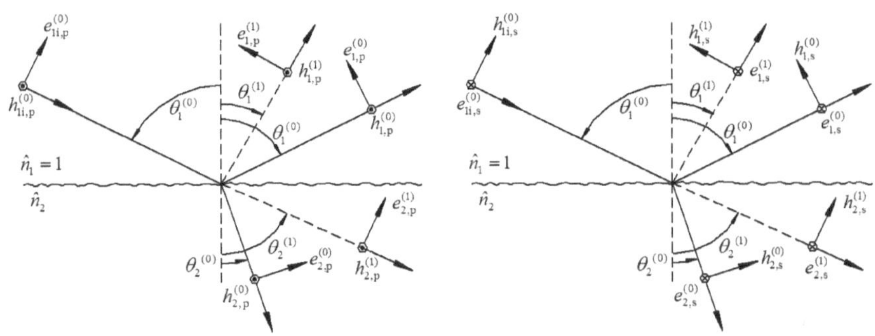

FIGURE 2.14 Scattering from a rough surface. Only the in-plane geometry is shown, but for most of the scattered directions, there are also angles describing the azimuthal orientation of the propagation direction. The diagrams show the electric and magnetic fields for (left) p-polarized and (right) s-polarized incident light. The solid rays indicate the specular and regularly transmitting rays, while the dashed rays indicate scattered rays.

where $k_0 = k\sin\theta_1^{(0)}$ is the projection of the incident wavevector onto the mean surface, and $\mathbf{e}_j^{(0)}$ and $\mathbf{h}_j^{(0)}$ contain the respective z-dependences of the fields. The incident wavevector is assumed to be contained in the x–z plane. For the first-order fields, we assume that we can use a Fourier expansion

$$\mathbf{E}_j^{(1)} = \frac{1}{4\pi^2}\int \mathbf{e}_j^{(1)}(\mathbf{k}, z)\exp(i\mathbf{k}\cdot\boldsymbol{\rho})\mathrm{d}^2k,$$
$$\mathbf{H}_j^{(1)} = \frac{1}{4\pi^2}\int \mathbf{h}_j^{(1)}(\mathbf{k}, z)\exp(i\mathbf{k}\cdot\boldsymbol{\rho})\mathrm{d}^2k, \tag{2.213}$$

where $\boldsymbol{\rho} = (x, y, 0)$ is the projection of the position vector onto the mean surface plane. The sign of the z-component of the wavevector, $k_{z,j} = \left(\tilde{n}_j^2\omega^2/c^2 - k_x^2 - k_y^2\right)^{1/2}$, must be chosen so that $\mathrm{Re}k_{z,1} + \mathrm{Im}k_{z,1} > 0$, since reflected waves must be upward propagating, and $\mathrm{Re}k_{z,2} + \mathrm{Im}k_{z,2} < 0$, since transmitted waves must be downward propagating. The Fourier expansion is rigorously only valid in a half-space, so it is strictly only valid in the region that is above the highest point on the surface in the reflected region and below the lowest point of the surface in the transmitted region. However, for very small amounts of roughness, we are going to assume that we can extend the fields into the valleys in the reflected region or into the peaks in the transmitted region and match the boundary conditions that exist between them. The total electric and magnetic fields at the boundary between the two media can be expanded in a Taylor series to first order in $\Delta z(x, y)$,

$$\mathbf{E}_j[x,y,\Delta z(x, y)] = \mathbf{E}_j^{(0)}(x,y,0) + \mathbf{E}_j^{(1)}(x,y,0) + \frac{\partial}{\partial z}\mathbf{E}_j^{(0)}(x, y, z)\Big|_{z=0}\Delta z(x,y),$$
$$\mathbf{H}_j[x,y,\Delta z(x, y)] = \mathbf{H}_j^{(0)}(x,y,0) + \mathbf{H}_j^{(1)}(x,y,0) + \frac{\partial}{\partial z}\mathbf{H}_j^{(0)}(x, y, z)\Big|_{z=0}\Delta z(x,y). \tag{2.214}$$

At this point, we are left with the relatively straightforward, but tedious, process of matching the boundary conditions at the surface $\Delta z(x, y)$. The reader who does not want to wade through the details can skip to the point after Eq. (2.224). Continuity of the electric and magnetic fields across the surface requires that

$$\hat{\mathbf{n}} \times \{\mathbf{E}_1[x,y,\Delta z(x, y)] - \mathbf{E}_2[x,y,\Delta z(x, y)]\} = 0,$$
$$\hat{\mathbf{n}} \times \{\mathbf{H}_1[x,y,\Delta z(x, y)] - \mathbf{H}_2[x,y,\Delta z(x, y)]\} = 0. \tag{2.215}$$

Using Eqs. (2.211) and (2.214) in Eq. (2.215) yields continuity relationships for the x and y components of $\mathbf{E}_i^{(1)}[x, y, 0]$ and $\mathbf{H}_i^{(1)}[x, y, 0]$,

$$\Delta E_x{}^{(1)}[x,y,0] = -\left[\frac{\partial \Delta z(x,y)}{\partial x}\right]\Delta E_z^{(0)}(x,y,0) - \Delta z(x,y)\Delta\left[\frac{\partial}{\partial z}\Delta E_x^{(0)}(x,y,0)\right],$$

$$\Delta E_y{}^{(1)}[x,y,0] = -\left[\frac{\partial \Delta z(x,y)}{\partial y}\right]\Delta E_z^{(0)}(x,y,0) - \Delta z(x,y)\Delta\left[\frac{\partial}{\partial z}\Delta E_y^{(0)}(x,y,0)\right],$$

$$\Delta H_x{}^{(1)}[x,y,0] = -\left[\frac{\partial \Delta z(x,y)}{\partial x}\right]\Delta H_z^{(0)}(x,y,0) - \Delta z(x,y)\Delta\left[\frac{\partial}{\partial z}\Delta H_x^{(0)}(x,y,0)\right],$$

$$\Delta H_y{}^{(1)}[x,y,0] = -\left[\frac{\partial \Delta z(x,y)}{\partial y}\right]\Delta H_z^{(0)}(x,y,0) - \Delta z(x,y)\Delta\left[\frac{\partial}{\partial z}\Delta H_y^{(0)}(x,y,0)\right], \tag{2.216}$$

where Δ applied to the fields indicate the difference between region 1 and 2. Inserting Eqs. (2.212) and (2.213) into (2.216), we find that

$$\begin{aligned}
\Delta e_x^{(1)}(\mathbf{k}) &= -\mathrm{i}k_x \Delta e_z^{(0)}(k_0)\Delta Z(\mathbf{k}_0 - \mathbf{k}),\\
\Delta e_y^{(1)}(\mathbf{k}) &= -\mathrm{i}k_y \Delta e_z^{(0)}(k_0)\Delta Z(\mathbf{k}_0 - \mathbf{k}),\\
\Delta h_x^{(1)}(\mathbf{k}) &= \mathrm{i}(\omega/c_0)\Delta d_y^{(0)}(k_0)\Delta Z(\mathbf{k}_0 - \mathbf{k}),\\
\Delta h_y^{(1)}(\mathbf{k}) &= -\mathrm{i}(\omega/c_0)\Delta d_x^{(0)}(k_0)\Delta Z(\mathbf{k}_0 - \mathbf{k}),
\end{aligned} \tag{2.217}$$

where

$$\Delta \mathbf{d}^{(0)} = \mathbf{e}_1^{(0)}(k_0) - \hat{n}_2{}^2\mathbf{e}_2^{(0)}(k_0) \tag{2.218}$$

and

$$\Delta Z(\mathbf{k}_0 - \mathbf{k}) = \int \Delta z(x,y)\exp[\mathrm{i}(\mathbf{k}_0 - \mathbf{k})\cdot\boldsymbol{\rho}]\mathrm{d}^2\rho. \tag{2.219}$$

The zero-order electric fields for s and p polarizations in region 1, evaluated at $z=0$, are

$$\begin{aligned}
\mathbf{e}_{1,s}^{(0)}\Big|_{z=0} &= -\hat{\mathbf{y}}e_{1,s}{}^{(0)}(1+r_s),\\
\mathbf{e}_{1,p}^{(0)}\Big|_{z=0} &= \hat{\mathbf{x}}e_{1,p}{}^{(0)}(1-r_p)\cos\theta_1{}^{(0)} + \hat{\mathbf{z}}e_{1,p}{}^{(0)}(1+r_p)\sin\theta_1{}^{(0)},\\
\mathbf{e}_{2,s}^{(0)}\Big|_{z=0} &= -\hat{\mathbf{y}}e_{1,s}{}^{(0)}t_s,\\
\mathbf{e}_{2,p}^{(0)}\Big|_{z=0} &= \hat{\mathbf{x}}e_{1,p}{}^{(0)}t_p\cos\theta_2{}^{(0)} + \hat{\mathbf{z}}e_{1,p}{}^{(0)}t_p\sin\theta_2{}^{(0)},
\end{aligned} \tag{2.220}$$

where the $r_{s,p}$ and $t_{s,p}$ are the reflection and transmission coefficients, given by Eqs. (2.146) and (2.149), evaluated at $\theta_1 = \theta_1^{(0)}$ and $\theta_2 = \theta_2^{(0)}$. Combining Eqs. (2.217) and (2.220) we have

$$\Delta e_x^{(1)}(\mathbf{k}) = -\mathrm{i}(\omega/c_0)\Delta Z(\mathbf{k}_0-\mathbf{k})e_{1,\mathrm{p}}^{(0)}\left[(1+r_\mathrm{p})\sin\theta_1{}^{(0)} - t_\mathrm{p}\sin\theta_2{}^{(0)}\right]\sin\theta_{1\mathrm{r}}\cos\phi,$$
$$\Delta e_y^{(1)}(\mathbf{k}) = -\mathrm{i}(\omega/c_0)\Delta Z(\mathbf{k}_0-\mathbf{k})e_{1,\mathrm{p}}^{(0)}\left[(1+r_\mathrm{p})\sin\theta_1{}^{(0)} - t_\mathrm{p}\sin\theta_2{}^{(0)}\right]\sin\theta_{1\mathrm{r}}\sin\phi,$$
$$\Delta h_x^{(1)}(\mathbf{k}) = -\mathrm{i}(\omega/c_0)\Delta Z(\mathbf{k}_0-\mathbf{k})e_{1,\mathrm{s}}^{(0)}(1+r_\mathrm{s}-\hat{n}_2{}^2 t_\mathrm{s}),$$
$$\Delta h_y^{(1)}(\mathbf{k}) = -\mathrm{i}(\omega/c_0)\Delta Z(\mathbf{k}_0-\mathbf{k})e_{1,\mathrm{p}}^{(0)}\left[(1-r_\mathrm{p})\cos\theta_1{}^{(0)} - \hat{n}_2{}^2 t_\mathrm{p}\cos\theta_{2\mathrm{i}}\right]. \tag{2.221}$$

The first-order electric fields, evaluated at $z=0$, can be expressed as

$$\mathbf{e}_{1,\mathrm{s}}^{(1)}\Big|_{z=0} = e_{1,\mathrm{s}}{}^{(1)}\left(\hat{\mathbf{x}}\sin\phi_1{}^{(1)} - \hat{\mathbf{y}}\cos\phi_1{}^{(1)}\right),$$
$$\mathbf{e}_{1,\mathrm{p}}^{(1)}\Big|_{z=0} = e_{1,\mathrm{p}}{}^{(1)}\left(-\hat{\mathbf{x}}\cos\theta_1{}^{(1)}\cos\phi_1{}^{(1)} - \hat{\mathbf{y}}\cos\theta_1{}^{(1)}\sin\phi_1{}^{(1)} + \hat{\mathbf{z}}\sin\theta_1{}^{(1)}\right),$$
$$\mathbf{e}_{2,\mathrm{s}}^{(1)}\Big|_{z=0} = e_{2,\mathrm{s}}{}^{(1)}\left(\hat{\mathbf{x}}\sin\phi_2{}^{(1)} - \hat{\mathbf{y}}\cos\phi_2{}^{(1)}\right),$$
$$\mathbf{e}_{1,\mathrm{p}}^{(1)}\Big|_{z=0} = e_{2,\mathrm{p}}{}^{(1)}\left(\hat{\mathbf{x}}\cos\theta_2{}^{(1)}\cos\phi_2{}^{(1)} + \hat{\mathbf{y}}\cos\theta_2{}^{(1)}\sin\phi_2{}^{(1)} + \hat{\mathbf{z}}\sin\theta_2{}^{(1)}\right). \tag{2.222}$$

Likewise, using Eq. (2.35), the first-order magnetic fields are

$$\mathbf{h}_{1,\mathrm{s}}^{(1)}\Big|_{z=0} = (\epsilon_0/\mu_0)^{1/2}e_{1,\mathrm{s}}{}^{(1)}\left(\hat{\mathbf{x}}\cos\theta_1{}^{(1)}\cos\phi_1{}^{(1)} + \hat{\mathbf{y}}\cos\theta_1{}^{(1)}\sin\phi_1{}^{(1)} - \hat{\mathbf{z}}\sin\theta_1{}^{(1)}\right),$$
$$\mathbf{h}_{1,\mathrm{p}}^{(1)}\Big|_{z=0} = (\epsilon_0/\mu_0)^{1/2}e_{1,\mathrm{p}}{}^{(1)}\left(\hat{\mathbf{x}}\sin\phi_1{}^{(1)} - \hat{\mathbf{y}}\cos\phi_1{}^{(1)}\right),$$
$$\mathbf{h}_{2,\mathrm{s}}^{(1)}\Big|_{z=0} = \hat{n}_2(\epsilon_0/\mu_0)^{1/2}e_{2,\mathrm{s}}{}^{(1)}\left(-\hat{\mathbf{x}}\cos\theta_2{}^{(1)}\cos\phi_2{}^{(1)} - \hat{\mathbf{y}}\cos\theta_2{}^{(1)}\sin\phi_2{}^{(1)} - \hat{\mathbf{z}}\sin\theta_2{}^{(1)}\right),$$
$$\mathbf{h}_{1,\mathrm{p}}^{(1)}\Big|_{z=0} = \hat{n}_2(\epsilon_0/\mu_0)^{1/2}e_{2,\mathrm{p}}{}^{(1)}\left(\hat{\mathbf{x}}\sin\phi_2{}^{(1)} - \hat{\mathbf{y}}\cos\phi_2{}^{(1)}\right). \tag{2.223}$$

For Eq. (2.221), we need discontinuities in the first-order electric and magnetic fields in the mean surface plane, which, from Eqs. (2.222), and (2.223), are

$$\Delta e_x^{(1)}(\mathbf{k}) = e_{1,\mathrm{s}}^{(1)}\sin\phi_1^{(1)} - e_{1,\mathrm{p}}^{(1)}\cos\theta_1{}^{(1)}\cos\phi_1^{(1)} - e_{2,\mathrm{s}}^{(1)}\sin\phi_2^{(1)} - e_{2,\mathrm{p}}^{(1)}\cos\theta_2{}^{(1)}\cos\phi_2^{(1)},$$
$$\Delta e_y^{(1)}(\mathbf{k}) = -e_{1,\mathrm{s}}^{(1)}\cos\phi_1^{(1)} - e_{1,\mathrm{p}}^{(1)}\cos\theta_1^{(1)}\sin\phi_1^{(1)} + e_{2,\mathrm{s}}^{(1)}\cos\phi_2^{(1)} - e_{2,\mathrm{p}}^{(1)}\cos\theta_2^{(1)}\sin\phi_1^{(1)},$$
$$\Delta h_x^{(1)}(\mathbf{k}) = (\epsilon_0/\mu_0)^{1/2}\Big(e_{1,\mathrm{s}}^{(1)}\cos\theta_1^{(1)}\cos\phi_1^{(1)} + e_{1,\mathrm{p}}^{(1)}\sin\phi_1^{(1)} + \hat{n}_2 e_{2,\mathrm{s}}^{(1)}\cos\theta_2^{(1)}\cos\phi_2^{(1)}$$
$$-\hat{n}_2 e_{2,\mathrm{p}}^{(1)}\sin\phi_2^{(1)}\Big),$$
$$\Delta h_y^{(1)}(\mathbf{k}) = (\epsilon_0/\mu_0)^{1/2}\Big(e_{1,\mathrm{s}}^{(1)}\cos\theta_2^{(1)}\sin\phi_1^{(1)} - e_{1,\mathrm{p}}^{(1)}\cos\phi_1^{(1)} + \hat{n}_2 e_{2,\mathrm{s}}^{(1)}\cos\theta_2^{(1)}\sin\phi_2^{(1)}$$
$$+\hat{n}_2 e_{2,\mathrm{p}}^{(1)}\cos\phi_2^{(1)}\Big). \tag{2.224}$$

Finally, combining Eqs. (2.221) and (2.224), solving for the first-order terms, using Eqs. (2.135), (2.146), and (2.149), and wading through a lot of algebra, we arrive at

$$\begin{aligned} e_{1,q}^{(1)} &= 2ik\Delta Z(\mathbf{k}_0 - \mathbf{k})\cos\theta_1{}^{(0)} r_{qp} e_p^{(0)}, \\ e_{2,q}^{(1)} &= 2ik\Delta Z(\mathbf{k}_0 - \mathbf{k})\cos\theta_1{}^{(0)} t_{qp} e_p^{(0)}, \end{aligned} \tag{2.225}$$

where we identify $\theta_{\rm i} = \theta_1^{(0)}$. The scattering reflection coefficients are

$$\begin{aligned} r_{\rm ss} &= \frac{(\tilde{n}^2 - 1)\cos\phi_{\rm r}}{\left(\cos\theta_{\rm i} + \sqrt{\tilde{n}_2{}^2 - \sin^2\theta_{\rm i}}\right)\left(\cos\theta_{\rm r} + \sqrt{\tilde{n}^2 - \sin^2\theta_{\rm r}}\right)}, \\ r_{\rm ps} &= \frac{(\tilde{n}^2 - 1)\sqrt{\tilde{n}^2 - \sin^2\theta_{\rm r}}\sin\phi_{\rm r}}{\left(\cos\theta_{\rm i} + \sqrt{\tilde{n}^2 - \sin^2\theta_{\rm i}}\right)\left(\tilde{n}^2\cos\theta_{\rm r} + \sqrt{\tilde{n}^2 - \sin^2\theta_{\rm r}}\right)}, \\ r_{\rm sp} &= \frac{(\tilde{n}^2 - 1)\sqrt{\tilde{n}^2 - \sin^2\theta_{\rm i}}\sin\phi_{\rm r}}{\left(\tilde{n}^2\cos\theta_{\rm i} + \sqrt{\tilde{n}^2 - \sin^2\theta_{\rm i}}\right)\left(\cos\theta_{\rm r} + \sqrt{\tilde{n}^2 - \sin^2\theta_{\rm r}}\right)}, \\ r_{\rm pp} &= \frac{(\tilde{n}^2 - 1)\left(\tilde{n}^2\sin\theta_{\rm i}\sin\theta_{\rm r} - \sqrt{\tilde{n}^2 - \sin^2\theta_{\rm i}}\sqrt{\tilde{n}^2 - \sin^2\theta_{\rm r}}\cos\phi_{\rm r}\right)}{\left(\tilde{n}^2\cos\theta_{\rm i} + \sqrt{\tilde{n}^2 - \sin^2\theta_{\rm i}}\right)\left(\tilde{n}^2\cos\theta_{\rm r} + \sqrt{\tilde{n}^2 - \sin^2\theta_{\rm r}}\right)}, \end{aligned} \tag{2.226}$$

where we let $\tilde{n} = \tilde{n}_2$ and identify $\theta_{\rm r} = \theta_1^{(1)}$, and $\phi_{\rm r} = \phi_1^{(1)}$. The scattering transmission coefficients are

$$\begin{aligned} t_{\rm ss} &= \frac{(\tilde{n}^2 - 1)\cos\phi_{\rm t}}{\left(\cos\theta_{\rm i} + \sqrt{\tilde{n}^2 - \sin^2\theta_{\rm i}}\right)\left(\tilde{n}\cos\theta_{\rm t} + \sqrt{1 - \tilde{n}^2\sin^2\theta_{\rm t}}\right)}, \\ t_{\rm ps} &= \frac{-(\tilde{n}^2 - 1)\sqrt{1 - \tilde{n}^2\sin^2\theta_{\rm t}}\sin\phi_{\rm t}}{\left(\cos\theta_{\rm i} + \sqrt{\tilde{n}^2 - \sin^2\theta_{\rm i}}\right)\left(\cos\theta_{\rm t} + \tilde{n}\sqrt{1 - \tilde{n}^2\sin^2\theta_{2\rm r}}\right)}, \\ t_{\rm sp} &= \frac{(\tilde{n}^2 - 1)\sqrt{\tilde{n}^2 - \sin^2\theta_{\rm i}}\sin\phi}{\left(\tilde{n}^2\cos\theta_{\rm i} + \sqrt{\tilde{n}^2 - \sin^2\theta_{\rm i}}\right)\left(\cos\theta_{\rm t} + \sqrt{1 - \tilde{n}^2\sin^2\theta_{\rm t}}\right)}, \\ t_{\rm pp} &= \frac{(\tilde{n}^2 - 1)\left(\sqrt{\tilde{n}^2 - \sin^2\theta_{\rm i}}\sqrt{1 - \tilde{n}^2\sin^2\theta_{\rm t}}\cos\phi + \tilde{n}\sin\theta_1\sin\theta_{\rm t}\right)}{\left(\tilde{n}^2\cos\theta_{\rm i} + \sqrt{\tilde{n}^2 - \sin^2\theta_{\rm i}}\right)\left(\cos\theta_{\rm t} + \tilde{n}\sqrt{1 - \tilde{n}^2\sin^2\theta_{\rm t}}\right)}, \end{aligned} \tag{2.227}$$

where we identify $\theta_{\rm t} = \theta_2^{(1)}$ and $\phi_{\rm t} = \phi_2^{(1)}$ The radiance associated with the roughness is

$$L_1 - \frac{k^4}{2\pi^2} Z_0 \langle |Z(\mathbf{k}_0 - \mathbf{k})|^2 \rangle \cos^2\theta_i \cos\theta_r |r_{qp}|^2 \langle |e_p^{(0)}|^2 \rangle,$$
$$L_2 = \frac{k^4}{2\pi^2} \tilde{n}^2 Z_0 \langle |Z(\mathbf{k}_0 - \mathbf{k})|^2 \rangle \cos^2\theta_i \cos\theta_t |t_{qp}|^2 \langle |e_p^{(0)}|^2 \rangle, \quad (2.228)$$

while the incident irradiance is

$$E = \frac{\cos\theta_i}{2} Z_0 \left\langle |e_p^{(0)}|^2 \right\rangle. \quad (2.229)$$

Thus, the BRDF, given by $f_r = L_1/E$ is given by

$$f_r = \frac{k^4}{\pi^2} \cos\theta_i \cos\theta_r |r_{pq}|^2 \langle |Z(\mathbf{k}_0 - \mathbf{k})|^2 \rangle, \quad (2.230)$$

and the BTDF, given by $f_t = L_2/E$ is given by

$$f_r = \frac{k^4}{\pi^2} \tilde{n}^2 \cos\theta_i \cos\theta_t |t_{pq}|^2 \langle |Z(\mathbf{k}_0 - \mathbf{k})|^2 \rangle. \quad (2.231)$$

The quantity $\langle |Z(\mathbf{k}_0 - \mathbf{k})|^2 \rangle$ is often referred to as the power spectral density (PSD) of the surface height function. It describes the power (the square of the amplitude) contained, per unit bandwidth in spatial frequency space, in the roughness function. The spatial angular frequency of the roughness, $\mathbf{k}_0 - \mathbf{k}$, can be evaluated from the incident and scattering angles,

$$(\mathbf{k}_0 - \mathbf{k})_x = (2\pi/\lambda)(\sin\theta_i - \sin\theta_r \cos\phi_r),$$
$$(\mathbf{k}_0 - \mathbf{k})_y = (2\pi/\lambda)(-\sin\theta_r \sin\phi_r). \quad (2.232)$$

Total diffuse reflectance and total diffuse transmittance can be obtained by integration of Eq. (2.230) or (2.231), respectively. For large period roughness, $\theta_i \cong \theta_r$ for most scattering angles that contribute to significant scatter. Furthermore, $r_{ss} \cong r_s$, so that

$$\rho_s = \frac{k^2}{\pi^2} \cos^2\theta_1 |r_s|^2 \sigma_z^2, \quad (2.233)$$

where σ_z is the root-mean-square roughness, given by

$$\sigma_z^2 = \int \langle |Z(\mathbf{k})|^2 \rangle \, d^2k. \quad (2.234)$$

REFERENCES

[1] CIE, ILV: International lighting vocabulary, CIE, Vienna, Austria, 2011, Ref Type: Report.
[2] F. Hengstberger, Absolute Radiometry, Academic Press, Boston, MA, 1989.
[3] D.J. Griffiths, Introduction to Electrodynamics, Prentice Hall, Upper Saddle River, New Jersey, 1998.

[4] J.D. Jackson, Classical Electrodynamics, John Wiley & Sons, New York, NY, 1975.
[5] J.R. Reitz, F.J. Milford, R.W. Christy, Foundations of Electromagnetic Theory, Addison-Wesley, Reading, 2008.
[6] C.F. Bohren, D.R. Huffman, Absorption and Scattering of Light by Small Particles, Wiley, New York, NY, 1983.
[7] W.H. Venable Jr., J.J. Hsia, Describing spectrophotometric measurements, National Bureau Standards, Gaithersburg, MD, 1974, Ref Type: Report.
[8] F.E. Nicodemus, J.C. Richmond, J.J. Hsia, I.W. Ginsberg, T. Limperis, Geometrical Considerations and Nomenclature for Reflectance, National Bureau of Standards, Gaithersburg, 1977.
[9] F.E. Nicodemus, Reflectance nomenclature and directional reflectance and emissivity, Applied Optics 9 (1970) 1474–1475.
[10] R.J. Potton, Reciprocity in optics, Reports on Progress in Physics 67 (2004) 717–754.
[11] R.A. Chipman, Polarimetry, in: Michael Bass (Ed.), Handbook of Optics, McGraw-Hill, New York, NY, 1995, pp. 1–37.
[12] D.H. Goldstein, Polarized Light, CRC Press, Boca Raton, Florida, 2010.
[13] P. Kubelka, F. Munk, Ein Beitrag zur Optik der Farbanstriche, Z. Technische Physik 12 (1931) 593–601.
[14] B. Hapke, Theory of Reflectance and Emittance Spectroscopy, Cambridge University Press, Cambridge, 1993.
[15] D.E. Barrick, Rough surface scattering based on the specular point theory, IEEE Trans. Ants. Prop. AP-16 (1968) 449–454.
[16] K.E. Torrance, E.M. Sparrow, Theory of off-specular reflection from roughened surfaces, Journal of the Optical Society of America 57 (1967) 1105–1114.
[17] B.G. Smith, Geometrical Shadowing of a Random Rough Surface, IEEE Trans. Ants. Prop. AP-15 (1967) 668–671.
[18] L. Rayleigh, On the Dynamical Theory of Gratings, Proceedings of the Royal Society of London A 79 (1907) 399.
[19] S.O. Rice, Reflection of Electromagnetic Waves from Slightly Rough Surfaces, Communications on Pure and Applied Mathematics 4 (1951) 351–378.
[20] D.E. Barrick, W.H. Peake, A Review of Scattering from Surfaces with Different Roughness Scales, Radio Science 3 (1968) 865–868.

Chapter 3

Dispersive Methods

Arnold A. Gaertner*, Howard W. Yoon† and Thomas A. Germer†
*National Research Council Canada, Ottawa, Ontario, Canada
†National Institute of Standards and Technology, Gaithersburg, Maryland, USA

Chapter Outline

3.1 Introduction 67
3.2 General Description 69
3.2.1 Bandpass Function 70
3.2.2 Gratings 74
3.2.3 Prisms 78
3.2.4 Resolving Power 81
3.3 Spectral Analyzer Design 82
3.3.1 Monochromator Design 82
3.3.2 Polychromator Design 84
3.4 Wavelength Calibration 84
3.5 Stray Light 87
3.6 Optical Radiation Sources 88
3.7 Optical Radiation Detectors 90
References 94

3.1 INTRODUCTION

A key feature of spectrophotometric measurements is that the measurements of reflection, transmission, absorption, scattering, fluorescence, and emittance are performed as a function of wavelength and, in some cases, are spectral concentrations. These concepts are discussed in detail in Chapter 2. For example, reflectance ρ is a function of wavelength, for example, $\rho=\rho(\lambda)$. Fluorescence is not only a function of the excitation and emission wavelengths, μ and λ, respectively, but also a spectral concentration in the emission wavelength. That is, the radiance L emitted by an illuminated surface is spread out over wavelength, so that the spectral radiance is given by

$$L_\lambda=\frac{\mathrm{d}L}{\mathrm{d}\lambda}. \tag{3.1}$$

Thus, it is important that one not only be able to measure quantities as functions of wavelength but also be able to measure spectral concentrations.

There are two primary methods for obtaining spectral information in optical measurements: one can disperse the different wavelengths in different

Experimental Methods in the Physical Sciences, Vol. 46. http://dx.doi.org/10.1016/B978-0-12-386022-4.00003-0

directions or locations, or one can perform an interferometric measurement. In this chapter, we concentrate on the former method, while Chapter 4 will discuss the interferometric methods.

The spectral analyzer is often buried as a "black box" within the body of the spectrophotometer. Nevertheless, the characteristics of this black box will have a significant influence upon the type and accuracy of the spectral information obtained by the spectrophotometer. The spectral analyzer may be used for either, or both, of two purposes in the spectrophotometer. In the *monochromatic mode*, the spectral analyzer is placed in the radiation path before the sample, thus being used for the purpose of providing a known, narrow spectral bandwidth of radiation that is then used to measure sample properties. In the *polychromatic mode*, the sample is first irradiated with polychromatic radiation and the radiation that exits the sample is then spectrally analyzed. Figure 3.1 graphically illustrates these two principal methods.

In many cases, it is not critical which of the two different sample illumination modes is used. However, there are several special circumstances where the choice is critical. Thermochromic or heat-sensitive samples, for example, should be measured in the monochromatic mode, since the large amount of radiation that is incident upon the sample in polychromatic mode may impact the measurement. Fluorescent samples are often measured using a combination of both modes, where two spectral analyzers are used, one for monochromatic irradiation and one for spectral analysis of the emitted radiation. Fluorescent color measurements are often performed in the polychromatic mode, where the polychromatic radiation source is filtered to simulate one of the standard illuminants.

In this chapter, we describe the fundamental operation of a spectral analyzer, which can either be a monochromator or a polychromator (sometimes referred to as a multichannel spectrometer). Features common to all of these devices will be first discussed, including spectral issues associated with

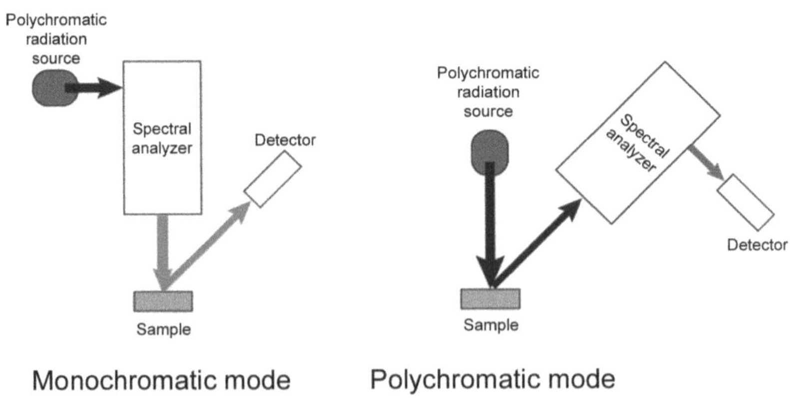

FIGURE 3.1 Spectral analyzer modes.

resolution and bandwidth, and characteristics of focal length and aperture. The two specific types of dispersive elements, gratings and prisms, which are used in these devices, and their characteristics will be described. Methods for wavelength calibration of these systems will be given. Nonideal aspects of these systems, including stray light and the effects of image aberrations, are then discussed. Finally, we discuss some of the light sources and detectors that are commonly used in these systems. Other useful texts on these topics can be found in Refs. [1–7].

3.2 GENERAL DESCRIPTION

Figure 3.2 shows a general schematic of a spectral analyzer. The principal element in a spectral analyzer is the dispersive element, which is usually either a grating or a prism. This element serves to take collimated radiation and redirect that radiation into directions that depend upon the wavelength. Thus, we generally have a relationship that the deviation angle θ is a function of the wavelength λ: $\theta = \theta(\lambda)$. It is not important from where θ is measured. Gratings, both reflective and transmissive, and prisms each have coordinate systems unique to their application. However, what is important is the dispersion, $d\theta/d\lambda$, which indicates how that angle depends upon the wavelength.

The input optic of a spectral analyzer, be it a monochromator or a polychromator, usually serves to take an image of an input aperture, or a slit,

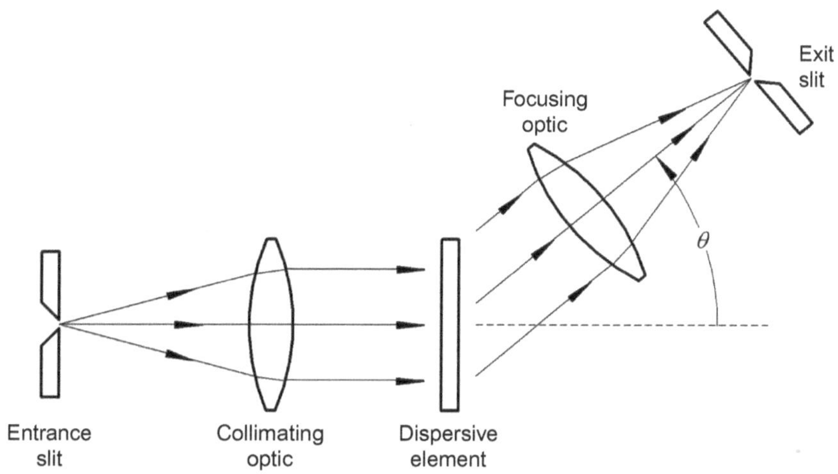

FIGURE 3.2 A general schematic of a spectral analyzer, showing the entrance slit, a collimating optic, a dispersive element, a focusing optic, and an exit slit. The dispersive element serves to change the direction of the light by an angle θ, which is a function of wavelength λ. The collimating and focusing optics are usually curved mirrors instead of lenses.

and focus it at infinity, so that the light is approximately collimated. Some spectral analyzers use the dispersive element as part of the focusing optics, but we will limit our discussion right now to the simpler case. A second optic placed after the dispersive element then takes a point at infinity and focuses it on either an exit aperture, or a slit, in the case of a monochromator, or onto a position-sensitive detector, in the case of a polychromator. Assuming that the focal lengths of the two optics are the same and given by f, the linear dispersion of the focus point is related to the angle dispersion by

$$D = \frac{\mathrm{d}x}{\mathrm{d}\lambda} = f\frac{\mathrm{d}\theta}{\mathrm{d}\lambda}. \tag{3.2}$$

One also sometimes refers to the reciprocal linear dispersion D^{-1}, which is

$$D^{-1} = \frac{\mathrm{d}\lambda}{\mathrm{d}x} = \left(\frac{\mathrm{d}x}{\mathrm{d}\lambda}\right)^{-1} = \frac{1}{f(\mathrm{d}\theta/\mathrm{d}\lambda)}. \tag{3.3}$$

The details of the relationship given in Eq. (3.2) depend upon the type of dispersive element and how it is used.

3.2.1 Bandpass Function

A monochromator images the entrance slit onto the exit slit, when the wavelength λ is the center wavelength λ_0. For wavelengths near λ_0, the image of the entrance slit is shifted on the exit slit by the distance $D\Delta\lambda$, where $\Delta\lambda = \lambda - \lambda_0$ is the shift in wavelength. That is, for every wavelength λ, a point on the entrance slit, $(x_{\text{enter}}, y_{\text{enter}})$, is imaged onto a point on the exit slit, $(x_{\text{exit}}, y_{\text{exit}})$, according to

$$\begin{aligned} x_{\text{exit}} &= x_{\text{enter}} + D\Delta\lambda, \\ y_{\text{exit}} &= y_{\text{enter}}. \end{aligned} \tag{3.4}$$

Illuminating the entrance slit, there is a spectral radiance $L_\lambda(\lambda)$, which we assume is uniform in position and direction. We further assume that the transmittance of the entrance slit is given by a function $\tau_{\text{enter}}(x_{\text{enter}}, y_{\text{enter}})$, which is unity inside the slit and zero outside it. The spectral emittance that is captured by the imaging system is then

$$E_{\lambda,\text{enter}}(x_{\text{enter}}, y_{\text{enter}}, \lambda) = \tau_{\text{enter}}(x_{\text{enter}}, y_{\text{enter}}) L_\lambda(\lambda) \Omega_{\text{proj}}, \tag{3.5}$$

where the projected solid angle (see Eq. 2.9) collected by the monochromator is

$$\Omega_{\text{proj}} = \frac{\pi}{\left[1 + 4(f/d)^2\right]}. \tag{3.6}$$

The quantity f/d, where d is the diameter of the collection optic, is the f-number of the system. From a point on the entrance slit having area $\mathrm{d}x_{\text{enter}}\mathrm{d}y_{\text{enter}}$, the radiant power passing through the exit slit is

$$\begin{aligned} d\Phi_{exit}(x_{exit}, y_{exit}, \lambda) &= e(\lambda)\tau_{exit}(x_{exit}, y_{exit})E_{\lambda,enter}(x_{enter}, y_{enter}, \lambda)\, dx_{enter}dy_{enter}d\lambda \\ &= e(\lambda)\tau_{exit}(x_{exit}, y_{exit})\tau_{enter}(x_{enter}, y_{enter}) \\ &\quad L_{\lambda}(\lambda)\Omega_{proj}dx_{enter}dy_{enter}d\lambda, \end{aligned} \tag{3.7}$$

where $e(\lambda)$ denotes the combined efficiency (including reflectance of the mirrors and the absolute efficiency of the dispersive element) of the internal monochromator components, and $\tau_{exit}(x_{exit}, y_{exit})$ is the transmittance of the exit slit, which is unity inside the slit and zero outside it. Making the change in variables given in Eq. (3.4), Eq. (3.7) can be rewritten as

$$\begin{aligned} d\Phi_{exit}(x_{exit}, y_{exit}, \lambda) &= e(\lambda)\tau_{exit}(x_{exit}, y_{exit})\tau_{enter}(x_{exit} - D\Delta\lambda, y_{exit}) \\ &\quad L_{\lambda}(\lambda)\Omega_{proj}dx_{exit}dy_{exit}d\lambda. \end{aligned} \tag{3.8}$$

The radiant power passing through the exit slit is then obtained by integrating over the spatial variables,

$$\begin{aligned} d\Phi_{exit}(\lambda) &= e(\lambda)L_{\lambda}(\lambda)\Omega_{proj} \\ &\quad \left[\iint \tau_{exit}(x_{exit}, y_{exit})\tau_{enter}(x_{exit} - D\Delta\lambda, y_{exit})dx_{exit}dy_{exit}\right]d\lambda. \end{aligned} \tag{3.9}$$

The quantity in the brackets can be seen to be a convolution between the entrance slit function τ_{enter} and the exit slit function τ_{exit} and is referred to as the spectral bandpass function,

$$W(\Delta\lambda) = \iint \tau_{exit}(x_{exit}, y_{exit})\tau_{enter}(x_{exit} - D\Delta\lambda, y_{exit})dx_{exit}dy_{exit}, \tag{3.10}$$

which has units of area. Thus, the total power passing through the monochromator is

$$\Phi_{exit} = \Omega_{proj}\int e(\lambda)L_{\lambda}(\lambda)W(\Delta\lambda)d\lambda. \tag{3.11}$$

In most cases, the combined efficiency of the monochromator is a relatively slowly varying function and can be evaluated at λ_0 and taken out of the integral. However, some dispersive elements can have sharp spectral features, which necessitate keeping $e(\lambda)$ in the integral or at least including its effects in an analysis of uncertainties.

The bandpass function can be calculated either by directly evaluating Eq. (3.10) or by recognizing that the convolution is simply the overlap area between the two slit functions as illustrated in Fig. 3.3. For a rectangular entrance slit of width Δx_{enter} and height Δy_{enter} and rectangular exit slit of width Δx_{exit} and height Δy_{exit}, the spectral bandpass function $W(\Delta\lambda)$ has a trapezoidal shape with a peak value of (see Fig. 3.3, left)

$$W_{max} = \min(\Delta x_{enter}, \Delta x_{exit})\min(\Delta y_{enter}, \Delta y_{exit}), \tag{3.12}$$

a top width of

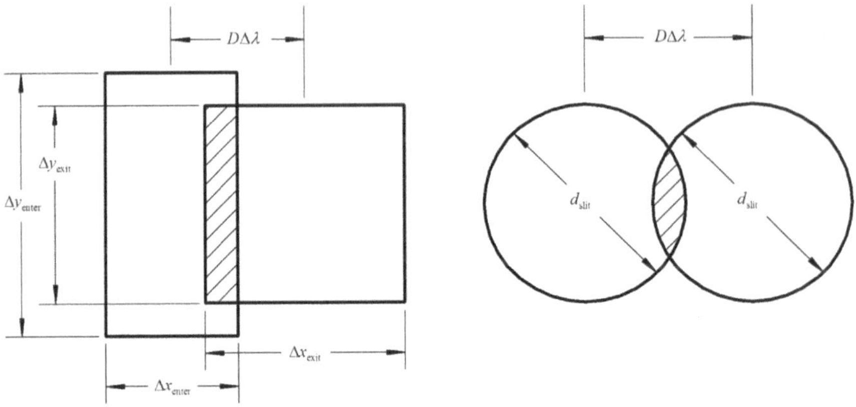

FIGURE 3.3 Calculation of the spectral bandpass function. On the left are two rectangular apertures, and on the right are two circular apertures. The spectral bandpass function is the overlap (hatched areas) as a function of $\Delta\lambda$.

$$\delta\lambda_{top} = \frac{|\Delta x_{enter} - \Delta x_{exit}|}{D}, \tag{3.13}$$

a full base width of

$$\delta\lambda_{base} = \frac{\Delta x_{enter} + \Delta x_{exit}}{D}, \tag{3.14}$$

and a full width at half maximum (FWHM) of

$$\delta\lambda_{FWHM} = \frac{(\delta\lambda_{base} + \delta\lambda_{top})}{2}. \tag{3.15}$$

The integral of the bandpass function is

$$\int W(\Delta\lambda)\mathrm{d}\lambda = W_{max}\delta\lambda_{FWHM}. \tag{3.16}$$

Equation (3.16) is useful for calculating the total power passing through the monochromator, given a known radiance at the entrance slit. If the spectral dependence of $e(\lambda)$ and $L_\lambda(\lambda)$ are slowly varying, the total radiant power exiting the monochromator is

$$\Phi_{exit} \approx \Omega_{proj} e(\lambda) L_\lambda(\lambda) \int W(\Delta\lambda)\mathrm{d}\lambda. \tag{3.17}$$

When the two slits are of the same width, the bandpass function is triangular.

For two circular apertures of diameter d_{slit}, the function $W(\Delta\lambda)$ is given by (see Fig. 3.3, right)

$$W(\Delta\lambda) = \frac{d_{slit}^2}{2}\left[\cos^{-1}\left(\frac{|\Delta\lambda|D}{d_{slit}}\right) - \frac{|\Delta\lambda|D}{d_{slit}^2}\sqrt{d_{slit}^2 - (D\Delta\lambda)^2}\right]. \tag{3.18}$$

The bandpass function given by Eq. (3.18) is close to, but not actually, triangular. The peak value of the bandwidth function is $W_{max} = \pi d_{slit}^2/4$, the area of the slit. The full base width is

$$\delta\lambda_{base} = \frac{2d_{slit}}{D}, \tag{3.19}$$

the FWHM is

$$\delta\lambda_{FWHM} = \frac{0.808d_{slit}}{D}, \tag{3.20}$$

and the integral of the bandwidth function is

$$\int W(\Delta\lambda)\mathrm{d}\lambda = \frac{\pi d_{slit}^3}{12D}. \tag{3.21}$$

Consider the case of a monochromator with rectangular slits of equal width Δx. If one considers Eq. (3.8), one can see that at one edge of the exit slit, say $x_{exit} = -\Delta x/2$, only wavelengths from $\Delta\lambda = -\Delta x/D$ to $\Delta\lambda = 0$ contribute to the radiant power, while at the opposite edge of the exit slit, say $x_{exit} = \Delta x/2$, only wavelengths from $\Delta\lambda = 0$ to $\Delta\lambda = \Delta x/D$ contribute to the radiant power. This can sometimes be a problem, because there is dispersion in the wavelengths across the aperture. One way to remove this effect is to use two monochromators in *subtractive mode*. The two monochromators are configured so that the dispersion in one of them is the opposite of the other. Such a monochromator would have a spectral bandwidth function that would be determined by the first monochromator, but the second monochromator would remove the spatial dispersion across the slit. In addition, the use of two monochromators substantially lowers the effects of stray light, so that stray radiant energy (SRE), discussed later, is significantly reduced.

A second monochromator can also be used in *additive mode*. In that case, the second monochromator serves to further reduce the bandwidth, as if the dispersion term in Eq. (3.4) were doubled. In addition, as for the monochromator being used in subtractive mode, there is a substantial reduction in the SRE.

The analysis above for the spectral line width relies upon the monochromator performing in an ideal fashion. Since the source is generally focused onto the entrance slit of the monochromator, the radiance at the entrance slit is rarely uniform. The collimating and focusing optics in the monochromator are usually operated in an off-axis geometry, so aberrations will alter the imaging of the entrance slit on the exit slit. Thus, as the slit width increases, the instrument spectral function deviates from the ideal shape defined by the above analysis. For these reasons, it is important in critical applications to measure this lineshape function. This can generally be accomplished using a second monochromator. The output of the monochromator under test is best collected by an integrating sphere before being sampled by the second monochromator. One can also use narrow-band laser sources to measure the spectral bandwidth function. Again, one should use integrating spheres on both the illumination and collection sides of the monochromator. The laser method, however, will only account for the internal aspects of the monochromator and not for the source spatial distribution on the entrance slit.

3.2.2 Gratings

When optical radiation is incident upon a repetitive structure with a period Λ, such as that shown in Fig. 3.4, diffraction from the structure can occur [1,2,5,6]. Gratings are usually specified by their inverse period $G = 1/\Lambda$, usually in units of lines or grooves per millimeter. If radiation is collimated and incident upon the grating at an angle ι, the grating will diffract radiation into specific angles θ_m that satisfy the diffraction equation

$$m\lambda G = \sin\iota - \sin\theta_m. \tag{3.22}$$

The integer m is the diffraction order. When the diffraction order is $m = 0$, the effect is the same as the specular reflection, where $\iota = \theta_0$. We can rewrite Eq. (3.22) in terms of the half angle between the incident and diffraction directions,

$$\alpha = \frac{\iota + \theta_m}{2}, \tag{3.23}$$

and the orientation of the grating normal with respect to the centerline between the incident and diffraction directions,

$$\beta = \frac{\iota - \theta_m}{2}. \tag{3.24}$$

Equation (3.22) can then be written in terms of α and β as

$$\lambda = \frac{2}{mG}\sin\beta\cos\alpha. \tag{3.25}$$

The form of Eq. (3.25) is useful, because monochromators are usually designed so that there is a fixed, small angle 2α between the entrance optics

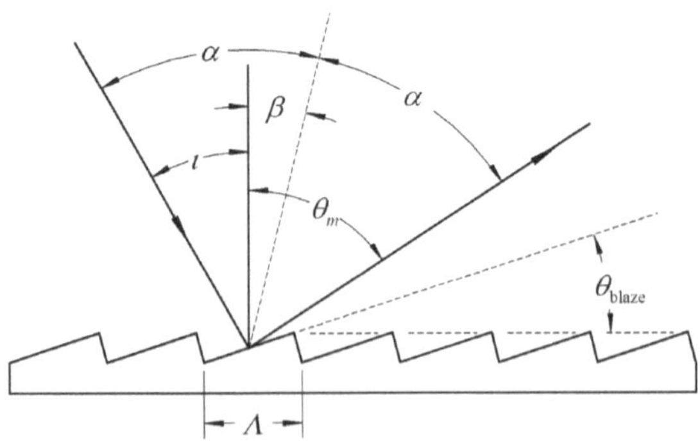

FIGURE 3.4 Plane reflection grating variables.

and the exit optics, and the grating is rotated to tune the monochromator to a specific wavelength. When $\alpha=0$, the grating is said to be operated in the Littrow geometry. The angular dispersion is found by differentiating Eq. (3.22), yielding

$$\frac{\mathrm{d}\theta_m}{\mathrm{d}\lambda}=-\frac{Gm}{\cos\theta_m}. \tag{3.26}$$

Since $\theta_m=\alpha-\beta$, we can write Eq. (3.26) as

$$\frac{\mathrm{d}\theta_m}{\mathrm{d}\lambda}=-\frac{Gm}{\cos(\alpha-\beta)}. \tag{3.27}$$

When α is small, which is often the case, the dispersion is approximately

$$\frac{\mathrm{d}\theta_m}{\mathrm{d}\lambda}\approx-\frac{2Gm}{\sqrt{4-G^2\lambda^2m^2}}. \tag{3.28}$$

If one considers all of the expressions, Eqs. (3.22)–(3.28), one finds that the wavelength λ is always combined with the diffraction order m as the product $m\lambda$. For this reason, when a diffractive spectral analyzer is configured for wavelength λ and order $m=1$, the wavelengths $\lambda/2$, $\lambda/3$, etc., will also be passed through the analyzer. It is usually necessary that one remove these extraneous orders in a process referred to as *order sorting*. Order sorting is usually accomplished by including one or more, possibly interchangeable longpass filters into the entrance or exit of the device. For example, a diffractive spectral analyzer intended to be operated from 200 to 2500 nm needs interchangeable filters that block wavelengths shorter than 1250, 625, and 313 nm. These filters are usually contained in a filter wheel just outside the device, usually on the side where their presence would have the least negative impact on any imaging optics: the source side of a monochromator or the detector side of a polychromator. A prism predisperser, which does not pass higher orders, can be also be used for order sorting.

Gratings can be characterized by their grating efficiency, which is a measure of the power radiated in a diffraction order compared to the radiant power incident upon it. The *absolute grating efficiency* for the *m*th order is given by

$$\eta=\frac{\Phi_m}{\Phi_\mathrm{i}}, \tag{3.29}$$

where Φ_m is the radiant power in the *m*th order and Φ_i is the incident radiant power. The term *relative grating efficiency* is also used and represents the absolute grating efficiency compared to the specular reflectance ρ of a smooth flat surface with the same coating,

$$\eta_\mathrm{rel}=\frac{\eta}{\rho}. \tag{3.30}$$

It should be recognized that the grating efficiency is dependent upon wavelength, polarization, and incident angle. For many wavelength regions, the dependence on wavelength may be relatively smooth. However, in the presence of Wood's anomalies [8], one of the diffraction orders, usually not the one being used, propagates along the surface, that is, with angle $\theta_m = \pm 90°$. In that case, there can be considerable structure in the grating efficiency spectrum, especially for the polarization state where the magnetic field is parallel to the lines of the grating (transverse magnetic, TM). When the electric field is parallel to the lines of the grating (transverse electric, TE), the grating efficiency spectrum is usually smoother. Grating efficiency curves are often quoted for the Littrow condition. Figure 3.5 shows example diffraction efficiency curves for two aluminum gratings. Sharp features are observed at the long wavelength side of each of the Wood's anomalies. Since the Wood's anomalies are observed primarily in the TM polarization and not the TE polarization, large changes in the relative polarization behavior of the spectral analyzer can occur at these wavelengths. As an aside, the convention used for polarization on gratings is the opposite of that used in reflection from surfaces, as described in Chapter 2. For gratings, it is often said that p-polarization is TM and s-polarization is TE. That is, the parallel (p) and perpendicular (s) directions refer to the lines of the grating, not the plane of incidence.

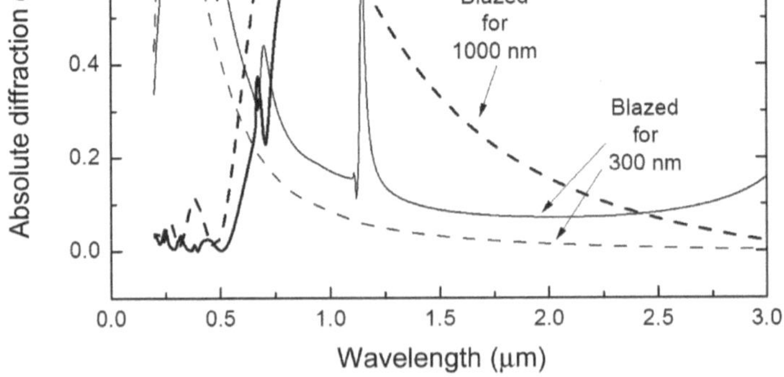

FIGURE 3.5 Absolute first-order diffraction efficiencies in the Littrow geometry for two 600 lines mm^{-1} blazed aluminum gratings, one blazed for 300 nm, and one blazed for 1000 nm. The efficiencies for the two polarizations are shown. Also shown near the top are vertical marks showing the expected locations of Wood's anomalies.

There are a number of types of gratings to consider. A common style of reflective grating is the mechanically ruled blazed grating, an example of which is shown in Fig. 3.6. The angle of the steps is chosen to optimize specular reflection into a specific order, usually $m=1$, for a specific wavelength. In the Littrow geometry, $\theta_m = -\imath$, and we see from Eq. (3.22) that the surfaces need to be tilted by the angle

$$\theta_{\text{blaze}} = \sin^{-1}\left(\frac{m\lambda G}{2}\right). \tag{3.31}$$

The blazing tends to approximately maximize the efficiency of the grating at the design wavelength. One can see this in Fig. 3.5, which shows the diffraction efficiency for gratings blazed for 300 and 1000 nm. Notice that if the grating blazed for 1000 nm were operated at 300 nm, there would be very little throughput of the spectral analyzer. A general rule of thumb is that a blazed grating will have a useful range from about 2/3 the blaze wavelength to 3/2 the blaze wavelength. For example, the grating blazed at 300 nm will be useful from 200 to 450 nm, while the grating blazed for 1000 nm will be useful from about 700 to 1500 nm. Spectral analyzers are often constructed to house multiple gratings on a turret, so that the diffraction efficiency can be optimized for each wavelength.

Blazed gratings are usually constructed by mechanical ruling. For that reason, they are subject to higher levels of scattered light, which result from errors in the ruling process, including chatter in the tool. Gratings can also be constructed using holographic processes. Two carefully tailored plane waves are allowed to interfere on a substrate coated with a photoresist. The nascent grating is then chemically processed to yield a relatively smooth, nearly sinusoidal profile. They are often then coated with a metallic layer. Holographic gratings tend to have significantly lower levels of scattered light, but they suffer from not necessarily having the optimum profile for maximizing diffraction efficiency. Since the amount of scatter from a surface is roughly inversely proportional to the square of the wavelength

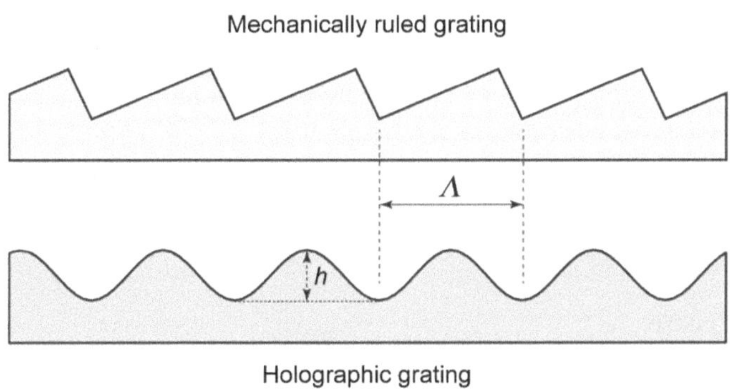

FIGURE 3.6 Groove profile for ruled and holographic gratings.

(e.g., see Chapter 2, Eq. 2.232), achieving low levels of scatter is particularly important for ultraviolet optics. Furthermore, it is easier with holographic gratings to achieve the higher groove densities needed for ultraviolet applications. While it is possible to blaze holographic gratings, it is not common. The curvature of the grooves in holographic gratings can be adjusted to improve the spectral imaging required for flat-field spectrographs, such as those using flat detector arrays where many wavelengths are measured simultaneously.

The properties of holographic diffraction gratings are often discussed with reference to the modulation $\mu = h/\Lambda = hG$, where h is the modulation height of the grating. These vary from very low modulation, $\mu < 0.05$, to very high modulation, $\mu > 0.4$. When used in reflection with a metallic overcoat, they show similar polarization effects to those noted above for plane ruled gratings. There is no simple difference between the efficiency curves for holographic gratings compared with ruled gratings, and the user is recommended to study the data available from the manufacturer for the gratings considered.

Diffraction gratings can also be designed as part of the focusing elements in the spectral analyzer. Focusing can be accomplished either by ruling the lines onto a curved surface or by having a line spacing that varies across the grating. These types of gratings are significantly more expensive, but can offer significant advantages in some applications. Fewer optical elements are needed, for example, in a spectral analyzer that combines the focusing and the dispersion in the same optic.

There are many variations of the above themes available for both ruled and holographic gratings. There are reflection gratings, transmission gratings, concave gratings, and even grisms, a combination of grating and prism in one piece. Echelle gratings are coarse gratings that are optimized for high incident angle and high diffraction orders and are used in conjunction with a second grating to disperse wavelengths along two dimensions. It should also be noted that most gratings in use are replicas of original master gratings that are ruled or made from photoresist (holographic).

Many monochromators are configured such that the wavelength λ is changed by rotating the grating about an axis coincident with its central ruling. In this case, the directions of the incident and the diffracted radiation beams remain constant in the monochromator and define the angle α in Eq. (3.25). For a high-resolution monochromator with focal length approximately 1 m, the angle between these directions, 2α, for one manufacturer is approximately 15°. This small angle causes the angle dispersion to vary slowly with wavelength. The small variation in the dispersion is one of the main advantages of a grating disperser system over a prism disperser system.

3.2.3 Prisms

Operation of a prism is based upon the refraction of light, governed by Snell's law, and the dispersion of the index of refraction $n(\lambda)$ with wavelength.

Snell's law, introduced in Section 2.6.1 of Chapter 2, relates the angles of propagation inside and outside a material,

$$n_{air}\sin\theta_{air} = n(\lambda)_{prism}\sin\left[\theta(\lambda)_{prism}\right], \tag{3.32}$$

where n_{air} is the index of refraction of the air, which is close to unity, θ_{air} is the angle of propagation in air, $n(\lambda)_{prism}$ is the wavelength-dependent index of refraction of the prism, and $\theta(\lambda)_{prism}$ is the wavelength-dependent angle of propagation in the prism. The wavelength dispersing properties of prism-based monochromators depend upon the material dispersion of the prism: the variation in the refractive index of the prism material with wavelength. In transparent media, the index of refraction is usually very smooth, following the Cauchy relationship

$$n(\lambda) = A_1 + \frac{A_2}{\lambda^2} + \frac{A_3}{\lambda^4}, \tag{3.33}$$

where A_1, A_2, and A_3 are constants specific to each material. Another common expression for the dispersion of materials is the Sellmeier equation

$$n^2(\lambda) = A + \frac{B_1\lambda^2}{\lambda^2 - C_1} + \frac{B_2\lambda^2}{\lambda^2 - C_2}, \tag{3.34}$$

where, again, A, B_1, C_1, B_2, and C_2 are constants specific to the material. Figure 3.7 shows the dispersion of silica glass along with its derivative, $dn/d\lambda$ [9].

Figure 3.8 shows rays passing through a prism with an apex angle of α. The angle of incidence on the left face is β. Applying Snell's law to the two interfaces, we can derive an expression for the deviation angle θ,

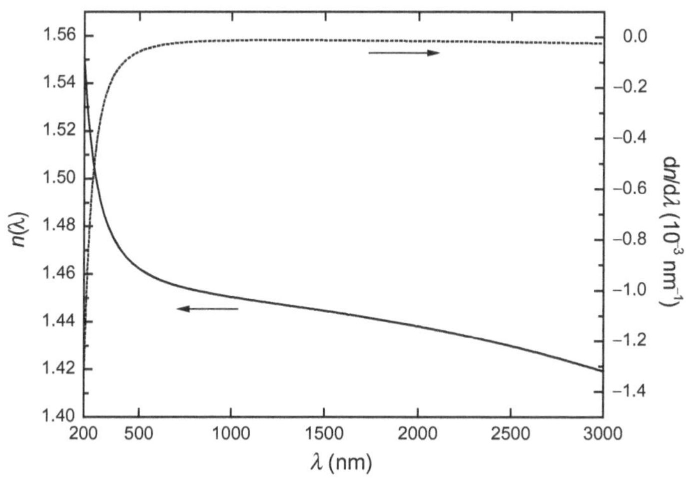

FIGURE 3.7 The refractive index and the index dispersion of fused silica.

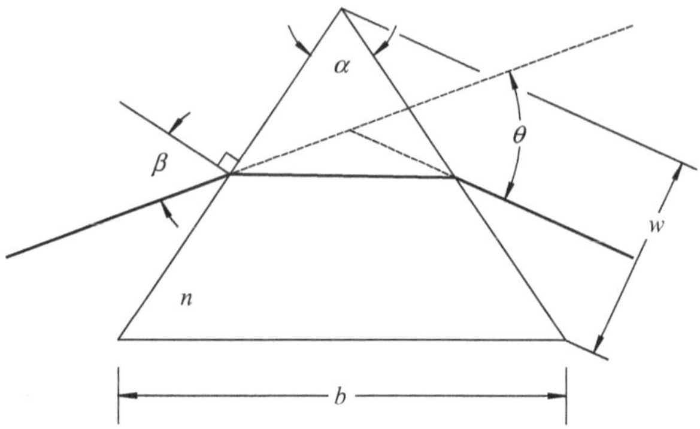

FIGURE 3.8 Optical rays passing through a prism. The angle of incidence on the left face is β, the apex angle of the prism is α, and the deviation angle is θ.

$$\theta = \beta - \alpha + \sin^{-1}\left\{ n \sin\left[\alpha - \sin^{-1}\left(\frac{\sin\beta}{n}\right)\right]\right\}. \tag{3.35}$$

Equation (3.35) is the prism equivalent of Eq. (3.22) for gratings. We can differentiate this expression with respect to the index of refraction,

$$\frac{d\theta}{d\lambda} = \frac{n\sin\alpha}{\sqrt{n^2 - \sin^2\beta}\sqrt{1 - n^2\sin^2\left[\alpha - \sin^{-1}\left(\frac{\sin\beta}{n}\right)\right]}}\frac{dn}{d\lambda}. \tag{3.36}$$

Although Eqs. (3.35) and (3.36) are not the most elegant expressions, they form the basis for the prism spectral analyzer. The minimum deviation angle $\theta_{\min}$ is often used for measuring the index of refraction of a transparent material and occurs when the rays are symmetric and satisfies

$$n\sin\frac{\alpha}{2} = \sin\left(\frac{\alpha + \theta_{\min}}{2}\right). \tag{3.37}$$

In this configuration, the incident angle is $\beta = (\alpha + \theta_{\min})/2$. Prism spectral analyzers are often configured so that they operate near this angle for one of the wavelengths in the spectral range of interest. Near this angle, the dispersion can be approximated by

$$\frac{d\theta}{d\lambda} = \frac{b}{w}\frac{dn}{d\lambda} = \frac{2\sin(\alpha/2)}{\cos\beta}\frac{dn}{d\lambda}, \tag{3.38}$$

where b is the base width of the prism and w is the projection of the outgoing side in the outgoing direction (see Fig. 3.8).

Figure 3.9 shows the angular dispersion of a prism spectrometer configured to be in the minimum deviation angle for $\lambda = 250$ nm. It can be seen that the angular dispersion is a strong function of wavelength and has dispersion near or above 1 mrad nm^{-1} in the ultraviolet. In the visible range, however,

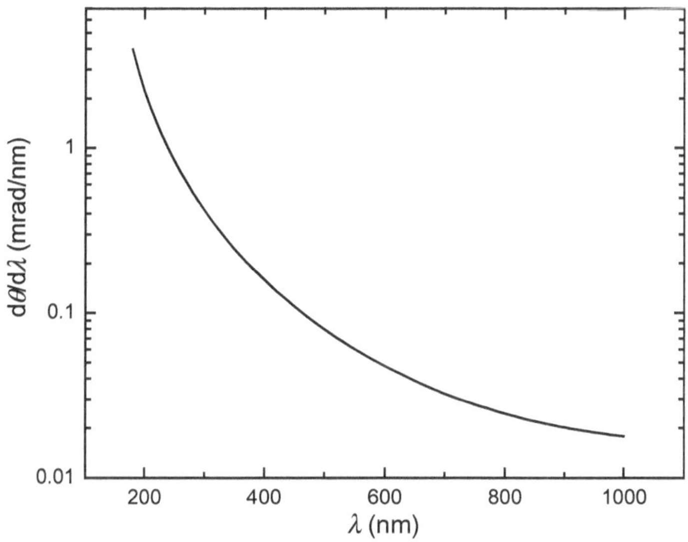

FIGURE 3.9 The angle dispersion of a prism spectrometer using a fused silica prism with apex angle $\alpha = 60°$ and configured to be in the minimum deviation angle $\lambda = 250$ nm. The minimum deviation angle is $\theta_{min} = 37.83°$, and the incident angle is $\beta = 48.91°$.

the dispersion and thus the resolution and bandwidth are significantly poorer. The strong dispersion in the ultraviolet is what makes prism-based spectral analyzers ideal for that wavelength range.

A prism can also be operated in the reflection mode. In this case, the prism has half the apex angle α, and one face of the prism has a metallic coating. This is akin to folding the diagram shown in Fig. 3.8 in half, with the deviation angles and angular dispersion continuing to be given by Eqs. (3.35) and (3.36), respectively.

Another useful application of prism spectral analyzers is as a premonochromator for a grating monochromator. Using this combination can remove the need to have order sorting filters, while also decreasing the stray light common with grating monochromators.

3.2.4 Resolving Power

The resolution of a spectral analyzer does not approach zero when the slits are infinitesimally small. This is due to diffraction and the finite size of either the grating or the prism. Consider the transverse extent w of the grating or prism as observed from the outgoing direction (see Fig. 3.8 for the case of the prism). In either case, we can consider waves emanating from the extreme sides of the beam. When one side has a path length that is longer by a single wavelength, there will be, over the extent of the beam, destructive interference. At this angle, given by $\Delta\theta = \lambda/w$, the first minimum in the diffraction from the entire beam will occur. Thus, we have an inherent angle dispersion of

$$\frac{d\theta}{d\lambda} \approx \frac{\Delta\theta}{\Delta\lambda} \approx \frac{\lambda}{w\Delta\lambda}. \tag{3.39}$$

Equating Eq. (3.39) with Eq. (3.26) and recognizing that the length of the grating is $L = w/\cos\theta_m$, we find that

$$\frac{\lambda}{\Delta\lambda} = mGL. \tag{3.40}$$

The expression in Eq. (3.40) is referred to as the *resolving power*. A similar argument applies to the prism. Combining Eq. (3.39) with Eq. (3.38), we have

$$\frac{\lambda}{\Delta\lambda} = b\frac{dn}{d\lambda}. \tag{3.41}$$

Given spectral analyzers with similar dimensions, so that L for the grating is comparable to b of the prism, and letting $m = 1$ in Eq. (3.40), we are left comparing G of the grating with $dn/d\lambda$ of the prism material. Viewing the prism dispersion in Fig. 3.9 and comparing it to a 600 lines mm^{-1} (6×10^{-4} nm^{-1}) grating, it can be seen that the grating will have better resolving power, except in the ultraviolet.

In practice, the theoretical resolving power is rarely achieved. Any wavefront error encountered by the radiation tends to reduce the resolving power. Furthermore, in spectrophotometric measurements, one is often balancing resolving power with instrument throughput. The limiting resolution of the instrument is only achieved when the slits are small and the throughput is small.

3.3 SPECTRAL ANALYZER DESIGN

3.3.1 Monochromator Design

The basic design concepts for prism-based and grating-based spectral analyzers are very similar. Unless the dispersive element is acting as a focusing element, it is usually irradiated with collimated radiation. Since the input and output slits are usually fixed in position, the selection of the desired wavelength at the output slit is usually performed by rotation of the dispersive element. There are many configurations for the arrangement of the collimating and focusing optics with respect to the dispersive element. In the case of gratings, the grating itself can be constructed to be concave, so that it serves as both the dispersive and the focusing functions in the monochromator.

The common mount types of monochromator are the Czerny–Turner design, the Ebert–Fastie design, and the Littrow design. These types are

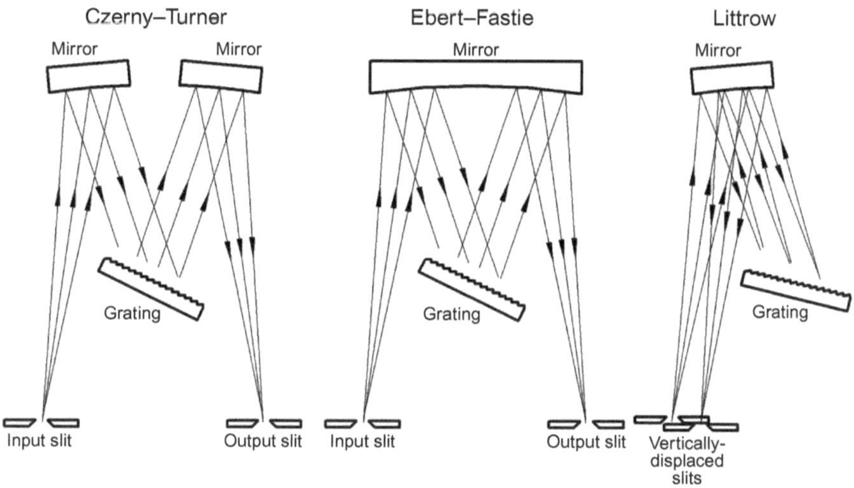

FIGURE 3.10 The Czerny–Turner, Ebert–Fastie, and Littrow monochromator designs.

illustrated in Fig. 3.10 with planar reflective grating dispersive elements. Similar designs can be used for prism and transmission grating monochromators.

The Czerny–Turner mount uses two separate focusing mirrors for collimating the radiation from the entrance slit and focusing the radiation from the dispersive element onto the output slit. The typical design schematic using a plane grating is shown in Fig. 3.10. The Ebert–Fastie mount is a modification of the Czerny–Turner mount with a single large concave mirror performing both the collimating and focusing functions. This feature reduces the number of components in the monochromator, but the cost of one large optic may be larger than two smaller optics. In addition, stray light and aberrations are more difficult to control in the Ebert–Fastie design. The Littrow mount is also called an autocollimation mount, since the dispersed radiation is reflected back in the same direction as the incident light. The input and output beams for the Littrow mount are shown displaced in Fig. 3.10 for illustration. In general, however, the two slits will be displaced in the third dimension, perpendicular to the page, one above the optical plane and the other below it. Reflections may pose more of a problem in the Littrow mount since the input and output beams pass through the same space in the housing. The Littrow mount also places the entrance and exit slits close to one another, not leaving much room for order sorting filter wheels and source optics. Many more details concerning mount types, design, and construction may be found in Refs. [2,4,7].

Reflective mirrors are almost always used in monochromators, because they do not exhibit chromatic aberration. However, all of the designs shown in Fig. 3.10 use reflective optics in an off-axis geometry. While spherical optics are generally used in low-cost monochromators, they have focal lengths that differ in the spectral direction (the tangential focus) than in the transverse direction (the sagittal focus). The slits should be placed at the tangential focus, in order to maintain the spectral resolution of the instrument. There will be a loss of power, however, if the slit is not very large in the vertical direction. Improved designs will have toroidal mirrors instead of spherical mirrors. In that case, the tangential and sagittal foci can coincide.

3.3.2 Polychromator Design

Polychromators, commonly referred to as array spectrometers, capture a large part of the spectrum simultaneously. Rather than have a fixed output slit and a rotating dispersive element, they place an array detector where the output slit would normally be. Due to the use of a multichannel detector, they can significantly improve the speed of many spectrophotometric measurements. Their dynamic range can be limited by their need to capture the entire spectrum. Figure 3.11 shows polychromators of the crossed Czerny–Turner and uncrossed Czerny–Turner designs. These devices can be quite compact. The gratings are often curved (not shown in Fig. 3.11) and designed to be a part of the focusing system, which can either further reduce their size or allow for improved imaging. Array spectrometers can be more susceptible to stray light, but the numerical processing methods described in Zong have proved successful in reducing these effects [10].

3.4 WAVELENGTH CALIBRATION

Since, from Eq. (3.25), the wavelength tuned by the monochromator is proportional to the sine of the grating angle, it is common for monochromators to use a sine bar mechanism, which translates linear motion into the sine of the rotation angle. Nominally, the wavelength in such a monochromator is linearly related to the motion of the lead screw. It is important to calibrate the

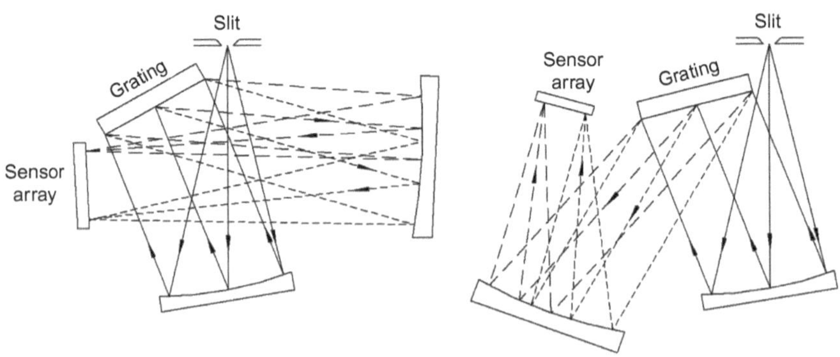

FIGURE 3.11 Two styles of polychromator design: (left) the crossed Czerny–Turner and (right) the uncrossed Czerny–Turner designs. Solid rays indicate polychromatic radiation, short dashed lines indicate the shortest wavelength, and long dashed lines indicate the longest wavelength.

wavelength scale, at least insofar as obtaining the slope and intercept from the graph of wavelength versus linear position. It is best to calibrate the scale using one of a number of reference methods. Low-pressure elemental emission lamps, such as Hg, Ar, Kr, Ne, and D_2, can be used as intrinsic sources with uncertainties of 1.0 pm [11,12]. These lamps should be placed in integrating spheres so that the radiation from the sphere fills the input optics and the internal optics of the monochromator. If the signals are not intense enough, then the lamps can be directly imaged onto the entrance slit. However, this latter technique can lead to higher uncertainties, since the output of the lamps will not be Lambertian and the method relies upon correct alignment of the source. The exit aperture of the integrating sphere should be placed at the exact position of the broadband source used in the monochromator; otherwise, the wavelength scale may be shifted. The peaks used for the calibrations should be well-separated spectral lines and have the expected shape, that is, triangular for two equal-sized entrance and exit rectangular slits, as discussed above in Section 3.2.1. If they are not, then the user should suspect that something is wrong and correct it: either the monochromator is not well focused or the method used to illuminate the input slit is poor. The wavelength positions can be determined by using the centroid method. If the signal $S(x_s)$ is measured over the entire bandpass for a spectral line as a function of the linear translator coordinate x_s, the centroid can be found from

$$\bar{x}_s = \frac{\int x_s S(x_s) dx_s}{\int S(x_s) dx_s}. \tag{3.42}$$

Alternatively, the peak position is often found by fitting the four topmost data points to a quadratic function,

$$S(x_s) = a + bx_s + cx_s^2. \tag{3.43}$$

The peak value of Eq. (3.43) is $x_s^{peak} = -b/(2c)$. The centroid values $\bar{x}_s$ (or the peak values x_s^{peak}) for all of the spectral lines are then graphed, as illustrated in Fig. 3.12. A simple straight line fit to the data shown in Fig. 3.12 yielded a standard deviation of 0.14 nm, but showed a significant systematic trend across the wavelength range. Therefore, the data were fit to successively higher-order polynomials until no significant systematic trend was observed. The residuals shown in Fig. 3.12, using a fifth-order polynomial, have a standard deviation of 0.023 nm, which is significantly below the nominal bandwidth of the monochromator, which was set to 3 nm.

For spectrophotometers where the atomic emission lamps cannot be placed at the same positions as the lamp sources, material wavelength standards can be used [13]. These include didymium [14] and holmium oxide [15] glass filters and the McCrone wavelength standard [16], as well as rare-earth doped glasses containing mixtures of Ho_2O_3, Sm_2O_3, and Yb_2O_3 [17,18]. Cuvettes

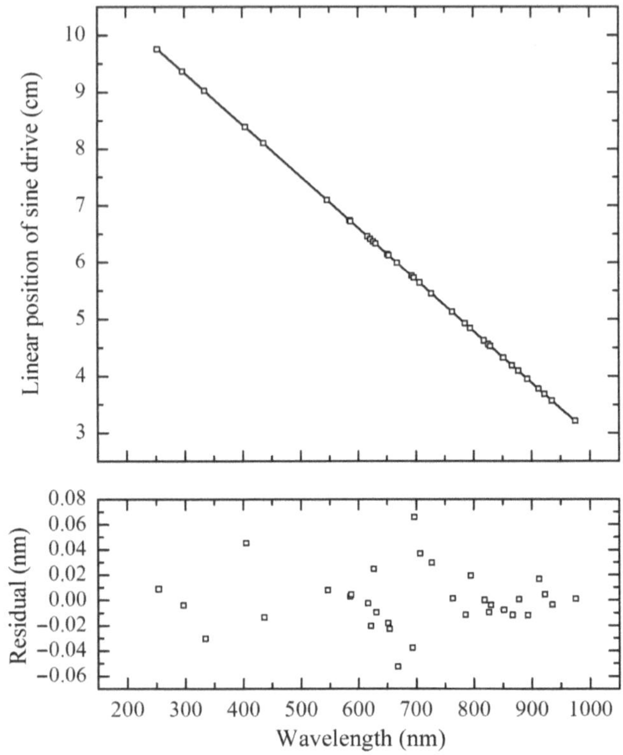

FIGURE 3.12 Results from a wavelength calibration of a grating monochromator. The top graph shows the (symbols) centroid of the linear position for each of the spectral lines as a function of the standard wavelengths and (curve) a fit to a fifth-order polynomial. The lower graph shows the wavelength residual of the predicted values.

containing holmium oxide solutions can also be used [19]. The visible and the infrared transmittances of the NIST SRM 2035a are shown in Fig. 3.13. The wavelengths at which the absorption features occur in these materials have been carefully measured and are traceable to the International System of Units. The doped-glass standard is placed into the sample compartment of the spectrophotometer and the wavelengths can be scanned for calibrating wavelengths, as was done above with the much narrower atomic spectral lines. However, rather than using the centroid method for determining the peak position, the data near each peak should be fit to a smooth function and the maximum or minimum values, as appropriate, used [20]. The positions of these transmission minima (reflection maxima) depend critically upon the instrument bandpass. Complementary wavelength scale calibration data can be obtained from the inflection points of these material standards, which are less dependent on instrument bandpass. These positions are calculated from the zero crossings of the second-derivative spectrum.

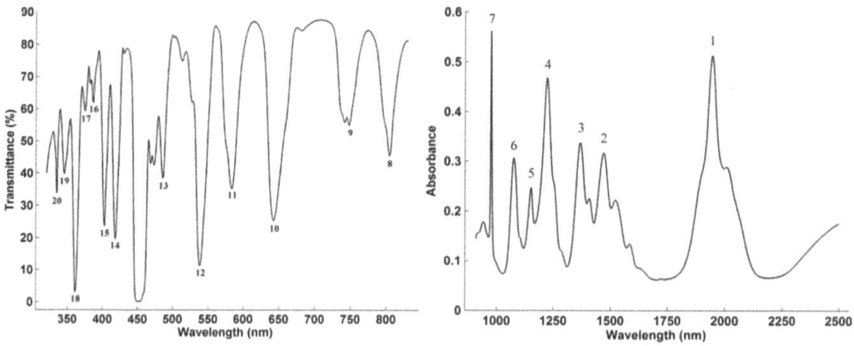

FIGURE 3.13 The transmittance (left) and the absorbance (right) measured for the NIST SRM® 2035a rare-earth doped standard reference material. Twenty-numbered features have assigned wavelengths with uncertainties. *From Refs. [17,18].*

Due to the complicated relationship between deflection angle and wavelength (e.g., see Eqs. 3.33 and 3.35), it is particularly important to perform wavelength calibrations of prism-based spectral analyzers. Furthermore, since the wavelength dependence of the index of refraction can follow the Cauchy relationship, Eq. (3.33), it is common to make fits to polynomials in $1/\lambda^2$ rather than λ.

3.5 STRAY LIGHT

Stray light is undesired or spurious signal that contributes to the measured signal. The SRE is the ratio of the measured spurious radiation to the detected primary radiation. Its effects are generally greatest when there is significant scattering in the instrument or at wavelengths of decreased primary signal, for example, near the source energy and detector response limits, as well as when the grating efficiency is low.

Stray light can be particularly problematic with polychromators that use charge-coupled devices (CCDs) or photodiode arrays (PDAs), because they detect all wavelengths simultaneously. A small percentage of stray light originating in the red portion of the spectrum, where silicon detector sensitivity is usually high, will have a very large impact at the shorter wavelengths in the ultraviolet and blue region, where the primary radiation level is low. This situation also exists in spectrophotometers which use broadband photomultiplier tubes (PMTs) or Si photodiodes as the sensor. Methods have been developed by Zong *et al.* to correct for such stray light errors in CCD-based instruments [10]. This method involves using monochromatic radiation, from either a tunable laser source or a monochromator, at discrete wavelengths across the range of interest, and measuring the system response to obtain a spectral stray light correlation matrix. These matrix methods have since been incorporated in several commercial array-based spectroradiometers to reduce their spectral stray light error. However, it should be noted that these methods are essentially deconvolutions, which can increase the noise level in the spectrum.

For a conventional spectrophotometer, SRE can be described as either isochromatic or heterochromatic in origin. *Isochromatic* SRE is radiation of the same wavelengths as incident on the sample, that is, within the instrument bandpass, but which reaches the detector without actually interacting with the sample. *Heterochromatic* SRE is radiation of wavelengths outside the instrument bandpass setting.

The heterochromatic stray radiant energy (HSRE) is primarily caused by scattering from inside the monochromator from the grating and the optics and also higher orders in a grating instrument. This can be reduced by matching the *f*-number of the input optics to the internal optics of the spectrophotometer, thus reducing the amount of extraneous radiation inside the spectrophotometer.

A second source of HSRE is reentry spectra. This phenomenon is sometimes called double refraction and occurs when light diffracts back to the collimator, and returns to the grating and diffracts again, eventually emerging from the exit slit at constant offset with respect to the set wavelength of the monochromator. Since the zeroth order specular reflection is especially intense, spectrophotometers must be well designed to suppress this source of reentry spectra. If the precise wavelength difference of this double refracted light is known, it is possible to reduce this directional stray light by inserting additional bandpass filters in the optical path of the instrument.

The measurement of stray light can be performed using broadband radiation with a cut-on filter, in a method described in Ref. [21]. The filter is placed between the monochromator and the source, and the monochromator is scanned in the wavelength region where the cut-on filter should be suppressing any radiation from being transmitted. The ratio of the signal in the wavelength region that is blocked to that in the wavelength region that is transmitted is the stray light suppression ratio. It should be noted that the heterochromatic stray light so determined is a function of the particular absorber used. For this reason, specifications for spectrophotometer performance always report the relevant wavelength and the particular absorber that was used for this stray light measurement (e.g., stray light $<0.00008\%$ at 370 nm (50 mg L^{-1} $NaNO_2$)). An intense laser source can also be used to determine the stray light. Single-grating monochromators and spectrographs typically have SREs of 10^{-3} or 10^{-4}, while double monochromators can decrease the SRE to 10^{-7} to 10^{-8}.

3.6 OPTICAL RADIATION SOURCES

Accurate spectrophotometry poses a few requirements on the light sources that are used. Generally, one needs to have light sources available that cover the spectral range of interest. As discussed above in Section 3.4, the center wavelength accuracy of a monochromator can be calibrated to be significantly better than its bandwidth. Thus, the spectral radiance should be relatively smooth over the intended spectral range, so that the spectral shape does not impart wavelength shifts that can be as large as the spectral bandwidth. In addition, the light source must have sufficient radiant output to meet the particular measurement needs. In this case, it is usually the radiance of the source that is most important, since, from Chapter 2, Section 2.2.1, we saw that radiance is conserved in an imaging system, and from Section 3.2.1, we find that the radiance $L_\lambda(\lambda)$ at the entrance slit ultimately determines the output of the monochromator. Finally, stability of the radiation source is also important for accurate spectrophotometric measurements.

Figure 3.14 shows the radiance for four different sources commonly used in spectrophotometry applications. Quartz tungsten halogen (QTH) lamps are essentially blackbody radiators operating at a temperature near 3300 K. The spectrum is very smooth, but because of its limited temperature, it cuts off in the ultraviolet. Their peak radiance is near 880 nm and is approximately 0.6 mW mm^{-2} nm^{-1} sr^{-1}. Xenon arc lamps reach temperatures 5000–7000 K, which allow them to provide significant radiance in the ultraviolet. Their radiance through the visible region is significantly higher, around 10 mW mm^{-2} nm^{-1} sr^{-1}, than QTH lamps. Unfortunately, they suffer from having significant atomic line structure in the infrared wavelengths, which makes their use difficult for the highest accuracy measurements in that spectral region. Furthermore, the arc tends to wander about the electrodes, creating problems with stability. Because the arc is imaged onto the entrance slit of the monochromator, any motion of the arc will cause variations in the spectral

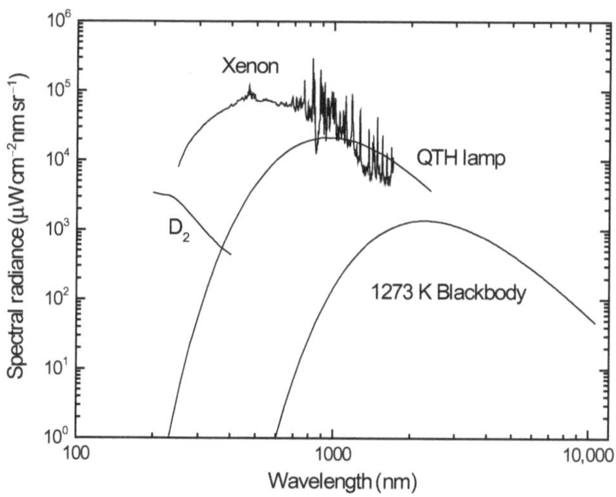

FIGURE 3.14 The spectral radiance for four sources: a D_2 lamp, a Xe arc lamp, a QTH lamp, and a 1273 K (1000 °C) blackbody.

content within the bandpass function, which are difficult to eliminate by simply monitoring the total output power. Deuterium lamps have significantly lower radiance (about 0.05 mW mm^{-2} nm^{-1} sr^{-1} at 200 nm) but are one of the few compact broadband sources available in the deep ultraviolet. Because deuterium lamps tend to be extended sources (extending a few millimeters in size), and because their radiance is low, they are often used in conjunction with QTH lamps in a pass-through configuration. That is, they are placed close to the entrance slit to the monochromator, transmitting the radiation from the QTH lamp through it. Figure 3.14 also shows a 1273 K blackbody source, which is common in infrared applications and is known for its stability and spatial uniformity. In the infrared, it is also common to use glow bars or Nernst glowers. These devices are ceramic-coated tungsten wires which are electrically heated. The temporal stability of cavity blackbody sources is much better than the other sources due to the thermal mass of the blackbody as compared to the Nernst glower. Furthermore, the cavity blackbody is spectrally smoother and has a higher emissivity.

Finally, there are some novel light sources that may find applications in spectrophotometry in the future. The maximum operating temperature of a Xe arc lamp is limited by the ability for the electrodes to withstand the arc conditions. New Xe plasma sources, pumped by a high power laser, have the advantage that they can reach much higher temperatures (10,000–20,000 K), not have issues with electrode degradation, and have plasma sizes close to the focused spot size of the pump laser [22]. The laser-pumped Xe plasma sources, however, have the same spectral issues in the ultraviolet as their arc lamp cousins. Supercontinuum fiber sources use self-phase modulation to convert short-pulse, high-repetition rate laser pulses into broadband radiation [23]. They will often have a spectral peak from the pump laser, often at 1064 nm, and have moderate structure elsewhere. However, their spectral range has a sharp cutoff between 400 and 500 nm at one end of the spectrum, but extends to about 2400 nm. Both of these novel sources are not extended sources, like the QTH, Xe arc, or deuterium lamps, but rather are for uses in spectrophotometry that need point sources.

3.7 OPTICAL RADIATION DETECTORS

The types of detectors used in spectrophotometers can range from PMTs to various semiconductor photovoltaic or photoconductive devices. PMTs have very high sensitivity and rely upon the photoelectric effect, followed by cascaded secondary electron amplification by a dynode structure. The spectral range of PMTs is dependent upon the photocathode material, and extends from the deep ultraviolet to the near infrared. Their responsivity depends upon the details of their construction (size and shape) and operating parameters (e.g., temperature and voltages applied to the photocathode and dynode elements). They can be operated in a single photon counting mode or as a conventional continuous detector. Of the photovoltaic or photoconductive detectors, silicon is the most popular, since it has high sensitivity, large area, and uniformity, and for some types can have very high linearity, but it is limited in its spectral response to wavelengths less than 1100 nm. For longer wavelengths, it is common to use InGaAs detectors, which operate between 800 and 1700 nm, and extended InGaAs detectors, which operate between 800 and 2500 nm. PbS detectors, which are photoconductive devices and used from 900 to 3000 nm, and germanium detectors, which operate from 800 to 1800 nm, are still used in commercial spectrophotometers but are falling out of favor, since their performance is generally superseded by regular and extended InGaAs detectors. Beyond this spectral region, thermal detectors such as pyroelectric and thermopile detectors are used. The infrared sensors generally need to be cooled or temperature stabilized, either thermoelectrically or with a cryogen. The infrared sensors are usually used with chopped radiation sources, due to their sensitivity to background radiation, and because many are inherently ac devices.

One- and two-dimensional array detectors are commonly constructed from the semiconductor materials. CCDs, PDAs, and complementary metal oxide semiconductor (CMOS) arrays fabricated from silicon find themselves used in a wide variety of consumer applications, including digital cameras and flat-bed scanners. As a result, they are relatively inexpensive. All of these detectors share the fundamental absorption properties of silicon, but differ in how they are addressed and read out. CCDs and CMOS sensors integrate charge that has accumulated at a junction, while PDAs are addressable photodiodes and measure photocurrent. For this reason, CCDs and CMOS sensors can have an extremely wide dynamic range, since the integration time can be varied accurately from milliseconds to many minutes. Array detectors using the other semiconducting materials are usually in the form of PDAs.

Detectors are often classified according to their *noise equivalent power* (NEP), with units of W $Hz^{-1/2}$. When multiplied by the measurement bandwidth, the NEP indicates the radiant power level at which the signal-to-noise ratio would be unity. Because a detector has a responsivity that is wavelength dependent, the NEP also depends upon wavelength, but is usually quoted at a single wavelength. There are two dominant noise sources in solid-state detectors: Johnson noise and shot noise. Johnson noise results from the thermal motion of electrons in a resistor and is proportional to the square root of the device's resistance. Shot noise is due to the Poisson statistics associated with the random arrival time of the discrete electrons in the detector. For most solid-state detectors, the NEP is dominated by Johnson noise and thus is proportional to the square root of the area A of the detector. Thus, it is common to characterize detectors by a *specific detectivity*, given by

$$D^* = \frac{\sqrt{A}}{\mathrm{NEP}}. \tag{3.44}$$

The higher a detector's D^*, generally, the more sensitive it is. For further information, the reader should consult with the technical literature from specific manufacturers.

There are a few important considerations for detectors used in spectrophotometry. Spatial and angle of illumination uniformity is important if the detector is used directly behind or at the image of the exit slit of a monochromator. PMTs usually have a wire grid in front of the photocathode, which imparts strong spatial variations across its sensitive area. The use of a diffuser in front of the detector can improve the effects of nonuniformity. Even better, but usually at the loss of some signal, an integrating sphere can be used with a detector to give very good uniformity.

A very important consideration in all detector applications is linearity. That is, the signal recorded should be proportional to the radiant quantity incident upon the detector. It is important to verify the linearity of the detectors being used in spectrophotometry. Linearity should be considered with the detector and its associated electronics taken together as a package or detection system. The radiant quantity is radiant power if the beam underfills the detector aperture and is irradiance if the beam overfills this aperture. Usually, with modern electronics design, the low power end of detector responsivity can be extremely linear. However, all detectors have a level at which they saturate, usually due to the space charge that accumulates in its active region. As a result, detectors often exhibit nonlinearity that can either be a function of local irradiance (power density) rather than total radiant power. In cases where it is shown that the detection system nonlinearity is reproducible, the effects of this nonlinearity can be corrected [24].

There are a number of ways of checking for linearity. These methods can be divided into dependent methods and independent methods. The dependent methods are based upon performing spectrophotometric measurements on a set of known samples, such as a solution of different concentrations of an absorbing compound in a solvent or a set of independently calibrated optical filters, and testing for linearity in the measured result. For the method that uses absorbance measurements on different solutions, which is based on the Beer–Lambert law discussed in Chapter 2, it does not detect nonlinearity due to absorbance readings being off by a common factor; thus, it needs to be used in combination with a separate check for linearity [25].

The independent methods assess the linearity of the system, without the use of a standard. These can be broadly classified into three different methods: superposition, attenuation, and differential, which are described below.

The *superposition method* relies on exposing the detector to two independent sources of radiation [26–29]. One measures each of the two signals, S_1 and S_2, and then both at the same time S_{1+2}. If the detector is linear, then $S_1+S_2=S_{1+2}$. This test is performed over a range of S_{1+2} to assess the range over which this condition holds true. The measurements from the superposition method that sums light over two paths can be combined with a technique where each path has more than one attenuation setting. Such a device can be used to determine the nonlinearity function using a least-squares fit to a model nonlinearity [26,28,29].

A variant of this method, the *flux doubling method*, uses two apertures in the beam, opening and closing one or both of them to read the three signal levels. Figure 3.15 shows a schematic of a device that can be placed in the sample position of a spectrophotometer. Although the flux doubling method produces sparse, exponentially spaced data with cumulating uncertainties, it is a recommended method for checking the linearity of a spectrophotometer.

The *attenuation method* uses a set of unknown absorbing filters [29]. If the transmittances are $\tau_1, \tau_2, \ldots, \tau_N$, then combinations of filters, say τ_1 and τ_2, should yield the product $\tau_1\tau_2$. By performing a sufficient number of these measurements, using not just two, but every combination of the filters, one can determine the values of each of the transmittances and the linearity in the system over a wide signal range. This method is fairly easy to perform in a spectrophotometer, since it uses the established setup for which it is designed. However, one should be careful to avoid interreflections between the filters, which can cause the total transmittance to not follow the simple product rule. One way to minimize interreflections is to slightly tilt these filters in opposite directions.

The *differential method* applies a small, constant ac fluctuation to the lamp power and measures the ac fluctuation of the detected signal [30]. If the detector is linear, the ratio of the ac fluctuation to the dc value will remain constant. Different filters can be placed in the beam to change the overall power level. It may seem logical to perform this with a single sample having spectral features covering a wide range of transmittances (see, e.g., Fig. 3.13). However, the ac modulation of the lamp radiance will depend upon wavelength. This method is more difficult to perform in a standard spectrophotometer, because one does not necessarily have direct access to the lamp power.

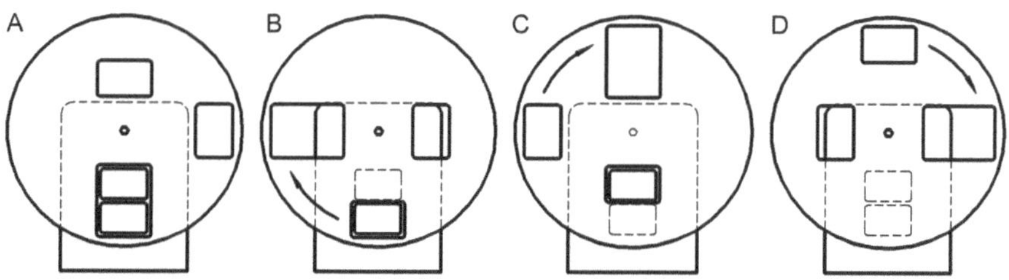

FIGURE 3.15 Schematic of a device used for measuring the linearity of a spectrophotometer or a detector. The beam passes through the two apertures in the back plate. In (a), both apertures are open, in (b) the bottom aperture is open, in (c) the top aperture is open, and in (d) neither aperture is open. Position (d) is not necessary for linearity testing, but it is useful for isochromatic SRE measurement.

REFERENCES

[1] M. Born, E. Wolf, Principles of Optics, Pergamon, Oxford, 1980.

[2] S.P. Davis, Diffraction Grating Spectrographs, Holt Rinehart and Winston, New York, 1970.

[3] F. Grum, R.J. Becherer, Radiometry, Academic Press, New York, 1979.

[4] J.F. James, R.S. Sternberg, The Design of Optical Spectrometers, Chapman and Hall, London, 1969.

[5] F.A. Jenkins, H.E. White, Fundamentals of Optics, second ed., McGraw Hill, New York, 1950.

[6] C. Palmer, E. Loewen, Diffraction Grating Handbook, sixth ed., Richardson Grating Laboratory, Rochester, New York, 2005.

[7] R.A. Sawyer, Experimental Spectroscopy, Dover, New York, 1963.

[8] R.W. Wood, On a remarkable case of uneven distribution of light in a diffraction grating spectrum, Philos. Mag. 4 (1902) 396–402.

[9] R. Kitamura, L. Pilon, M. Jonasz, Optical constants of silica glass from extreme ultraviolet to far infrared at near room temperature, Appl. Optics 46 (2007) 8118–8133.

[10] Y. Zong, S.W. Brown, B.C. Johnson, K.R. Lykke, Y. Ohno, Simple spectral stray light correction method for array spectroradiometers, Appl. Optics 45 (2006) 1111–1119.

[11] C.J. Sansonetti, M.L. Salit, J. Reader, Wavelengths of spectral lines in mercury pencil lamps, Appl. Optics 35 (1996) 74–77.

[12] J.E. Sonsonetti, W.C. Martin, Handbook of basic spectroscopic data, J. Phys. Chem. Ref. Data 34 (2005) 1559–2259.

[13] C. Burgess, J. Hammond, Wavelength standards for the near-infrared spectral region, Spectroscopy 22 (2007) 40–47.

[14] W.H. Venable Jr., K.L. Eckerle, Didymium glass filters for calibrating the wavelength scale of spectrophotometers—SRM 2009, 2010, 2013, and 2014, National Bureau of Standards, Washington, DC, 1979, Special Publication 260–266.

[15] D.W. Allen, Holmium oxide glass wavelength standards, J. Res. Natl. Stand. Technol. 112 (2007) 303–306.

[16] O.D.D. Soares, J.L.C. Costa, Spectrophotometer spectral bandwidth calibration with absorption bands crystal standard, Appl. Optics 38 (1999) 2007–2013.

[17] S.J. Choquette, J.C. Travis, L.E. O'Neal, C. Zhu, D.L. Duewer, A rare earth oxide glass for the wavelength calibration of near infrared dispersive and Fourier transform spectrometers, Spectroscopy 16 (2001) 14–19.

[18] S.J. Choquette, J.C. Travis, D.L. Duewer, SRM 2035: a rare earth oxide glass for the wavelength calibration of near infrared dispersive and Fourier transform spectrometers, Proc. SPIE 3425 (1998) 94–102.

[19] J.C. Travis, J.C. Acosta, G. Andor, J. Bastie, P. Blattner, C.J. Chunnilall, S.C. Crosson, D. L. Duewer, E.A. Early, F. Hengstberger, C.-S. Kim, L. Liedquist, F. Manoocheri, F. Marcader, L.A.G. Monard, S. Nevas, A. Mito, M. Nilsson, M. Noël, A.C. Rodriguez, A. Ruíz, A. Schirmacher, M.V. Smith, G. Valencia, N. van Tonder, J. Zwinkels, Intrinsic wavelength standard absorption bands in holmium oxide solution for UV/visible molecular absorption spectrophotometry, J. Phys. Chem. Ref. Data 34 (2005) 41–56.

[20] J.C. Travis, J.C. Zwinkels, F. Mercader, A. Ruíz, E.A. Early, M.V. Smith, M. Noël, M. Maley, G.W. Kramer, K.L. Eckerle, D.L. Duewer, An international evaluation of holmium oxide solution reference materials for wavelength calibration in molecular absorption spectrophotometry, Anal. Chem. 74 (2002) 3408–3415.

[21] ASTM E387-04, Standard test method for estimating stray radiant power ratio of dispersive spectrophotometers by the opaque filter method, ASTM, West Conshohocken, PA, 2009.

[22] H. Zhu, P. Blackborow, LDLS sheds light on analytical-sciences applications, Laser Focus World 47 (12) (2011) 53.

[23] R.R. Alfano, The Supercontinuum Laser Source, second ed., Springer, New York, 2006.

[24] J. Zwinkels, D. Gignac, Automated high precision variable aperture for spectrophotometer linearity testing, Appl. Optics 30 (1991) 1678–1687.

[25] A. Reule, Testing spectrophotometer linearity, Appl. Optics 7 (1968) 1023–1028.

[26] A. Thompson, H.-M. Chen, Beamcon III, a linearity measurement instrument for optical detectors, J. Res. Natl. Inst. Stand. Technol. 99 (1994) 751.

[27] K.D. Mielenz, K.L. Eckerle, Spectrophotometer linearity testing using the double-aperture method, Appl. Optics 11 (1972) 2294–2303.

[28] R.D. Saunders, J.B. Shumaker, Automated radiometric linearity tester, Appl. Optics 23 (1984) 3504–3506.

[29] L. Coslovi, F. Righini, Fast determination of the nonlinearity of photodetectors, Appl. Optics 19 (1980) 3200–3203.

[30] R.G. Frehlich, Estimation of the nonlinearity of a photodetector, Appl. Optics 31 (1992) 5926–5929.

Chapter 4

Fourier Transform Methods

Simon G. Kaplan* and Manuel A. Quijada†
*National Institute of Standards and Technology, Gaithersburg, Maryland, USA
†NASA Goddard Space Flight Center, Code 551, Greenbelt, Maryland, USA

Chapter Outline

4.1 Introduction: Ideal Michelson Interferometer 98
4.2 Real Fourier Transform Spectrometers 101
4.2.1 Finite Scan Length and Source Size: Spectral Resolution 101
4.2.2 Sampling the Interferogram: Spectral Bandwidth 106
4.2.3 Phase Error and Phase Correction 107
4.2.4 FTS Versus Dispersive Instruments 110
4.2.5 FTS Design Considerations 114
4.2.6 Instrument Design Examples 121
4.3 Sources of Uncertainty and Their Reduction 123
4.3.1 Noise 124
4.3.2 Interferometer Alignment Drift 124
4.3.3 Phase Error 125
4.3.4 Detector Nonlinearity 126
4.3.5 Interreflections 129
4.3.6 Nonsource Emission 130
4.3.7 Beam Geometry Errors 132
4.3.8 Polarization Effects 133
4.3.9 Stray Light 134
4.3.10 Atmospheric Absorption 134
4.3.11 Wavenumber Uncertainty 135
4.4 Measurement Applications 136
4.4.1 Measurement of Transmittance and Reflectance 136
4.4.2 Measurement of Refractive Index 137
4.4.3 Assessment of the Accuracy of FTS Measurements 137
4.5 Recommendations for Accurate FTS Measurements 138
4.5.1 Matching Instrument to Measurement Requirements 138
4.5.2 Instrument Software and Data Handling 139
4.5.3 Maintenance of Measurement Traceability 139
References 140

Experimental Methods in the Physical Sciences, Vol. 46. http://dx.doi.org/10.1016/B978-0-12-386022-4.00004-2

4.1 INTRODUCTION: IDEAL MICHELSON INTERFEROMETER

A stable source of spectrally defined optical radiation is a primary requirement for spectrophotometry. Major advances have been made in recent years in tunable lasers and other compact solid-state light generating devices, but most laboratory work in spectrophotometry is still done using broadband thermal sources and some type of opto-mechanical apparatus for spectral definition. The previous chapter described the use of dispersive spectrometers which separate the source light spatially into spectral bins. In this chapter, we describe a complementary method that uses scanning interferometry to generate the Fourier transform of the source spectrum. With a FTS, the light is not spatially dispersed, so the entire spectrum is observed simultaneously. This approach has advantages, but also leads to a somewhat different set of concerns in making accurate spectrophotometric measurements, as we discuss below.

While there are many different FTS designs used for spectrophotometry, most commercially available instruments operate on the principle of the Michelson interferometer. The concept of Michelson's original design [1] is shown in Fig. 4.1. Consider a source of monochromatic plane waves of wavelength λ incident from the right. The basic idea of the instrument is to split the incident wave into two nominally equal components, introduce a precisely measured phase difference between them, and then send the recombined output beam (bottom) to the detector. For simplicity, we consider an infinitesimally thin, nonabsorbing beamsplitter with Fresnel reflection and transmission coefficients r and t, respectively, assume perfectly reflecting mirrors, and also ignore polarization for the time being.

The time-dependent electric field at the input side of the beamsplitter can be represented in complex notation as $\mathbf{E}_0 e^{i\omega t}$, where $\mathbf{E}_0$ represents the electric

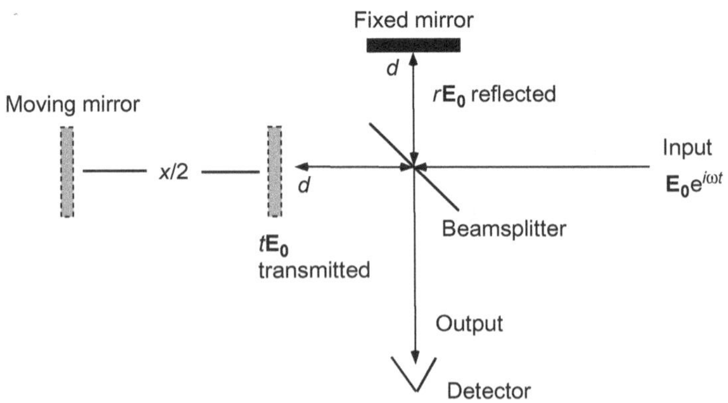

FIGURE 4.1 Optical layout of a Michelson interferometer, showing the input and output ports, the fixed and moving mirrors, and beamsplitter. Plane wave radiation incident from the input port is split into two components which are recombined and sent to either port.

field amplitude and $\omega = 2\pi c/\lambda$ is the radial frequency of the wave. One component of the split wave is first reflected upward, bounces off of the fixed mirror, and then is transmitted through the beamsplitter. Another component is first transmitted, reflects from the moving mirror, and then is reflected from the beamsplitter toward the detector. The sum of these two amplitudes, $\mathbf{E}_{\text{det}}$, is given by:

$$\begin{aligned}\mathbf{E}_{\text{det}} = \mathbf{E}_{\text{ref}} + \mathbf{E}_{\text{trans}} &= t\left(r\mathbf{E}_0 e^{i(\omega t - 2\pi\sigma 2d)}\right) + r\left(t\mathbf{E}_0 e^{i(\omega t - 2\pi\sigma(2d+x))}\right)\\ &= rt\mathbf{E}_0 e^{i(\omega t - 2\pi\sigma 2d)} + tr\mathbf{E}_0 e^{i(\omega t - 2\pi\sigma(2d+x))},\end{aligned} \tag{4.1}$$

where $\sigma = 1/\lambda$ is the wavenumber of the incident light, d is the distance from the center of the beam splitter to either of the mirrors in "zero-path-difference" (ZPD) configuration, and x is the optical path difference (OPD) introduced by displacing the moving mirror by $x/2$. The additional terms in the exponentials represent the phase change in the transmitted and reflected components upon making their round trips back to the beamsplitter. We can apply Eq. (2.39) for the time-averaged Poynting vector to derive the irradiance at the detector plane of the recombined plane wave as

$$E = \frac{1}{2} Z_0 \left\langle \left| rt\mathbf{E}_0 e^{i(\omega t - 2\pi\sigma 2d)} + tr\mathbf{E}_0 e^{i(\omega t - 2\pi\sigma(2d+x))} \right|^2 \right\rangle, \tag{4.2}$$

where Z_0 is the impedance of free space as defined in Chapter 2. We have taken the refractive index of the propagation medium to be 1. Equation (4.2) then reduces to

$$\begin{aligned}E &= \frac{1}{2} Z_0 \rho\tau |\mathbf{E}_0|^2 \left(2 + e^{-2\pi i\sigma x} + e^{2\pi i\sigma x}\right)\\ &= \frac{1}{2}\eta S_0 (1 + \cos(2\pi\sigma x)),\end{aligned} \tag{4.3}$$

where $\rho = |r|^2$, $\tau = |t|^2$, $\eta = 4\rho\tau$ is the beamsplitter efficiency, and $S_0 = (1/2)Z_0|\mathbf{E}_0|^2$ is the incident beam irradiance [2,3].

The irradiance at the detector plane thus shows a sinusoidal dependence upon x, as shown below in Fig. 4.2. At $x = 0$, the two components of the split wave add up in phase and if $\eta = 1$ then all the incident light is transmitted through to the detector port. This also occurs where x is any integer multiple of the source wavelength, while destructive interference leads to zero transmitted light at half-integer multiples of λ. At these points, the light is completely returned to the input port. In general, we are not concerned with this back-reflected radiation unless some of it comes through again to eventually reach the detector and lead to measurement error, as discussed in Section 4.3.

It is clear from Eq. (4.3) and Fig. 4.2 that a Michelson interferometer is an excellent device for measuring the wavelength of light with high precision. In fact, its original application was to determine whether the wavelength of light varied (and thus, the speed of light varied) relative to the uniform motion of

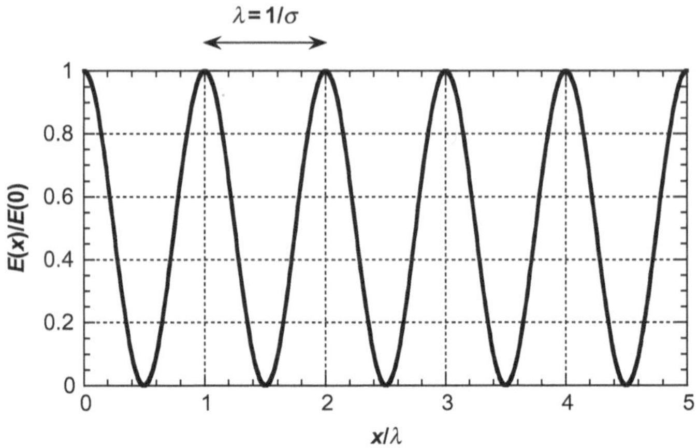

FIGURE 4.2 Modulation of collimated monochromatic light by an ideal Michelson interferometer as a function of OPD. Measurement of the modulated signal versus x leads to a precise determination of λ.

the earth through space. It did not and therefore the hypothesis of stationary ether was shown to be incorrect [4]. Eventually, interferometric measurements of the wavelength of lasers tied to atomic frequency standards became so precise that the meter itself was redefined in terms of the speed of light in vacuum and the second [5]. Interferometers with various designs are routinely used today for high-accuracy measurements of laser or atomic emission line wavelengths. However, the real utility of interferometers in spectrophotometry comes with broadband input sources (a point that was apparently realized by Michelson very early on Ref. [6]). Equation (4.3) can be extended to apply to a broadband input source as

$$E(x) = \frac{1}{2}\eta \int_0^\infty S_\sigma (1 + \cos(2\pi\sigma x)) d\sigma, \tag{4.4}$$

where we have replaced the Poynting flux/unit area of the monochromatic input wave with an integral over the input spectral irradiance denoted by S_σ. Equation (4.4) can then be rewritten as

$$I(x) = E(x) - \frac{1}{2}\eta|\mathbf{S}_0| = \frac{1}{2}\eta \int_0^\infty S_\sigma \cos(2\pi\sigma x) d\sigma. \tag{4.5}$$

With the x-independent term subtracted from both sides, the right-hand side can be recognized as the cosine Fourier transform of the source spectral irradiance. In practice, the quantity on the left hand side of Eq. (4.5) is called the *interferogram*, and S_σ is referred to simply as the *spectrum*. In most working FTS instruments, the detector output is ac-coupled as the moving mirror is scanned to produce audio-frequency modulation of the signal. Any residual dc signal from the input amplifier is subtracted before processing the interferogram, $I(x)$. However, it should be remembered that for a broadband source, on average, approximately half of the incident radiation is present as a dc background on the detector. This can have an important effect on the linearity of the measurements, as discussed in Section 4.3.

In order to recover the spectrum from the recorded interferogram, one performs an inverse cosine transform

$$S_\sigma = \frac{1}{\pi\eta}\int_{-\infty}^{\infty} I(x)\cos(2\pi\sigma x)\mathrm{d}x, \tag{4.6}$$

where now the integral is over all values of x. The relationship between the interferogram and spectrum is illustrated in Fig. 4.3, which shows the spectra and the central portion of the respective interferograms of a hypothetical source centered near 2000 cm^{-1} of irradiance 47 $\mathrm{mW/cm}^2$ with lorentzian line shape and three different line widths. The narrow band source shown in Fig. 4.3a produces a nearly sinusoidal modulated signal in the interferometer, while the broader sources in (b) and (c) produce interferograms which are progressively more peaked near $x=0$. In each case, constructive interference yields a maximum signal at $x=0$, but the phase differences in the different spectral components of the broader sources yield destructive interference and rapidly diminishing modulation as the path difference is increased. In general, it can be seen that the central part of the interferogram contains information about the overall shape of the spectrum, while the outlying parts contain the high-resolution information.

Equations (4.5) and (4.6) represent the mathematical description of an ideal FTS instrument that features a perfectly collimated light beam, infinite extent for the mirror drive, and infinitesimal spatial resolution in sampling the modulated signal. Any real spectrometer of course uses a finite-sized source (and hence has nonzero beam divergence) and has limits on the mirror motion and fineness of the sampling grid. We see in the next section how these physical constraints determine the resolution and spectral bandwidth of the instrument.

4.2 REAL FOURIER TRANSFORM SPECTROMETERS

4.2.1 Finite Scan Length and Source Size: Spectral Resolution

Most modern FTS instruments are designed to scan the moving mirror symmetrically about the ZPD point, at least over part of their full range of motion. In this case, the OPD covers the range $-L<x<L$. Thus, the interferogram is only measured over a finite range of values for x, and the inverse cosine transform integral in Eq. (4.6) must be truncated at $\pm L$. Another way of looking at the truncated integral makes use of the fact that the Fourier transform of the product of two functions is equal to the *convolution* of their individual Fourier

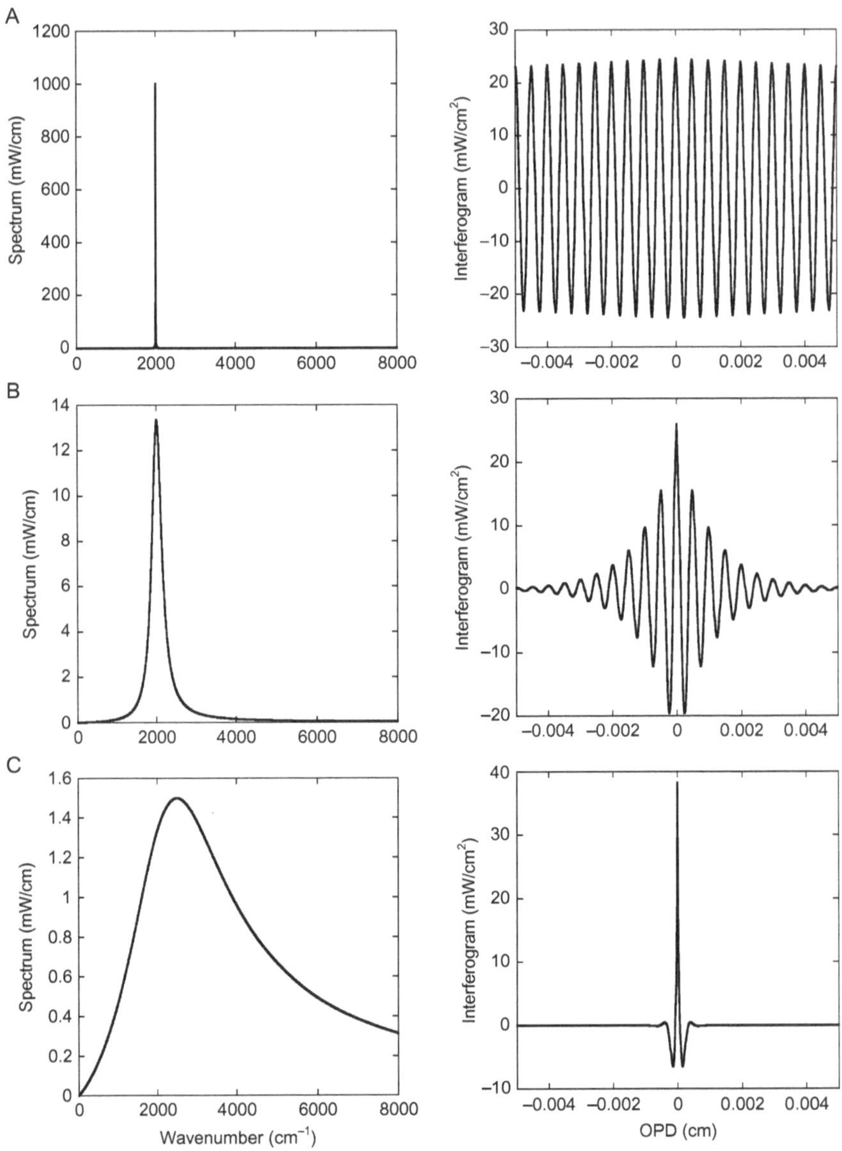

FIGURE 4.3 Spectral irradiance (left) and modulated interferogram irradiance (right) of a hypothetical source with a Lorentzian line shape centered near 2000 cm^{-1} with a line width of (a) 2 cm^{-1}, (b) 150 cm^{-1}, or (c) 1500 cm^{-1}.

transforms. We may consider the ideal interferogram in Eq. (4.6) to be multiplied by the rectangular function $\Pi(x/L)$, where $\Pi(y)=1$ if $|y|\leq 1$ and 0 otherwise:

$$I_{\text{obs}}(x)=I(x)\Pi\left(\frac{x}{L}\right). \tag{4.7}$$

The observed spectrum S_{obs} will therefore be the convolution of the ideal S_σ from Eq. (4.6) and the Fourier transform of the Π function [2], which is just a sinc function

$$S_{\text{obs}}(\sigma) = \int_{-\infty}^{\infty} S_{\sigma-\sigma'} 2L \frac{\sin(2\pi L\sigma')}{2\pi L\sigma'} d\sigma'. \tag{4.8}$$

The results of Eqs. (4.7) and (4.8) are illustrated for a monochromatic source in Fig. 4.4. The infinitely extending cosine interferogram would produce a delta function spectrum through Eq. (4.6). Upon truncation with the rectangular function in Eq. (4.7), we see the spectral response changing to a sinc function (right frame). The width of the central lobe of this curve is given by

$$\partial\sigma = \frac{1}{L}, \tag{4.9}$$

which is therefore the spectral resolution of an FTS with maximum OPD of L. The form of Eq. (4.8) may be recognized as the general expression that defines the *instrument line shape* (ILS) of a spectrometer. In the case of an FTS with an abruptly truncated interferogram, the ILS is thus a sinc function.

While there is nothing intrinsically "wrong" with this ILS, the strong ringing and even negative ordinate values that are seen in the vicinity of sharp, incompletely resolved spectral features can be a cause of concern in practice. One could take a model expression for the sample optical property (say, transmittance) that one is measuring and run it through Eq. (4.8) before comparing to the FTS data. Then, line width or absorption cross-section data could be extracted without loss of spectral information. Often such a model

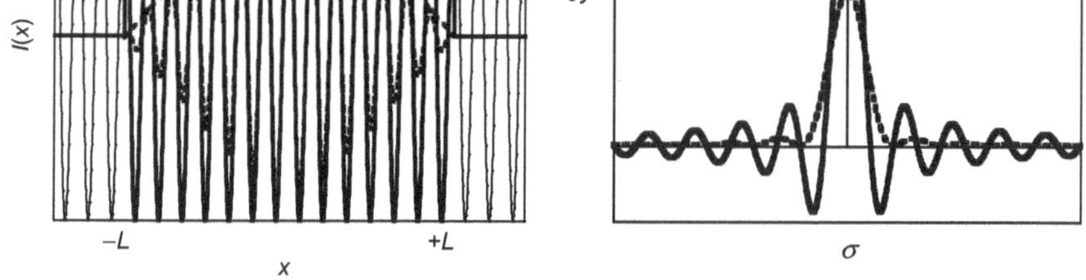

FIGURE 4.4 Effect on the recorded spectrum (right frame) of truncating the interferogram of a monochromatic source (left frame) in an FTS instrument. The thin curves show the response of an ideal FTS, while the thick and dashed curves show the results of rectangular truncation or triangular apodization of the interferogram, respectively.

expression is not readily available or would have too many parameters to be of practical use. It therefore is common practice to multiply the recorded interferogram by a smooth function equal to one at $x=0$ that goes to zero at $x=\pm L$ before applying Eq. (4.6) to retrieve the spectrum, and thereby reduce the amplitude of the spectral side lobes to an acceptable level. This procedure is called *apodization* (from Latin for "remove the foot") and is equivalent to replacing the sinc function ILS in Eq. (4.8) with the Fourier transform of the apodization function.

The simplest apodization function is the triangular form shown as the dashed line in Fig. 4.4 which yields a sinc^2 ILS, the same as a monochromator with rectangular slits and diffraction-limited resolution. The apodization procedure can be seen to significantly reduce the amplitude of the side lobes and produce a strictly nonnegative theoretical spectrum so any negative values can then be attributed to noise or some other error source. However, this smoothing comes at the cost of about 50% degradation in spectral resolution. Spectroscopists have come up with many different apodization functions, and the reader is referred to literature on the subject [3] for further information. There is a trade-off between side lobe suppression, resolution degradation, and distortion of the spectral line shape. Popular general-purpose functions include Happ-Ganzel and variations of Norton-Beer and are typically included with the software packages that come with commercial FTS instruments. As a general rule, it is a good idea to use one of these apodization functions if the spectral data in question contain features that are not completely resolved, but if everything is well-resolved then simple rectangular truncation of the interferogram in Eq. (4.7) (sometimes called "boxcar" apodization) is fine. If you find yourself in a situation where you cannot decide if a spectral feature is real or not, and its behavior depends strongly on the choice of apodization function, the best course may be to simply increase the resolving power σL if possible and average longer.

A second limitation on the resolving power of a real FTS comes from the finite source size and consequent lack of perfect beam collimation in the interferometer. Typically, the input radiation for an FTS originates from a blackbody or lamp irradiated aperture at the focal point of a collimating mirror. The combination of aperture size and collimator focal length determines the range of angles relative to the central optical axis that are present in the interferometer beam. Imagine a bundle of rays that pass through the interferometer at an angle α, and bounce off the moving mirror as shown in Fig. 4.5. Considering the distance between the perpendicular phase fronts, we can see that the OPD on the round trip between the beamsplitter and the moving mirror will be *shorter* than the on-axis rays and in fact is given by $x\cos(\alpha)$ instead of x. Thus, the mirror will have to move farther to produce the same phase difference as the on-axis rays, and the off-axis light will be recorded at a *longer* wavelength in the spectrum, leading to an error that needs to be corrected for.

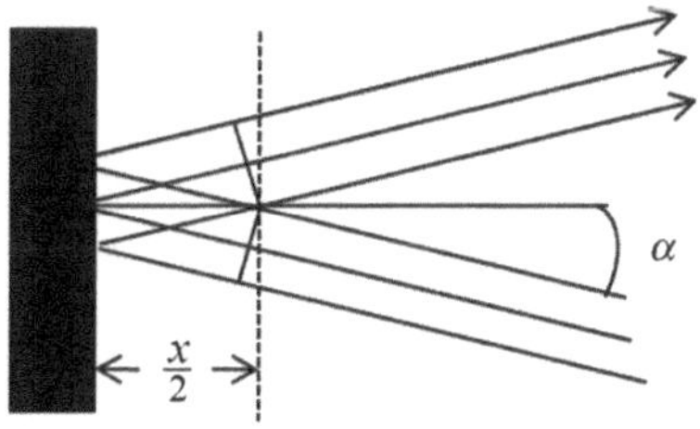

FIGURE 4.5 Rays that strike the moving mirror of an FTS at an angle α experience an OPD of $x\cos(\alpha)$ as opposed to x for on-axis rays.

Extending this argument to a range of off-axis angles, it is clear that there will be a spread of wavelengths recorded for a given input wavelength, and thus a loss in spectral resolving power in addition to a net wavelength bias in the instrument. In fact, for an azimuthally symmetric beam the modulation efficiency of the interferometer is directly dependent upon the beam divergence. Assuming a uniform distribution of α from 0 to $(\Omega_m/\pi)^{1/2}$ where Ω_m is the solid angle of the beam extent, it can be shown [2] that Eq. (4.5) is generalized to:

$$I(x) = \frac{1}{2}\eta\int_0^\infty \Omega_m \mathrm{sinc}\left(\frac{\sigma x \Omega_m}{2\pi}\right) S_\sigma \cos\left(2\pi\sigma x\left(1-\frac{\Omega_m}{4\pi}\right)\right)d\sigma. \tag{4.10}$$

Comparing Eqs. (4.10) and (4.5), we can see there are two modifications due to the integration over off-axis rays. First, there is the prefactor "envelope function" $\Omega_m \mathrm{sinc}(\sigma x\Omega_m/2\pi)$ before the spectrum. For a maximum OPD of L, the first zero of the sinc function will occur at $\sigma L = 2\pi/\Omega_m$ and the amplitude of the side lobes will decrease after that. Therefore, from Eq. (4.9), to retain the full resolving power of the instrument, it is necessary to impose a limit on the maximum solid angle of the beam:

$$\Omega_m \le \frac{2\pi\delta\sigma}{\sigma}. \tag{4.11}$$

Any rays beyond this maximum angular extent will experience reduced or even reversed phase modulation. As the mirror is scanned to longer OPD, the ideal cosine variation in irradiance for a given wavenumber component of the spectrum will be diminished by the envelope function in Eq. (4.10) and thus the effective spectral resolution will be decreased. This effect is sometimes called "self-apodization" of the interferogram and determines the degree of beam collimation required for a given resolving power in an FTS instrument. The second difference between Eqs. (4.5) and (4.10) is the factor of $1-(\Omega_m/4\pi)$ inside the cosine function in Eq. (4.10). This can be thought of as a scale factor on the wavenumber σ, and produces a relative wavenumber shift

$$\frac{\sigma' - \sigma}{\sigma} = -\frac{\Omega_m}{4\pi} = -\frac{1}{2\mathfrak{R}_m}, \tag{4.12}$$

where $\mathfrak{R}_m = (2\pi/\Omega_m)$ is the maximum resolving power of the instrument. This wavenumber error is half the maximum resolution for a given instrument solid angle or *f*-stop. It is a straightforward matter to correct for it, as some commercial instrument software does automatically. In general, one has to keep careful track of the optical setup when using an FTS for a spectrophotometric measurement in order to know what actually limits the instrument beam divergence and make the appropriate wavenumber correction, as well as to report the correct spectral resolution. In practice, this information can be obtained by measuring atomic discharge lamps, or narrow atmospheric absorption lines, similar to what one would do to calibrate the wavelength scale and resolving power of a dispersive spectrometer.

4.2.2 Sampling the Interferogram: Spectral Bandwidth

The OPD of an FTS may be scanned continuously or in finite steps. In either case, the interferogram must be sampled with fine enough spatial resolution that the inverse cosine transform of Eq. (4.6) may be performed to retrieve the spectrum. Typically for a continuously scanning system, this is done using a single-mode HeNe laser in parallel with the source beam of the FTS to track the OPD from the recorded cosine interferogram of the laser and trigger the collection of source beam data at regular spatial intervals. In newer designs, the laser metrology beam and source beam may be sampled asynchronously at very high rates, and the x-positions in the source interferogram interpolated from the laser signal. This helps to correct for changes in mirror speed during the scan.

However, after the data are collected and processed, in the end one has a discretely sampled signal in the spatial domain that will be transformed to a spectrum in the wavenumber domain using a discrete form of Eq. (4.6). This usually makes use of the inverse fast Fourier transform (iFFT) algorithm, which assumes a regular sampling grid. Thus, for an interferogram sampled at intervals of Δx, from $-L$ to $+L$, the iFFT will yield a spectrum of values at regular intervals of $1/(2L)$ from 0 to $1/(2\Delta x)$. From Nyquist's theorem, any signal present at wavenumbers larger than $\sigma_N = 1/(2\Delta x)$ will be "folded" back into the recorded spectrum at the wrong wavenumber. For most spectrophotometry applications in the visible through infrared spectral regions at moderate resolutions of $\geq 0.1\ \mathrm{cm}^{-1}$, it is better to err on the side of caution and decrease the sampling interval Δx so that this folding frequency is comfortably above the spectral region of interest. Even over sampling by a factor of two or more is usually not a strain on the storage capacity or processing power of contemporary microcomputers.

It is a good idea to limit the actual electrical and optical bandwidths of the instrument as much as possible to only record data in the desired wavenumber region, with the use of electrical and optical low-pass filters. In this way, any high-frequency signals that are under sampled will not end up contributing noise in the spectral region of interest. Continuously scanning systems have a direct relationship between the modulation frequency f of the detected signal and the wavenumber of corresponding spectral component: $f = \sigma v$, where v is the rate of change in OPD. The minimum sampling interval is thus limited by v for a given analog-to-digital conversion system bandwidth. The optimum scan rate is usually constrained by $1/f$ noise and mechanical stability on the slow end, and by mirror inertia and driving force on the fast end. For typical bench top FTS instruments, it is in the 0.1–5 cm/s range. Thus, an analog-to-digital converter (ADC) with 1 MHz bandwidth will probably be adequate. The scan rate and sampling interval can be adjusted to optimize stability and signal-to-noise ratio (SNR), as discussed in Section 4.3.

4.2.3 Phase Error and Phase Correction

The model of FTS response represented in Eqs. (4.5) and (4.10) will always produce a symmetric interferogram, with $I(x) = I(-x)$ and a maximum at $x = 0$. This simply reflects the setup from Fig. 4.1, where the phase difference between the two split wave components is entirely due to the difference in geometrical path length in the two arms of the interferometer. In practice, there are other possible sources of phase shift.

Most beamsplitters for the mid- to near-infrared spectral region are constructed from thin film coatings on crystalline substrates, where the coating is designed to give close to 50% reflectance and transmittance values over the wavenumber range where the substrate is transparent. Generally, the substrate needs to have a thickness on the order of a centimeter to achieve flatness to within a small fraction of a wavelength across the beam. Considering again the optical layout in Fig. 4.1, if the reflective coating is on the input side of the beamsplitter, one can see that the component reflected from the fixed mirror will be transmitted only once through the substrate, while the moving mirror component will go through three times. The extra optical path length for the moving mirror component would result in a phase shift, so that the ZPD point for a given input wavenumber will not generally occur at $x = 0$. In fact, dispersion in the dielectric substrate material will produce a wavenumber-dependent phase shift, and the resulting interferogram for a broadband source will be "chirped" as well as shifted in x. This effect is usually minimized by including on the reflective side of the beamsplitter a "compensator" plate of the same thickness and material as the substrate. However, small errors in manufacture and alignment will leave behind some residual uncompensated phase shift between the two arms of the interferometer.

Another source of wavenumber-dependent phase shift for continuously scanning FTS systems comes from the frequency-dependent response of the detector/amplifier. In addition to the natural roll-off and phase shift of a typical current or voltage amplifier at higher frequency, one usually adds a low-pass filter to reduce the effect of inadequately sampled high-frequency noise, as discussed in the previous section. Thus, each wavenumber component of the spectrum will produce a sinusoidal signal with a different phase shift, producing a similar effect to beamsplitter dispersion. There is a trade-off between aggressively filtering out high-frequency noise, and introducing large phase shifts into the higher wavenumber components of the spectrum.

Finally, as discussed in the previous section, the sampling of the interferogram cannot be continuous, so one will not in general end up with a set of data points that is symmetrically distributed about $x=0$. While this may seem to be a trivial issue, in practice, it is often the largest contributor to phase error, especially at longer wavelengths where the dispersion and electrical bandwidth effects are smaller. The net result of these effects is that the recorded interferogram will in general be asymmetric, and an inverse Fourier transform analysis will yield a spectrum with both real and imaginary components. Since the actual optical power spectrum recorded by the detector must be a real quantity, one needs to somehow account for the extraneous phase shifts before using the data to derive spectrophotometric quantities.

Mathematically, the most efficient way to describe the wavenumber-dependent phase shift in an FTS is to first rewrite Eq. (4.5) in complex form:

$$I(x)=\frac{1}{2}\eta\int_{-\infty}^{\infty} S_\sigma e^{i2\pi\sigma x}d\sigma, \tag{4.13}$$

where now the integral runs over negative as well as positive wavenumbers, so the imaginary component cancels out. It is then a simple matter to include a wavenumber-dependent phase factor $\phi(\sigma)$ as

$$I_{\text{obs}}(x)=\frac{1}{2}\eta\int_{-\infty}^{\infty} S_\sigma e^{i(2\pi\sigma x+\varphi(\sigma))}d\sigma. \tag{4.14}$$

Except in the vicinity of very strong spectral features in the beamsplitter or instrument electronics, $\phi(\sigma)$ should be a smooth, slowly varying function of wavenumber. It can therefore generally be derived from a low-resolution interferogram that is sampled over roughly equal intervals on both sides of the peak signal.

The basic outline of one method of phase correction can be seen in Fig. 4.6. The recorded interferogram $I_{\text{obs}}(x)$ in frame (a) is shifted so that a point near the maximum signal becomes $x=0$. Then, the interferogram is apodized with a smooth function that goes to zero after a relatively small number of sample intervals (often ± 128 or ± 256 points). An inverse FFT is applied to yield the low-resolution complex spectrum $S'(\sigma)$ shown in frame (b). The phase shift $\phi(\sigma)$ is then calculated as

$$\varphi(\sigma)=\tan^{-1}\left[\frac{\text{Im}(S'(\sigma))}{\text{Re}(S'(\sigma))}\right], \tag{4.15}$$

as shown in frame (c). This function is smoothly interpolated to the same frequency grid as the high-resolution complex spectrum $S'(\sigma)$ (which could be

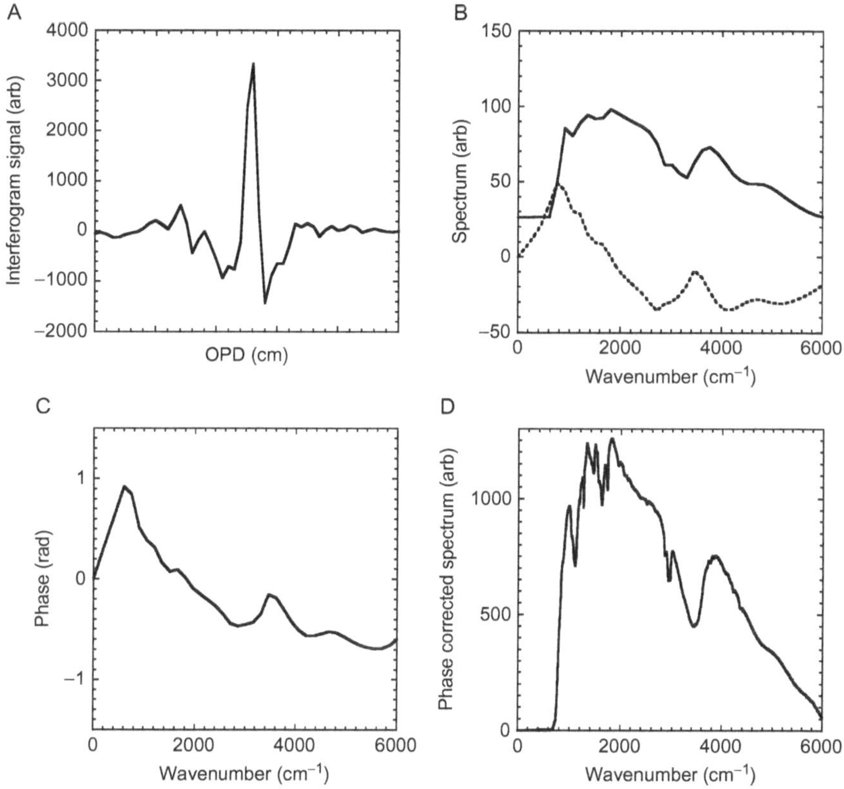

FIGURE 4.6 Phase correction of an asymmetric interferogram to yield a real spectrum. Frame (a) shows the central portion of the interferogram, (b) the real (solid line) and imaginary (dashed line) parts of the low-resolution complex spectrum, (c) the phase function, and (d) the final phase corrected spectrum.

effected by simply leaving zeroes in the feet of the apodized interferogram from step (a)) and used to "rotate" the complex spectrum back to the real axis as shown in frame (d)

$$S_\sigma = \mathrm{e}^{-i\varphi(\sigma)} S'(\sigma). \qquad (4.16)$$

At this point, one can examine the resulting spectrum to see if it makes sense. In the case illustrated in Fig. 4.6, the results in frame (d) look reasonable. The spectrum goes to zero at wavenumbers lower than about 600 cm^{-1}, consistent with the band gap of the photoconductive HgCdTe detector that was used to collect these data. The spectrum is nonnegative and has the correct general shape based on the roughly Planckian output of the approximately 1200 K source and known absorption features in the optics such as OH^- near 3500 cm^{-1}. The phase function shown in frame (c) does vary smoothly as a function of wavenumber and stays within the range of $\pm\pi/2$. If $\phi(\sigma)$ goes through many windings of π, this indicates that the $x=0$ point was chosen poorly and the analysis should be repeated with a new starting point, or possibly there is an alignment issue with the FTS which needs to be addressed.

In general, one may not have roughly equal sampling of the interferogram on both sides of ZPD. If the interferogram is mostly one sided, then the weighting of the FFT must be done appropriately; the analysis is either performed in interferogram or spectrum space. Popular methods incorporated into commercial FTS software include those of Mertz [7] and Forman *et al.* [8]. For spectrophotometric applications, it is important that the FTS phase function be properly handled for both the "sample" and "reference" spectra used to derive spectral transmittance, for instance. This means either that the phase does not change appreciably between these two measurements or that any drift in the FTS phase is accounted for. Some of the ways in which this can go wrong are discussed below in Section 4.3.

One may ask why it is necessary to go through the process described in Eqs. (4.14)–(4.16) at all in the case of spectrophotometry. One could simply use the modulus of $S'(\sigma)$ and ignore the phase completely in calculating spectral ratios. There are several problems with this approach. First of all, the modulus spectrum includes extra noise from the imaginary component and is "rectified" such that the spectrum will not average to zero as it should in wavenumber ranges where the sample is black. This can produce large relative errors where the transmittance or reflectance of the sample is small. Also, in practice it is often necessary to account for detected signal which does not originate from the source, but from other components such as the sample or optics. These signals will in general have different phase shifts that need to be kept in the calculation up to the last step. Correcting the phase in FTS measurements is somewhat analogous to doing proper phase sensitive detection with a lock-in amplifier.

4.2.4 FTS Versus Dispersive Instruments

4.2.4.1 Spectral Multiplexing

Consider an FTS instrument using a source with spectral radiance L_σ (W/cm^2/sr/cm^{-1}) and optical throughput $\pi a^2\Omega$, where a is the beam radius and Ω is the solid angle of accepted rays in the interferometer, and interferometer modulation efficiency η. Typically, in the mid-infrared spectral region, the SNR of an FTS instrument using a moderate temperature (approximately 1200 K) thermal source is limited by noise in the detector rather than in the source. In this case, the SNR for a given spectral interval $\delta\sigma$ and averaging time Δt is given by

$$\mathrm{SNR} = \frac{L_\sigma(\pi a^2\Omega)(\eta)(\delta\sigma)(\Delta t^{1/2})}{\mathrm{NEP}}, \tag{4.17}$$

where NEP is the noise equivalent power of the detector ($W/Hz^{1/2}$). The key feature of this otherwise generic expression is that for an FTS, all of the spectral bins are averaged for time Δt simultaneously, as opposed to a dispersive instrument where the averaging is done sequentially, and each interval is only observed for time $\Delta t/N$, where N is the number of spectral bins. Thus, the FTS in principle has a SNR advantage compared to a dispersive instrument given by the square root of the number of spectral bins in the measurement. This is sometimes called the *multiplexing* advantage or *Fellgett's* advantage [9].

Equation (4.17) assumes that the beam solid angle is small enough that self-apodization effects can be ignored and the spectral resolution is only limited by the OPD in the FTS. This is often true in practice as the limiting aperture for the source is typically set to smaller than required for a given resolution because of the need to limit the flux on the detector or because of sample or detector size limitations. If the FTS is operated at the limit of its spectral resolving power, then we can use Eq. (4.11) to substitute for Ω in Eq. (4.17) and arrive at

$$\mathrm{SNR} = \frac{L_\sigma(2\pi^2 a^2)(\eta)(\delta\sigma)^2(\Delta \mathrm{t})^{1/2}}{(\sigma)\mathrm{NEP}}, \tag{4.18}$$

where the SNR is now seen to be proportional to resolution squared, as in the case of a dispersive instrument where both the entrance and exit slits are reduced in width to achieve finer spectral resolution. This expression also indicates the importance of using largest possible value for $\delta\sigma$, which does not cause significant spectral errors due to convolution of the sample or reference spectra with the ILS. A typical rule of thumb is that $\delta\sigma$ should be equal to 1/5 of the smallest line width of interest in the spectrum [3].

The multiplexing advantage of an FTS relative to a dispersive spectrometer of similar wavenumber coverage and resolution assumes that both instruments are using single-element detectors. Some dispersive spectrometers make use of one-dimensional array detectors configured to average over multiple spectral bins at the same time, thus achieving the same advantage as a single-detector FTS. Of course, FTS instruments can also use multiple detectors, in particular two-dimensional arrays to achieve imaging capability, or enhanced spectral resolution. Many different designs have been realized for both types of instruments that trade-off spectral and spatial degrees of freedom for various applications.

In general, there are other significant sources of noise in addition to thermally induced electrical fluctuations in the detector. In particular, in the near-infrared, visible, and near-ultraviolet spectral region, where low-noise Si or InGaAs photoconductive detectors are available, photon shot noise in the source may be the dominant contributor. The denominator of Eq. (4.17) would then be proportional to $(L_\sigma)^{1/2}$ for an FTS instrument but only $(L_\sigma/N)^{1/2}$ for a dispersive instrument. The factors of $1/N^{1/2}$ in the numerator and denominator would cancel for the dispersive case, and there would be no difference in SNR between the FTS and dispersive spectrometers. In this case, a dispersive instrument with an array detector used for spectral averaging would have a SNR advantage of $N^{1/2}$ compared to the single-element FTS.

Another limiting case occurs when the noise in the system is dominated by fluctuations in overall source output. This can be caused for instance by instability in the source power supply or mechanical vibrations. In this case, the noise per spectral bin is proportional to L_σ for an FTS but only L_σ/N for a dispersive spectrometer, and the dispersive design SNR wins by a factor of $N^{1/2}$ even for a single-detector element. These SNR considerations, in addition to the difficulty and expense of making an interferometer work well at very short wavelengths, explain the predominance of dispersive spectrometers over FTS instruments in use for the 250–2500 nm spectral region. However, at longer infrared wavelengths for spectrophotometry over a wide spectral range, the FTS advantage from Eq. (4.17) does apply and is realized in superior performance in practice as long as other sources of error (Section 4.3) are kept under control.

4.2.4.2 *Throughput and Instrument Size*

Another advantage of FTS over dispersive designs is related to the instrument geometry. Dispersive instruments based on prisms or linear diffraction gratings typically use rectangular slits at the input and output to define the spectral resolution. The optics can be expanded in one spatial direction to collect more light, but in the other direction are restricted by the desired resolving power. In the FTS case, both spatial dimensions perpendicular to the beam can be expanded without affecting the resolving power up to the self-apodization limit of Eq. (4.11). For a given beam diameter, the FTS will have a higher optical throughput. This is sometimes called *Jacquinot's* advantage [3].

From Eq. (4.11), we can see that for circular optics with beam radius a the maximum optical throughput of an FTS is given by

$$\Theta_{\mathrm{FTS}} = \frac{2\pi\delta\sigma}{\sigma}\left(\pi a^2\right). \tag{4.19}$$

This expression can be compared to the throughput for a grating-based dispersive spectrometer which in first order is given by Ref. [3]

$$\Theta_{\mathrm{D}} = \frac{h\delta\sigma}{f\alpha\sigma^2}(A_{\mathrm{G}}), \tag{4.20}$$

where h is the slit height, f is the focal length of the collimating mirror, α is the grating constant, and A_G is the illuminated area of the grating. For equivalent beam areas, it can be seen that the resolution-limited throughput of the FTS relative to the grating instrument will tend to increase at larger wavenumbers. Putting in values for typical mid-IR dispersive instruments with spectral coverage from 400 to 5000 cm^{-1} and 2 cm^{-1} resolution, the FTS throughput is higher by a factor of 10–300.

In practice, the multiplexing and throughput advantages of infrared FTS relative to dispersive instruments are somewhat offset by other factors such as the interferometric modulation efficiency, losses in the beamsplitter, the need to block part of the beam for a metrology laser path, and frequency dependence of the detectors used in continuously scanning systems. However, with typical designs for 2 cm^{-1} resolution in the mid-infrared spectral region, it is fairly common to have approximately 100 mW of modulated optical power incident on the detector in an FTS system versus <1 mW for a dispersive system. This in turn can lead to issues with nonlinearity of the detector response which can be dealt with in various ways, but most simply by attenuating the beam with filters or stops, negating some of these advantages. Strategies for making optimal use of the dynamic range of an FTS are discussed in Section 4.3.

Despite the loss of the multiplexing advantage of the FTS at shorter wavelengths, the throughput advantage is sometimes exploited to build instruments of very high spectral resolution (less than 0.03 cm^{-1}) of much smaller size than the equivalent dispersive instrument would be. Such instruments have been used for measurement of atomic discharge spectra in the visible and UV [10] as well as for measurement of refractive index of materials [11].

4.2.4.3 Wavelength Scale

Another difference between FTS and dispersive instruments that is worth noting is how they derive the wavelength scale of the spectrum. In an FTS, the wavelength scale comes naturally from the Fourier transform of the interferogram which is sampled over a grid of OPDs x. These values of x are determined by reference to the cosine interferogram generated by a collinear single-mode laser beam. Differences in the average value of x between the metrology laser and infrared beams due to beam divergence (Eq. 4.12) or misalignment produce errors in the wavelength scale which are typically less than 0.03% for reasonably well-constructed instruments. Moreover, the correction factor is independent of wavenumber and thus can be determined by measurement of relatively few atmospheric absorption lines at high resolution. Higher order errors due to dispersion of the gas within the FTS optical path (for nonvacuum systems), diffraction, and laser wavelength drift tend to be small, and the first-order correction is usually good enough to reduce the residual wavelength error to about 0.0003% for mid-infrared through ultraviolet spectrophotometry.

In a grating instrument, the wavelength scale is nominally derived from the physical dimensions of the grating, focal length of the collimating mirror, and gear ratios of the grating drive mechanism. Atomic discharge lamps are measured over a number of orders to derive correction factors which depend upon wavelength due to imperfections in the grating drive. With considerable work, it is possible to achieve the same level of wavelength accuracy as an FTS, but this procedure must be repeated for different gratings, filters, and slit assemblies. One final note which is occasionally overlooked is that FTS instruments inherently measure vacuum wavelength (except for the small atmospheric dispersion effect) whether operated under purge or vacuum conditions, whereas dispersive instruments measure wavelength in the instrument atmosphere.

4.2.5 FTS Design Considerations

Figure 4.7 shows a schematic optical layout for a generic two-beam Michelson-type FTS. Here, we discuss the major components of an FTS system and how their design choices reflect the trade-offs of wavelength coverage, spectral resolution, throughput, sensitivity, speed, stability, radiometric accuracy, and cost. In many commercial FTS instruments, some of these elements are designed to be interchangeable to cover different spectral regions or radiometric power ranges. Most "research-grade" instruments also provide ports and repositionable mirrors to enable users to add their own external sources or detectors.

4.2.5.1 Source

The ideal source for most applications in the mid-infrared spectral region would be a blackbody cavity, for its high emissivity, stability, and Lambertian output quality. However, the bulk, expense, and power requirements of such a source lead most manufacturers of FTS systems for spectrophotometry to opt

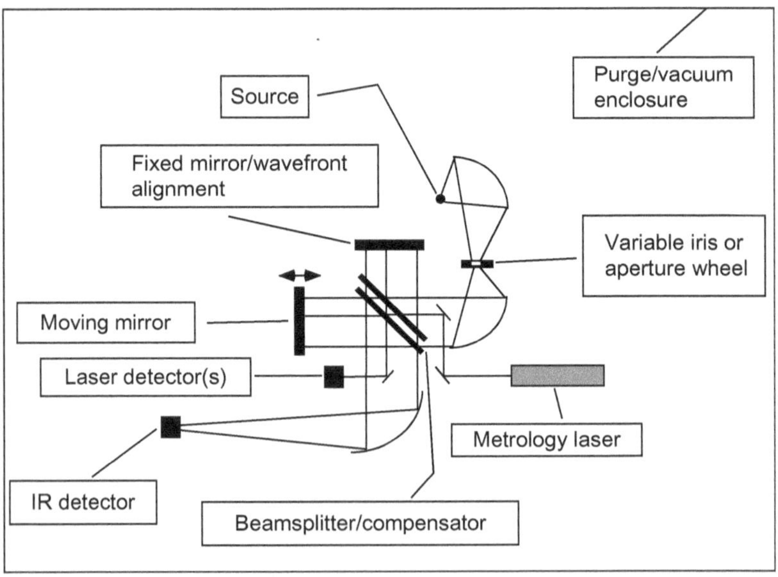

FIGURE 4.7 Schematic optical layout of a Michelson-type FTS showing the major components of the system.

for a small ceramic-coated heater coil element "globar" source instead. Typically, these sources are approximately 1 cm in diameter, with emissivity approximately 0.8. They are run in constant current mode with a temperature of about 1300 K and peak power output near 2500 cm^{-1}, with a usable range of 200–6000 cm^{-1}. Temperatures much higher than this require either vacuum or purge with a high-purity inert gas such as Ar. The lifetime of most globar sources is 5–10 years.

For larger wavenumbers in the near-infrared spectral region, the most common spectral source is the simple quartz tungsten halogen (QTH) light bulb (in fact, most manufacturers use commonly available automotive headlamp bulbs). The effective blackbody temperature of the tungsten element is approximately 3000 K, giving peak output near 6000 cm^{-1}, with usable range from about 2900 cm^{-1} (below which the quartz envelope is opaque) to 20,000 cm^{-1}. QTH bulbs typically last approximately 1 year of fairly constant use but are inexpensive and easy to replace. For FTS instruments operating into the ultraviolet spectrum, some manufacturers provide deuterium arc lamps, which provide some continuum coverage up to 50,000 cm^{-1}.

Far-infrared spectra below 200 cm^{-1} are usually acquired with the use of a high-pressure HgXe arc lamp source. The quartz envelope of the lamp is transparent below 150 cm^{-1} so that the plasma inside, with effective temperature of approximately 3000 K and reasonably high emissivity (at least compared to tungsten wire) can be viewed. At these low wavenumbers, the output of any Planckian source is only linearly dependent upon temperature, so even Ar arc sources with an effective temperature of about 5000 K are hardly worth the considerable effort to make them work. All of the thermal sources that have been mentioned are typically 50–100 W and are usually enclosed in water-cooled jackets to prevent overheating of the electrical contacts as well as to shield the rest of the instrument from all but the desired angular output range of the source.

Nontraditional sources for FTS could include free-electron or supercontinuum lasers which can provide coherent broadband output with high radiance. Synchrotron light sources are also used, with several facilities around the world having been set up with FTS systems spanning the far-infrared to vacuum ultraviolet spectral regions [10,12]. By tuning the synchrotron beam energy, it is possible to produce a broadband source that exceeds the radiance of a 3000 K blackbody by several orders of magnitude. The high brightness of these sources makes them ideally suited to applications in microscopy, which requires the smallest possible focus spot on the sample, or refractometry and ellipsometry, which require highly collimated sample beams for the best accuracy.

4.2.5.2 Entrance Aperture

Most research-grade FTS instruments that are designed for spectral resolution less than 2 cm^{-1} include fore-optics that collect light from the source, focus it onto a variable sized field stop, and then recollimate the beam going into the interferometer. Typically, the source and field stop are located at the foci of an ellipsoidal mirror, and a paraboloidal mirror is used to recollimate the beam. The entrance aperture, commonly called the *Jacquinot stop*, limits the angular range of rays entering the interferometer and thus the minimum spectral resolution that can be achieved through Eq. (4.11). For a Jacquinot stop of diameter d_J and collimating mirror focal length f_c, the maximum solid angle Ω_m of rays entering the interferometer is given approximately by $\pi d_J^2/4f_c^2$, corresponding to a minimum spectral resolution of

$$\delta\sigma = \frac{\sigma d_J{}^2}{8 f_c{}^2}. \qquad (4.21)$$

Putting in numbers for a typical mid-IR FTS, we can see that for 0.5 cm^{-1} resolution at 4000 cm^{-1} (a resolving power of 8000) Eq. (4.12) gives $\Omega_m = 7.85 \times 10^{-4}$ sr, or an *f*-number of 31.6 for the collimated beam. For a 1-mm entrance aperture, the required collimator focal length from Eq. (4.21) is 3.16 cm. A system designed with 5 cm diameter optics for the interferometer would then have an optical throughput of 1.54×10^{-2} cm^2 sr. The Jacquinot stop mechanism is usually either a variable iris or a wheel with several different precision apertures that can be rotated into position. In either case, the area of the aperture must be known well enough that Eq. (4.12) can be used to make the proper wavenumber correction, assuming the entrance aperture actually defines the solid angle limit of the whole optical system, and not the sampling or detector optics.

4.2.5.3 Beamsplitter

The chief design considerations for the beamsplitter are that it is as close as possible to 50% reflecting and 50% transmitting over the spectral range of interest, and that it is optically flat to within approximately 1/10 of a wave for the highest wavenumber light it will be used with. For the mid-infrared through UV spectral regions, the flatness requirement necessitates the use of fairly thick (approximately 1 cm) substrates of transparent crystalline materials like KBr, SiO_2, or CaF_2, that can be polished to the required degree. These low index materials are usually given a thin optical coating to enhance the reflectivity. The spectral range is limited by lattice vibration (phonon) absorption at the low wavenumber end, and either the coating properties or optical nonflatness at the high wavenumber end.

As discussed in Section 4.2.3, it is usually necessary to add an uncoated compensator plate of the same material and thickness as the beamsplitter substrate in order to produce the same path length difference for rays in either arm of the interferometer. The two plates are wedged slightly to eliminate standing-wave interference in the substrate and carefully aligned to minimize deviation of the beam through the interferometer. Typically, there are no antireflection coatings on the back of the beamsplitter or either side of the compensator, so other reflected beams are present within the FTS. If you look at the refocused source image in the sample compartment of a commercial FTS setup for the near-infrared, you will usually see a bright central spot with satellite spots on either side from the outside surfaces of each plate. These extra beams are not interferometrically aligned so do not contribute to the modulated FTS response, but do add a dc component to the detector signal.

For wavenumbers below 600 cm^{-1}, pellicle beamsplitters made from mylar or coated mylar are typically used. In uncoated designs, the efficiency η is peaked at the wavenumber determined by the first-order maximum reflectance due to interference within the thin film, with additional higher order peaks at larger wavenumber. Coatings such as Si or Ge have been used to enhance the reflectance and cover a wider spectral range without regions where η goes to zero [3]. These kinds of beamsplitters are thin enough (less than 100 μm) that a compensator plate is not necessary.

Most FTS designs allow beamsplitters to be interchanged for different spectral regions, usually by manual swapping, although there are examples of automated beamsplitter wheels that allow change at the push of a button. The beamsplitters are held in kinematic mounts to facilitate the change, but either manual or automatic alignment tweaking is usually needed when changing from one to another. Some part of the beamsplitter (usually the center) is used by the collinear metrology laser for mirror position and tilt tracking. For most spectral ranges, this will mean the center part of the beamsplitter has a different coating from the part used for the signal beam. For far-infrared pellicle beamsplitters, there is usually a hole in the middle filled with a small visible splitter that must be carefully aligned parallel to the outer annular portion. Some instruments provide storage within the purge enclosure for extra beamsplitters, but otherwise any hygroscopic materials like KBr or CsI must be stored in a dry box when not in use.

4.2.5.4 Moving Mirror Mechanism

Both flat mirrors in a conventional Michelson FTS must meet the same strict 1/10 wave flatness specification as the beamsplitter for optimum performance. In addition, they must be maintained perpendicular to the incident beam over the range of motion of the moving mirror to a similar tolerance. The motion must be highly uniform to avoid adding noise to the modulated optical signal. An instrument with 0.1 cm^{-1} spectral resolution requires 10 cm OPD, or 5 cm travel on each side of ZPD, 10 cm total travel for a symmetric design best suited for high radiometric accuracy. In order to operate in the near-infrared at a wavelength of 1 μm ($\sigma = 10{,}000\ cm^{-1}$), 1/10 of a wave of OPD over a beam with diameter $D = 5$ cm corresponds to a tilt angle $\theta = \tan^{-1}\left(\frac{1}{2}\frac{\lambda/10}{D}\right) = 1\,\mu\text{rad}$. A few mechanical approaches have been developed to support smooth motion with the required alignment over such a range.

These include air-bearing and porch-swing designs. This kind of performance can also be achieved by using corner-cube or cats-eye reflectors in place of the plane mirrors. Such a design greatly reduces the wavefront tilt error in the interferometer at the expense of adding shear between the two beams. An air-bearing translation control system can easily keep the shear under several micrometers in amplitude, which will produce no appreciable loss in modulation for a 5-cm diameter beam.

The drive for the moving mirror can be provided by a voice-coil, a motor-belt system, or even a linear motor. In any case, the motion must be tracked with an uncertainty of a tiny fraction of a wavelength (less than 1 nm for mid-infrared) to not add appreciable sampling noise to the interferogram and must be smoothly reversible without losing track of the position.

4.2.5.5 Fixed Mirror/Wavefront Alignment Mechanism

In order to relax the requirements on tilt control of the moving mirror (and hence the size, weight, and cost of the motion control system), many FTS designs incorporate a compensating tilt adjustment mechanism on the fixed mirror, or sometimes on an additional folding mirror in the moving arm of the interferometer in order that the fixed mirror truly does not move. The idea is to effectively tilt the fixed mirror during the scan through the same angular trajectory as the moving mirror, thus keeping the two beams parallel to each other. Most often this is done using either piezoelectric transducers or small voice coils configured to tilt the mirror around two axes perpendicular to the beam direction. Thus, it becomes possible to allow approximately 30–100 μrad tilt on the moving mirror and still keep the two beams parallel. Tracking of the moving mirror tilt trajectory is done using the metrology laser system as discussed below.

4.2.5.6 Metrology Laser System

Nearly all commercial FTS designs use a single-mode polarization selected and thermally stabilized HeNe gas laser operating at 632.8 nm wavelength (15,800 cm^{-1}) to track the moving mirror position. Usually, the center of the beamsplitter is reserved for this laser beam, which is made collinear with the main signal beam using small pickoff mirrors or folding prisms. The cosine interferogram is recorded using a Si photodiode detector, and the digitized signal is processed to determine the average OPD versus time. In older designs, this was done using zero-crossings of the interferogram, which occur twice per laser wavelength, limiting the Nyquist frequency to 15,800 cm^{-1} although this could be extended with interpolation schemes based on tracking the mirror velocity over time. This signal would be used to trigger an acquisition of the main signal interferogram. In some newer designs, the entire digitized signal is used to produce a position trajectory with less than 1/1000 wave precision that can be synchronized to the main signal interferogram in postprocessing.

In addition to tracking the change of OPD during a mirror scan, the metrology system must also produce a stable absolute reference point near the center burst of the signal interferogram that can be used as the ZPD point input to the phase-correction algorithms discussed in Section 4.2.3. In some systems, this is done by performing several short scans near the ZPD position and picking the point where maximum amplitude is seen in the signal interferogram as the ZPD point. As long as the system continues to scan and track the OPD, this point can remain stable over thousands of scans, so the signal interferograms can be coadded to improve the SNR. An alternative system uses a small collimated white light source fed through the interferometer right next to the laser beam, which produces a centerburst interferogram independent of the main signal beam. The maximum of this interferogram can then be used as a ZPD reference point, which can be sampled on each scan or just during a short prescan alignment routine as above.

A problem arises when the moving mirror reaches the end of its scan range and turns around. The OPD velocity must go briefly through zero and mechanical vibrations may produce a fringe shift that looks like motion of the mirror; in general, it is not possible using one metrology signal to determine which direction the mirror is moving. The common solution is to insert a quarter-wave retarder plate into half of the metrology beam, producing two simultaneous cosine interferograms sampled by two detectors, which differ by 90° in phase. The leading signal becomes the lagging signal when the OPD velocity changes sign.

Finally, it is also common to expand the metrology laser beam and sample it at several points with detectors near the center of the beam as well as near the edge in two perpendicular directions to track the tilt of moving mirror as discussed above. Typically, a servo control loop is set up to keep the phase of the center and two offset detector signals constant during the scan by using an error voltage generated from each phase difference to produce a corrective offset in the wavefront alignment drive. Contemporary commercial bench top infrared FTS instruments incorporating air-bearing or porch-swing moving mirror drives along with such a "dynamic alignment" tilt compensation system are capable of producing spectral output in the signal beam which is stable to less than 0.1% variation over several hours of observation time.

4.2.5.7 Detector

Optical detectors used with FTS systems fall generally into two classes: (a) thermal detectors, in which the absorbed radiation produces a change in temperature of the detector element that is recorded by electrical transduction, or (b) photon detectors, which directly produce a current or voltage signal due to a change in the electronic quantum state of the detector material upon absorbing a photon [13]. To be useful with a continuously scanning FTS in the mid- to near-infrared spectral range, a detector must have an electrical bandwidth of at least 1 kHz. Thus, thermocouple or thermopile detectors commonly used in older monochromator instruments are not applicable. Step-scan instruments can make use of slower detectors.

The most commonly used thermal detectors for mid-infrared are ambient temperature deuterated triglycine sulfate (DTGS) pyroelectric elements with 1 or 2 mm diameter. These have NEP of approximately 2×10^{-9} W/Hz$^{1/2}$ but produce a fairly linear response at the flux levels typically present in an FTS instrument. For wavenumbers below 300 cm^{-1}, where the available flux from thermal sources is very low, it is common to use liquid He cooled composite bolometer detectors of 2 mm diameter with NEP of 10^{-12} W/Hz$^{1/2}$ or less. These detectors can exhibit some nonlinearity at higher flux levels and also need to have cold long-pass filters to limit the view of ambient temperature background flux.

Commonly used photon detectors include liquid N_2 cooled HgCdTe and InSb for mid- and near-infrared up to 10,000 cm^{-1} and ambient or thermoelectrically cooled Si, Ge, or InGaAs for higher wavenumbers. The Johnson noise-limited NEP of photon detectors is generally proportional to the square root of the detector area, so most FTS manufacturers use detectors as small as possible, often 1 mm or less in diameter. The peak NEP of a 1 mm diameter HgCdTe detector designed for mid-infrared use is approximately 3×10^{-12} W/Hz$^{1/2}$. The responsivity of any photon detector drops off roughly linearly with increasing wavenumber above the band gap and goes to zero exponentially for wavenumbers approaching the band gap from above. Clearly, this kind of detector can be used to expand the dynamic range of an FTS by several orders of magnitude below the pyroelectric noise limit. However, they are generally nonlinear at flux levels above approximately 1 mW, so it is usually necessary to somehow attenuate the FTS beam to make use of them. Generally, one needs to underfill the detector element for accurate spectrophotometry measurements to reduce the sensitivity to sample alignment issues. Thus, for some applications it may be wise to trade-off an increase in NEP for larger detector area.

An FTS instrument must digitize the sampled interferogram signal recorded by the detector with high fidelity to avoid introducing spectral artifacts. Many manufacturers use an ADC with 18 or even 20 bits of dynamic range to accommodate the SNR at the centerburst of the interferogram, which can be as high as 10^5 for an optimized system. Many systems incorporate gain-ranging to avoid saturating the ADC near the ZPD point and boost the gain to avoid digital noise in the lower signal wings of the interferogram. In general, the amplifier, band pass filter, and ADC electronics produce a frequency (and hence wavenumber) dependent phase shift in the signal. This means that the phase characterization of the interferogram described in Section 4.2.3 must be redone for different scan velocities and in some cases, change in amplifier gain.

4.2.5.8 Optical Bench and Purge/Vacuum Enclosure

For accurate spectrophotometry measurements in the laboratory, an FTS generally needs to be kept in a thermally stable, mechanically and electrically quiet environment. A commercial bench top instrument that is usually stable to less than 0.1% over a 1 h scan can drift by several percent if the room temperature changes by more than about 2 °C. Generally, the FTS components are mounted to an optical plate of about 2 cm thickness to provide mechanical rigidity and stability to the interferometer. Although not shown in Fig. 4.7, most laboratory instruments include a compartment with an intermediate focal point for holding samples to be measured.

For work in the infrared, atmospheric carbon dioxide and water vapor are strong absorbers, whereas oxygen becomes a limiting factor in the ultraviolet above 50,000 cm^{-1}. The best solution would be to operate the FTS under purge with dry nitrogen gas boiled off from a liquid N_2 dewar. This is not practicable in most laboratories (including ours) so it is common to use a compressed air source and a cleaner unit that removes dust, oil, carbon dioxide, and water vapor. Water vapor in particular is continuously evolved from surfaces that have been exposed to the ambient laboratory air unless they are baked under vacuum. However, for most bench top FTS systems, a dry air purge of 20 min to 1 h is enough to reduce the residual atmospheric absorption to a low, stable level, which is sufficient for the spectral features to divide out in spectrophotometric ratio measurements. High-resolution (less than 0.1 cm^{-1}) or far-infrared (less than 300 cm^{-1}) measurements are usually best carried out under vacuum conditions. Several commercial manufacturers build instruments with heavy-duty enclosures that can be evacuated to less than 30 Pa, which is sufficient for most measurements of solid samples.

4.2.6 Instrument Design Examples

One common variation of the conventional Michelson FTS is a four-beam design that uses corner-cube reflecting mirrors and includes two input ports and two output ports as shown in Fig. 4.8. Such an instrument is often used in radiometric applications because it provides a direct method to compare the output of two sources placed say at ports 1 and 4 by using a detector at port 2 or port 3. Conversely, the responsivity of two detectors placed at ports 2 and 3 can be directly compared by observing a source at port 1 or port 4. Advantages for applications in spectrophotometry include the ability to easily subtract ambient background flux contributions originating from the sample or detector (one of the two source inputs can just be a room temperature plate). In addition, this design eliminates first-order interreflections between a sample or detector in one of the output paths and the interferometer, sending the reflected beam back to the input ports. Also, two detectors can be used to simultaneously collect spectra over different spectral regions or to collect

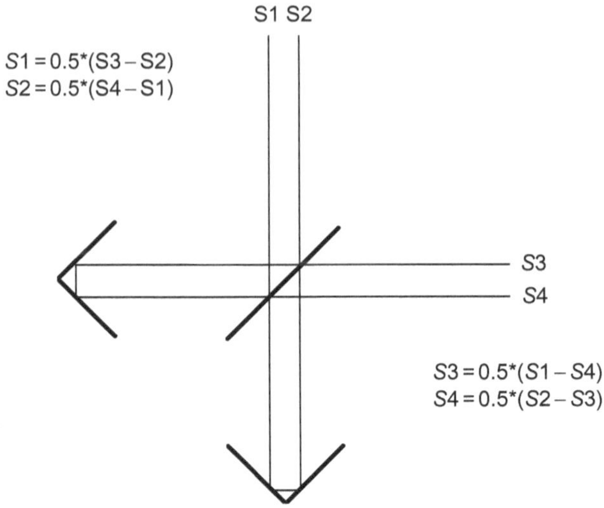

FIGURE 4.8 Schematic optical layout of a four-beam interferometer design featuring two input and two output ports as described in the text.

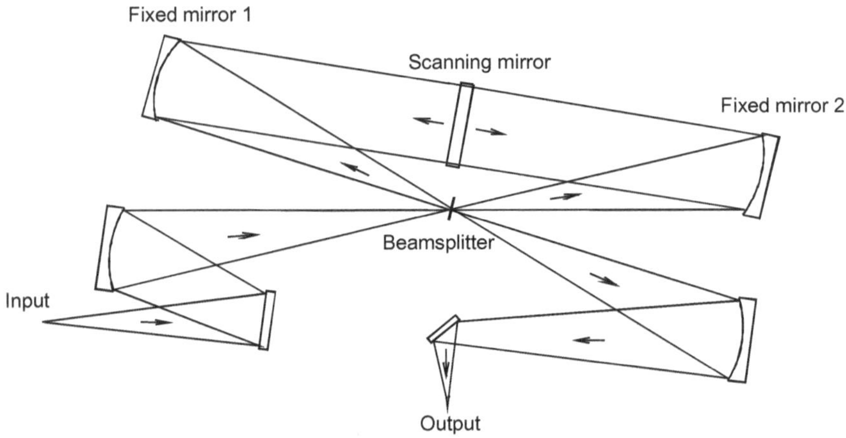

FIGURE 4.9 Schematic optical layout of a Genzel interferometer design showing the input and output ports as well as the scanning mirror as described in the text.

sample and reference spectra. The trade off is a reduction of throughput compared to a conventional design with the same size optics.

A second variation of the conventional Michelson FTS that has been used in far-infrared applications is based on a Genzel interferometer [14] as shown in the schematic diagram of Fig. 4.9. Contrary to a standard Michelson interferometer where the light is collimated at the beamsplitter and is never focused until it emerges, the Genzel interferometer has the light focused onto the beamsplitter. The transmitted and reflected components pass to two collimating mirrors and then to a double-sided moving mirror. As the mirror moves, the path in one arm increases and the path in the other arm decreases. Another characteristic of Genzel interferometer is that beamsplitters can have a small size (≈ 2 cm in diameter). This allows for several beamsplitters to be mounted on a wheel and easily interchanged without breaking the instrument vacuum. The reduction in size significantly reduces the amplitude of the drum-head vibrations commonly associated with large-diameter thin dielectric beamsplitters.

Another feature of the Genzel interferometer is that the angle of incidence at the beamsplitter (15°) is much smaller than a standard Michelson interferometer. This has the advantage of reducing the difference in efficiency between the s- and p-polarized beam components. For a displacement, x, of the moving mirror, an OPD of $4x$ is generated by the Genzel interferometer. Although this has been claimed as an advantage for the Genzel over the Michelson interferometer, any tilt introduced on moving the mirror leads to a doubling of the error in comparison to the error introduced in a standard Michelson interferometer with the same tilt. There is also a consideration to maintain a very stable temperature in the optical bench to obtain a stable alignment. Thus, these are not considered to be major advantages. Nonetheless, the Genzel interferometer does have several significant advantages for far-infrared spectrophotometry in terms of a compact design that yields an instrument with a relatively high spectral resolution (0.03 cm^{-1}).

Other methods can be used to achieve the wavefront division that is done with a conventional beamsplitter in a Michelson or Genzel FTS design. In the far-IR spectral region, the efficiency of thin dielectric beamsplitters falls off at very low wavenumbers (below 30 cm^{-1}) and splitter designs based on lamellar gratings or polarizers are sometimes used [3]. It is also possible to build FTS systems which produce a range of OPDs without any moving parts, for instance by passing the two polarization components of the input beam through a prism and onto a linear detector array [15].

4.3 SOURCES OF UNCERTAINTY AND THEIR REDUCTION

There are a large number of potential sources of uncertainty in spectrophotometric measurements performed with an FTS [16,17]. While many of the effects discussed below are common to measurements made with any type of spectrometer, their magnitudes and the ways in which they manifest themselves can be quite different in FTS spectrophotometry. We try to provide guidance in detecting the presence of various systematic errors, ways to reduce them, and estimate the residual uncertainties. The reduction of one error may lead to an increase in another (e.g., detector nonlinearity vs. noise), so some judgment must be applied in deciding the optimum instrument configuration for a given measurement requirement.

4.3.1 Noise

It is straightforward to estimate the uncertainty due to random noise from the standard uncertainty of the mean of repeated instrument scans. In the ideal case in the infrared, the only significant contributor to noise in an FTS measurement would be the Johnson noise in the detector/amplifier system, and the multiplexing advantage would be fully realized. Of course random noise from the source due to fluctuations or photon shot noise will decrease this advantage, but still can be evaluated in a simple way. As discussed in Section 4.2.2, it is important to filter out any electrical noise above the Nyquist frequency determined by the sampling interval of the interferogram. In practice, however, there are other sources of noise, which may not be completely random in nature.

Coherent noise sources such as electrical pickup from the 60 Hz power line or electronics, or mechanical vibrations, will not usually be synchronized with the FTS scan cycle. Their contribution will therefore tend to average out over time and decrease in amplitude as $1/N_c^{1/2}$, where N_c is the number of coadded scans. However, they can be fairly large in amplitude. A noise source with frequency f will show up in the processed spectrum as a peak at wavenumber $\sigma = f/v$ (where v is the OPD scan speed in cm/s) with a half-width probably equal to the spectral resolution $\delta\sigma$. These features could be confused as originating from the sample. If they cannot be sufficiently reduced by electrical shielding or vibration isolation systems, then they may be minimized or least identified as spurious by changing the scan velocity of the FTS. Increasing v may push the noise peak to a wavenumber below the spectral region of interest.

Noise may also be introduced by errors in the positions at which the interferogram is sampled. If these errors are random, it will look like detector noise and will limit the SNR to $4/(\varepsilon_{RMS}\sigma_{max})$, where ε_{RMS} is the root mean square error in OPD and σ_{max} is the maximum wavenumber present in the spectrum [18]. For a SNR of 10^5 in the mid-infrared region, this means the positional error must be less than 1 nm, which is actually accomplished by commercial FTS instruments. Nonrandom errors in the mirror motion, particularly sinusoidal errors in position or velocity, will give rise to sidebands on sharp spectral features, but fortunately these effects are very small in most modern FTS systems.

4.3.2 Interferometer Alignment Drift

Misalignment of the interferometer reduces the spectral modulation efficiency. Variations in alignment on a time scale longer than one scan show up as drift in the baseline spectral response of the FTS. Variations that occur between the sample and reference scans directly lead to errors in the spectrophotometric measurement. The change in optical phase difference for a given mechanical misalignment of the interferometer scales linearly with wavenumber, so baseline errors due to this effect increase with wavenumber. A common way of evaluating the size of these errors is to perform repeated sets of spectral scans at regular time intervals, usually several minutes at a time. The ratios of these spectra are colloquially known as "100% lines" and give a statistical measure of the stability of the instrument that includes other effects (like source temperature stability and atmospheric absorption) as well. The component of baseline variation that varies linearly with wavenumber is most likely due to alignment changes in the interferometer.

Many bench top FTS systems achieve baseline variations of less than 0.1% at 4000 cm^{-1} over a period of hours, but some may show up to several percent drift over a few minutes. For a less stable instrument, it may be necessary to decrease N_c and perform many groups of scans switching between sample and reference to reduce the uncertainty from alignment drift.

4.3.3 Phase Error

There are a number of ways in which the phase-correction process discussed in Section 4.2.3 can go wrong. Generally, the phase function $\phi(\sigma)$ is calculated from a smoothly truncated low-resolution version of the sample interferogram. However, this can be problematic if the signal in a given spectral region is close to the noise level. In this case, the phase corrected spectrum can change sign rapidly and exhibit the rectified noise effects that phase correction is supposed to eliminate. One solution to this problem is to calculate $\phi(\sigma)$ from a reference configuration of the instrument (e.g., the empty beam with no sample in place) and then use this function to correct the sample spectrum. An example of this procedure is shown below in Fig. 4.10. The phase function (a) derived from the noisy sample interferogram becomes undefined above 2200 cm^{-1}, leading to excess noise and negative signal in the phase

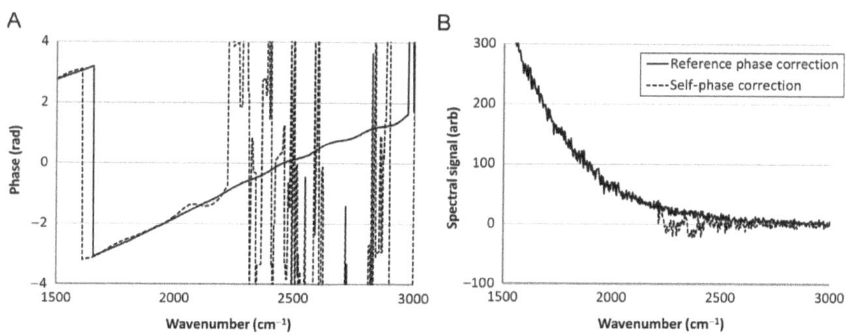

FIGURE 4.10 FTS phase correction of a noisy spectrum using the recorded sample signal ("self-phase," dashed lines) or a reference measurement with five times higher signal (solid line). Frame (a) shows the phase functions computed from the reference spectrum and sample spectrum, while (b) shows the phase corrected sample spectrum using the two different phase functions.

corrected spectrum (b). Using the phase function derived from the reference measurement eliminates this excess noise.

In order for the reference phase function to be used to correct the sample spectrum, the phase of the FTS must not change significantly between the two measurements. This may not always be true for less stable instruments. In particular, the method used by some FTS systems to find the nominal ZPD position may not be reproducible from one scan to the next. This can also lead to large radiometric errors if interferograms with offset sampling grids are accidentally averaged together. It may be possible to recenter the interferograms with a fitting procedure, or individually phase correct each one if the SNR is large enough. Comparing the results of self-phase correction and reference phase correction may help to determine whether this kind of error is occurring, but it may be difficult to fix without addressing the underlying stability issue with the FTS system.

4.3.4 Detector Nonlinearity

If the modulated signal recorded by the detector in an FTS is not strictly proportional to the change in received optical power with OPD, then the distortion of the interferogram will be distributed throughout the spectrum by Eq. (4.6). The resulting spectral errors will vary depending on the nature and magnitude of the nonlinear response variation [19]. As an exaggerated example, consider the effect upon the hypothetical spectra shown in Fig. 4.3 of a detector whose responsivity is proportional to $I(x)^{0.9}$. These results are illustrated in Fig. 4.11. The effect of the nonlinear response is most pronounced near the peak of the interferogram, producing an error whose Fourier transform yields a roughly constant background for the broadband spectrum in (c). The distortion of the cosine modulation from the spectral peak at 2000 cm^{-1} produces harmonics at integer multiples of the true wavenumber, of either sign, whose amplitude decrease with increasing wavenumber.

While this is a somewhat artificial case, it demonstrates several characteristic signatures of nonlinear detector response in an FTS system: false apparent signal in spectral regions where there should be none (especially at wavenumbers below the cutoff of the detector or beamsplitter), harmonics of strong spectral lines, and the possibility of negative response for strong spectral features. In practice, it is quite easy to drive cooled photon detectors like HgCdTe and InSb into their nonlinear regime with the unattenuated modulated flux from a moderate-resolution (2 cm^{-1}) infrared FTS and generate relative errors of more than 10%. Attenuating screens, neutral density filters, or beam-waist reducing apertures can be used to reduce the incident flux on the detector and yield a linear response. An estimate of the residual nonlinearity uncertainty for a transmittance or reflectance measurement can be made by repeating the reference and sample scans with varying levels of beam attenuation and looking at the change in their ratios.

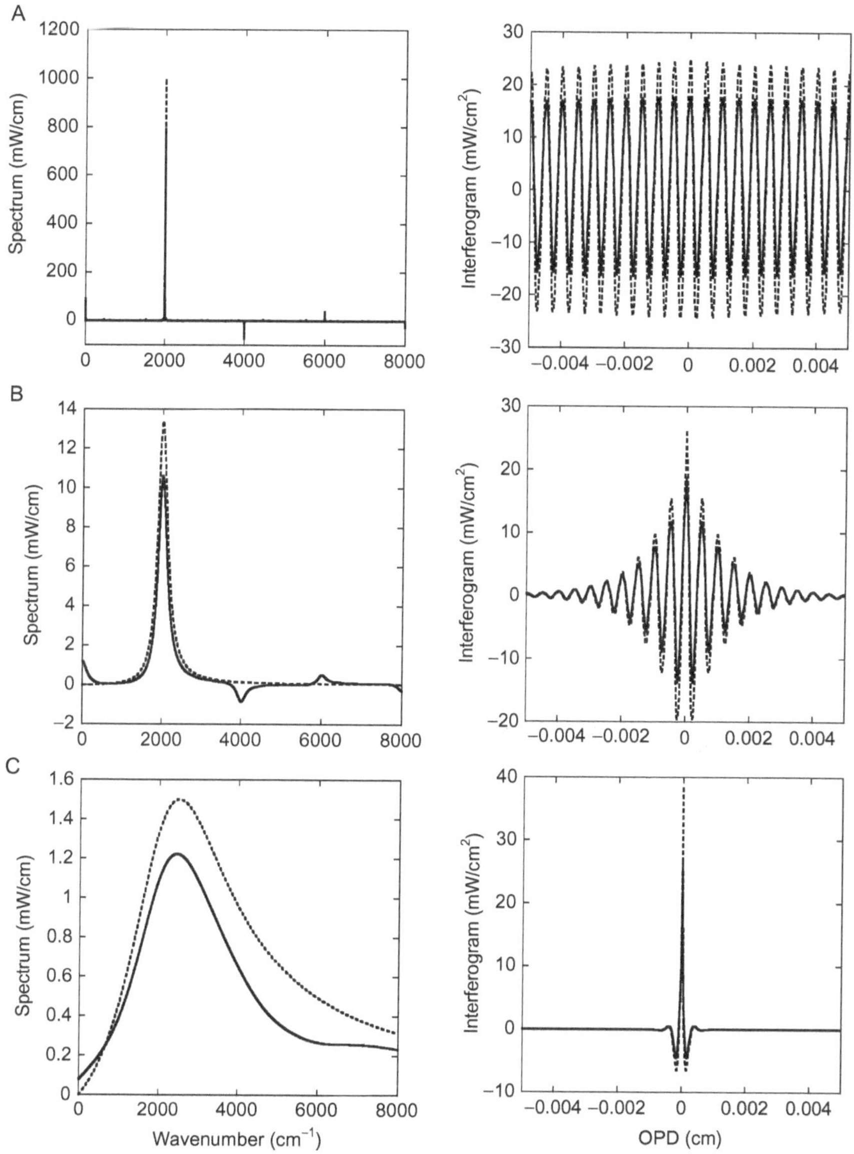

FIGURE 4.11 Effect on the hypothetical interferograms and spectra from Fig. 4.3 of a detector response that is proportional to $I(x)^{0.9}$. The original signals (dashed lines) are recorded as distorted data (solid lines). Frames (a), (b), and (c) show spectra with linewidths of 2, 150, and 1500 cm^{-1}, respectively.

It is also possible to correct for reasonably small levels of nonlinearity using a model for the detector responsivity versus incident power, and a fitting routine to vary the nonlinear parameter(s) and minimize the below-cutoff spectral response. Variations of this approach are sometimes used in commercial instruments, and detector amplifiers can even be modified to linearize the output to the ADC. It is important to record the dc voltage due to unmodulated flux on the detector from the background or less than ideal FTS modulation, as this affects the response linearity.

Pyroelectric detectors generally have a small nonlinearity (less than 0.3%) at the power levels encountered in an infrared FTS. However, their responsivity does depend upon their base operating temperature. The change in average flux reaching the detector between the sample and reference measurements can change the base temperature of the detector and hence its responsivity. This effect is sometimes called *detector nonequivalence* [20]. An example for a transmittance measurement of Si in the mid-infrared is shown in Fig. 4.12. Here, a series of measurements of the sample and reference are made, with the incident flux level successively reduced with filters and screens by a factor of 10 below the original output of the FTS. The detector voltage level recorded at the peak of the interferogram of the empty reference beam is noted in the legend. In the initial configuration, the measured transmittance is low compared to the curve calculated from handbook values [21] of n and k of Si by about 0.02. As the beam is attenuated, the difference in detector responsivity between the sample and reference measurements is reduced, and the measured values approach the expected level to within

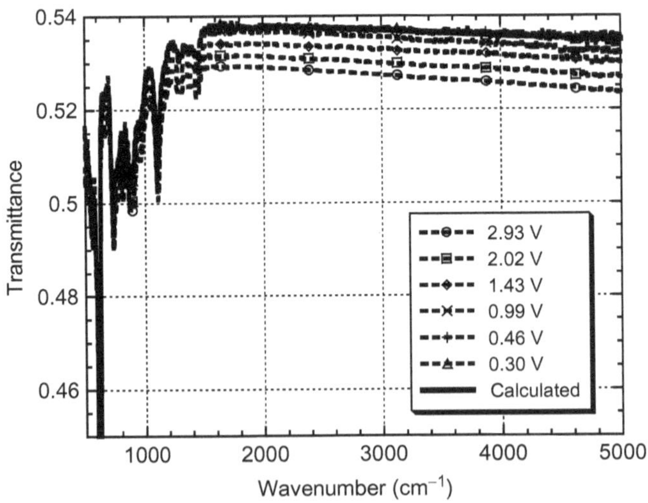

FIGURE 4.12 Measured transmittance of a 0.25-mm thick Si wafer using a mid-infrared FTS with a DTGS detector. The flux entering the sample compartment of the instrument is attenuated with filters and screens to reduce the detector voltage signal at ZPD to the indicated levels.

0.002. The overall effect of detector nonequivalence is to introduce a relative error in the transmittance that is roughly a constant multiplicative factor, independent of wavenumber, but not to distort spectral features.

4.3.5 Interreflections

Interreflections between the sample and the detector, spectrometer, or other optical elements in a spectrophotometric measurement system introduce additive error terms to the measurand. In particular, light that is reflected back to an FTS by the sample or other optical element can pass through the beamsplitter back to the interferometer mirrors, be modulated a second time by the scanning mirror, recombine, and then pass through the sample again to the detector. For a given wavenumber σ, any part of the optical power that is modulated twice by the FTS will experience twice the path length change and thus be recorded at 2σ. Successive interreflections can generate higher harmonics in the spectrum [22]. Other interreflections not involving the FTS directly will still add an error to the measurement that can include spectral features of other components in the system. This situation is illustrated for a transmittance measurement in Fig. 4.13.

In a focused sample geometry, one way to eliminate interreflections is to insert semicircular beam blocks before and after the sample, as shown in Fig. 4.13. The reflection of the half-cone beam incident on the sample will thus be blocked from reentering the FTS, and similarly any light reflected from the detector will not return to the sample or to the spectrometer. An example of the effects of interreflections on a transmittance measurement (normal incidence, *f*/3 focusing geometry) of a 0.25-mm thick Si wafer is shown in Fig. 4.14.

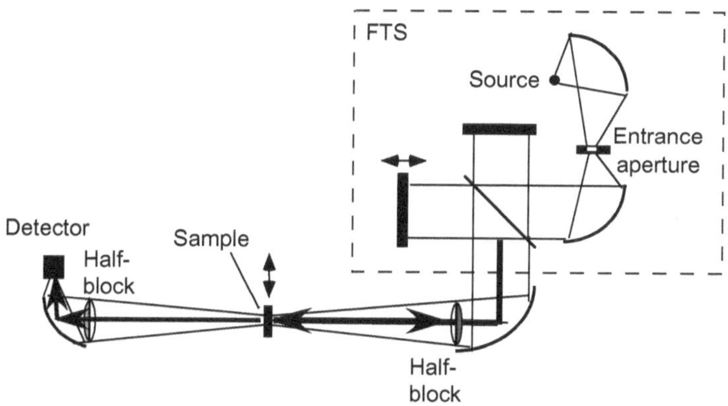

FIGURE 4.13 Schematic layout of a sample transmittance measurement with an FTS showing possible interreflections and their elimination through the use of semicircular beam blocks.

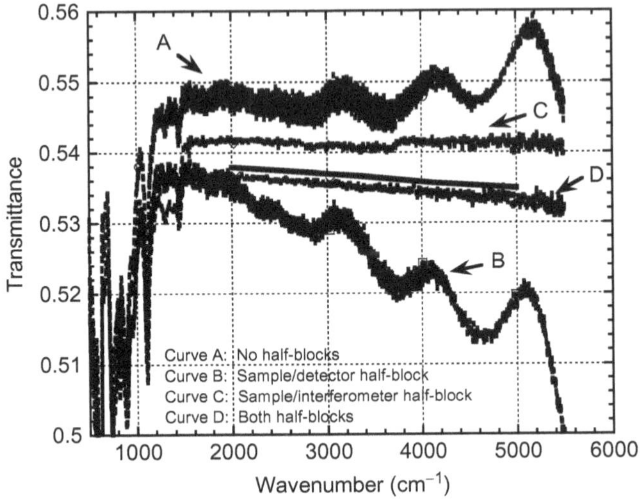

FIGURE 4.14 Effects of interreflections on a normal incidence transmittance measurement of a 0.25 mm Si wafer. The lettered curves show the results of inserting half-blocks illustrated in Fig. 4.13 compared to the expected transmittance (solid curve).

Without any half-blocks in the beam, curve A shows large deviations from the expected transmittance curve. These include both wavenumber-doubled spectral features originating from the spectral irradiance of the incident FTS beam and high-frequency fringes due to doubling of the etalon spectrum of the Si wafer sample. Inserting a half-block between the sample and detector (curve B) lowers the error at low wavenumbers, but does not get rid of the interreflections with the FTS. Curve C with the half-block before the sample eliminates the sample-FTS interreflection, while curve D with both half-blocks gives a result close to the expected transmittance spectrum.

The use of half-beam blocks, or half-field stops, can effectively eliminate interreflection errors at the cost of 50% reduction in signal, and some increase in sensitivity to beam geometry errors as discussed below. Other options include filters between the sample and FTS, or simply tilting the sample, which may be the simplest approach if knowledge of the precise incident beam geometry is not critical to the measurement. Si represents close to the worst case, being approximately 50% transmitting; lower index materials will have less uncertainty. Finally, one may also consider interreflections on the source side of the FTS. These in principle can lead to errors in the reference spectrum that can propagate to the measured transmittance.

4.3.6 Nonsource Emission

Spectrophotometric measurements in the infrared may be affected by ambient thermal emission from the spectrometer, sample, detector, or other optics in the system. A typical way in which these errors are avoided is to modulate the source with a mechanical chopper and use synchronous lock-in detection of the transmitted or reflected light. However, this is not practical with standard continuously scanning FTS instruments which already modulate the source radiation at audio frequencies with the time-varying OPD in the interferometer. The source would have to be modulated at very high frequency (approximately 100 kHz) and the detector signal then demodulated before performing the Fourier transform, which is difficult to do reliably over a wide spectral range. A source chopper can be used with step-scan FTS systems.

As a result, infrared FTS spectrophotometric measurements are often subject to background thermal emission errors that need to be corrected for [23]. If we again consider a simple transmittance measurement, we can see that the spectra recorded by the detector, S_1 for the reference measurement and S_2 for the sample measurement, will contain contributions from the detector itself and from the sample as well as from the source:

$$S_1 = S_{\text{source}} - S_{\text{det}}, \tag{4.22}$$

$$S_2 = \tau S_{\text{source}} - \tau S_{\text{sample}} - \tau^2 S_{\text{det}}, \tag{4.23}$$

where τ is the sample transmittance. Note that the signals that originate from the output side of the interferometer are negative since their modulation is out of phase by π with the interferogram originating from the source. The extra terms in Eqs. (4.22) and (4.23) will cause an error if the transmittance is calculated simply as S_2/S_1. However, they can be subtracted out by performing an additional measurement for both the sample and reference with the source off, or blocked, assuming that this does not affect the temperature of the sample or detector. The background corrected transmittance can then be derived from

$$\tau = \frac{S_2(\text{source on}) - S_2(\text{source blocked})}{S_1(\text{source on}) - S_1(\text{source blocked})}. \tag{4.24}$$

For an FTS instrument working at room temperature, the error terms in Eqs. (4.22) and (4.23) will be significant for wavenumbers below 1500 cm^{-1} if either the sample or detector is far from room temperature. An example of this effect is shown in Fig. 4.15 which displays the measured transmittance of a metallic neutral density filter at 77 K before and after correction using Eq. (4.24). In this case, a room temperature DTGS pyroelectric detector was used, whose black coating makes it an emitter with peak output near 700 cm^{-1}. The emission of the 77 K sample is very small in this spectral region. One can see that the extra factor of τ in the last term of Eq. (4.23) will cause a positive error relative to Eq. (4.22) in the ratio S_2/S_1, in agreement with the observed data. After correction, the transmittance spectrum shows the flat behavior expected for this filter from 800 to 2000 cm^{-1}. The relative error due to background thermal emission for an infrared FTS measurement can be as large as 20% at wavenumbers below 700 cm^{-1}.

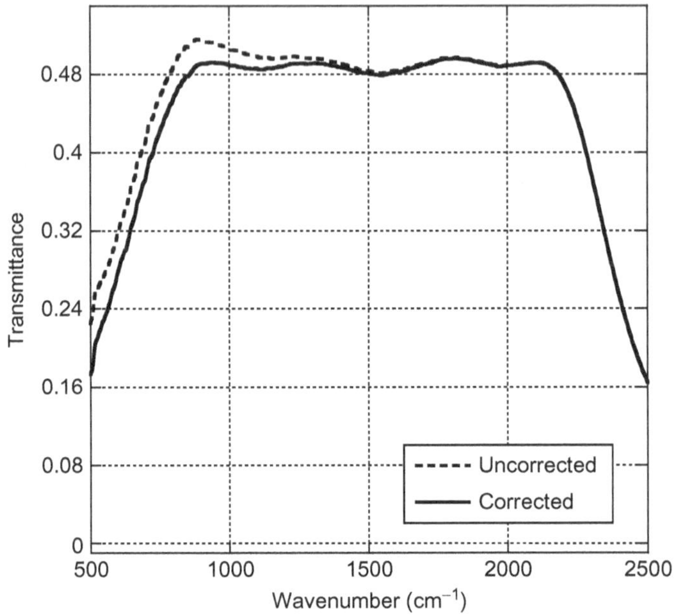

FIGURE 4.15 Measured transmittance of a metal-coated neutral density filter at 77 K using an infrared FTS before (dashed line) and after (solid line) correction for background thermal emission error.

The entrance aperture of an FTS as shown in Fig. 4.7 can also act as a secondary source of radiation, especially if heated significantly above ambient by the source beam that is focused onto it. The effective source diameter at lower wavenumbers can thus be much larger than the opening of the entrance aperture. This can cause errors due to overfilling of the sample, which normally is placed at an intermediate image plane of the source. It also can produce wavenumber errors since now the angular range of the beam in the interferometer is wavenumber dependent and the limiting resolution (Eq. 4.11) and wavenumber correction (Eq. 4.12) may become poorly defined. For these reasons, it is common practice to place a field stop at a convenient intermediate source image plane somewhere after the exit of the FTS to block radiation from the heated aperture.

4.3.7 Beam Geometry Errors

Changes in beam geometry at the detector plane between the sample and reference configurations are a common cause of errors in spectrophotometry whether using an FTS or other type of wavelength selective source. Displacement, deviation, or focus-shifting of the beam is produced by thick or wedged samples in a transmittance measurement, or by misalignment of the sample plane in a reflectance measurement. The resulting motion or change in shape or size of the irradiated spot on the detector can cause significant errors in the sample/reference ratio. In the worst case, the beam may start to move off the detector. While not specific to FTS systems, this error source is exacerbated by the tendency of instrument manufacturers to use the smallest possible detector element in order to optimize speed, SNR, and cost. Some detectors (such as Si, InSb, and some pyroelectrics) have good spatial uniformity across their active areas, showing less than 1% variation in responsivity. HgCdTe detectors on the other hand, can have as much as 80% spatial variation, with the maximum in responsivity unfortunately occurring near the edge on some detectors.

Transmittance measurements performed in the sample compartment of commercial FTS instruments will often show relative errors of several percent or more for sample thicknesses of greater than 2 mm due to beam geometry effects. Errors in specular reflectance measurements can be larger than this. Although this type of uncertainty is conceptually simple, it can be the most difficult and costly to reduce to the tenths of a percent level. Strategies used to reduce beam geometry-related errors include (1) larger, more spatially uniform detectors (carbon nanotube-coated pyroelectrics are a promising contemporary candidate), (2) adjusting the detector position between sample and reference measurements, (3) use of a spatial averaging device such as a diffuser or integrating sphere on the detector, (4) use of a nonuniformity corrected array detector to spatially average the beam as it moves between sample and reference measurements, and (5) better handling of sample alignment for reflectance, often with precision motorized stages and the use of laser/quadrant detector setups. Of course, improvements in source uniformity help as well. And the lowest possible uncertainty can end up being limited by the spatial uniformity of the sample itself.

A related type of spectrophotometric error arises from imprecise knowledge of the incident or exiting beam geometries for samples which are very sensitive to these parameters. For FTS systems, the uncertainty in beam geometry is increased by the use of off-axis focusing mirrors as well as by the annular beam shape due to the central portion of the optics being used for the metrology laser. Measurements of multilayer coated samples such as filters and dielectric mirrors in the infrared can be limited in accuracy by these factors.

4.3.8 Polarization Effects

Measurements of isotropic materials at off-normal incidence, or of anisotropic materials in general, are sensitive to beam polarization. If knowledge of the polarization state is critical, then polarizing optics are used to define the incident and/or exit beam states. Polarizing optics should have broadband coverage to be most useful with an FTS; it is not practical to tune a retarder to different wavelengths within the spectral region of interest and acquire multiple spectra. For many spectrophotometric measurements, polarization is not critical but can be a minor contributor to uncertainty.

While FTS instruments do not generally exhibit the strongly wavelength-dependent polarization that is seen with diffraction gratings, the multiple 45° mirror reflections and especially the coated beamsplitter typically result in 10–15% degree of linear polarization in the output of the spectrometer. This will usually result in less than 0.1% error for near-normal measurements of isotropic samples if it were assumed that the incident light was unpolarized. For measurements at greater than about 15° incidence, one should measure with two perpendicular polarizations and average the resultant spectra if the desired result is specified for unpolarized incident light.

4.3.9 Stray Light

Modulated radiation that is scattered or diffracted outside of the FTS and reaches the detector without having interacted with the sample in the proper way leads to spectrophotometric error. This generally results in a nonzero spectral baseline the same as would be seen with any type of spectrometer. However, light from the source that is scattered such that it does not pass properly through the interferometer is not modulated by the scanning mirror. It therefore just shows up as an increase in the dc background signal of the detector and does not produce an error in the spectral measurement.

This is in contrast to the situation with dispersive spectrometers, where source light diffusely scattered by the grating or other optics is recorded as signal at the wrong wavelength. It is thus generally easier to make high optical density measurements as low as 10^{-6} transmittance with an FTS system as long as nonlinearity and other error sources are well controlled. Such low-level measurements must however be checked against a configuration that should yield zero spectral signal (for instance an opaque reference sample) in order to assess and possibly correct for stray light contributions.

4.3.10 Atmospheric Absorption

As discussed in Section 4.2.5, atmospheric absorption lines do not usually cause significant spectrophotometric errors (>0.1%) in well-purged or evacuated FTS instruments in the mid- to near-infrared spectral regions. The variation in spectral baseline over time can be evaluated in the neighborhood of the strong CO_2 and H_2O absorption lines in order to assess their contribution to the overall uncertainty of the measurement. Some reduction in SNR and spectral distortions can occur over these wavenumbers. Measurement of sample spectral lines of similar width to strong atmospheric lines in the same spectral region should performed under vacuum conditions.

Stable residual carbon dioxide or water vapor absorption can produce a bias in the results if there is a significant difference in average optical path length between the sample and reference measurements, as for samples with large optical thickness. For setups where there is a large mechanical motion of optical elements between the sample and reference measurements, it is important that the atmosphere within the purge enclosure be well mixed and uniform.

4.3.11 Wavenumber Uncertainty

The wavenumber scale of an FTS measurement needs to be corrected for beam divergence and difference in alignment between the metrology laser and the signal beam. For measurements where the wavenumber uncertainty must be less than or equal to the limiting resolution from Eq. (4.11), it is wise to check the scale for both the sample and reference configurations by measuring residual atmospheric absorption lines within the instrument beam path. In principle, this should probe the same beam path as used in the spectrophotometric measurement and provide a valid wavenumber correction. An example of this procedure is shown in Fig. 4.16 for measurement of carbon dioxide absorption lines near 2350 cm^{-1} at 0.25 cm^{-1} spectral resolution. The observed positions are seen to vary linearly with the expected positions from handbook data [24] with a correction factor of 0.99987 and residuals of less than 0.01 cm^{-1}. It is thus fairly easy to establish that the wavenumber scale of an instrument is sufficiently accurate in principle for most spectrophotometry of solid materials.

However, the wavenumber uncertainty for a given sample spectral feature in FTS spectrophotometry is usually dominated not by the intrinsic scale of

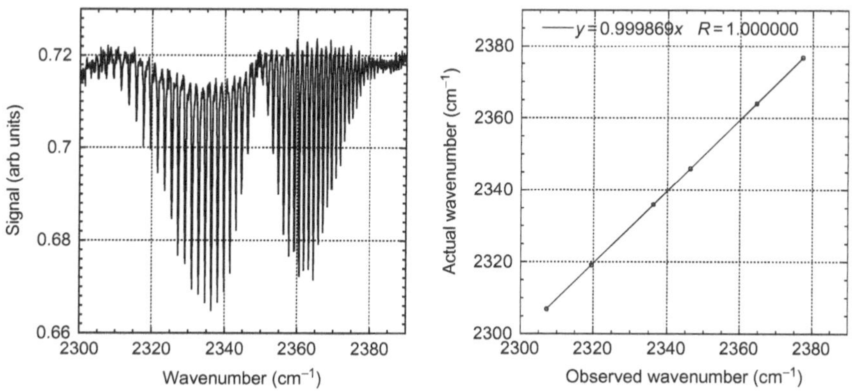

FIGURE 4.16 Measurement of CO_2 absorption lines in the mid-infrared (left frame) using an FTS at 0.25 cm^{-1} spectral resolution. A linear fit of the observed line positions to handbook values (right frame) yields a wavenumber correction factor of 0.99987 with residuals of less than 0.01 cm^{-1}.

the instrument but by other things such as spectral distortion due to detector nonlinearity, phase error, noise, alignment drift, Fabry–Perot effects, other nearby spectral features, or sensitivity of the sampled quantity to beam geometry, temperature, or other physical factors. Considerable work has been done in assessing the accuracy of spectral definition that can be achieved in practice. A general conclusion [25] is that the wavenumber accuracy for a given measurement should be checked using a reference sample with similar width features in the same spectral region as the sample under test. A very popular reference material for the mid-infrared is polystyrene film, which can be obtained as a calibrated reference standard [26] designed to fit the sample compartment of most commercial FTS systems.

4.4 MEASUREMENT APPLICATIONS

4.4.1 Measurement of Transmittance and Reflectance

Many commercially available accessories fit into the sample compartment of an FTS to enable the user to make transmittance or reflectance measurements in various regular or diffuse geometries. Reflectance accessories can be designed for either relative measurements, using a reference mirror in substitution mode, or absolute measurements, with a mirror that moves between two configurations for the sample and reference spectra. Some manufacturers have also modified microscopes with all-reflective optics to enable sampling areas of less than 100 μm, and even have options for adding focal plane array HgCdTe or other multi-element detectors to enable hyperspectral imaging of samples in the infrared. While the less expensive (<$5K) of these assemblies are useful for qualitative spectroscopic work, they tend to lack the precise, repeatable control of sample and reference positioning that is needed to avoid large beam geometry errors as discussed in the previous section. Relative uncertainties due to beam geometry error with these devices are often several percent or more.

One option for high-accuracy infrared transmittance and reflectance measurements with an FTS is the use of integrating spheres, even for specular sample geometries. The low throughput of the sphere (less than 0.5% for many designs in the infrared) does reduce the available dynamic range. However, this attenuation of the input flux enables a 77 K HgCdTe detector to be used in its linear response regime. The spatial averaging characteristics of the sphere greatly reduce beam geometry errors between the sample and reference measurements. A side-mount 150 mm diameter sphere is used at NIST [27] for transmittance and reflectance measurements with less than 0.3% expanded uncertainty. Variable-angle infrared reflectance measurements using a goniometric instrument may not be able to afford the throughput loss of a sphere on the detector and can be limited in accuracy by the spatial uniformity of available detectors [28]. A key element of these high-accuracy instruments is precise alignment of the sample and other optics using motorized translation stages with less than 25 μm uncertainty and rotation stages with less than 0.005° uncertainty. Such setups are expensive ($50K or more) and usually too big to fit inside a typical FTS sample compartment.

Measurements of low-level infrared τ or ρ ($\leq$0.0001) at moderate spectral resolution with an FTS, making use of the multiplexing advantage, are also possible as long as the detector nonlinearity error is controlled. One way to do this is with neutral density filters whose transmittance can be measured accurately with a linear, low sensitivity detection system, and then used as reference samples for higher optical density measurements with a cooled photon detector. Out-of-band filter infrared transmittance can then be measured over six decades of dynamic range with appropriate blocking filters in place [29].

4.4.2 Measurement of Refractive Index

The refractive index of a transparent material can be determined from a high-resolution transmittance measurement of an etalon sample, which exhibits Fabry–Perot fringes with interference maxima at wavenumbers

$$\sigma_m = \frac{m}{2dn(\sigma_m)}, \tag{4.25}$$

where m is the fringe order, d is the sample thickness, and $n(\sigma_m)$ is the refractive index at the fringe maximum. High-resolution measurements over a wide spectral range such as this are good applications of the FTS advantages described in Section 4.2.4, and there is a long history of FTS refractive index measurements from the far-infrared through the vacuum ultraviolet. With samples of $d \approx 5$ mm, it is possible to combine precision thickness measurements with high-resolution spectral measurements in a well-collimated beam to obtain uncertainties in n of order 0.00001 [11,30].

The drawbacks of this method compared to traditional prism refractometry include the need for $\lambda/10$ sample flatness, correction for thermal expansion, and determination of the integer fringe order m. Various methods for determining m from spectral fitting work for some materials, especially in the far-infrared where n approaches a constant at wavenumbers below the lattice vibrations and clearly $m=0$ at $\sigma_m=0$. However, in general, it is necessary to know n at least for one value of σ_m in the spectral range of interest to determine m, which often means using data from an absolute prism refractometry measurement at this wavenumber. Industry continues to favor the more direct approach of minimum-deviation prism refractometry for the most critical work.

4.4.3 Assessment of the Accuracy of FTS Measurements

Despite the plethora of sources of uncertainty described above, it is possible to perform infrared spectrophotometric measurements with an FTS that

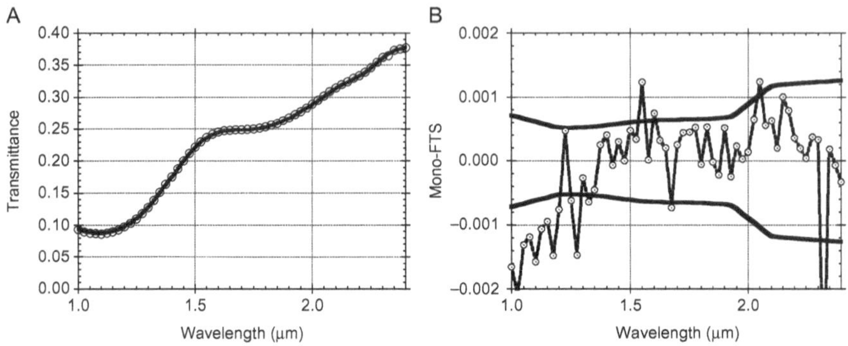

FIGURE 4.17 Comparison of regular transmittance measurements of a 2-mm thick absorbing glass filter made with an FTS/integrating sphere system and a monochromator-based system. The monochromator data are shown as the open circles in frame (a) and the FTS data as the solid line. The solid lines in frame (b) show the combined expanded uncertainties ($k=2$) of the two measurements, and the symbols the difference between the two measurements.

approach the uncertainty level that has traditionally been claimed by National Metrology Institutes (NMIs) using dispersive instruments, that is, about 0.1%. An example of an intercomparison of regular transmittance measurements in the 1–2.5 μm spectral region performed with both an FTS/integrating sphere system and a grating monochromator system at NIST [31] is shown in Fig. 4.17. The sample is a 2-mm thick absorbing glass filter designed for 10% transmittance in the visible. Most of the data points in the difference graph shown in Fig. 4.17b fall within the combined expanded uncertainties of the two measurements. Similar results have been obtained for both transmittance and reflectance of other specular samples in the 0.99–0.01 range. Comparisons have also been made with other laboratories that tend to confirm the FTS system uncertainty [32].

4.5 RECOMMENDATIONS FOR ACCURATE FTS MEASUREMENTS

4.5.1 Matching Instrument to Measurement Requirements

The chief criteria in selecting an FTS system are spectral range, spectral resolution, stability, and flexibility of input/output options. If vacuum is needed for far-infrared or very high-resolution measurements, then this will necessarily drive the user to select a "research-grade" instrument. However, for lower resolution measurements over the mid-infrared spectral region, the main deciding factors may be stability, ability to use an external sample and detector setup, and cost. A number of available systems are capable of meeting these basic requirements and the final uncertainty that can be achieved will depend upon how the measurements are configured. Below is a brief list of recommendations for making accurate FTS spectrophotometry measurements that summarizes many of the points discussed above.

1. Select a detector for appropriate wavelength coverage and dynamic range. Pyroelectrics are good for general use, when tau or rho are in the range 1–99%. Infrared photoconductors are generally nonlinear with standard FTS configuration, but can be used to expand the dynamic range with appropriate beam attenuation or nonlinearity correction. Consider detector size and spatial uniformity relative to the expected variations in incident beam geometry that need to be accommodated.
2. Eliminate interreflections among the sample and other instrument components.
3. Test for thermal background effects, especially when sample or detector is far from ambient temperature.
4. Always correct and check the phase. Use double-sided interferograms when possible. Avoid unnecessary apodization of the interferogram—this leads to distorted line shapes and loss of resolution.
5. Find the best measurement scheme to optimize SNR with given system stability. Sometimes, this means lots of repeated scans, sample-reference-sample, etc.
6. Test linearity and overall system accuracy with appropriate reference samples—for example, glass, Si, Ge, gold mirror, etc.
7. Check the wavenumber scale with atmospheric absorption lines, if resolution is sufficient. Otherwise, polystyrene, rare-earth-doped glass or other artifact wavelength standards are useful.

4.5.2 Instrument Software and Data Handling

Commercially available FTS systems typically come with a microcomputer and proprietary software for automating the spectrometer, setting spectral scan parameters, and collecting the data. Various options for processing and displaying the data are also usually included. While many software packages are fairly comprehensive, some users may want to export the data and use their own analysis routines. For this reason, it is important to have access to the raw, unprocessed interferograms. It is also helpful if the software keeps a good record of all the instrument settings for a given measurement, and what processing was done to the data. Finally, most programs allow some kind of real-time display of the interferogram or spectrum as the instrument is scanning. This is very helpful for alignment and general troubleshooting of a measurement.

4.5.3 Maintenance of Measurement Traceability

Like any piece of precision laboratory equipment, FTS systems need periodic maintenance including alignment of the optics and occasional replacement of sources, the metrology laser, beamsplitter, or electronics. Mid-infrared beamsplitters are usually made of soft, hygroscopic materials like KBr or CsI. Repeated removal and reinsertion of these optical elements can eventually cause stress damage and even delamination of the coatings. Also, the material will become foggy over years of use unless kept in an extremely dry purge environment or vacuum at all times.

Although accurate spectrophotometry does not rely on stability of the absolute radiometric response of an instrument over a time scale of years, changes in FTS system responsivity can indicate alignment problems or other issues that need to be resolved so they do not add additional uncertainty to the measurements. It is a good idea to keep records of the detector ZPD voltages for all relevant settings of entrance aperture, scan speed, source, beamsplitter, and detector to have a baseline to compare against if problems are found. Spectra should be collected and archived whenever a beamsplitter is replaced with a new or refurbished one. The performance of an alkali halide beamsplitter at high wavenumbers (above $1500\ cm^{-1}$) will usually degrade over a period of years, and if it drops by more than 50% it may be time to replace it.

It is a good idea to keep a small set of reference samples on hand for periodic checking of the instrument performance, especially before and after important sets of measurements which require uncertainties close to the best that can be achieved with that instrument.

REFERENCES

[1] A.A. Michelson, Visibility of interference fringes in the focus of a telescope, Philos. Mag. 31 (190) (1891) 256–259.

[2] S.P. Davis, M.C. Abrams, J.W. Brault, Fourier Transform Spectrometry, Academic Press, San Diego, 2001.

[3] P.R. Griffiths, J.A.D. Haseth, Fourier Transform Infrared Spectrometry, second ed., Wiley, Hoboken, 2007.

[4] A.A. Michelson, E.W. Morley, On the relative motion of the Earth and of the luminiferous ether, Sidereal Messenger 6 (1887) 306–310.

[5] K.M. Evenson, J.S. Wells, F.R. Peterson, B.L. Danielson, G.W. Day, R.L. Barger, J.L. Hall, Speed of light from direct frequency and wavelength measurements of the methane-stabilized laser, Phys. Rev. Lett. 29 (19) (1972) 1346–1349.

[6] A.A. Michelson, On the application of interference-methods to spectroscopic measurements, Philos. Mag. 31 (191) (1891) 338–346.

[7] L. Mertz, Auxiliary computation for Fourier spectrometry, Infrared Phys. 7 (1) (1967) 17–23.

[8] M.L. Forman, H. Steel, G.A. Vanasse, Correction of asymmetric interferograms obtained in Fourier spectroscopy, J. Opt. Soc. Am. 56 (1) (1966) 59–61.

[9] P. Fellgett, Theory of multiplex interferometric spectrometry, J. Phys. Radium 19 (1958) 187–191.

[10] U. Griesmann, R. Kling, J.H. Burnett, L. Brasatz, R.A. Gietzen, The NIST FT700 vacuum ultraviolet Fourier transform spectrometer: applications in ultraviolet spectrometry and radiometry, Proc. SPIE 3818 (1999) 180–188.

[11] R. Gupta, S.G. Kaplan, High accuracy ultraviolet index of refraction measurements using a Fourier transform spectrometer, J. Res. Natl. Inst. Stand. Technol. 108 (2003) 429–437.

[12] G.P. Williams, C.J. Hirschmugl, E.M. Kneedler, E.A. Sullivan, D.P. Siddons, Y.J. Chabal, F. Hoffman, K.D. Moeller, Infrared synchrotron measurements at Brookhaven, Rev. Sci. Instrum. 60 (7) (1989) 2176–2178.

[13] A. Rogalski, History of infrared detectors, Opto-Electron. Rev. 20 (3) (2012) 279–308.

[14] P.R. Griffiths, C.C. Homes, Instrumentation for Far-Infrared Spectroscopy, John Wiley and Sons, Chichester, UK, 2006.

[15] M.J. Padgett, A.R. Harvey, A static Fourier-transform spectrometer based on Wollaston prisms, Rev. Sci. Instrum. 66 (4) (1995) 2807–2811.

[16] J.R. Birch, F.J.J. Clarke, Fifty categories of ordinate error in Fourier transform spectrometry, Spectrosc. Europe 7 (1995) 16–22.

[17] S.G. Kaplan, L.M. Hanssen, R.U. Datla, Testing the radiometric accuracy of Fourier transform infrared transmittance measurements, Appl. Opt. 36 (1997) 8896–8908.

[18] T. Hirschfeld, Quantitative IR: a detailed look at the problems involved, in: Fourier Transform Infrared Spectroscopy: Applications to Chemical Systems, vol. 2, Academic Press, New York, 1979, p. 193.

[19] D.B. Chase, Nonlinear detector response in FT-IR, Appl. Spectrosc. 38 (1984) 491–494.

[20] M.I. Flik, Z.M. Zhang, Influence of nonequivalent detector responsivity on FT-IR photometric accuracy, J. Quant. Spectrosc. Radiat. Transf. 47 (1992) 293–303.

[21] E.D. Palik, Handbook of Optical Constants of Solids, Academic Press, Orlando, 1985.

[22] H.W.H.M. Jongbloets, M.J.H.V.d. Steeg, E.J.C.M.V.d. Werf, J.H.M. Stoelinga, P. Wyder, Spectrum distortion in far-infrared Fourier spectroscopy by multiple reflections between sample and Michelson interferometer, Infrared Phys. 20 (3) (1980) 185–192.

[23] J.R. Birch, E.A. Nicol, The removal of detector port radiation effects in power transmission or reflection Fourier transform spectroscopy, Infrared Phys. 27 (1987) 159–165.

[24] G. Guelachvili, K.N. Rao, Handbook of Infrared Standards II: With Spectral Coverage of 1.4 um-4 um and 6.2 um-7.7 um, Academic Press, San Diego, 1993.

[25] L.M. Hanssen, C. Zhu, Wavenumber standard for mid-infrared spectrometry, in: Handbook of Vibrational Spectroscopy, John Wiley and Sons, Chichester, 2002.

[26] D. Gupta, L. Wang, L.M. Hanssen, J.J. Hsia, R.U. Datla, Standard Reference Material: Polystyrene Films for Calibration of the Wavelength Scale of Infrared Spectrometers—SRM 1921, NIST Special Publication 260–122, Government Printing Office, Washington, DC, 1995.

[27] L.M. Hanssen, Integrating-sphere system and method for absolute measurement of transmittance, reflectance, and absorptance of specular samples, Appl. Opt. 40 (19) (2001) 3196–3204.

[28] S.G. Kaplan, L.M. Hanssen, Silicon as a standard material for infrared reflectance and transmittance from 2 to 5 μm, Infrared Phys. Tech. 43 (6) (2002) 389–396.

[29] Z.M. Zhang, L.M. Hanssen, R.U. Datla, High-optical-density out-of-band spectral transmittance measurements of bandpass filters, Opt. Lett. 20 (9) (1995) 1077–1079.

[30] D.F. Edwards, E. Ochoa, Infrared refractive index of silicon, Appl. Opt. 19 (1980) 4130–4131.

[31] S.G. Kaplan, L.M. Hanssen, E.A. Early, M.E. Nadal, D. Allen, Comparison of near-infrared transmittance and reflectance measurements using dispersive and Fourier transform spectrophotometers, Metrologia 39 (2) (2002) 157.

[32] C.J. Chunnilall, F.J.J. Clarke, M.P. Smart, L.M. Hanssen, S.G. Kaplan, NIST–NPL comparison of mid-infrared regular transmittance and reflectance, Metrologia 40 (1) (2003) S55.

Chapter 5

Regular Reflectance and Transmittance

Peter A. van Nijnatten
OMT Solutions BV, Eindhoven, Netherlands

Chapter Outline

5.1 Introduction 144
5.2 Relevant Background Information 144
5.2.1 Measurement Geometry 144
5.2.2 Measurement Equation 145
5.2.3 Evaluation of Measurement Uncertainty 146
5.2.4 The Role of Integrating Spheres 147
5.2.5 Avoiding Spectral Artifacts 148
5.3 Measurements Near-Normal Incidence 150
5.3.1 Typical Configurations for Near-Normal Reflection Measurements 151
5.3.2 Measuring the Transmittance and Reflectance of Coated Substrates 153
5.3.3 Measuring Very Low Reflectance Materials 153
5.3.4 Measuring Specular Reflectance with an FTIR 154
5.3.5 Tools for Absolute Reflectance and Transmittance Measurements 156
5.4 Measurements at Oblique Incidence 158
5.4.1 Relevant Issues 159
5.4.2 Directional Transmittance Measurements 163
5.4.3 Relative Directional Reflection Measurements 166
5.4.4 Absolute Directional Reflection Measurements 169
5.5 Measuring the Reflectance of Highly Reflecting Materials 171
5.5.1 VW Method 172
5.5.2 IV Method 175
References 177

Experimental Methods in the Physical Sciences, Vol. 46. http://dx.doi.org/10.1016/B978-0-12-386022-4.00005-4

5.1 INTRODUCTION

Reflectance and transmittance measurements of non-scattering samples are probably the most common measurements in optical spectroscopy. Applications are found in the glass and coating industry [1,2], as well as in characterization of optical elements, like filters and mirrors [3,4], and optical materials for display technology, solar cells, and optical solar collectors. Regular reflectance and transmittance measured near-normal incidence form a basis for many standardized procedures for calculating solar energy and daylight properties [5,6].

In recent years, the need for reliable measurement techniques for directional optical properties is growing. Fortunately, in the past two decades, considerable progress has been made in identifying and overcoming the sources of error in measurement of transmittance and reflectance as a function of the angle of incidence, wavelength, and polarization. Variable angle spectrophotometry provides a means for analysis that yields the thickness and optical constants of the individual layers in multilayer coatings, as well as other parameters that can be related to optical material properties [7,8].

This chapter offers a state-of-the-art overview and discusses the various methods and associated measurement problems and commercially available solutions in detail. A topic that is given special attention is the overview of the relatively new methods that have been developed for the calibration of reference mirrors and characterization of laser mirrors at oblique incidence.

5.2 RELEVANT BACKGROUND INFORMATION

5.2.1 Measurement Geometry

Generally speaking, the regular, specular reflectance of a sample refers to the mirror-like surface reflection of light from a single incoming direction into a single outgoing direction in which both these directions are under the same angle with respect to the surface normal of the sample. The regular transmittance refers to the transmission of rays from a single incoming direction into a single outgoing direction in accordance with Snell's law of refraction. In case of a plan-parallel sample, the outgoing and incoming directions are the same, although the optical path of the transmitted beam is shifted due to refraction in the sample.

In a commercial spectrophotometer, the beam that interacts with the sample is usually not very well collimated, and a beam divergence on the order of 2–3° is not uncommon. The area illuminated by the beam lies on the order of 50 mm^2 and may vary depending on wavelength and instrument settings. Under these conditions, the angle of incidence is the average of incident directions spanning a total solid angle and the illuminated area on the sample. So, if we say that we are measuring the transmittance at 0° incidence, we actually obtain the transmittance for an average angle of incidence that is probably around 1–2°, depending on how the energy in the beam is distributed over the various directions in the beam. Fortunately, this small deviation from zero angle of incidence generally results in a negligible difference in the normal transmittance of specular samples. For example, the difference between the unpolarized transmittances at 0° and 2° of a transparent sample with a refractive index of 1.5, typical for window glass, is less than 0.001%.

The geometry of the reflected and transmitted beams in terms of detected solid angle and sample area corresponding to the reflectance and transmittance is generally not the same as the geometry of the incident beam. The detected solid angle is only exactly the same if the sample surface is perfectly flat and, in case of transparent samples, if both back and front surfaces are perfectly flat and parallel. If the sample is transparent, the interreflections increase the area on the sample from which the reflected and transmittance beams emerge.

The biconical configuration with small solid angles according to the generalized statement given by Eq. (2.76) in Chapter 2 is accurate for all cases and can be simplified only for flat plan-parallel samples. The practical importance of such a statement lies in its usefulness to help us to understand the outcome of a measurement and, more importantly, the differences that typically occur when measuring with different instruments (with different shape and size of the beam and detector properties) or even with different detectors in the same instrument.

5.2.2 Measurement Equation

The measurement equation is a function describing the measurement process in terms of all quantities that can contribute a significant uncertainty to the measurement result [9]. It forms the basis for the evaluation of measurement uncertainty as we discuss in Section 5.2.2.

To determine the transmittance τ of a sample with a spectrophotometer, three different instrument readings are combined and corrected according to the measurement equation

$$\tau(\lambda) = C_A(\lambda) C_N(\lambda) T(\lambda), \tag{5.1}$$

where

$$T(\lambda) = \frac{M_S(\lambda) - M_0(\lambda)}{M_{100}(\lambda) - M_0(\lambda)} \tag{5.2}$$

is the measured sample transmittance, C_A and C_N are correction factors related to alignment and detector nonlinearity, respectively, M_S is the instrument reading obtained with the sample placed in the beam, M_{100} is the reading without the sample (100% transmittance), and M_0 is the reading measured with the sample beam blocked (0% transmittance) to determine the stray radiation level.

The misalignment correction factor C_A accounts for the difference in the geometry of the detected beam with and without the presence of the sample. In practice, the multipliers C_A and C_N are assumed to have a best estimated value of one with a given standard uncertainty (u_A and u_N, respectively), and T (or an average of a number of measurements of T) provides the best estimate of the sample transmittance. The readings M_{100} and M_0 are often taken only once or twice a day to determine the so-called instrument background or baseline correction. Sample measurements are automatically corrected after each reading of M_S by the software controlling the instrument to obtain the transmittance. To complete the measurement equation, other multipliers ("correction factors") can be added to represent other influences that may apply, for example, the uncertainties in wavelength and polarization.

A similar procedure is used for reflectance measurements. In the case of a relative measurement, M_{100} is replaced by an instrument reading M_R obtained on a reference mirror and the measurement equation becomes

$$\rho(\lambda) = C_A(\lambda) C_N(\lambda) R(\lambda), \tag{5.3}$$

where

$$R(\lambda) = \frac{M_S(\lambda) - M_0(\lambda)}{M_{100}(\lambda) - M_0(\lambda)} \rho_{ref}(\lambda) \tag{5.4}$$

is the measured sample reflectance, and $\rho_{REF}(\lambda)$ is the reflectance spectrum of the reference mirror, which has been obtained by calibration.

5.2.3 Evaluation of Measurement Uncertainty

The most commonly used procedure for determining measurement uncertainty is described in the ISO document "Guide to the Expression of Uncertainty in Measurement" (GUM) [9]. This procedure starts with determining the contributions of the various corrections and instrument readings to the uncertainty of the measured property. The measurement equation plays a key role in this and is only complete if it contains all variables that contribute to the measurement result. It is for this reason that correction factors like C_A and C_N are introduced. Even though an expectation value equal to one is assumed for these factors, which does not affect the outcome of the measurement, the uncertainty in this value does contribute to the measurement uncertainty.

In the evaluation of the measurement uncertainty, each contribution is represented by a standard uncertainty, which can be considered to be an estimate of its standard deviation. Similarly, the square of a standard uncertainty can be considered as an estimate of its variance. By propagating the variances of the components through the measurement equation, the combined measurement uncertainty u_C is given as the square root of the resulting variance,

$$u_C = \sqrt{(u_M)^2 + (u_A)^2 + (u_N)^2 + \ldots} \tag{5.5}$$

In Eq. (5.5), u_M, u_A, and u_N are the standard uncertainties related to the instrument readings, misalignment, and detector nonlinearity, respectively. More contributions related to other effects, such as wavelength uncertainty, angular uncertainty, polarization uncertainty, and reference mirror calibration uncertainty, can be added when applicable.

For the standard uncertainty related to the instrument readings, generally the best estimate of the standard deviation of the mean is taken, which is calculated from the standard deviation $s(x)$ of N readings of the measured property x (in our case to be replaced with T or R) using

$$u_M = t_p(N-1)\frac{s(x)}{\sqrt{N}}, \tag{5.6}$$

where $t_p(N-1)$ is the Student-t factor for $N-1$ degrees of freedom and a confidence level of 68.27%.

Detector nonlinearity produces a systematic uncertainty component that, in principle, can be corrected [3,10]. This correction requires a thorough investigation of the instrument in use. Investigation of high-end commercial spectrophotometers has shown that, when a measurement is made with a proper baseline correction, to obtain a scale with values between 0% and 100%, the nonlinearity error in a measured property m with values between 0 and 1 (like τ or ρ) depends on the value of m and can be approximated by the parabolic equation [3]

$$u_N(m) = 4m(1-m)u_{PH}, \tag{5.7}$$

where u_{PH} is equal to half the photometric accuracy as specified by the instrument manufacturer.

The combined standard uncertainty of Eq. (5.5) is multiplied by a coverage factor k to obtain the expanded uncertainty U. The expanded uncertainty provides an interval $\tau - U$ to $\tau + U$ about the result τ, within which the value of τ can be asserted with a high level of confidence. The value $k=2$, corresponding to a confidence level of approximately 95%, is commonly used.

5.2.4 The Role of Integrating Spheres

Integrating spheres are typically intended to measure diffuse samples and are therefore discussed in detail in Chapter 6. However, it is also common to use an integrating sphere to measure coated and uncoated glass samples that do not have diffusing properties. Putting a glass sample into the beam of a spectrophotometer changes the optical geometry of the beam after interaction with the sample. As a result, the spatial energy distribution and shape of the image on the detector are also changed. This is also true for perfectly flat samples with perfectly parallel surfaces under normal incidence where the change in the beam after the sample is created mostly by the change in optical path length, caused by the refractive index of the glass being higher than that of air.

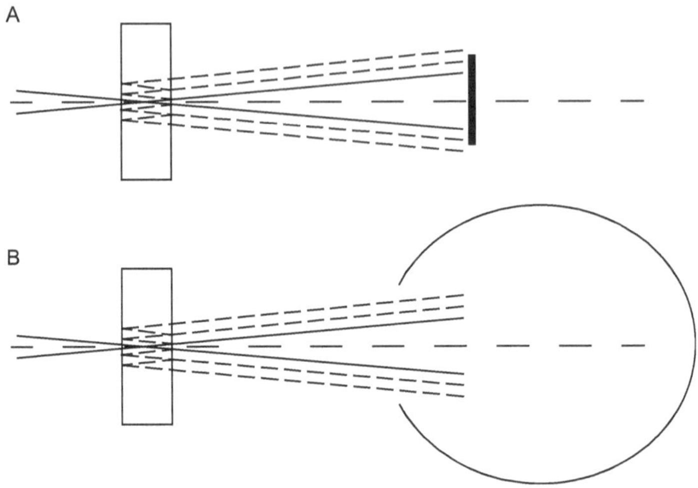

FIGURE 5.1 Effect of the choice of the type of detector on the measurement of transmittance. (a) Using a standard detector, secondary beams (dashed) are usually only partially measured. (b) Using an integrating sphere, all beams are properly collected.

Furthermore, secondary beams occur due to multiple reflections in the sample between the front and back surface. These secondary beams project progressively enlarged multiple images on the detector surface, which can cause systematic errors [11,12]. This is illustrated in Fig. 5.1. In reflection, these multiple images are progressively shifted, thereby increasing the error.

In the case of uncoated glass with a refractive index close to 1.5, these secondary beams represent only about 0.2% of the total signal, but their contribution for coated glass can be in the order of 1% for single-sided coated glass and several percent for double-sided coated glass. In the case of drawn sheet glass and curved samples, additional systematic errors occur by beam distortion and deflection as a result of non-flatness, surface topology (non-smoothness), and the opposite surfaces not being parallel. By using an integrating sphere, almost all transmitted (or reflected) radiation is collected and measured, and the remaining error due to these effects can become negligible.

Although specular samples can be positioned directly against the reflection or transmission port of an integrating sphere in the same way as diffusing samples, they are often placed in a separate sample compartment with additional optics between the sample and the integrating sphere, so that the sphere is only used as a uniform underfilled detector to capture the beam after sample interaction.

5.2.5 Avoiding Spectral Artifacts

Spectral artifacts are features in the measured spectrum that are not related to the measured sample property but mostly due to the instrument settings, instrument properties, and measurement geometry. In this section, we discuss some of the most common spectral artifacts and how to avoid them.

5.2.5.1 Changeover Points

In dispersive spectrophotometers, a scan over a wide wavelength range typically contains several wavelength points at which the instrument exchanges one or more components, such as filters, gratings, or detectors. If a step or jump occurs in a measured spectrum that cannot be contributed to the properties of the sample, the first thing, we have to check is whether the wavelength at which this step or jump occurs coincides with a changeover point.

A well-known changeover point in an ultraviolet/visible/near-infrared (UV/Vis/NIR) spectrophotometer is the wavelength at which the instrument switches from a NIR detector, typically a PbS or InGaAs photodetector, to a UV/Vis detector, typically a Si photodetector or photomultiplier tube (PMT). The detector changeover point is often at a wavelength for which both detectors are at the end of their usable wavelength range and have low sensitivity. For a PbS/PMT combination, the changeover point is typically set around 860 nm. For an InGaAs/PMT combination, it can be set to shorter wavelengths.

A typical problem is a jump at the detector changeover point due to the poor sensitivity of the PMT for longer wavelengths. This can be caused by a high gain setting of the detector electronics needed to cope with the lower signal but also amplifies stray-light. The same amount of stray-light should be measured when the beam is blocked (0% transmittance). Therefore, a proper baseline correction should include the 0% transmittance or reflectance measurement to avoid this jump.

Another important changeover point is the grating changeover wavelength, which for a UV/Vis/NIR spectrophotometer is usually chosen to coincide with the detector change. A grating change can lead to a large change in the beam polarization, as well as wavelength resolution. Both can have an impact on the response of the sample to the beam and lead to spectral artifacts. This effect we discuss in more detail in Sections 5.2.5.2 and 5.2.5.3.

5.2.5.2 Slit and Aperture Settings

Both Fourier transform infrared (FTIR) and dispersive spectrophotometers have apertures that determine the amount of radiant flux in the beam, the spectral resolution, or the shape of the area of the beam projected on the sample. Typically, the groove density of a UV/Vis grating is four times higher than a NIR grating. In that case, a wavelength bandwidth of 5 nm in the UV/Vis region corresponds to the same physical slit width as a wavelength bandwidth of 20 nm in the NIR region. Since the slit is optically imaged onto the sample, the ratio of the bandwidth in the UV/Vis and NIR regions should be 1:4 in order to keep the spot size constant during the measurement.

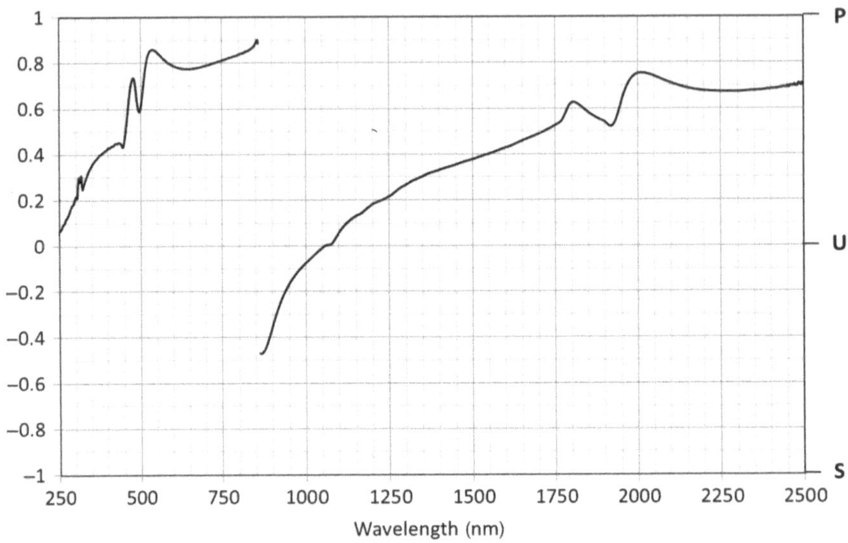

FIGURE 5.2 Degree of polarization $(\Phi_p - \Phi_s)/(\Phi_p + \Phi_s)$, where Φ_p and Φ_s are the radiant flux in p- and s-polarization, respectively, as function of the wavelength for a typical double monochromator UV/Vis/NIR spectrophotometer.

This means that if the requirement for maximum wavelength bandwidth is, for instance, 10 nm over the whole region, the resolution in the NIR should be set at 10 nm and the UV/Vis resolution to 2.5 nm to keep the spot size constant over the entire wavelength range.

5.2.5.3 Polarization

Without precautions, the light incident upon the sample in a dispersive spectrophotometer may be strongly polarized, with the degree of polarization dependent upon wavelength. This behavior is illustrated in Fig. 5.2, which shows the degree of polarization for a typical instrument as a function of wavelength. For most wavelengths, p-polarization is dominant, except at the low end of the NIR region where there seems to be a larger degree of s-polarization. What is clear from Fig. 5.2 is that, during the grating change-over, there is a large jump in polarization. If samples are measured that are polarization sensitive, we can therefore expect a jump in the spectrum and significant measurement errors, if the beam is not depolarized first.

5.3 MEASUREMENTS NEAR-NORMAL INCIDENCE

There is a wide range of spectrophotometer accessories available for the measurement of reflectance near-normal incidence. In this section, we discuss various applications and the accessories that are available to measure reflectance and transmittance near-normal incidence. Near-normal incidence, we assume that the angle of incidence is small enough that we can neglect the effect of polarization. Generally, angles of incidence up to 10° are regarded as near-normal. For FTIR spectrophotometers, the beam tends to be much less polarized than that of dispersive instruments, and angles of incidence up to 15° are often regarded as near normal for them.

5.3.1 Typical Configurations for Near-Normal Reflection Measurements

Figure 5.3 shows an overview of typical configurations for the measurement of specular reflectance at near-normal incidence. The most basic tool for this purpose is a small accessory that uses a mirror to redirect the sample beam of the spectrophotometer toward the sample and a second mirror to direct the reflected beam back into the original direction toward the detector of the instrument. This tool, schematically depicted in Fig. 5.3a, has no moving parts and it provides a relative measurement using a calibrated mirror as reference. Ideally, such a device should have a three-point sample support to make sure that both sample and reference have their front surfaces in exactly the same plane during measurement to eliminate alignment errors. If properly designed, this tool is often the most reliable and accurate, with practically zero alignment error, and its accuracy is mostly dependent on the calibration uncertainty of the reference mirror.

Although integrating spheres, schematically depicted in Fig. 5.3b, are originally developed as a tool for reflectance measurements of diffuse samples, they are widely used for specular samples as well. This is also a relative measurement for which a reference mirror has to be used for calibrating the ordinate scale.

The configuration shown in Fig. 5.3c uses a moving detector to provide for an absolute measurement. Ideally, the detector is an integrating sphere to compensate for the difference in spatial radiant flux distribution on the detector between the reference and sample measurement. This difference is caused by the fact that the beam during the sample measurement is flipped over with respect to that during the reference measurement, thereby providing a mirrored image on the detector. Since the beam of a commercial spectrophotometer is far from uniform, a uniform detector is crucial for obtaining accurate results. The same configuration can be used for directional reflectance and transmittance measurements, which we discuss later in this chapter.

The last configuration in Fig. 5.3d is known as the VN configuration, named after the typical V and N shapes of the optical path during reference and sample measurements. This accessory also provides an absolute measurement with a mirrored image on the detector during the sample measurement with respect to that during the reference measurement. To switch from reference to sample mode, one mirror is rotated and another mirror moved to a

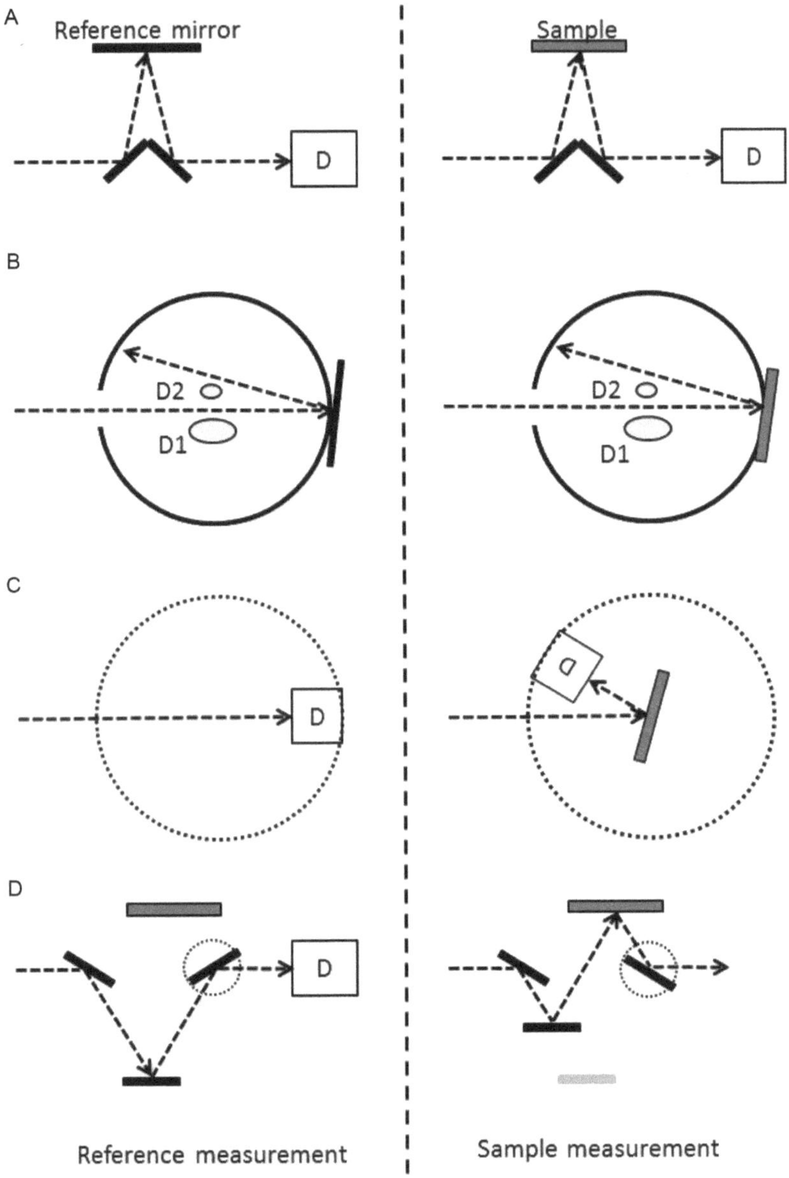

FIGURE 5.3 Typical configurations for the measurement of specular reflectance at near-normal incidence: (a) relative reflectance accessory, (b) integrating sphere, (c) absolute reflectance accessory with moving detector, and (d) VN-type absolute reflectance accessory.

position that ensures the same path length in the two configurations. Since this configuration has two moving parts, it is usually more sensitive to misalignment.

Other configurations for absolute reflectance measurements, known as VW and IV configurations, measure the square of the reflectance. They are used for determining the specular reflectance of highly reflecting samples and discussed in Section 5.5.

5.3.2 Measuring the Transmittance and Reflectance of Coated Substrates

In the glass industry, it is common practice to use an integrating sphere for measuring the transmittance and reflectance of coated glass [13,14]. Interlaboratory comparison has demonstrated that spectral transmittance measurements performed on the same samples by different industrial laboratories using spectrophotometers equipped with an integrating sphere are in good agreement [15]. A similar result is obtained with reflectance measurements when a standardized procedure is used and a calibrated reflectance standard having optical characteristics similar to the samples [15]. When measuring specular samples, one should use a specular reference for calibration and not the diffuse reference that is commonly supplied with an integrating sphere. If a diffuse reflectance standard is used instead, the substitution error alone can be as high as 0.5% for a 150 mm sphere and even 6% for a 60 mm sphere with the same port size!

5.3.3 Measuring Very Low Reflectance Materials

The main difficulty in measuring the reflectance of transparent samples with very low reflection, like glass substrates with antireflection coatings, lies in avoiding the influence of stray-light. Measuring these samples against an integrating sphere, as shown in Fig. 5.3b, is generally not a good idea, because the stray-light produced by the cover behind the sample can contribute significantly to the signal. The best approach to minimize the influence of stray-light is illustrated in Fig. 5.4. A specular reflectance accessory is separated from the detector, allowing the detector to see only the radiation reflected directly by the sample.

A calibrated mirror can be used as a reference sample, in spite of the large difference in reflectance, provided the spectrophotometer has a large dynamic range covering several decades in reflectance and with sufficient linearity. If not, reference samples that have a reflectance closer to that of the sample (but preferably not lower![1]) can be used, for example, a thin fused silica plate.

---

1. Reflectance values higher than the reference require extrapolation of the calibrated scale which results in higher uncertainty.

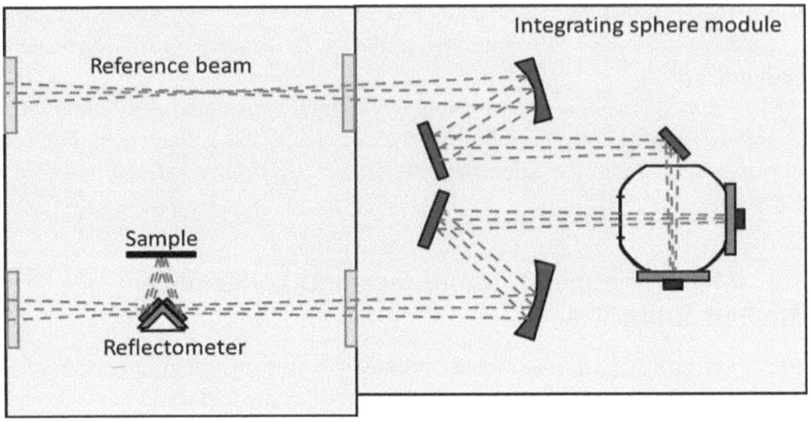

FIGURE 5.4 Reflection setup with a fixed angle specular reflectance accessory.

An opaque glass plate like a glass neutral density filter with less than 1% transmittance has a negligible contribution from the back reflectance and after calibration provides a suitable reference sample with reflectance around 0.04.

5.3.4 Measuring Specular Reflectance with an FTIR

Spectral reflectance measurements in the infrared are widely used in determining the emissivity of coated glazing to establish the energy-saving capability of the coating. Traditionally, these measurements where made using a dispersive (grating) spectrophotometer equipped with a specular reflectance accessory. Manufacture of dispersive IR spectrophotometers ceased in the 1990s and has been replaced by more modern and less expensive FTIR spectrophotometers, which have many advantages over the dispersive IR spectrophotometers. Since FTIR spectrometers are single beam instruments, a proper measurement procedure should take into account for the effect of drift [16].

This procedure should involve multiple measurements according to a specified sequence of reference mirror and sample(s) and correction of measured curves according to a time-weighted average of the reflectance measured for the reference mirrors. That is, we assume that the incident radiant flux varies linearly between measurements of the reference mirror. The measurement sequence must be completed in its entirety without interruption. The recommended measurement procedure consists of a number of scans that contains reference scans as a first and last scan, as well as intermediate reference scans that allow samples measured in between to restrict the effect of drift. The reference scans are used to compensate for drift in the source signal.

Table 5.1 gives an overview of a recommended measurement sequence for different numbers of samples. Each sample Sn (where $n=1,\ldots,N$ enumerates each of the N samples) is measured twice. This allows for checking of the

TABLE 5.1 Recommended Scan Sequence [16] Depending on the Number of Samples to be Measured (R, Reference Mirror)

Scan	Number of Samples											
Nr	*1*	*2*	*3*	*4*	*5*	*6*	*7*	*8*	*9*	*10*	*11*	*12*
1	**R**	**R**	**R**	**R**	**R**	**R**	**R**	**R**	**R**	**R**	**R**	**R**
2	S1	S1	S1	S1	S1	S1	S1	S1	S1	S1	S1	S1
3	**R**	S2	S2	S2	S2	S2	S2	S2	S2	S2	S2	S2
4	S1	**R**	S3	S3	S3	S3	S3	S3	S3	S3	S3	S3
5	**R**	S1	**R**	S4	S4	S4	S4	S4	S4	S4	S4	S4
6		S2	S1	**R**	S5	S5	**R**	**R**	S5	S5	S5	S5
7		**R**	S2	S1	**R**	S6	S5	S5	**R**	**R**	S6	S6
8			S3	S2	S1	**R**	S6	S6	S6	S6	**R**	**R**
9			**R**	S3	S2	S1	S7	S7	S7	S7	S7	S7
10				S4	S3	S2	**R**	S4	S8	S8	S8	S8
11				**R**	S4	S3	S1	**R**	S9	S9	S9	S9
12					S5	S4	S2	S1	**R**	S10	S10	S10
13					**R**	S5	S3	S2	S1	**R**	S11	S11
14						S6	S4	S3	S2	S1	**R**	S12
15						**R**	**R**	S4	S3	S2	S1	**R**
16							S5	**R**	S4	S3	S2	S1
17							S6	S5	S5	S4	S3	S2
18							S7	S6	**R**	S5	S4	S3
19							**R**	S7	S6	**R**	S5	S4
20								S4	S7	S6	S6	S5
21								**R**	S8	S7	**R**	S6
22									S9	S8	S7	**R**
23									**R**	S9	S8	S7
24										S10	S9	S8
25										**R**	S10	S9
26											S11	S10
27											**R**	S11
28												S12
29												**R**

reliability of the measurements and for obtaining an estimate of the reproducibility that can be used in the evaluation of the measurement uncertainty. If we assume that the appropriate reference value can be determined by linear interpolation between the measured reference measurements from the time the measurement was taken and assume that the measurements are equally spaced in time, two reflectance values are calculated for each sample according to

$$R_{Sn}(\lambda) = \frac{(N+1)M_{Sn}(\lambda)}{(N+1-n)M_{R1}(\lambda) + nM_{R2}(\lambda)}\rho_R(\lambda), \tag{5.8}$$

where M_{R1} and M_{R2} are the reference measurements before and after the respective sample measurement M_{Sn}, respectively, and $\rho_R(\lambda)$ is the reflectance of the reference mirror. The best estimate of the sample reflectance at each wavelength λ (or, alternatively, at each wavenumber ν) is given by the average of the two independent reflectances $R_{Sn,1}$ and $R_{Sn,2}$,

$$\rho_{Sn} = \frac{R_{Sn,1} + R_{Sn,2}}{2}. \tag{5.9}$$

The standard uncertainty associated with this average is the estimated standard deviation of the mean using Eq. (5.6),

$$u_S = 1.84\frac{|R_{Sn,1} - R_{Sn,2}|}{\sqrt{2}} = 1.30|R_{Sn,1} - R_{Sn,2}|. \tag{5.10}$$

The factor of 1.84 is the Student-t factor for one degree of freedom and a 68.27% confidence level.

The combined standard calibration uncertainty in the sample reflectance is obtained by summing the variances,

$$u_C = \sqrt{(u_S)^2 + (u_R)^2 + (u_A)^2 + (u_O)^2}, \tag{5.11}$$

where u_R is the standard uncertainty of the reflectance of the reference mirror (usually half the reported expanded calibration uncertainty), u_A accounts for differences in alignment between substituting samples and reference mirror, and u_O represents the combined uncertainty due to ordinate errors caused by detector nonlinearity, double modulation, and digitization errors.

The angle of incidence for IR reflectance accessories usually lies around 10–12°, to take into account the beam divergence. At these angles, polarization effects in the measurement with a FTIR spectrophotometer can usually be neglected.

5.3.5 Tools for Absolute Reflectance and Transmittance Measurements

An accessory capable of performing absolute reflection and transmittance measurements is shown in Fig. 5.5. This accessory is designed for a FTIR

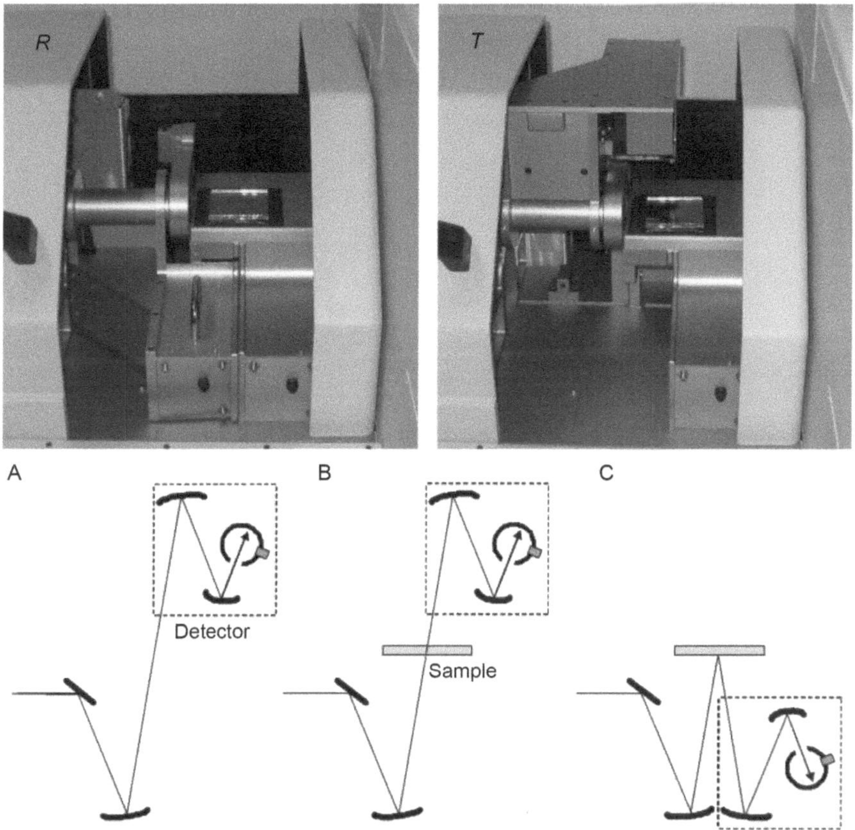

FIGURE 5.5 Absolute 10° reflectance/transmittance accessory [2] for an FTIR spectrophotometer. The schematic shows the beam path in the (a) reference mode, (b) transmittance mode, and (c) reflectance mode.

spectrophotometer and uses a deuterated triglycine sulfate detector in combination with a 25-mm rough-gold-coated integrating sphere to reduce the sensitivity to misalignment. The sample placement is horizontal and the measurement principle is based on the configuration of Fig. 5.3b. Special features of this tool are (a) the sample is placed horizontally, (b) the position and size of the measurement spot is the same for transmittance and reflectance, and (c) the angle of incidence is the same for transmittance and reflectance. These features make this tool ideal for analyzing the properties of thin films and determining absorption based on $\alpha = 1 - \tau - \rho$, which is generally only valid for the same measured sample area and the same angle of incidence for reflectance and transmittance. For high accuracy, a measurement sequence similar to that given in Table 5.1 should be applied for both transmittance and reflectance. This tool was tested by measuring the reflectance of a certified

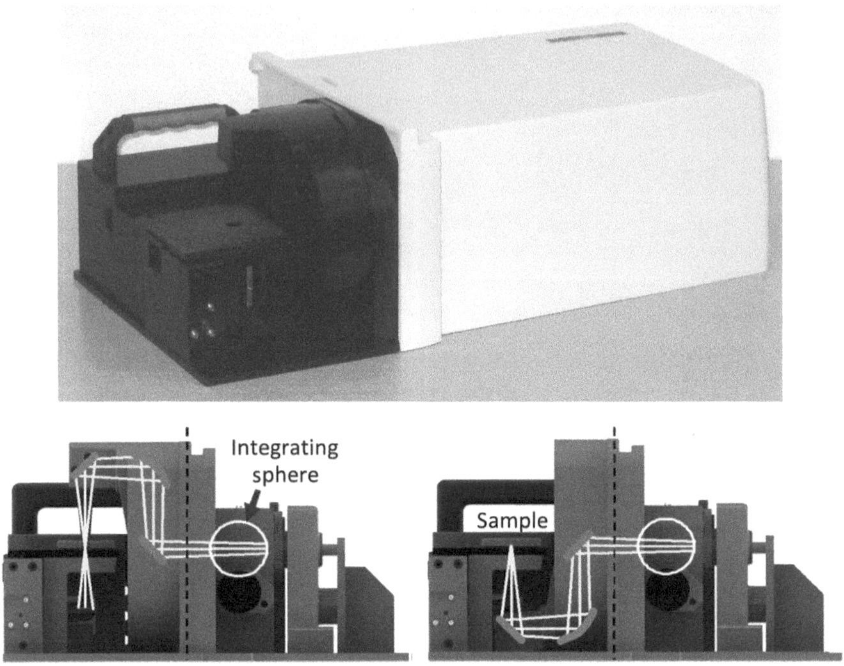

FIGURE 5.6 Absolute 8° reflectance/transmittance accessory [2] for a UV/Vis/NIR spectrophotometer. The optical path of the beam between sample and detector sphere is shown below in transmittance (left) and reflectance (right).

gold mirror with calibration uncertainty of 0.3%. The tests demonstrated an agreement with the certified values within 0.2% for the investigated wavelength range of 2.5–20 μm [2].

A similar tool developed for a UV/Vis/NIR spectrophotometer is shown in Fig. 5.6. Here, a 60 mm pressed polytetrafluoroethylene integrating sphere is used in combination with a PMT and InGaAs detector to cover the wavelength range of 200–2500 nm. The accuracy was proven to be 0.2% T in transmittance mode and 0.5% R in reflectance mode [2].

5.4 MEASUREMENTS AT OBLIQUE INCIDENCE

Performing directional optical measurements is not a simple task. A laboratory intercomparison organized in the early 1990s clearly demonstrated the lack of understanding and capability of the industry in performing such measurements at that time [17]. Since then, much work has been done to obtain a better understanding of the problems involved. A new laboratory intercomparison on measurements at oblique incidence was organized in 1999 and the results published in 2001 already showed significant improvement [18].

Research into the measurement problems encountered with commercial spectrophotometers has resulted in a new generation of commercial spectrophotometer accessories that are capable of performing accurate measurements at oblique incidence. The results of a recent laboratory intercomparison, which was organized to investigate the state-of-the-art, showed [19] that the total measurement uncertainty obtained on coated glass samples now typically lies in the range of 0.2–1% depending on the wavelength, angle of incidence, and polarization.

5.4.1 Relevant Issues

5.4.1.1 Polarization

An issue we must deal with when measuring at oblique incidence is polarization. At oblique incidence, the optical properties of a specular sample depend on polarization, as shown in Fig. 5.7 for a sample of clear window glass. The so-called p- and s-polarization states represent two extremes.

Natural daylight consists of rays of electromagnetic radiation and each of these rays has a different state of polarization (direction of the electric field). Since the effect is random, the combination of all these rays gives the average, which represents unpolarized radiation. The curves R_u and T_u in Fig. 5.7 are the averages that represent natural (unpolarized) daylight. For measurements at oblique incidence, we use a polarizer and measure at p- and s-polarization separately. For a dispersive spectrophotometer, which produces predominantly p-polarized radiation, a depolarizer can be used to produce equal amounts of s- and p-polarized radiation before the beam enters the polarizer, improving the noise level at s-polarization. Using only the depolarizer to directly measure, the reflectance or transmittance for unpolarized radiation is generally

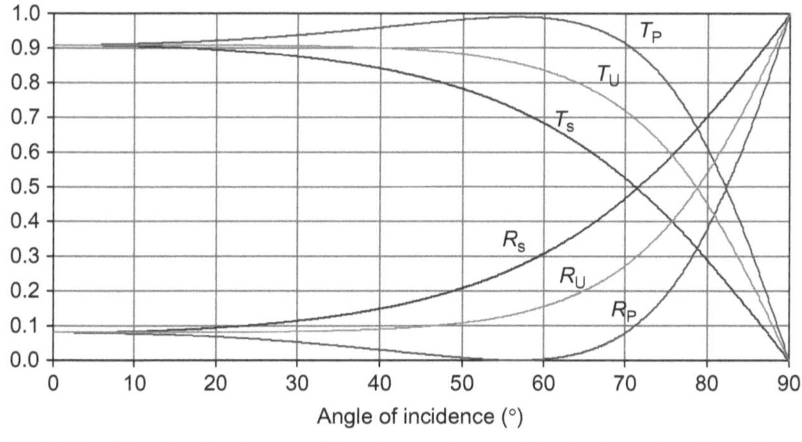

FIGURE 5.7 Directional reflectance (R) and transmittance (T) of a 4 mm clear float glass at a wavelength of 500 nm. The subscripts s, p, and u refer to the polarization state.

not a good idea, unless no other optical elements are positioned in the beam between the depolarizer and the sample, and the beam is also depolarized directly behind the sample.

A simple technique for measuring the transmittance for unpolarized radiation at oblique incidence without the use of polarizing elements is to take the average of the transmittance obtained from two orientations of the plane of incidence. This requires a tool that maintains the angle of incidence while the plane of incidence formed by the direction of the beam and the normal of the sample surface can be rotated 90° about the axis of the beam. Thus, we obtain two orientations in which the direction of s-polarization in one orientation is the same as the direction of p-polarization in the other orientation and vice versa.

A polarizer is an optical element that ideally passes radiation of only one polarization. Measurement errors associated with the use of a polarizer are due to the unwanted radiation that is passed from polarization states other than the one selected. Two types of error have to be considered, one type is associated with the extinction ratio of the polarizer, which is the ratio of unwanted and wanted transmitted radiation. The extinction ratio e is typically 10^{-2}–10^{-3} for wire grid polarizers and 10^{-4}–10^{-6} for Glan–Thompson prism polarizers. The standard uncertainty related to the systematic error caused by a nonzero value of e can be estimated by

$$u_{\mathrm{PE}} = \frac{e\left|m_{\mathrm{s}} - m_{\mathrm{p}}\right|}{2\sqrt{3}}, \tag{5.12}$$

where m_{s} and m_{p} are the values of the optical property (τ or ρ) for s- and p-polarization, respectively, and we assume a rectangular probability distribution for e.

Another error is associated with the polarizer angle φ that controls the polarization by determining the orientation of the polarizer with respect to the plane of incidence and optical axis. A deviation in this angle results in a systematic error in the measured property. The standard uncertainty in the measured property, $u_{\mathrm{P}\phi}$, related to this error is given by

$$u_{\mathrm{P}\phi} = u_{\phi}\left|m_{\mathrm{s}} - m_{\mathrm{p}}\right|, \tag{5.13}$$

where u_{ϕ} is the standard uncertainty in the polarizer angle in radians. The total standard uncertainty related to the polarizer errors is then given by

$$u_{\mathrm{P}} = \sqrt{\left(u_{\mathrm{PE}}\right)^2 + \left(u_{\mathrm{P}\phi}\right)^2}. \tag{5.14}$$

5.4.1.2 Multiple Reflections

Another issue that complicates measurements under oblique incidence is due to the secondary shifted beams resulting from multiple reflections inside the sample. This effect is illustrated in Fig. 5.8. The lateral distance, x, between

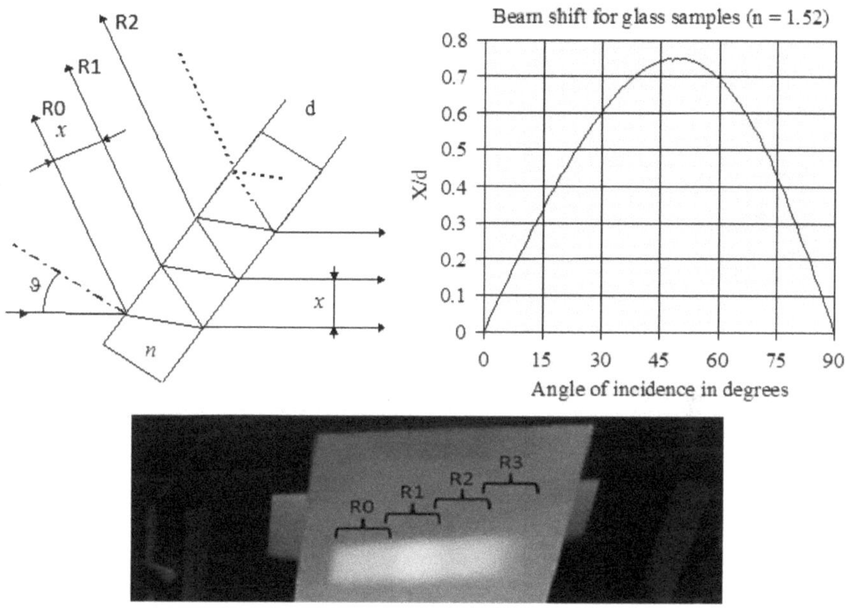

FIGURE 5.8 Multiple shifted beams at oblique incidence. The photograph demonstrates multiple beams projected on a paper target in front of a rectangular integrating sphere port, by a 7 mm glass sample with coating on the back side, illuminated at 60° with s-polarized light.

the secondary beams depends on the angle of incidence θ, the sample thickness d, and its refractive index n according to

$$x = 2d\cos\theta\tan\left\{\sin^{-1}[\sin(\theta)/n]\right\}. \tag{5.15}$$

A graph of x/d versus θ for a sample with a refractive index of 1.52 (regular window glass) is also shown in Fig. 5.8. The maximum occurs at about 50°. At this angle, x becomes approximately 75% of the sample thickness. The detector should be large enough to capture not only the primary reflected or transmitted beam but also a sufficient number of the secondary beams.

For example, assume that the beam width at the sample is 4 mm, the sample thickness is 8 mm, and that the illuminated area at the sample position is imaged on the detector. In this case, the total width of the light spot at the detector, taking into account only the primary beam and the first two secondary beams has a worst-case value (at 50° incidence) of $4\,\text{mm} + 2 \times 0.75 \times 8\,\text{mm} = 20\,\text{mm}$. In practice, the best and simplest solution is to use an integrating sphere detector with a large enough entrance port.

5.4.1.3 Angular Accuracy

An important requirement for accurate directional measurements is that the setting of the angle of incidence has to be accurate. Especially for angles of incidence $\theta > 45°$, where the angular dependence of the optical properties is

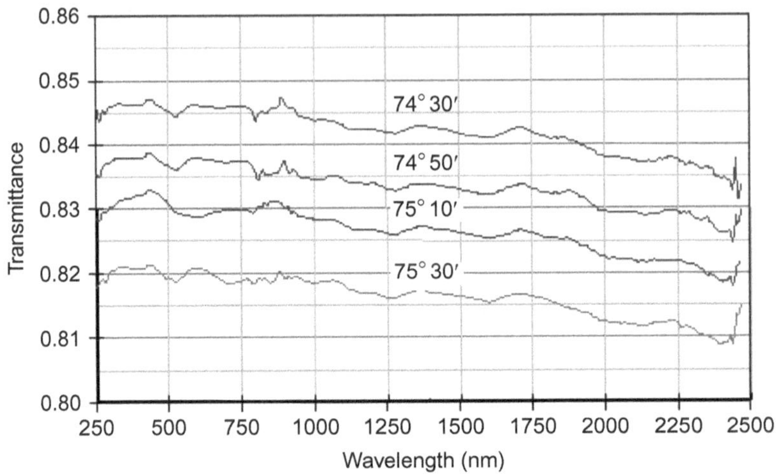

FIGURE 5.9 The spectral transmittance of a 2 mm fused silica sample for oblique incidence as measured around 75° (p-polarization) [25].

usually much greater than at normal incidence, inaccuracy in the angle of incidence can lead to significant measurement errors. This is demonstrated by the example in Fig. 5.9, showing that a change of 1° in θ causes the transmittance to change by 3%.

Apart from scale inaccuracies in the angle of the sample rotation stage, we can identify two additional sources of error in the angle of incidence. One source of error is caused by the direction of the average energy of a grating spectrophotometer's light beam being slightly wavelength dependent. Although the geometrical direction of the center of the beam may be constant, a spectrophotometer's beam is never perfectly collimated and the angular distribution of energy within the beam is generally not uniform. As a result, even with a perfectly accurate sample rotation stage, measurements performed under a positive angle $+\theta$ and a negative angle $-\theta$ do not generally give the same results at all wavelengths [1,20].

The second source of error is the offset in the zero angle of the rotation stage. If the angle of incidence θ' indicated by the scale of the rotation stage deviates from the true angle of incidence θ by a systematic error of $\Delta\theta$, the positive and negative angles in this situation are $\theta_+ = \theta' + \Delta\theta$ and $\theta_- = -\theta' + \Delta\theta$, respectively.

Both types of systematic error can be compensated by performing two scans at positive and negative angles, respectively and taking the average. For example, if the positive and negative angular settings are, respectively, +45° and −45° and the systematic error is −1.3°, the true angles of incidence will be, respectively, +43.7° and −46.3°. As a result, different values for the transmittance are obtained. However, if the systematic error is not too large, and if the optical property being measured is linear, the average of the two measurements will yield the transmittance at 45.0°.

5.4.1.4 Recommendations for Variable Angle Spectroscopy

The discussion in the previous sections leads to the following recommendations for accurate measurements of regular reflectance and transmittance at oblique incidence:

- Use polarizing elements and perform the measurements for p- and s-polarization separately.
- Use a large enough uniform detector to capture not only the direct but also sufficient secondary beams from interreflections and back reflection.
- Use accurate rotation states to obtain the angle of incidence with an accuracy of at least 0.1°.
- Average over positive and negative directions for the angle of incidence to compensate for systematic errors.

5.4.2 Directional Transmittance Measurements

A simple setup for accurate measurement of directional transmittance is shown in Fig. 5.10. The most time efficient procedure for measuring directional transmittance with this setup uses a sequence of scans with alternating p- and s-polarization, starting with the stray-light scans ($M_{0\%,\mathrm{p}}$ and $M_{0\%,\mathrm{s}}$, for p- and s-polarization, respectively). These are followed by the sample scans at all positive angles ($M_{\theta,\mathrm{p}}$ and $M_{\theta,\mathrm{s}}$, respectively). Then, 100% transmittance reference scans ($M_{100\%,\mathrm{p}}$ and $M_{100\%,\mathrm{s}}$, respectively) without the sample are performed. Finally, all of the sample scans at negative angles are performed so that the opposing angle is equal distance in time from the 100% transmittance measurement. In this way, the 100% transmittance measurement is performed half way in time between the positive and

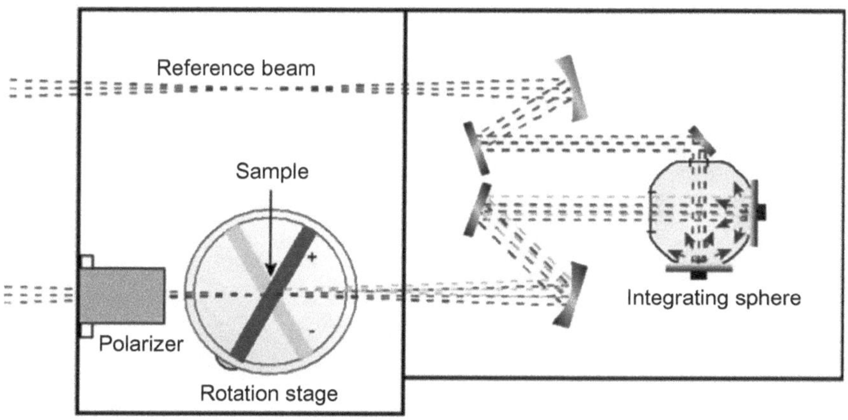

FIGURE 5.10 Setup for measuring directional transmittance with a double beam UV/Vis/NIR spectrophotometer [25].

negative angles, compensating for any linear drift in the system. An example of a measurement sequence for 30°, 60°, and 75° would thus be measured in the following sequence:

$$M_{0\%,\mathrm{p}},\ M_{0\%,\mathrm{s}},\ M_{75^\circ,\mathrm{p}},\ M_{75^\circ,\mathrm{s}},\ M_{60^\circ,\mathrm{p}},\ M_{60^\circ,\mathrm{s}},\ M_{30^\circ,\mathrm{p}},\ M_{30^\circ,\mathrm{s}},$$
$$M_{100\%,\mathrm{p}},\ M_{100\%,\mathrm{s}},\ M_{-30^\circ,\mathrm{p}},\ M_{-30^\circ,\mathrm{s}},\ M_{-60^\circ,\mathrm{p}},\ M_{-60^\circ,\mathrm{s}},\ M_{-75^\circ,\mathrm{p}},\ M_{-75^\circ,\mathrm{s}}.$$

If the transmittance at 0° needs to be included, two extra p and s scans can be added at the end of the procedure. Although the transmittance at 0° is the same for p and s, it is better for the operator to stick with the alternating p and s scans in order to avoid mistakes. Alternatively, the 0° scans can be included in the middle of the sequence, replacing the 100% reference scans which are then put at the end instead. To save time, the 0% and 100% scans can be omitted assuming the instrument produces results that are already properly corrected. Nevertheless, including them in the measurement procedure generally improves the baseline correction, so it is good practice to measure them separately.

The measured sample transmittance T for angle of incidence θ and polarization $q=\mathrm{p}$ or s is determined from these scans from

$$T_{\theta,q}(\lambda)=\frac{\frac{1}{2}M_{+\theta,q}(\lambda)+\frac{1}{2}M_{-\theta,q}(\lambda)-M_{0\%,q}(\lambda)}{M_{100\%,q}(\lambda)-M_{0\%,q}(\lambda)}. \tag{5.16}$$

If we repeat the measurement sequence N times, we will be able to obtain the standard measurement reproducibility u_{M} according to Eq. (5.6).

In case of only one measurement sequence, we can replace this by a standard uncertainty given from Eq. (5.10) by

$$u_{\mathrm{S},\theta,q}(\lambda)=1.30\frac{|M_{+\theta,q}(\lambda)-M_{-\theta,q}(\lambda)|}{M_{100\%,q}(\lambda)+M_{0\%,q}(\lambda)}. \tag{5.17}$$

Equation (5.17) overestimates the standard uncertainty unless the systematic difference between the results obtained at positive and negative angles is negligible compared to the reproducibility of the measurement.

In case of 0° incidence, the measured sample transmittance is given by averaging over p- and s-polarization according to

$$T_{0^\circ}(\lambda)=\frac{M_{0^\circ,\mathrm{p}}+M_{0^\circ,\mathrm{s}}-M_{0\%,\mathrm{p}}-M_{0\%,\mathrm{s}}}{M_{100\%,\mathrm{p}}+M_{100\%,\mathrm{s}}-M_{0\%,\mathrm{p}}-M_{0\%,\mathrm{s}}}, \tag{5.18}$$

with the standard uncertainty given by

$$u_{\mathrm{S},0^\circ}(\lambda)=1.30\frac{|M_{0^\circ,\mathrm{p}}-M_{0^\circ,\mathrm{s}}|}{M_{100\%,\mathrm{p}}+M_{100\%,\mathrm{s}}-M_{0\%,\mathrm{p}}-M_{0\%,\mathrm{s}}}. \tag{5.19}$$

In Eq. (5.19), we consider values obtained for p- and s-polarization to follow the same normal distribution with no effect of the polarization.

A suitable measurement equation for the transmittance of q polarized light is

$$\tau_{\theta,q}(\lambda) = C_A(\lambda)C_N(\lambda)C_P(\lambda)C_\theta(\lambda)C_W(\lambda)T_{\theta,q}(\lambda), \quad (5.20)$$

where $T_{\theta,q}$ is the measured sample transmittance and C_A, C_N, C_P, C_θ, and C_W are correction factors for errors in the alignment, ordinate linearity, polarization, angle of incidence, and wavelength, respectively. All correction values are assumed to have an expectation value of one with a 68% confidence interval given by its standard uncertainty. A complete evaluation of the combined standard uncertainty in the sample transmittance τ is then given by adding the variances,

$$u_C = \sqrt{(u_S)^2 + (u_A)^2 + (u_N)^2 + (u_P)^2 + (u_\theta)^2 + (u_W)^2}, \quad (5.21)$$

where u_S is determined by Eq. (5.17) or (5.19). The standard uncertainty related to misalignment strongly depends on the capability of the detector to cope with the change in the beam properties due to sample (beam shift and distortion). When an integrating sphere is used with large enough entrance port, u_A usually lies in the order of 0.1% or less.

Transmittance spectra obtained on a clear window glass sample at 0°, 30°, and 60° are shown in Fig. 5.11. Figure 5.12 shows an example of the spectral uncertainty of one of these spectra, $\tau_{60°,s}$, showing the various contributions.

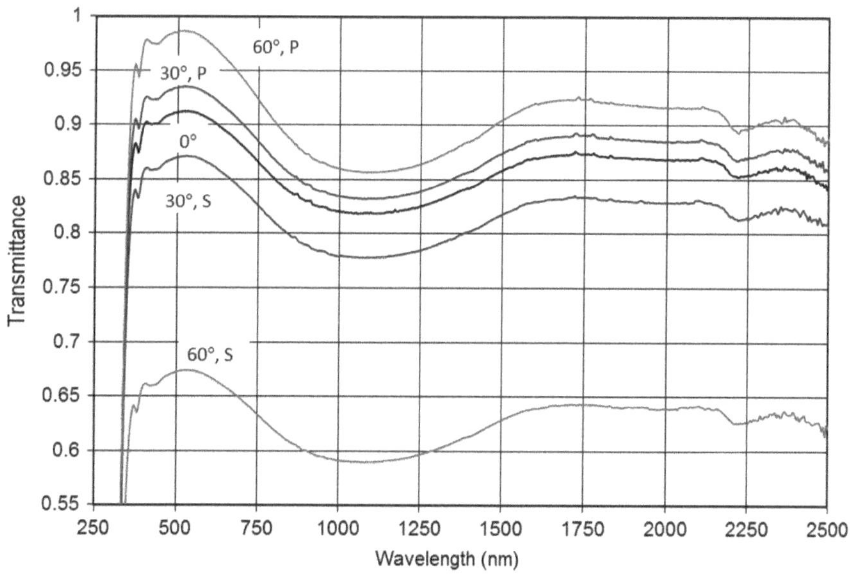

FIGURE 5.11 Transmittance spectra obtained on a clear window glass sample.

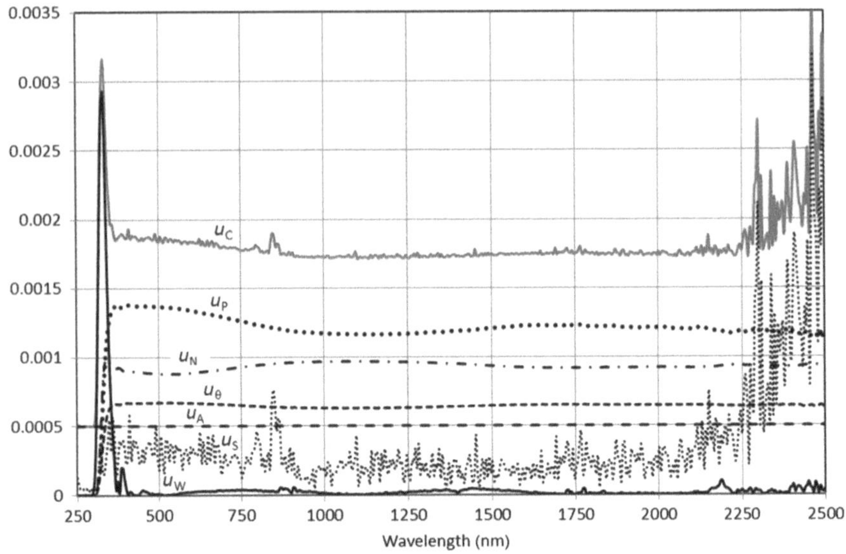

FIGURE 5.12 Spectral standard uncertainty components and the combined uncertainty for the transmittance spectrum $\tau_{60°,s}$ shown in Fig. 5.11.

5.4.3 Relative Directional Reflection Measurements

A popular tool for directional reflection measurements in specular samples is the retro-mirror pair reflectometer shown schematically in Fig. 5.13. This reflectometer is based on the principle that a pair of flat reflecting surfaces mounted on a rotation stage at a fixed mutual angle always reflects a light beam into the same direction, independent of the angle of the rotation stage. One condition is that the line of intersection of the two mirrors forming the retro-mirror pair lies exactly along the axis of rotation.

The version shown in the photograph of Fig. 5.13 was designed by the author [8] to fit in a dispersive IR spectrophotometer. The rotation stage has an angular accuracy of 0.08° and the retro-mirror can be rotated to switch between incident angles of opposite sign. Sample and retro-mirror are mounted with a mutual angle of 90°, although a right angle is not a requirement for the measurement principle. Harrick [21] demonstrated that the only requirement is that the reflected direction remains at a constant angle to the incident direction and used a 100° angle for his reflectometer.

The procedure for reflection measurements is basically identical to the procedure used for making transmission measurements. The major difference lies in the fact that the reference measurements are performed with the calibrated reference mirror. For each angle of incidence that is required for the sample measurement, we now also need to have calibration values of the

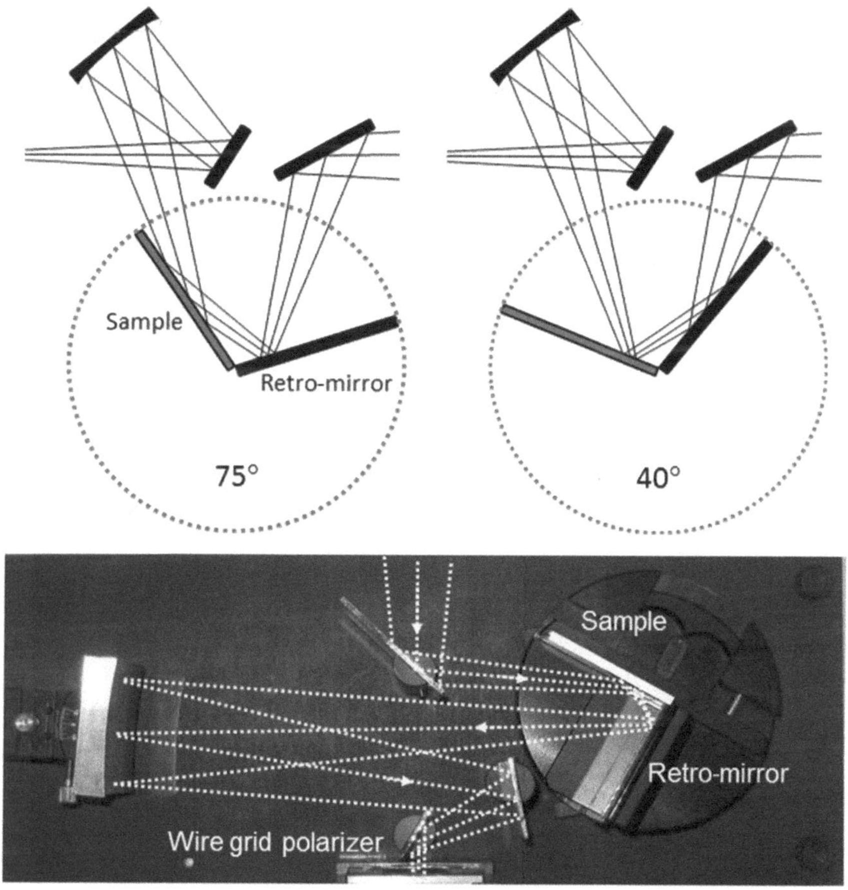

FIGURE 5.13 Optical layout of a Harrick-type retro-mirror pair reflectometer for relative directional reflection measurements shown at 75° and 40° angle of incidence. The photograph shows a different configuration with a mutual angle of 90° between sample and retro-mirror.

reference mirror. Changing the angle of incidence on the sample also changes the angle of incidence on the retro-mirror and as a result also the throughput of the reflectometer. This requires that the reflectometer is calibrated for each angle of incidence. Fortunately, we do not need to perform this calibration for each sample. It will be enough to select only one angle and polarization, and determine the change in throughput of the reflectometer for each other angle and polarization using the reference mirror that has known reflectance at the same angles and polarization [22].

A proper procedure for reflection measurements starts the sample scans at all negative angles, starting with the highest angle, followed by only one stray-light scan (0%) without sample present with p-polarized radiation

($M_{0\%,\mathrm{p}}$) and one reference scan with p-polarized radiation on the reference mirror at 45° ($M_{\mathrm{REF},+45°,\mathrm{p}}$) and finally the sample scans at positive angles starting with the smallest incident angle. An example of a sample measurement sequence for 30°, 60°, and 75° following this procedure would be

$$M_{-75°,\mathrm{p}},\ M_{-75°,\mathrm{s}},\ M_{-60°,\mathrm{p}},\ M_{-60°,\mathrm{s}},\ M_{-30°,\mathrm{p}},\ M_{-30°,\mathrm{s}},\ M_{\mathrm{REF},45°,\mathrm{p}},\\ M_{0\%,\mathrm{p}},\ M_{30°,\mathrm{p}},\ M_{30°,\mathrm{s}},\ M_{60°,\mathrm{p}},\ M_{60°,\mathrm{s}},\ M_{75°,\mathrm{p}},\ M_{75°,\mathrm{s}}.$$

The measured sample reflectance R for angle of incidence θ and polarization $q=\mathrm{p}$ or s is determined from these scans by

$$R_{\theta,q}(\lambda)=\frac{\frac{1}{2}M_{+\theta,q}(\lambda)+\frac{1}{2}M_{-\theta,q}(\lambda)-M_{0\%,\mathrm{p}}(\lambda)}{M_{\mathrm{REF},45°,\mathrm{p}}(\lambda)-M_{0\%,\mathrm{p}}(\lambda)}F_{\theta,q}(\lambda), \tag{5.22}$$

where

$$F_{\theta,q}(\lambda)=\frac{M^{*}{}_{\mathrm{REF},45°,\mathrm{p}}(\lambda)-M^{*}{}_{0\%,\mathrm{p}}(\lambda)}{\frac{1}{2}M^{*}{}_{\mathrm{REF},45°,q}(\lambda)+\frac{1}{2}M^{*}{}_{\mathrm{REF},-45°,q}(\lambda)-M^{*}{}_{0\%,\mathrm{p}}(\lambda)}\rho_{\mathrm{REF},\theta,q}(\lambda), \tag{5.23}$$

and $\rho_{\mathrm{REF},\theta,q}(\lambda)$ is the reflectance of the reference mirror. Equation (5.23) gives the correction necessary to calibrate for the varying throughput of the reflectometer. The asterisks on the values in Eq. (5.23) are used to point out that these scans do not belong to the same measurement series as that of the sample. This procedure requires the correction (5.23) to be determined only once using the same measurement sequence for the reference mirror as for the sample.

In a similar manner as Eq. (5.17), the standard uncertainty in the measured sample reflectance is given by

$$u_{\mathrm{S},\theta,q}(\lambda)=1.30\frac{|M_{+\theta,q}(\lambda)-M_{-\theta,q}(\lambda)|}{M_{\mathrm{REF},45°,\mathrm{p}}(\lambda)+M_{0\%,\mathrm{p}}(\lambda)}F_{\theta,q}(\lambda). \tag{5.24}$$

A complete evaluation of the combined standard uncertainty u_{C} in the sample reflectance ρ is given by

$$u_{\mathrm{C}}=\sqrt{(u_{\mathrm{S}})^2+(u_{\mathrm{R}})^2+(u_{\mathrm{A}})^2+(u_{\mathrm{N}})^2+(u_{\mathrm{P}})^2+(u_{\theta})^2+(u_{\mathrm{W}})^2}, \tag{5.25}$$

where u_{R} is the standard calibration uncertainty in the measurement related to the calibration uncertainty $u_{\mathrm{REF},\theta,q}$ of the reference mirror according to

$$u_{\mathrm{R}}=\frac{R_{\theta,q}(\lambda)}{\rho_{\mathrm{REF},\theta,q}}u_{\mathrm{REF},\theta,q} \tag{5.26}$$

and the other uncertainties are discussed earlier. Reflectance spectra obtained on a clear window glass sample at 8°, 30°, and 60° are shown in Fig. 5.14. Figure 5.15 shows an example of the spectral uncertainty of one of these spectra, showing the various contributions. For these measurements, the tool shown in Fig. 5.16 was used.

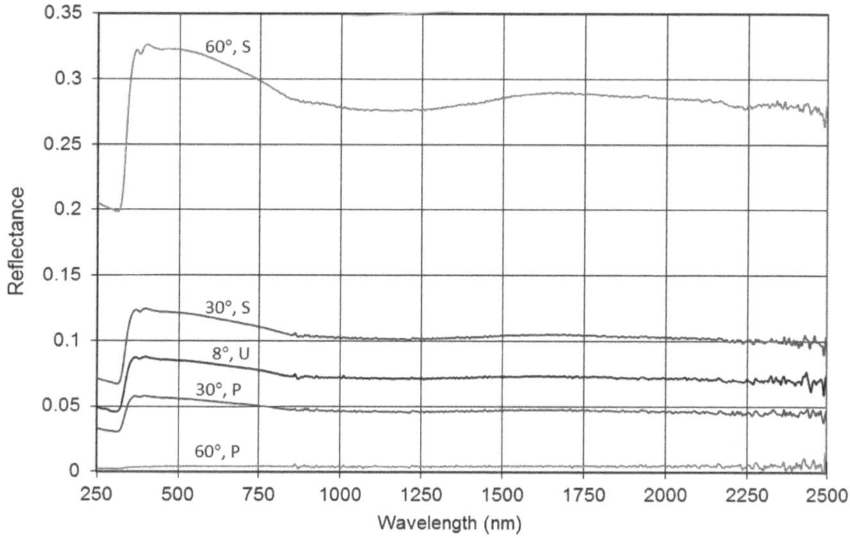

FIGURE 5.14 Reflectance spectra obtained on a clear window glass sample.

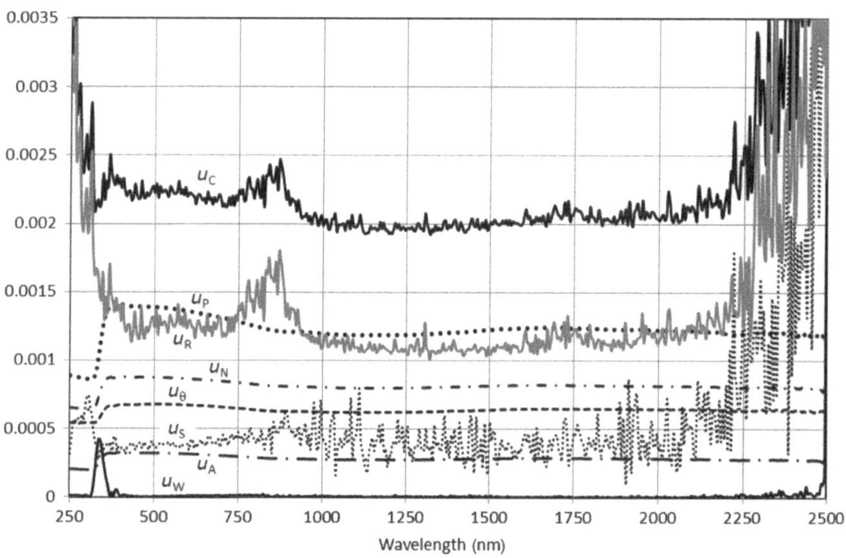

FIGURE 5.15 Spectral standard uncertainties and combined uncertainty for the reflectance $\rho_{60°,s}$ shown in Fig. 5.14.

5.4.4 Absolute Directional Reflection Measurements

With directional measurements, angles and polarization have to be changed and, if necessary, the alignment adjusted between scans. Avoiding mistakes usually requires the full attention of an experienced operator during a long

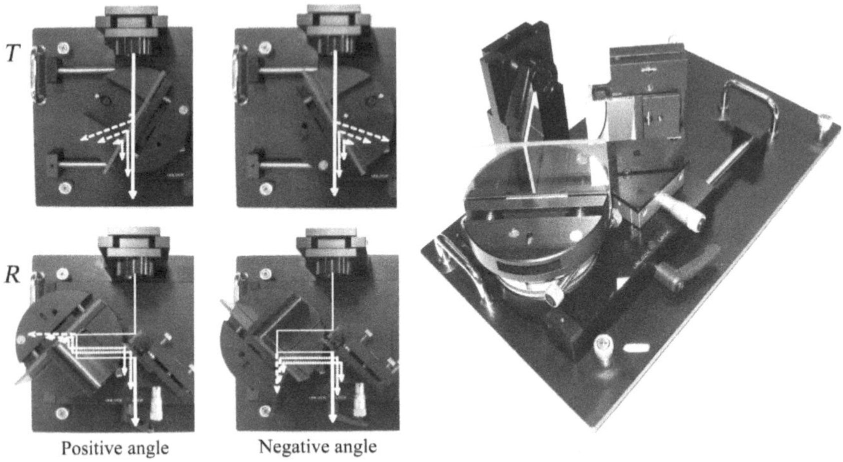

FIGURE 5.16 UV/Vis/NIR spectrophotometer accessory for directional reflectance and transmittance measurements [22].

period of time when many scans are required in transmittance and reflectance on both sides of the same sample. Automation not only removes the burden of performing the time-consuming measurements manually but also offers more reliability.

Goniometer-based accessories for double beam UV/Vis/NIR spectrophotometers have been made commercially available since the beginning of this century [2,20]. These accessories are stepper motor-driven tools that use an integrating sphere with a detector and are capable of recording transmittance and reflectance spectra as functions of the angle of incidence and polarization. Several measurement geometries are shown in Fig. 5.17. The sample is positioned on a motorized rotation stage in the center and the integrating sphere detector driven by a second rotation stage is either positioned behind the sample (at 180°) for transmittance measurements or in front of the sample for reflectance measurements (at twice the angle of incidence).

The goniometer is calibrated in the so-called autozero mode without sample and with the detector in the 180° position to capture the spectrophotometer's beam. For transmittance, the sample is inserted in the beam and the angle of incidence set with the sample rotation stage and the detector position at 180°. For reflection measurements, the angle of the detector rotation stage is set at twice the angle of the sample stage.

The angular range of the detector shown in Fig. 5.17 is 10–350° (180° being the position directly behind the sample). The angular range for measurement of specular samples is 5–85° for reflectance and 0–85° for transmittance (depending on sample type and size).

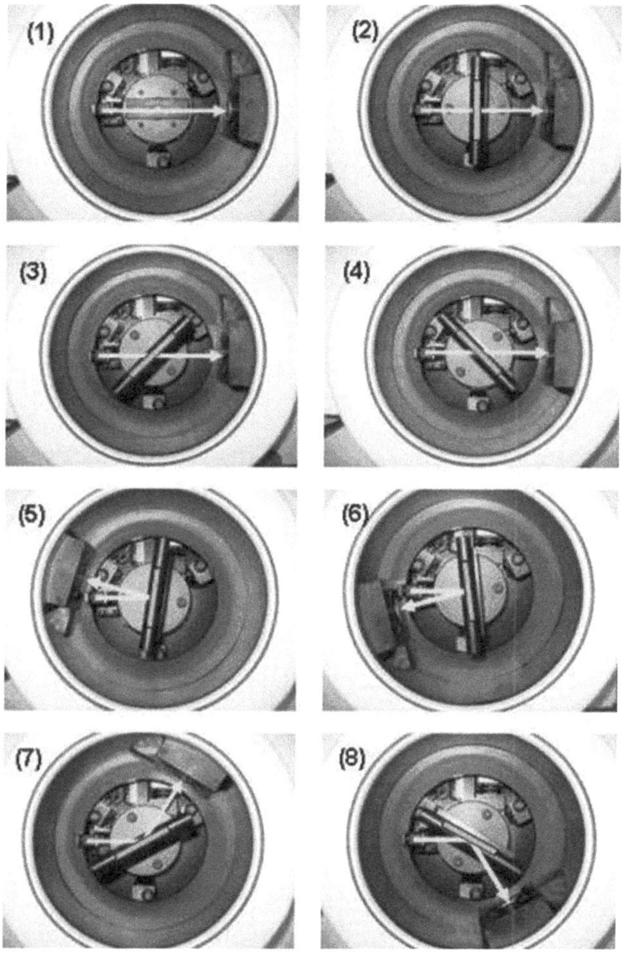

FIGURE 5.17 Top view of the sample chamber of an automated reflectance/transmittance goniometer [2] with different measurement geometries: (1) autozero mode (no sample installed), (2) transmission at 0°, (3) transmission at +45°, (4) transmission at −45°, (5) reflection at +8°, (6) reflection at −8°, (7) reflection at +60°, and (8) reflection at −60°.

5.5 MEASURING THE REFLECTANCE OF HIGHLY REFLECTING MATERIALS

Laser mirrors are multilayer coatings optimized to have absorption close to zero at the wavelength of interest. To quantify the absorption one needs to measure the reflection, a value close to one, extremely accurately. A similar demand for a highly accurate reflectance measurement lies in the calibration of reference mirrors used as reflectance standard in spectrophotometry. The reflectometers discussed below feature a double reflection on the sample so that the square of the reflectance is measured. This reduces the measurement errors in the reflectance by 50% which is a great advantage when high accuracy is important. The standard uncertainty in the reflectance can be as low as 0.0005 with these reflectometers. Only cavity ringdown reflectometers, which are beyond the scope of this chapter, are capable of measuring with a lower uncertainty [23].

5.5.1 VW Method

A device commonly used for the calibration of reflectance standards is the so-called VW reflectometer by Strong [24]. The principle of this VW reflectometer as designed to fit in a spectrophotometer's sample compartment is shown in Fig. 5.18. In the so-called V-mode, the instrument beam is interacting with three mirrors (M1, M2, and M3). The angle of incidence on mirror M2 must be as small as possible but large enough to make sure that in the W-mode the beam can pass this mirror on either side without vignetting. Typically, this angle is 8° and is determined by the position of the mirrors M1 and M3. M2 is a spherical mirror that compensates for the elongated optical path.

In the so-called W-mode, the beam additionally interacts twice with the sample. If the detector is linear and alignment is done perfectly, the image projected on the detector is identical for both V- and W-modes, and the ratio of the two readings produces exactly the square of the sample reflectance.

The VW reflectometer shown in Fig. 5.18 is designed to have directional reflectance capability for performing measurements at oblique incidence. An example of measurements at 60° with this accessory is shown in Fig. 5.19. In this case, two additional forms of the W-mode are possible, representing opposite angles of incidence. As discussed in Section 5.4.1.3, the ability to perform measurements at both positive and negative angles is important since it gives the user the possibility for compensating for systematic errors related

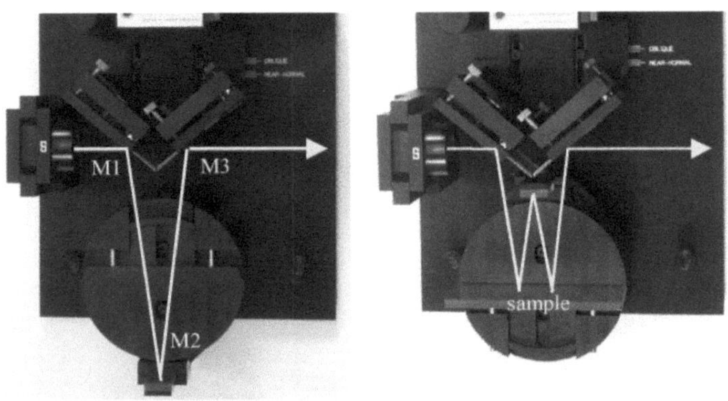

FIGURE 5.18 VW reflectometer accessory set for near-normal (8°) angle of incidence [1,4].

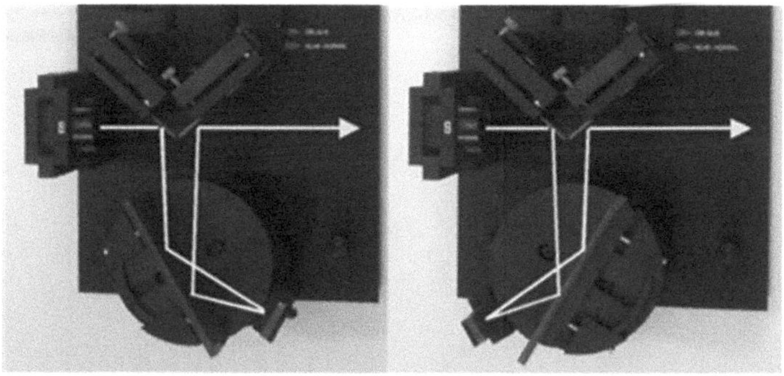

FIGURE 5.19 The directional VW accessory measuring at oblique incidence for positive angle of incidence (left) and for negative angle of incidence (right) [1,4].

to beam-shift effects and angular accuracy, by taking the average of these two types of measurements.

Mirrors M1 and M3 (see Fig. 5.18) are mounted on the same block and can be shifted backward (upward in Figs. 5.17 and 5.18) to get a smaller angle on M2. In the case of oblique incidence, the angle of incidence on mirror M3 must be as small as possible. In our case, this angle is 5°. Markers are engraved on the base plate of the accessory to indicate the proper position of the mirror block.

In the W-mode, there are always two reflections on the sample, one coming from mirror M1 (which is reflected toward M2) and one from mirror M2 (which is reflected toward M3). In the case of oblique incidence with the angle set to θ, the average angle of incidence of the beam coming from mirror M1 is $\theta_1 = \theta + 5°$, while for the second beam coming from mirror M2, the average is $\theta_2 = \theta - 5°$. This effect is shown in Fig. 5.20, where the angles are somewhat exaggerated to make the effect more visible. The systematic error related to this effect is given by

$$\Delta R = R(\theta) - \sqrt{R(\theta + 5°)R(\theta - 5°)}. \tag{5.27}$$

In the case of an aluminum-coated mirror, this systematic error typically is on the order of only 0.05–0.4%, depending on the wavelength, polarization, and θ.

A slight misalignment between the beams in the V- and W-modes of the VW accessory due to sample tilt and angular uncertainty can be expected. We deal with these alignment errors by using an integrating sphere with the detector.

In the case of a metallic mirror, the systematic error caused by the different angles for the two reflections on the sample has proven to be very small for angles below 60° and still quite acceptable at 75°. However, the effect can be disastrous in the case of multilayer coatings, often used for laser mirrors. The shape of the reflectance spectrum of such a coating is determined by

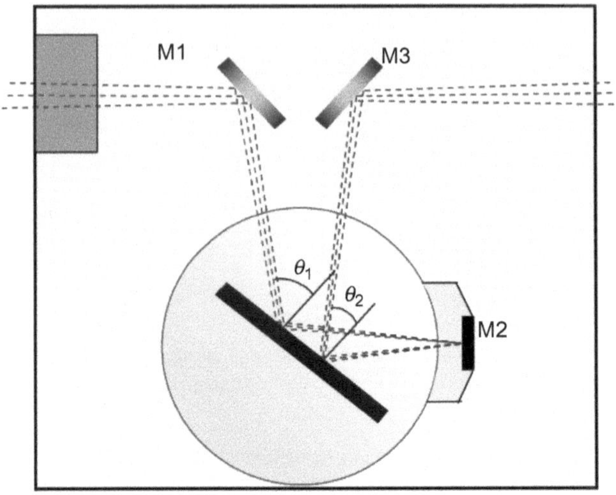

FIGURE 5.20 The directional VW features different angles of incidence of the two reflections on the sample at oblique incidence.

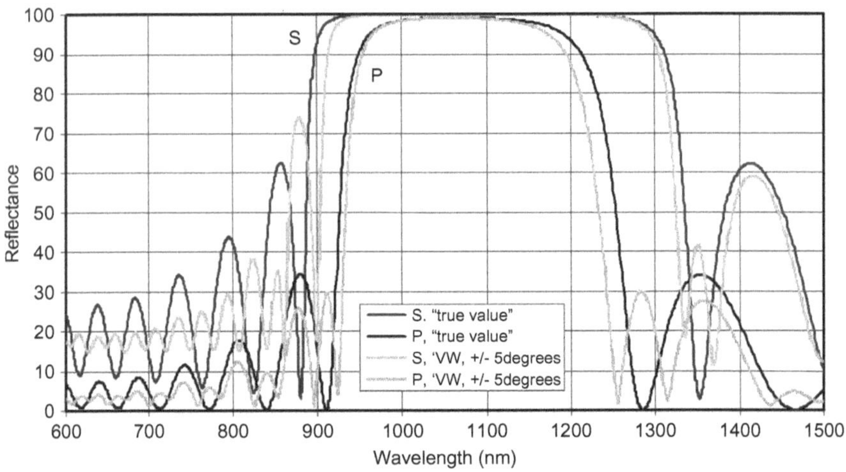

FIGURE 5.21 Simulated true values (ideal measurement with zero beam divergence) versus simulated VW measurements with reflections at 40° and 50° (average 45°).

constructive and destructive interference of radiation reflected at the layer interfaces, and this effect is sensitive to small changes in the angle of incidence. In Fig. 5.21, we compare the true reflectance of a laser mirror for an angle of incidence of 45° with the reflectance measured using the directional VW accessory. The results show a large difference between the two. In order to solve this measurement problem, a new technique was developed which we call the IV method [4].

5.5.2 IV Method

The difference between the IV and VW reflectometers is shown in Fig. 5.22, which gives a top view of the two accessories in the reference mode, which is the V-mode for the VW accessory and the I-mode for the IV reflectometer. In the IV reflectometer, the beam coming from the polarizer is reflected by three mirrors (M1, M2, and M3 in Fig. 5.22), before it is transmitted by a broadband beamsplitter (BS) and directed to the spherical mirror on the rotation stage. This mirror redirects the beam back toward the BS, where it is reflected 90° toward the exit window of the sample compartment to interact with the detector unit.

Figure 5.23 shows the IV reflectometer in the V-mode, which is the sample measurement mode for this reflectometer. Not only are the two reflections on the sample at the same spot (on the axis of rotation of the sample), but they also have the same angle of incidence for both reflections which makes the method suitable for multilayer dielectric coatings.

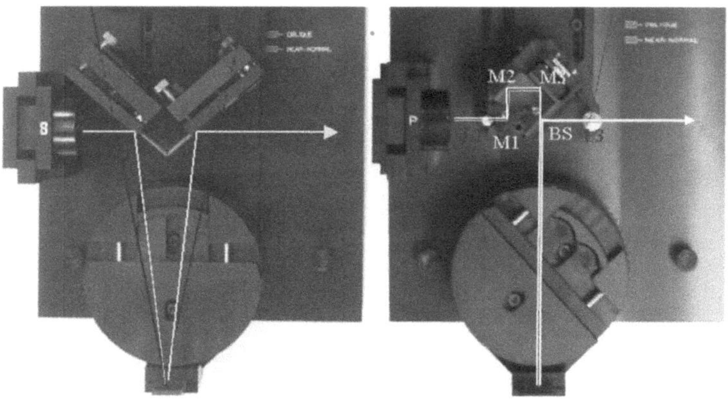

FIGURE 5.22 Top view of the VW accessory (left) and IV accessory (right) in the reference mode (without sample), showing the optical path of the sample beam [4].

FIGURE 5.23 Top view of the IV reflectometer in the V-mode, for (left) +60° incidence and (right) −60° incidence, showing the optical path of the sample beam [4].

An infrared version of the IV reflectometer connected to a FTIR spectrophotometer is shown in Fig. 5.24. This reflectometer was designed to handle sample sizes up to 250 mm diameter. Although this reflectometer was optimized for the wavenumber range of 3300–5000 cm^{-1}, where the combined standard uncertainty of the measurement is as low as 0.002, the workable range is 500–5000 cm^{-1} (wavelength range 2–20 μm). An example of results obtained on a commercial low-emissivity (single Ag) coated window glass measured at different angles of incidence with this setup is shown in Fig. 5.25.

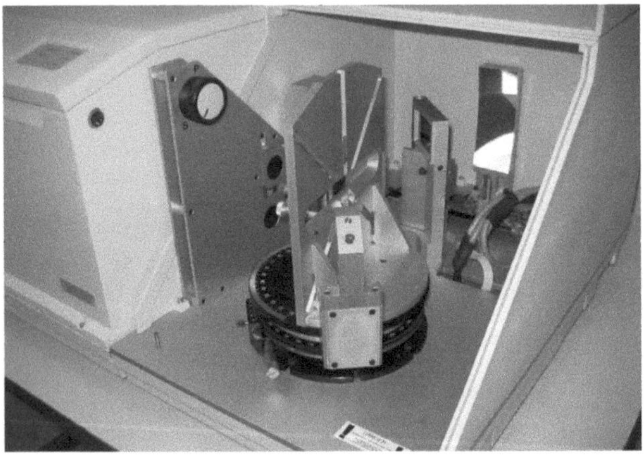

FIGURE 5.24 FTIR directional IV reflectometer for measuring absolute reflectance of highly reflecting materials at angles of incidence in the range of 10–80° [2].

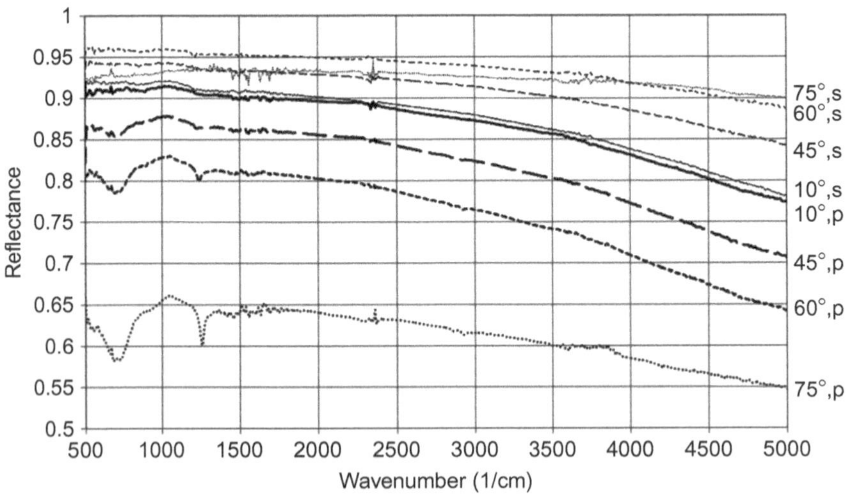

FIGURE 5.25 Reflectance of a low-emissivity coated window glass, measured with the FTIR IV reflectometer at different angles of incidence and polarization.

REFERENCES

[1] P.A. van Nijnatten, Measurement techniques for solar energy properties of glazing, in: Proceedings of the World Renewable Energy Congress VI, Brighton, 2000.

[2] P.A. van Nijnatten, J.M.C. de Wolf, I.J.E. Schoofs, Spectrophotometer accessories for thin film characterisation, in: 7th ICCG (International Conference on Coatings on Glass & Plastics), Breda, Netherlands, 2008.

[3] P.A. van Nijnatten, Calibration of neutral density glass filters to produce transmittance standards, in: 5th ESG Conference "Glass Science and Technology for the 21st Century," Prague, 1999.

[4] P.A. van Nijnatten, J.M.C. de Wolf, I.J.E. Schoofs, Directional reflection measurements on highly reflecting coatings, in: 7th International Conference on Coatings on Glass, Eindhoven, 2008.

[5] WG2, ISO/TC 160/SC 2, ISO 9050: glass in building—determination of light transmittance, solar direct transmittance, total solar energy transmittance, ultraviolet transmittance and related glazing factors, 2007.

[6] TC129, CEN, EN410: glass in building—determination of luminous and solar characteristics of glazing, 1998.

[7] A.V. Tikhonravov, T.V. Amotchkina, M.K. Trubetskov, R.J. Francis, V. Janicki, J. Sancho-Parramon, H. Zorc, V. Pervak, Optical characterization and reverse engineering based on multiangle spectroscopy, Appl. Optics 51 (2012) 245–254.

[8] P.A. van Nijnatten, Optical analysis of coatings by variable angle spectrophotometry, Thin Solid Films 516 (14) (2008) 4553–4557.

[9] ISO Guide to the Expression of Uncertainty in Measurement, ISO, Geneva, Switzerland, 1995.

[10] K.D. Mielenz, K.L. Eckerle, Spectrophotometer linearity testing using the double-aperture method, Appl. Optics 11 (1972) 2294–2303.

[11] F. Geotti-Bianchini, P. Polato, Error monitoring in the measurement of the spectral specular reflectance of flat transparent samples with commercial spectrophotometers as a function of plate thickness, Optik 71 (1985) 121–128.

[12] F. Geotti-Bianchini, P. Polato, Error monitoring in the measurement of the spectral specular reflectance of transparent samples, Optik 63 (1983) 239–246.

[13] G. Bonicatto, P. Polato, Reflectance measurements on coated architectural glasses using commercial spectrophotometers with integrating sphere accessories, Glass Technol. 28 (1987) 180–185.

[14] P. Polato, F. Geotti-Bianchini, P. Segato, Comparative specular reflectance measurements on coated flat glass with commercial spectrophotometers, Glass Technol. 27 (1986) 55–59.

[15] F. Nicoletti, P. Polato, J. Roucour, Interlaboratory comparison of solar range transmittance and reflectance of coated and uncoated flat glass, Glasstech. Ber. Glass Sci. Technol. 67 (1994) 31–44.

[16] P.A. van Nijnatten, M.G. Hutchins, N.B. Kilbey, A. Roos, K. Gelin, F. Geotti-Bianchini, P. Polato, C. Anderson, F. Olive, M. Köhl, R. Spragg, P. Turner, THERMES project, Final Report Contract no. G6RD-CT-2001-00658, European Commission—Cordis FP5, Brussels, 2005.

[17] M.G. Hutchins, P. Ageorges, Angular dependent spectral properties for architectural glazing: results of an inter-laboratory comparison, SPIE 2017 (1993) 13–24.

[18] M.G. Hutchins, A.J. Topping, C. Anderson, F. Olive, P.A. van Nijnatten, A. Roos, M. Rubin, Angle dependent optical properties of coated glass products: results of an inter-laboratory comparison of spectral transmittance and reflectance, Thin Solid Films 392 (2001) 269–275.

[19] M.G. Hutchins, N. Kilbey, C. Anderson, P.A. van Nijnatten, A. Roos, Angle dependent transmittance and reflectance measurements and associated colour values of glass products: a laboratory intercomparison, in: International Commission on Glass, TC10, Optical Properties of Glass, 2012.

[20] P.A. van Nijnatten, An automated directional reflectance transmittance analyser for coating analysis, Thin Solid Films 442 (2003) 74–79.

[21] N.J. Harrick, Determination of refractive index and film thickness from interference fringes, Appl. Optics 10 (1971) 2344–2349.

[22] P.A. van Nijnatten, A spectrophotometer accessory for directional reflectance and transmittance of coated glazing, Solar Energy 73 (3) (2002) 137–149.

[23] A. Duparré, D. Ristau, Optical interference coatings 2010 measurement problem, Appl. Optics 47 (13) (2008) C179–C184.

[24] J. Strong, Procedures in Experimental Physics, Prentice-Hall Inc., New York, 1938, p. 376.

[25] P.A. van Nijnatten, Measurement of directional optical properties of coated glazing products for building and automotive glazing, in: 6th International Conference on Architectural and Automotive Glass, Today and in the 21st Century, Glass Processing Days, Tampere, Finland, 1999.

Chapter 6

Diffuse Reflectance and Transmittance

Andreas Höpe
Physikalisch-Technische Bundesanstalt, Braunschweig, Germany

Chapter Outline

6.1 Introduction 180
6.2 Measurands 182
6.2.1 Reflectance 182
6.2.2 Reflectance Factor 182
6.2.3 Radiance Factor 183
6.2.4 Transmittance 183
6.2.5 Relationship Between Reflectance, Transmittance, and Absorptance 184
6.3 Notation of Diffuse Reflection Geometries 184
6.3.1 Recommended Integrating Sphere Geometries 186
6.4 Integrating Spheres 187
6.4.1 Historic Information 189
6.4.2 Setup 190
6.4.3 Construction Principles 190
6.4.4 Internal Coating Materials 191
6.4.5 Homogeneity 192
6.5 Absolute Sphere Methods for Measuring Diffuse Reflection 193
6.5.1 Basic Considerations 193
6.5.2 Methods of Taylor (First Taylor) and Benford 195
6.5.3 Method of Taylor (Third Taylor) 197
6.5.4 Method of Sharp-Little 198
6.5.5 Method of van den Akker 199
6.5.6 Method of Korte–Schmidt 200
6.5.7 Coblentz Sphere 202
6.5.8 Goniometric Method 203
6.6 Diffuse Reflection Standards 203
6.6.1 Requirements 204
6.6.2 Common Diffuse Reflection Standards 205
6.6.3 Fluorescence of Diffuse Reflection Standards 210
6.6.4 Handling Recommendations 211

Experimental Methods in the Physical Sciences, Vol. 46. http://dx.doi.org/10.1016/B978-0-12-386022-4.00006-6

6.7 Relative Sphere Methods for Measuring Diffuse Reflection 212
6.8 Diffuse Transmittance Measurements 214
References 216

6.1 INTRODUCTION

Diffuse reflectance is always present in our everyday surroundings, and reflection of light gives us the required stimulus by which our visual sense perceives the world around us [1,2]. Diffuse transmission may not play as prominent a visual role as reflectance, but nonetheless is important in lighting, whether it arises from the coatings on luminaires, the design of visual displays, or simply the coloring of stained glass windows. The technical relevance of diffuse transmission for the characterization of materials and suspensions is unquestionable [3].

Figure 6.1 gives an overview of the basic features in idealized reflection and transmission. For this illustration, it is convenient to use a representation where the reflected radiant flux is depicted by a radial vector in the appropriate angular direction θ, shown by dashed arrows in Fig. 6.1, with its length proportional to the radiant flux. These radial vectors trace out a polar curve (or surface in three dimensions), often referred to as the indicatrix [4]. For specular reflection, which results from the Fresnel–Huygens principle together with the presence of a smooth interface, we have just two directions with fixed angles obeying the law of reflection, where $\theta_r = -\theta_i$. In the case of $\theta_r = \theta_i$, we have retroreflection, where a portion of the incoming radiation is

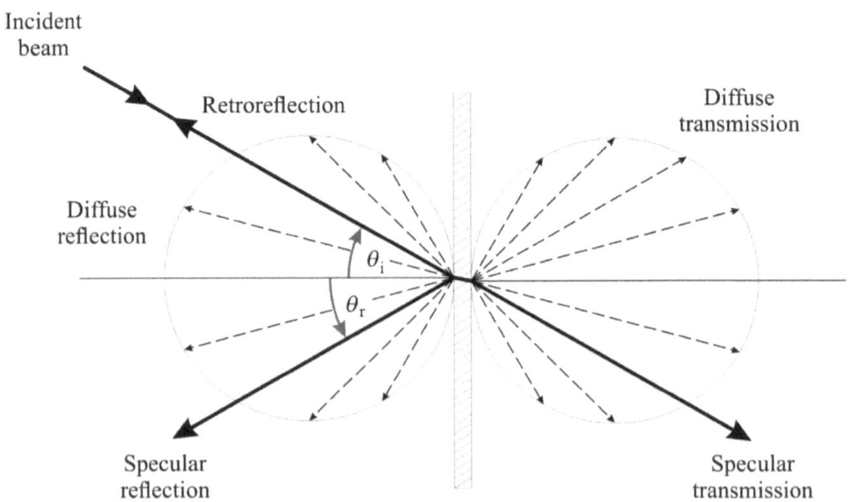

FIGURE 6.1 A schematic representation of reflection and transmission from a material.

returned to its source. In the transmitting direction, provided the incoming beam passes nearly unperturbed across the medium, only affected by Snell's law of refraction, we have normal, or specular, transmission.

In the ideal case of diffuse reflection, one gets a spherical angular distribution of the reflected radiation according to Lambert's cosine law. The same is true for ideal diffuse transmission through the material. If the radiance L of the reflected radiation is independent of the direction, one speaks of a Lambertian reflector.

In general, the reflection and transmission of light do not follow these simple, ideal cases. Generally applied, graphs of the indicatrix, some shown in Fig. 6.2, lead to intuitive representations of more typical behaviors of reflection and transmission. It can be seen, then, that to measure the reflectance or transmittance from arbitrary samples under test requires careful attention to ensure that the detection sensitivity is uniform over the appropriate half-sphere. In the case of specular reflection, one normally measures the reflectance of the sample as the ratio of the reflected flux to the incident flux in the given conditions, a relatively simple measurement provided the whole reflected radiation is detected. For diffuse reflection measurements, however, capturing all of the reflected or transmitted radiation uniformly can be a difficult procedure. This is the application where integrating sphere methods become important.

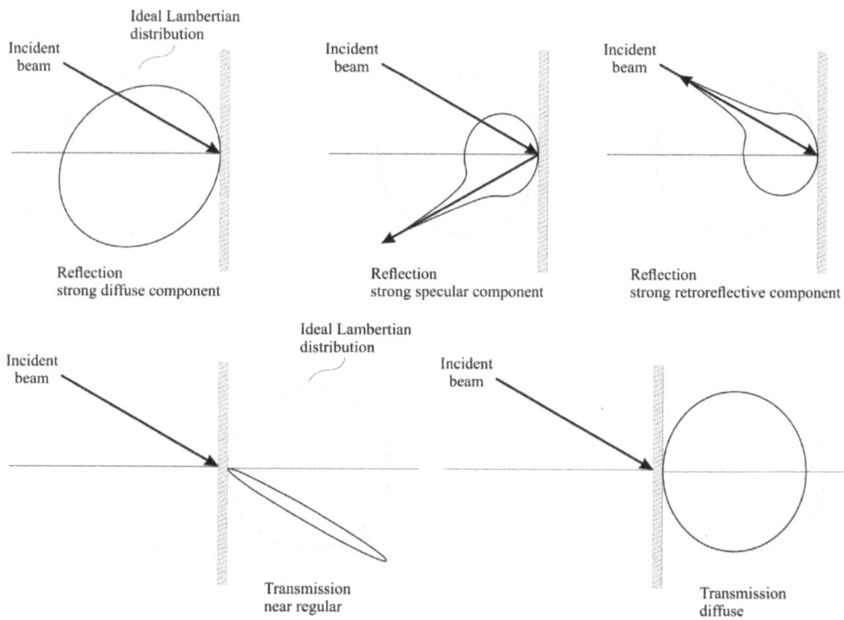

FIGURE 6.2 Different kinds of reflection and transmission characteristics of given materials.

6.2 MEASURANDS

The diffuse reflection and transmission of optical radiation from materials are best expressed by a number of different dimensionless classification values. These measurands are dimensionless, because they are ratios of physical quantities with identical dimension. These different measurands can differ in the range of values that they can accept, whether bounded in some range or not. The fact that these values are dimensionless, however, can lead to some confusion when using the appropriate measurand for a given application or measurement geometry. In this section, we will discuss these various measurands.

In many cases, reflection measurements on diffusely reflecting materials are carried out relative to a standard. For this purpose, the perfectly reflecting diffuser (PRD), which reflects the incoming radiation without losses, completely diffusely with a Lambertian directional distribution, and independent of wavelength, has been established as a primary standard [5]. The PRD, however, is only a theoretical concept and cannot be realized in practice. Since there is no material with these characteristics, the realization of the primary standard is carried out with physical methods, that is, by the measuring apparatus, which inherently measures the appropriate absolute quantity.

6.2.1 Reflectance

Reflectance is a dimensionless number defined by the ratio of the radiant flux Φ, expressed in Watts, of the reflected (index r) and the incident (index i) signals, written in a spectral dependent representation as

$$\rho(\lambda) = \frac{\Phi_\mathrm{r}(\lambda)}{\Phi_\mathrm{i}(\lambda)}. \tag{6.1}$$

From conservation of energy, it follows that reflectance for nonfluorescent materials may have values only in the interval 0–1, inclusive. The diffuse reflectance for real nonfluorescent materials is always less than unity. The reflectance depends upon the geometric illumination and collection conditions, which were discussed in Section 6.3.

6.2.2 Reflectance Factor

The reflectance factor R is a dimensionless number defined as the ratio of the radiant flux Φ_r reflected in the directions delimited by the given cone to that reflected in the same directions by the PRD irradiated in exactly the same way as the sample,

$$R(\lambda) = \frac{\Phi_\mathrm{r}(\lambda)}{\Phi_\mathrm{r}^{\mathrm{PRD}}}, \tag{6.2}$$

where Φ_r^{PRD} is the radiant flux from the PRD. It is important to realize that, unlike the reflectance ρ, the reflectance factor R is not bounded in the range from 0 to 1. For example, materials that exhibit some specularity can have a reflectance factor greater than unity, if the geometric conditions of the measurement are such that the specular direction, but not all directions, is included. Even nonspecular materials, if they are highly reflecting, can have reflectance factor greater than unity for some measurement conditions.

Because the total reflectance of the PRD is unity, when the solid angle Ω approaches 2π sr, the hemisphere above the material, the reflectance factor converges to the reflectance under the same conditions

$$R(\Omega=2\pi,\lambda)=\rho(\Omega=2\pi,\lambda). \qquad (6.3)$$

In this case, the reflectance factor, being the same as the reflectance, is in fact bounded in the interval 0–1.

6.2.3 Radiance Factor

The radiance factor β is a dimensionless number defined as the ratio of the radiance L_r (having a measurement unit of $W \cdot sr^{-1} \cdot m^2$) reflected by a sample surface to that which would be reflected into the same reflected-beam geometry by the PRD irradiated in exactly the same way as the sample,

$$\beta(\lambda)=\frac{L_r(\lambda)}{L_r^{PRD}}. \qquad (6.4)$$

The radiance factor can reach values much greater than unity. This can happen in gonioreflectometry (see Chapter 8) when measuring samples at arbitrary angles of incidence and reflection, even for diffuse reflection standards [6].

When the solid angle Ω of a radiance factor measurement approaches the limit of zero, the reflectance factor approaches the radiance factor,

$$\beta(\lambda)=\lim_{\Omega\to 0} R(\lambda). \qquad (6.5)$$

6.2.4 Transmittance

Transmittance is the transmission-mode analog to the reflectance. That is, transmittance is a dimensionless number defined by the ratio of the radiant flux Φ_t transmitted to the incident radiant flux Φ_i,

$$\tau(\lambda)=\frac{\Phi_t(\lambda)}{\Phi_i(\lambda)}. \qquad (6.6)$$

Like the reflectance, conservation of energy dictates that the transmittance for nonfluorescent materials have values in the interval 0–1, inclusive. The diffuse transmittance for real nonfluorescent materials is always less than unity.

The total transmittance τ_{total} of a medium can be separated into two parts, the regular transmittance τ_{reg} and the diffuse transmittance τ_{diff},

$$\tau_{total} = \tau_{reg} + \tau_{diff}. \tag{6.7}$$

If the incident radiant flux travels through the sample under test in such a way that the exit angle can be predicted from the entry angle according to Snell's law of refraction, the transmittance is referred to as regular. When the incident flux is scattered via its travel through the sample, for example, due to surface roughness or due to volume scattering, so that Snell's law no longer applies, the transmittance is referred to as diffuse.

We do not define transmissive analogs to the reflectance factor or the radiance factor, because it is impossible to have a perfectly transmitting diffuser, or even an approximation to one. Perhaps the best one could imagine would be to have a material that reflected half of the incident radiant flux uniformly in the reflection half-space and transmitted the other half of the incident radiant flux uniformly in the transmitting half-space. But, since no material exist that approximates this behavior, we generally do not speak of a transmittance factor.

6.2.5 Relationship Between Reflectance, Transmittance, and Absorptance

The terms of reflectance, transmittance, and absorptance are altogether defined in Chapter 2, in Eqs. (2.76), (2.82), and (2.89), respectively. These measurands can be combined together, since that which is reflected, transmitted, and absorbed must account for all of the incident flux. Thus, when the reflectance and transmittance includes all of the respective half-spaces, and in the absence of fluorescence and other inelastic scattering (such as Raman scattering), we have

$$\alpha(\lambda) + \rho(\lambda) + \tau(\lambda) = 1, \tag{6.8}$$

where α is the absorptance. Furthermore, Kirchhoff's radiation law states that the emissivity ε of a material is given by its absorptance. Thus, the emissivity is given by

$$\varepsilon(\lambda) = 1 - \rho(\lambda) - \tau(\lambda). \tag{6.9}$$

6.3 NOTATION OF DIFFUSE REFLECTION GEOMETRIES

As explained in Section 6.2, the diffuse reflection classification numbers are not fixed material-specific quantities. Rather, they depend on the geometric conditions of the measurement: the direction of irradiation and reflection, and the respective solid angles. Nicodemus [7] classified nine different measurement geometries for a reflection measurement. This scheme classifies the measurement by whether the incident and reflected radiation are each incident or collected over the entire hemisphere, within some cone, or along a specific direction. Any reflectance measurement must include some statement as to which geometry is used.

Table 6.1 lists these nine classifications of reflectance measurements, where the definitions are given with respect to the bidirectional reflectance distribution function (BRDF), $f_r(\theta_i, \phi_i; \theta_r, \phi_r)$, and where the solid angles for the incident and reflected conical geometries are Ω_i and Ω_r, respectively. In the table, the term directional refers to a differential solid angle $d\Omega$ in the direction specified by θ and ϕ, conical refers to a cone of finite extent centered in the direction of (θ, ϕ), and hemispherical refers to the full half-sphere above the sample. In the conical case, the solid angle Ω and its central direction (θ, ϕ) must be specified. The geometrical conditions are illustrated in Fig. 2.1 of Chapter 2. The nine types of reflectances of Table 6.1 can be visualized as all possible permutations of the different irradiation and detection possibilities, as can be seen in Fig. 6.3. The scheme can be also extended to nine goniometric

TABLE 6.1 Nomenclature for the Nine Types of Reflectance According to Nicodemus [7]

1.	Bidirectional reflectance	$d\rho(\theta_i, \phi_i; \theta_r, \phi_r) = f_r(\theta_i, \phi_i; \theta_r, \phi_r)\ d\Omega_{proj,r}$
2.	Directional–conical reflectance	$\rho(\theta_i, \phi_i; \Omega_r) = \int_{\Omega_r} f_r(\theta_i, \phi_i; \theta_r, \phi_r)\ d\Omega_{proj,r}$
3.	Directional–hemispherical reflectance	$\rho(\theta_i, \phi_i; 2\pi) = \int_{2\pi} f_r(\theta_i, \phi_i; \theta_r, \phi_r)\ d\Omega_{proj,r}$
4.	Conical–directional reflectance	$d\rho(\Omega_i; \theta_r, \phi_r) = (d\Omega_{proj,r}/\Omega_{proj,i}) \int_{\Omega_i} f_r(\theta_i, \phi_i; \theta_r, \phi_r)\ d\Omega_{proj,i}$
5.	Biconical reflectance	$\rho(\Omega_i; \Omega_r) = (1/\Omega_{proj,i}) \int_{\Omega_r}\int_{\Omega_i} f_r(\theta_i, \phi_i; \theta_r, \phi_r)\ d\Omega_{proj,i}\ d\Omega_{proj,r}$
6.	Conical–hemispherical reflectance	$\rho(\Omega_i; 2\pi) = (1/\Omega_{proj,i}) \int_{2\pi}\int_{\Omega_i} f_r(\theta_i, \phi_i; \theta_r, \phi_r)\ d\Omega_{proj,i}\ d\Omega_{proj,r}$
7.	Hemispherical–directional reflectance	$d\rho(2\pi; \theta_r, \phi_r) = (d\Omega_{proj,r}/\pi) \int_{2\pi} f_r(\theta_i, \phi_i; \theta_r, \phi_r)\ d\Omega_{proj,i}$
8.	Hemispherical–conical reflectance	$\rho(2\pi; \Omega_r) = (1/\pi) \int_{\Omega_r}\int_{2\pi} f_r(\theta_i, \phi_i; \theta_r, \phi_r)\ d\Omega_{proj,i}\ d\Omega_{proj,r}$
9.	Bihemispherical reflectance	$\rho(2\pi; 2\pi) = (1/\pi) \int_{2\pi}\int_{2\pi} f_r(\theta_i, \phi_i; \theta_r, \phi_r)\ d\Omega_{proj,i}\ d\Omega_{proj,r}$

The projected solid angle, discussed in Chapter 2, Section 2.2, is $d\Omega_{proj} = \cos\theta\, d\Omega$.

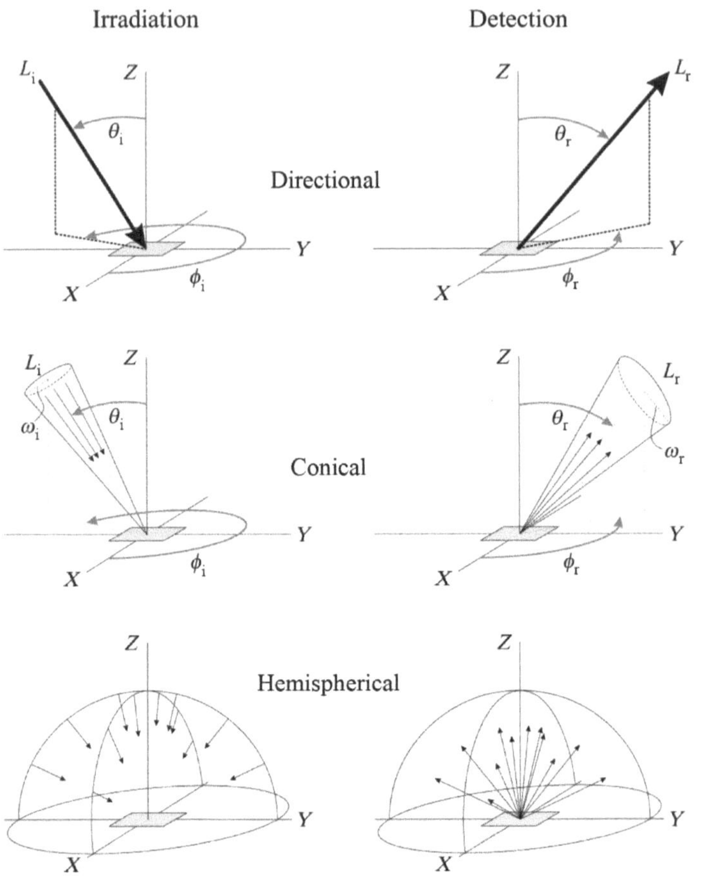

FIGURE 6.3 Designations for incident and reflected geometries, according to Nicodemus [7].

reflectance factors. Because radiance factor implies a directional reflected radiance, the directional–directional, conical–directional, and hemispherical–directional radiance factors can also be defined.

6.3.1 Recommended Integrating Sphere Geometries

The CIE (Commission Internationale de l'Éclairage, English for International Commission on Illumination) recommends a number of geometries for diffuse reflectance and transmittance measurements [8]. These geometries correspond to directional–hemispherical, hemispherical–directional, and hemispherical–hemispherical geometries described above in Table 6.1, with specific directions and one caveat. The geometries distinguish whether or not a small cone about the specular direction is included. The symbol used to indicate hemispherical is d (for diffuse, as opposed to 2π), while the inclusion or exclusion of the specular (or regular, in the case of transmittance) component is

TABLE 6.2 The CIE Recommended Integrated Reflectance and Transmittance Geometries

Reflectance geometries	
(di:8°)	Diffuse:8° geometry, specular component included
(de:8°)	Diffuse:8° geometry, specular component excluded
(8°:di)	8°:diffuse geometry, specular component included
(8°:de)	8°:diffuse geometry, specular component excluded
(d:0°)	Diffuse:normal geometry
(d:d)	Diffuse:diffuse geometry
Transmittance geometries	
(di:0°)	Diffuse:normal geometry, regular component included
(de:0°)	Diffuse:normal geometry, regular component excluded
(0°:di)	Normal:diffuse geometry, regular component included
(0°:de)	Normal:diffuse geometry, regular component excluded
(0°:0°)	Normal:normal geometry
(d:d)	Diffuse:diffuse geometry

indicated by i or e, respectively. Thus, a measurement of the reflectance $\rho(8°; 2\pi)$ would be indicated by (8°:di). For normal incidence, it is generally not possible to include the specular component, since it follows the same beam path as the incident light. Hence, the inclusion/exclusion specification is not needed for normal incidence reflectance measurements.

Table 6.2 lists the six CIE reflectance geometries and the six CIE transmittance geometries. Practical implementation of the measurements will be discussed later in this chapter. In these recommended geometries, is understood that for directional incidence, the angle between the normal to the sample and the optical axis of the incident beam must be within 10°, while the directional beams must be contained within 5° of the optical axis of that beam [3,8]. The set of transmittance geometries are very similar to the reflection geometries. Generally, reflection measurements, if the specular component is to be included, requires nonnormal incidence.

6.4 INTEGRATING SPHERES

An integrating sphere, sometimes known as an Ulbricht sphere, is an optical device for radiometric and photometric measurements consisting of a hollow sphere covered inside with a diffuse white reflective coating [9]. The sphere

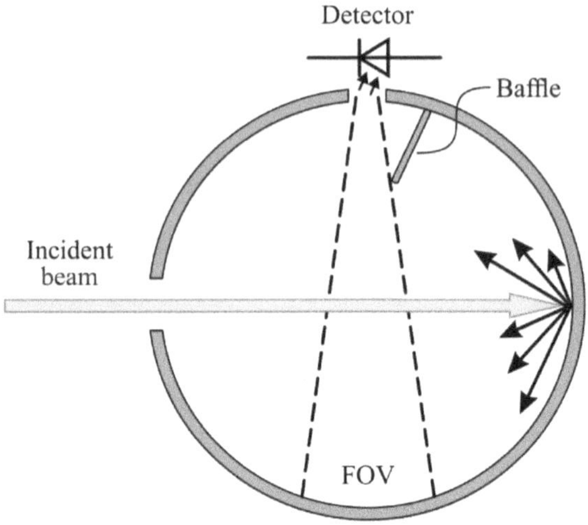

FIGURE 6.4 Typical setup of an integrating sphere measuring system.

setup typically has two or more small openings, denoted as ports, for introducing light or attaching an appropriate detector (see Fig. 6.4). The sphere is often configured with baffles that prevent the detector from directly viewing the light source or the area of the sphere wall directly illuminated by the light source. The arrangement causes many diffuse internal reflections of the entering light. The number of internal reflections, in the absence of loss of light through the ports, is typically on the order of

$$n \approx \frac{1}{1-\rho_w}, \tag{6.10}$$

where ρ_w is the wall reflectance. The value of n can be as high as 100 for a good coating (see also Fig. 6.5). The most relevant property of the integrating sphere is it uniform scattering or diffusing effect. If the sphere coating is Lambertian, then the irradiance of the sphere from light striking a point on the wall is uniform. Thus, it takes very few reflections before the light is completely uniform. An integrating sphere can be considered as a diffusing device which preserves power but destroys spatial information.

The radiance L on the wall of an integrating sphere generated by a flux Φ introduced into the sphere is given by

$$L = \frac{M\Phi}{\pi A_0}, \tag{6.11}$$

where Φ is the incident radiant flux, A_0 is the internal sphere wall area, M is a sphere multiplier quantity, which is equal to the average number n of the reflections within the sphere.

Ideally, the inner wall coating of the integrating sphere should have a high reflectance over the required wavelength range of the application and be nearly

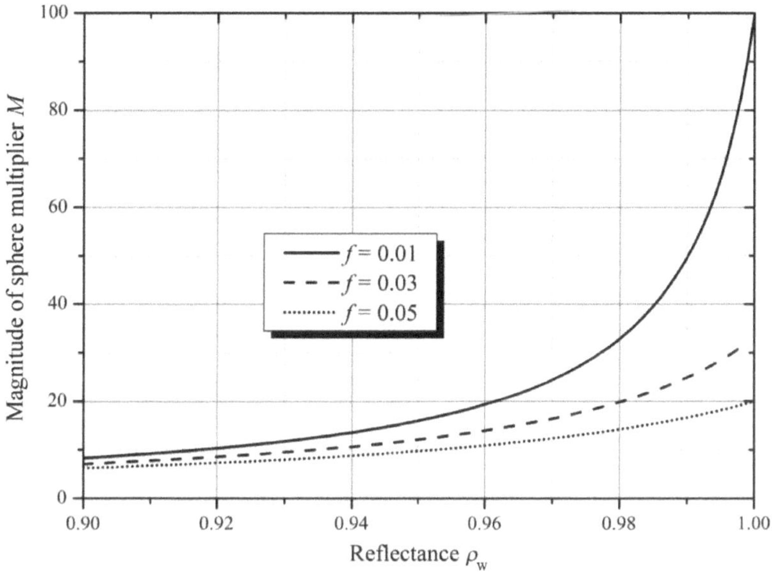

FIGURE 6.5 Dependence of the sphere multiplier quantity M on the sphere wall reflectance ρ_w and the port fraction f.

Lambertian. If the optical losses within the sphere and through the ports are low, the multiple reflections lead to a fairly high optical intensity inside the sphere (sphere multiplier) and consequently to a high optical efficiency,

$$M = \frac{\rho_w}{1 - \rho_w(1 - f)}, \tag{6.12}$$

where f is the fraction of the sphere wall area that is open to ports. Figure 6.5 illustrates the magnitude of the sphere multiplier M and its strong dependence on both the port fraction f and the sphere surface reflectance ρ_w. As a rule of thumb, it can be stated that for most integrating spheres with internal reflectances between $0.94 < \rho_w < 0.99$ and a port fraction between $0.02 < f < 0.05$ the value for the sphere multiplier M is in the range of 10–30.

6.4.1 Historic Information

The theoretical background and practical realization of the integrating sphere as a radiometric and photometric device was stated and demonstrated at the end of the nineteenth century. Two names have to be mentioned in this context, Sumpner and Ulbricht. The chronology and importance of the contributions to the subject may differ under a country specific point of view.

William Edward Sumpner (1865–1940) was an English mathematician and physicist, professor, and head of the Central Technical College of Birmingham. In a publication from 1892, he studied the influence of diffusing, unpolished surfaces on the illumination of ordinary rooms and open spaces [10]. Within this paper, he also introduced generally the concept of the integrating sphere to the field of light measurement.

Friedrich Richard Ulbricht (1849–1923) was a German engineer, professor at the technical university in Dresden, as well as general director of the royal Saxon railway company at this time. In 1892, Ulbricht began with preliminary studies for the construction of an electric power station for the railway station in Dresden. In the course of this work, he was also involved in the illumination of the station itself and the facility sites and offices. His photometric studies lead him, independent from Sumpner, who was working on that subject theoretically only, to the sphere device as an ideal case of uniform illumination by means of a point source. He developed from this insight in 1894 a measurement device consisting of two half spheres, which became the first sphere photometer. In the German-speaking parts of the world, this device is known in his honor as an Ulbricht sphere; otherwise, it is denoted just as an integrating sphere [11,12].

6.4.2 Setup

The design of an integrating sphere for a specific application involves the consideration of some basic parameters. For example, the selection of the optimum sphere diameter is based on the required number and size of the port openings. The selection of the internal sphere coating determines the spectral range as well as the performance [13]. Finally, the use of internal baffles with respect to the incident radiation and detector field-of-view must also be selected. In the following, the basic construction principles and appropriate coating materials are discussed.

6.4.3 Construction Principles

We consider an ideal integrating sphere, in order to show the basic principles of its operation. Such an ideal integrating sphere is characterized by the following properties:

- **i.** the entrance and exit ports are assumed to be infinitesimally small,
- **ii.** all installations inside the sphere, like light sources and baffles, are also considered to be infinitesimally small so that their influence on the optical radiation after the first reflection at the inner sphere surface can be neglected, and
- **iii.** the inner coating of the sphere is assumed to be a Lambertian PRD.

As will be explained in Section 6.5 about absolute sphere based methods, the light entering an integrating sphere should not directly irradiate an area of the sphere wall that the detector is viewing. In order to avoid this, baffles are often used in integrating sphere design. In doing so, however, certain inaccuracies are introduced, because the integrating sphere is no longer perfectly spherical, and light incident on a baffle does not uniformly irradiate the remainder of the sphere.

These considerations can be applied to real-world integrating spheres. As a rule of thumb, the number of baffles inside of the sphere should be minimized, and the surface area of openings (ports) should be less than 10% of the overall area. Intensity considerations pose a lower limit for the size of the entrance and exit ports, as the radiant power entering or exiting a sphere is proportional to the area of the respective port.

The wall reflectance ρ_w will also show a wavelength dependence. This dependence results in a wavelength dependent sphere multiplier M. Apart from these requirements, integrating spheres are also subject to temporal variations of their optical properties, which are primarily caused by degradation of their coating material.

6.4.4 Internal Coating Materials

When choosing a specific coating for an integrating sphere, two factors should be taken into account: its reflectance and its durability. Items located inside the sphere, like baffles or lamps will decrease the throughput of the sphere. This decrease in throughput is best avoided by coating all possible surfaces with a highly reflective coating. This behavior can be considered via the sphere multiplier magnitude M as discussed in Section 6.4.

Based on these conditions the ideal sphere coating can be defined to have the following attributes:

- **i.** high reflectance over the wavelength range of interest
- **ii.** highly diffuse, ideally Lambertian,
- **iii.** highly stable,
- **iv.** easily applied, and
- **v.** easily removable (if necessary for recoating).

Commercial relevant coatings include so-called sprayed-on coatings and packed or sintered polytetrafluoroethylene (PTFE) coatings.

6.4.4.1 Sprayed-On Coatings

Sprayed-on coatings are mostly based upon barium sulfate. The first commercial coating of this type was developed by Franc Grum and George William Luckey of Eastman Kodak, introduced in the 1970s [14,15]. This kind of coating is known as "Eastman 6080 White Reflectance Coating." This coating was the basis for other commercial barium sulfate coatings that later came on the market, like Spectraflect (Labsphere) or ODP97 (Gigahertz-Optik). The composition is made of barium sulfate powder, a polyvinyl acetate binder, water, alcohol, and possibly some coating additives. The advantages of the barium sulfate coatings are that they are highly reflective and highly diffuse, relatively close to Lambertian, easily applied, and easily removable. However, they can be easily damaged, and the spectral region where their reflectance is greater than 0.95 is limited to the range 400–1300 nm. More material depending aspects, including detailed wavelength dependencies, are discussed in Section 6.6.

6.4.4.2 Packed and Monolithic Coatings

Integrating sphere coatings can be produced by packing PTFE powder into the sphere shell with an electrically driven hammer device. In doing so, a constant density of approximately 1.0 $\mathrm{g\,cm^{-3}}$ can be achieved. This kind of internal sphere coating was first introduced at the National Bureau of Standards (NBS) in 1975 [16].

Monolithic integrating spheres are commonly constructed from sintered PTFE-based materials that come with different trade names like Spectralon® (Labsphere), Fluorilon® (Avian Technologies), OP.DI.MA® (Gigahertz-Optik), Zenith Polymer® (SphereOptics), and OptoPolymer (OptoPolymer). As a solid bulk material, they can be machined into a desired shape, including an integrating sphere. The advantage of these materials are that they have a reflectance greater than 0.95 from 250 nm to 2500 nm, are highly diffuse, and mechanically stable. However, their long-term stability is compromised by absorption of gasses from the air. They are relatively heavy, because they are highly translucent, requiring a thickness of at least 1 cm. Finally, they are relatively expensive, compared to the spray-on coatings.

6.4.5 Homogeneity

All integrating sphere techniques make substantial assumptions regarding the sphere's performance. The same also applies for the associated measurement procedure. For example, it is usually assumed that the wall reflectance ρ_w and its BRDF is uniform over the entire inner sphere wall. However, in practice, owing to the type of wall material and the process used to fabricate it, every sphere wall shows inhomogeneities. Furthermore, those inhomogeneities vary with wavelength.

So far, only a limited number of publications report on the performance of integrating spheres, taking into account the spatial and spectral dependence of the wall material [17–21]. Each of the common sphere techniques for the absolute calibration of reflection standards measures a ratio Q of reflection signals (see Sections 6.5.2–6.5.5) obtained when the light is incident upon reflectance standard under test to that obtained when the light is incident upon a specific point within the sphere wall. One important question pertaining to the calibration process is whether the respective point of the inner sphere wall is representative of the whole sphere. Moreover, the evaluation of that part of the uncertainty budget that is connected with the inhomogeneities is not trivial.

The homogeneity characteristics of integrating spheres can be determined using sphere scanning techniques. These techniques rely upon mirror-based rotating scanners to allow a probe to view different positions on the inner sphere wall [20] or light emitting diode (LED) sphere scanners [21] to illuminate different positions of the wall. The results of such investigations on integrating spheres show fractional variations of the sphere wall reflectance ranging from below 1% to several percent.

6.5 ABSOLUTE SPHERE METHODS FOR MEASURING DIFFUSE REFLECTION

Integrating spheres designed for absolute reflectance measurements play an important role in the calibration of diffuse reflections standards. These standards are required for distributing the absolute scale maintained at National Metrology Institutes (NMIs) to commercial reflectance instruments, which rely upon relative techniques. Eight absolute methods are briefly explained and discussed in the following sections. The first four absolute methods were, quite remarkably, all developed nearly simultaneously and published in the year 1920 [22–24].

6.5.1 Basic Considerations

In this section, we provide a theoretical analysis of an ideal integrating sphere and derive its throughput and efficiency. There are four general approaches that can be found in literature:

- **i.** energy balance technique [25], where we assume that the incident flux is the sum of the absorbed flux and that lost through its openings;
- **ii.** a summation of reflections, where the total flux is the sum of that reflected once, twice, etc. [26,27];
- **iii.** a matrix method, which solves a coupled set of rate equations accounting for the presence of a finite number of distinct regions of the sphere [28,29]; and
- **iv.** an integral method, which solves an integral equation of the sphere irradiance [30–32].

Methods (iii) and (iv) can account for the inhomogeneity of the sphere wall and the presence of nonideal barriers. However, for the purposes of illustrating the operation of an integrating sphere, we use the simpler technique (ii). We assume that the sphere is uniformly coated with a material having a Lambertian reflection characteristic, and that there are no baffles to obstruct the ray paths. The explicit spectral dependence of the reflectance quantities is omitted in order to simplify the notation.

The sphere, shown in Fig. 6.6, is assumed to have a radius r with an inner surface area of $A_0 = 4\pi r^2$. Letting the area of the entrance port be A_e, the area of the detector be A_d, and the area of the wall be A_w, we have

$$A_0 = A_e + A_d + A_w. \tag{6.13}$$

Or, expressed in terms of fractional areas,

$$f_e + f_d + f_w = 1, \tag{6.14}$$

where $f_e = A_e/A_0$ is the wall fraction of the entrance port, $f_d = A_d/A_0$ is the wall fraction of the detector, and $f_w = A_w/A_0$ is the wall fraction of the sphere wall. The average diffuse reflectance $\overline{\rho}$ of the sphere can be written in the form:

$$\overline{\rho} = f_e\rho_e + f_d\rho_d + f_w\rho_w, \tag{6.15}$$

where ρ_e, ρ_d, and ρ_w are the reflectances of the entrance port, the detector, and the wall, respectively. Since the entrance port and the detector presumably do not reflect light, $\rho_e = \rho_d = 0$, so that the average reflectance is

$$\overline{\rho} = f_w\rho_w. \tag{6.16}$$

At this point, we will distinguish between the (0°:d) reflectance of the wall, $\rho_{w,0}$, and the (d:d) reflectance of the wall, ρ_w. This distinction is made, because the incident light is directional, while further reflections from the wall arise from nominally diffuse illumination. If Φ_i is the radiant flux incident into the entrance port, the radiant flux incident upon the sphere wall after the first reflection is

$$\Phi_w^{(1)} = f_w\rho_{w,0}\Phi_i. \tag{6.17}$$

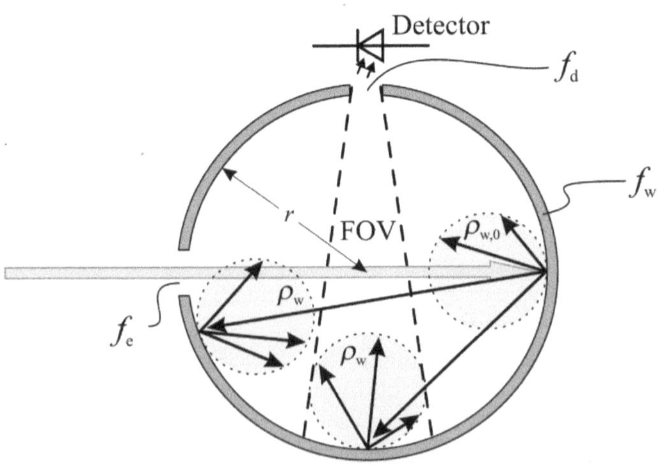

FIGURE 6.6 Graphical representation of the basic features of the summation technique.

The sphere radiant flux incident upon the sphere from the second reflection, is then

$$\Phi_{\mathrm{w}}^{(2)} = \overline{\rho}\,\Phi_{\mathrm{w}}^{(1)} = \overline{\rho} f_{\mathrm{w}} \rho_{\mathrm{w},0} \Phi_{\mathrm{i}}. \tag{6.18}$$

It is straightforward to see that the total radiant flux incident upon the wall is given by

$$\begin{aligned}\Phi_{\mathrm{w}}^{\mathrm{tot}} &= \Phi_{\mathrm{w}}^{(1)} + \Phi_{\mathrm{w}}^{(2)} + \Phi_{\mathrm{w}}^{(3)} + \cdots \\ &= f_{\mathrm{w}}\rho_{\mathrm{w},0}\Phi_{\mathrm{i}} + \overline{\rho} f_{\mathrm{w}}\rho_{\mathrm{w},0}\Phi_{\mathrm{i}} + \overline{\rho}^2 f_{\mathrm{w}}\rho_{\mathrm{w},0}\Phi_{\mathrm{i}} + \cdots \\ &= f_{\mathrm{w}}\rho_{\mathrm{w},0}\Phi_{\mathrm{i}}\left(1 + \overline{\rho} + \overline{\rho}^2 + \cdots\right) \\ &= \frac{f_{\mathrm{w}}\rho_{\mathrm{w},0}\Phi_{\mathrm{i}}}{1-\overline{\rho}}.\end{aligned} \tag{6.19}$$

The total radiant flux reflected by the wall surface is given by

$$\Phi_{\mathrm{r}}^{\mathrm{tot}} = \rho_{\mathrm{w}}\Phi_{\mathrm{i}}^{\mathrm{tot}} = \frac{\overline{\rho}\rho_{\mathrm{w},0}\Phi_{\mathrm{i}}}{1-\overline{\rho}}. \tag{6.20}$$

Finally, the total radiant flux incident upon the detector is that fraction of the reflected light the detector covers

$$\Phi_{\mathrm{d}}^{\mathrm{tot}} = f_{\mathrm{d}}\Phi_{\mathrm{r}}^{\mathrm{tot}} = \frac{f_{\mathrm{d}}\overline{\rho}\rho_{\mathrm{w},0}\Phi_{\mathrm{i}}}{1-\overline{\rho}}. \tag{6.21}$$

These preliminary results can be used to calculate basic equations for the different absolute sphere techniques discussed in the following sections.

6.5.2 Methods of Taylor (First Taylor) and Benford

In his paper from 1920, Albert Hoyt Taylor described five methods for the determination of reflectance [22]. Two of the described methods are absolute, in that they do not rely on the comparison with a previously calibrated standard. The approach for the so-called first Taylor method is shown in Fig. 6.7. The method determines the bihemispherical (d:d) reflectance $\rho(2\pi; 2\pi)$ of a material that is used for the inner coating of the integrating sphere, as well as that for a flat sample to be characterized. The sphere has three ports, namely an entrance, a sample, and a detector port. For the determination of the bihemispherical reflectance $\rho(2\pi; 2\pi)$, two distinct measurements have to be performed.

The first measurement is made with the sample mounted to the sample port. The radiation incident through the entrance port is irradiating a part of the sphere wall in a specific region. The detector is viewing the sphere wall through the detector port in another region, called the detector field of view (FOV). The second measurement is made with identical input and viewing geometry, except that the sample is removed, leaving the sample port open. The ratio Q of both measurements can be used to determine the absolute

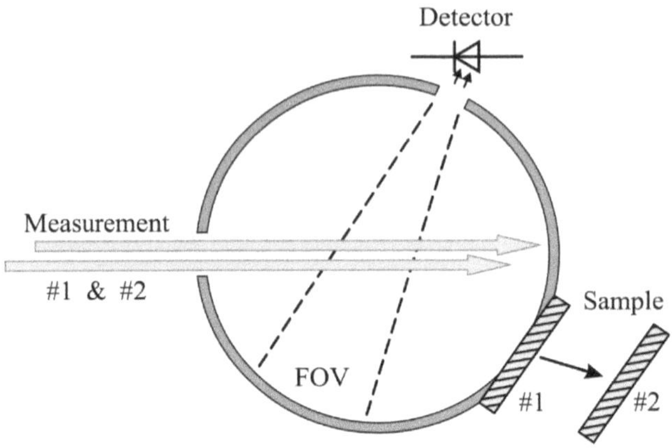

FIGURE 6.7 Graphical illustration of the sphere method according to Taylor (first Taylor) & Benford.

bihemispherical reflectance $\rho(2\pi;2\pi)$ of the sample. The reflectance of other sample materials distinct from the sphere wall coating can then be determined by additional relative measurements in the sphere.

The method of Frank Albert Benford [23], which was also published in 1920, is identical to the first Taylor method, with the minor, but important difference, that the sample is a section of the sphere wall itself, having spherical shape. If we let the sample port have an area A_s and wall fraction $f_s = A_s/A_0$, we can apply Eq. (6.21) for the first measurement, with $f_w \rightarrow f_w + f_s$, so that the radiant flux on the detector is

$$\Phi_{d,1} = \frac{f_d \rho_{w,0} (f_w + f_s) \rho_w}{1 - (f_w + f_s)\rho_w} \Phi_i, \tag{6.22}$$

where f_w is the now the wall fraction of the sphere, not including the sample port. For the second measurement, the radiant flux on the detector is

$$\Phi_{d,2} = \frac{f_d \rho_{w,0} f_w \rho_w}{1 - f_w \rho_w} \Phi_i. \tag{6.23}$$

The ratio of the radiant fluxes is then

$$Q = \frac{\Phi_{d,1}}{\Phi_{d,2}} = \frac{f_w + f_s}{f_w} \cdot \frac{1 - f_w \rho_w}{1 - (f_w + f_s)\rho_w}. \tag{6.24}$$

We can then solve Eq. (6.24) for the wall reflectance,

$$\rho_w = \frac{(1 + f_s/f_w) - Q}{(f_w + f_s)(1 - Q)}. \tag{6.25}$$

We can see that we have relied on the various assumptions used to derive Eq. (6.21). That is, the most accurate measurements using either of these two techniques require accurate modeling of the sphere performance. In particular, we have assumed that the sphere is, in fact, spherical. Thus, the use of flat samples, as described in the first Taylor method, requires a much more complicated description [33].

6.5.3 Method of Taylor (Third Taylor)

The second absolute method described by Taylor [22] in his fundamental publication from 1920 is schematically shown in Fig. 6.8. Interestingly, and leading to confusion, it is the second absolute method of Taylor, but the third method described in his paper (the second method being a measurement relative to a standard). Thus, this absolute method is normally referred to as the third Taylor method.

As in Section 6.5.2, two measurements are performed. In the first measurement, the incoming radiation is incident on the sample and the detector is viewing a separated sphere region, which is shaded from direct irradiation from the illuminated spot by means of an intervening baffle. Following the arguments used to derive Eqs. (6.19)–(6.21), excluding the term from the first reflected rays, we have

$$\begin{aligned}\Phi_{\mathrm{d},1}^{\mathrm{tot}} &= \rho_{\mathrm{w}} f_{\mathrm{d}}\left(\Phi_{\mathrm{w}}^{(2)} + \Phi_{\mathrm{w}}^{(3)} + \cdots\right)\\ &= \rho_{\mathrm{w}} f_{\mathrm{d}}\left(f_{\mathrm{w}}\overline{\rho}\rho_{\mathrm{s},0}\Phi_{\mathrm{i}} + f_{\mathrm{w}}\overline{\rho}^2\rho_{\mathrm{s},0}\Phi_{\mathrm{i}} + \cdots\right)\\ &= \rho_{\mathrm{w}} f_{\mathrm{d}}\left[f_{\mathrm{w}}\overline{\rho}\rho_{\mathrm{s},0}\Phi_{\mathrm{i}}\left(1 + \overline{\rho} + \overline{\rho}^2 + \cdots\right)\right]\\ &= \frac{f_{\mathrm{d}}\overline{\rho}^2\rho_{\mathrm{s},0}\Phi_{\mathrm{i}}}{1-\overline{\rho}}.\end{aligned} \tag{6.26}$$

In the second measurement, the detector has the same viewing geometry, but the sphere is rotated about the entrance port and incident radiation is

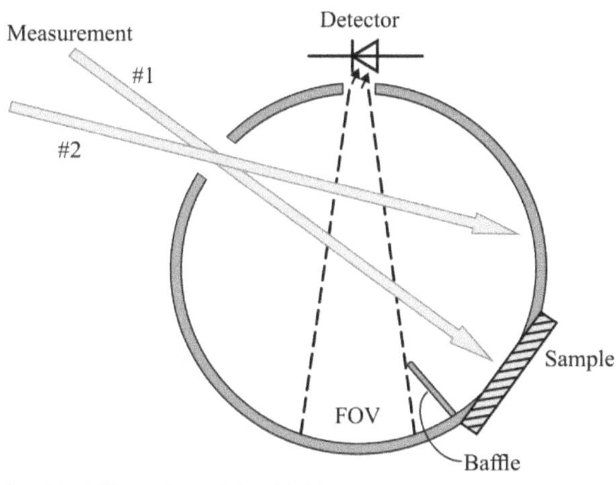

FIGURE 6.8 Graphical illustration of the third Taylor sphere method.

directed on a region that is not baffled anymore from the detector FOV. The signal is then given by Eq. (6.21),

$$\Phi_{\mathrm{d},2}^{\mathrm{tot}} = \frac{f_{\mathrm{d}}\overline{\rho}\rho_{\mathrm{w},0}\Phi_{\mathrm{i}}}{1-\overline{\rho}}. \tag{6.27}$$

The ratio between the two signals, with the assumption $\rho_{\mathrm{w}} = \rho_{\mathrm{w},0}$ is then

$$Q = \frac{\Phi_{\mathrm{d},1}^{\mathrm{tot}}}{\Phi_{\mathrm{d},2}^{\mathrm{tot}}} = \frac{\overline{\rho}\rho_{\mathrm{s},0}}{\rho_{\mathrm{w},0}} = \frac{f_{\mathrm{w}}\rho_{\mathrm{w}}\rho_{\mathrm{s},0}}{\rho_{\mathrm{w},0}} = f_{\mathrm{w}}\rho_{\mathrm{s},0}. \tag{6.28}$$

Thus, the directional–hemispherical reflectance $\rho(0°;2\pi)$ of the sample is

$$\rho_{\mathrm{s},0} = Q/f_{\mathrm{w}}. \tag{6.29}$$

6.5.4 Method of Sharp-Little

Also in 1920, in this apparently remarkable year of absolute integrating sphere method development, Clayton H. Sharp and William F. Little published a paper also describing various integrating sphere methods. The absolute Sharp-Little method [24] is an inversion of the third Taylor method of directional–hemispherical reflectance and is illustrated in Fig. 6.9. The sphere has the same arrangement as in Fig. 6.8, except that the positions of the irradiating source and the detector have been interchanged. In the sample measurement (#1), the detector directly views the sample, while the input beam is incident on a region of the sphere that is shaded by the baffle from direct illumination. For the reference measurement (#2), the detector views another sphere region that is not baffled from the directly illuminated region. Due to symmetry with

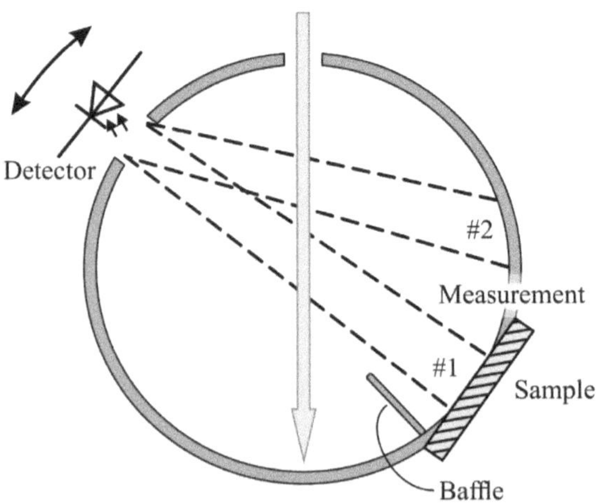

FIGURE 6.9 Graphical illustration of the Sharp-Little sphere method.

Taylors third method, the same equations apply, except that the Sharp-Little method measures the absolute hemispherical–directional reflectance factor $R(2\pi; 0°)$:

$$R(2\pi; 0°) = Q/f_w. \tag{6.30}$$

Differences occur in the methods of third Taylor and Sharp-Little if at the positions of the measurements #1 and #2 are different materials.

6.5.5 Method of van den Akker

In 1956, van den Akker *et al.* [34] published a short 21 line abstract about the "Evaluation of Absolute Reflectance for Standardization Purposes," where they refer to the use of two spheres. Ten years later, in 1966, they refined their work under the same title and gave more detailed information about their method and the apparatus [17]. The van den Akker, or double-sphere method, was used at the NBS in the United States to calibrate a standard reflectance material [35]. The setup of the method measures the bihemispherical reflectance $\rho(2\pi, 2\pi)$ of this material. The two integrating spheres are mounted together as shown in Fig. 6.10.

The van den Akker measurement relies on a measurement of the ratio of the directional–hemispherical reflectance of the sample (measured when it is in position #1) to the effective directional–hemispherical reflectance of a sphere with the sample at the back of that sphere (measured when the sample is in position #2). It makes the assumption that the efficiency of the left sphere does not change between the two measurements and that the signal is proportional to directional–hemispherical reflectance. Thus, the flux measured in position #1 is assumed to be proportional to the directional–hemispherical reflectance of the sample, $\rho_{s,0}$. Letting the average reflectance of the right sphere be

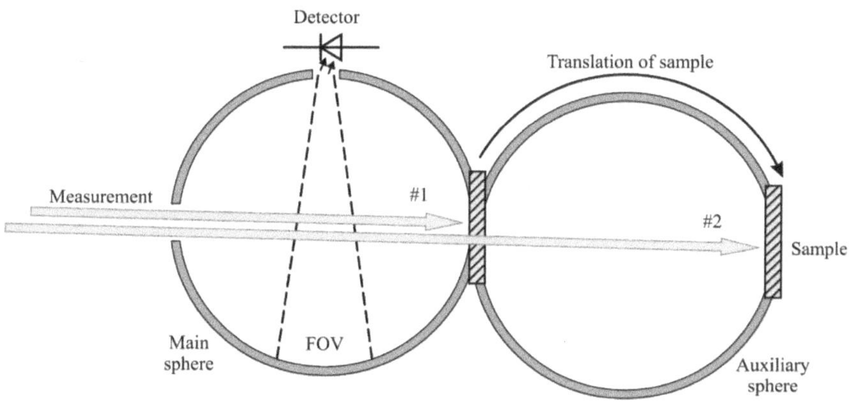

FIGURE 6.10 Graphical illustration of the van den Akker sphere method.

$$\overline{\rho} = f_s \rho_s + \left(1 - f_s - f_p\right) \rho_w, \tag{6.31}$$

where f_p is the port fraction of the entrance/exit port of the right sphere, and f_s is the port fraction of the sample port of the right sphere, the effective directional–hemispherical reflectance of the sphere, with the sample mounted on the back is

$$\rho_{\text{sphere},0} = \frac{f_p \overline{\rho} \rho_{s,0}}{1 - \overline{\rho}}. \tag{6.32}$$

Thus, the ratio of the two signals is given by

$$Q = \frac{\rho_{s,0}}{\rho_{\text{sphere},0}} = \frac{1 - \overline{\rho}}{f_p \overline{\rho}}. \tag{6.33}$$

If the sample material is the same as the sphere wall material, so that $\rho_s = \rho_w$, we have

$$\overline{\rho} = \left(1 - f_p\right) \rho_w. \tag{6.34}$$

The directional–hemispherical reflectance of the sphere wall and the sample is then given by

$$\rho_w = \frac{1 - Q f_p}{1 - f_p}. \tag{6.35}$$

A more complicated expression [35] is found when one includes the non-Lambertian behavior of the integrating sphere.

6.5.6 Method of Korte–Schmidt

The Korte–Schmidt method was developed in the 1950s by Heinrich Korte and Martin Schmidt at the Physikalisch-Technische Bundesanstalt (PTB) in Germany [36] and later revised to its current form by Wolfgang Erb in the 1970s [37,38]. Figure 6.11 illustrates this method. Distinct from the previously described methods, the sample to be measured is located at the center of the sphere, mounted on a holder in such a way that the surface of the sample is lying in the equatorial plane. The sphere itself can be opened and vertically divided into two identical halves for inserting of the sample to be measured. There is a lamp mounted at the back of the sample holder. A baffle between the lamp and the sample ensures that irradiation of the sample surface is only possible after a minimum of two inner wall reflections. Finally, the integrating sphere has two opposing ports allowing radiance measurements in two configurations: one directly viewing the sample, and one viewing the sphere wall at a location indirectly illuminated by the lamp.

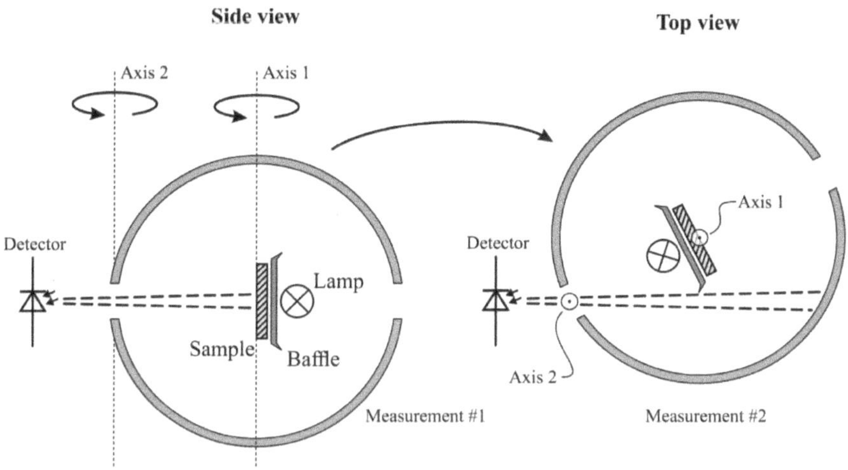

FIGURE 6.11 Graphical illustration of the Korte–Schmidt sphere method.

The radiance from the sphere wall is assumed to be uniform in regions that are not directly illuminated by the lamp. Due to the presence of the viewing port above the sample, the projected solid angle illuminating the sample is reduced. Using Eq. (2.9), the ratio of the projected solid angle illuminating the sample in the first configuration (see Fig. 6.11) to that for a complete half-sphere is given by

$$\frac{\Omega'_{\text{proj}}}{\Omega_{\text{proj}}} = \frac{r^2}{r^2 + r_{\text{p}}^2} = \frac{1}{1 - 4f_{\text{p}}}. \tag{6.36}$$

Recalling the definition of radiance factor β in Section 6.2.3, the radiance factor of the sample is given by

$$\beta = \frac{L_{\text{s}}}{L_{\text{PRD}}}. \tag{6.37}$$

The measurement of the sample radiance L_{s} is performed in the first configuration, while the measurement of the wall radiance L_{w} is performed in the second. Since the two radiance measurements are performed with the same incident radiance, but with different projected solid angles, the ratio of which is given by Eq. (6.36), and assuming the wall of the sphere is a PRD, the radiance factor is given by

$$\beta = \frac{1}{1 - 4f_{\text{p}}} \cdot \frac{L_{\text{s}}}{L_{\text{w}}}. \tag{6.38}$$

While this technique measures the radiance factor β, it is performed under conditions where the illumination is nearly uniform (with the exception of the port direction). Therefore, it is approximately equivalent to the hemispherical–direction reflectance, $\rho(2\pi; 0°)$.

It is typical to perform the wall radiance measurement at an angle of 25°. In this manner, the wall is not directly illuminated by the lamp, and there is negligible reduction of the projected solid angle from the ports. Furthermore, with the Korte–Schmidt method, other viewing angles, such as 8°, can be used, allowing measurement of $\rho(2\pi; 8°)$.

6.5.7 Coblentz Sphere

The Coblentz sphere was invented by William Weber Coblentz [39], an American physicist notable for his contributions to infrared radiometry and spectroscopy. He was working at NBS, where he spent his entire career and headed the radiometry section. The Coblentz sphere, illustrated in Fig. 6.12, is a device for measuring the directional–hemispherical reflectance and can be modified to measure total integrated scatter (TIS) or total scatter (TS). The term TIS is defined in a standard as the diffusely scattered flux (specular excluded), normalized by the hemispherical reflectance (specular included), and is often used in roughness measurements [40], while the term TS is defined in another standard as a particular specular excluded measurement, designed for optical surface testing [41]. The Coblentz sphere differs from the integrating sphere in that the inner surface consists of a hemispherical or hemi-ellipsoidal mirror rather than a diffuse surface coating. This setup is frequently used in the infrared spectral range, where it is difficult to manufacture a material that approximates a PRD. Instead, the inner sphere mirror surface compromises of a highly reflective gold coating. While an integrating sphere setup is usually easier to align, it tends to have a lower signal-to-noise ratio.

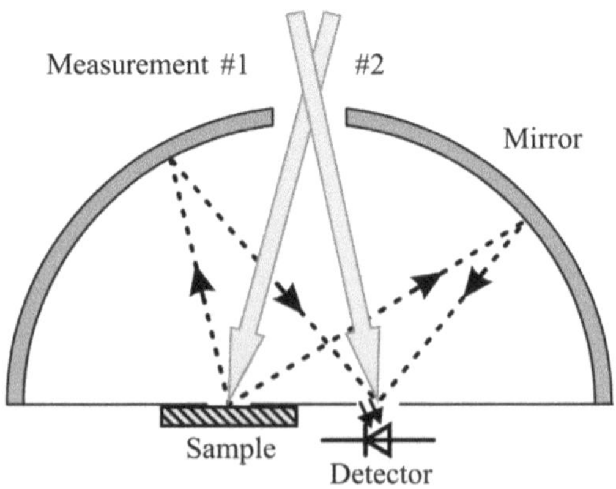

FIGURE 6.12 Graphical illustration of the Coblentz-sphere method.

The sample and detector are positioned at two closely spaced conjugate points near the center of curvature of the hemisphere. The incident radiation is first directed onto the detector (#1) for the measurement of the incident flux. Then the beam is switched to irradiate the sample (#2). The total flux reflected from the sample is collected by the hemispherical mirror and directed to the detector. In this technique, the reflectance ρ_m of the hemispherical mirror must be known. It may be determined if a flat sample of the mirror material is mounted at the sample port and a normal measurement is made. The resulting ratio of the two responses is the square of the mirror reflectance.

The Coblentz-sphere method delivers more signal to the detector; however, some of this signal is incident on the detector at very high angles. Thus, this approach tends to discriminate against high-angle scatter, which is not a problem for many samples. Scatter from rough surfaces, for example, often decays rapidly away from the specular direction.

6.5.8 Goniometric Method

A final method is based upon measurements of the BRDF and its subsequent integration. That is, to calculate the directional–hemispherical reflectance, one can perform the integral

$$\rho(\theta_i, \phi_i; 2\pi) = \int_{2\pi} f_r(\theta_i, \phi_i; \theta_r, \phi_r)\, d\Omega_{proj,r} \tag{6.39}$$

by approximating it with a summation. This method has the potential for being most accurate, since it makes no assumptions about the properties of an integrating sphere or its coatings. However, it can very time consuming, requiring hundreds, if not thousands of measurements. Section 8.2 of Chapter 8 describes an example of this measurement.

6.6 DIFFUSE REFLECTION STANDARDS

Until the CIE Session of 1959 in Brussels, colorimetric measurements of materials were referred not to the PRD but to smoked magnesium oxide [42]. The reflectance of the smoked MgO was assigned to a value of 100% for all wavelengths. Smoked MgO is an example of an artifact-based standard, albeit it is one that was intended to be intrinsic to a material and could be produced by anyone with access to the appropriate materials and methods. However, such standards suffer from a number of deficiencies. For example, it is difficult to control procedures between different laboratories, which are presumably following the same recipe. Furthermore, the reference standard may not be relevant to all of the applications, some of which may be unforeseen at the time the standard was developed.

In the current terminology, the definitions given in Eqs. (6.1), (6.2), and (6.4) replace the smoked MgO artifact-based terminology. In this case, the

PRD is the ultimate reference standard for the measurement, even though a physical artifact PRD is not available. The advantages of this terminology are numerous. The standard is tied to energy balance, so that the directional–hemispherical reflectance is related to how much light is absorbed. This fact will remain true forever, as different materials are developed as intermediate standards or different wavelengths are used. Improvements in measurement methods will reduce the uncertainty in subsequent measurements, rather than change the reference scale.

The transfer of the diffuse reflection scale to customers in research and industry is done by the calibration of commercially available diffuse reflection standards in the desired reflection geometry through a measurement chain, traceable to the PRD. The PRD is the ultimate standard in the field of radiometry and photometry with a radiance factor of 1 (unity) independent of the wavelength and geometric configuration. The PRD cannot be realized as an artifact. Since there is no material with these characteristics, the realization is carried out with an absolute measurement of reflectance, often performed at a NMI. The first material realization is considered a *primary standard*, and it is measured with respect to the PRD definition.

Transfer standards are used to deliver the reflectance scale to outside customers. It is impractical to measure every transfer standard using the absolute methods used to certify primary standards, so transfer standards are usually measured using relative methods, using the primary standard as a reference. Commercial vendors may provide transfer standards, provided they obtain a suitably certified primary standard (usually from an NMI). At last, *working standards* are often used for daily use. The working standard usually has an in-house calibration traceable to a transfer standard.

6.6.1 Requirements

The selection of a material for use as a standard reflection material must take into account not only the physical properties, but also the application. The requirements are similar, but more extensive, than those given in Section 6.4.4 above. The most important properties to be considered are the following:

- **i.** Reflectance: In most cases, a high reflectance is favored.
- **ii.** Surface: For many applications, a matte standard is required, while in other cases, a glossy surface is more preferable.
- **iii.** Opacity: A high opacity is usually desirable in order to avoid edge losses or significant diffusion during measurement.
- **iv.** Uniformity: The standard must be uniform across its surface in order to have a well-defined reflectance.
- **v.** Planarity: The surface of the standard should be flat, since most integrating sphere instruments are sensitive to small gaps between the sample surface and the plane of the sample port.

vi. Cleanability: Cleaning of working standards may be necessary, particularly in industrial use. In that case, cleaning shouldn't change its reflectance value.
vii. Stability: Long-term stability of the material with respect to radiation, temperature, atmosphere, etc., is required.
viii. Portability: Transfer standards must be mechanically stable and durable for shipping.
ix. Spectral flatness: It is highly desirable to have a spectrally flat reflectance. In this case, imperfections in the spectral recording during the measurement have less effect on the transfer of the reflectance value.
x. Inertness: Absence of fluorescence, thermochromism, photochromism, and hydrophilicity.

6.6.2 Common Diffuse Reflection Standards

Standards are often circular or rectangular with a diameter or edge length of 50 mm. For ceramic tiles, it is common to have rectangular samples with an edge length of 10 cm feed size. The most important wavelength range for diffuse reflection standards is the earth observation window, determined by the solar irradiance spectrum, which ranges from 250 to 2500 nm. It is a widely held belief in research and industry that the commonly used diffuse reflection materials only have minor deviations from the PRD in this range. This is a serious and false conclusion. Every standard reflection material existing today exhibits strong deviations from the idealized assumptions of the PRD [6], especially at extreme incident irradiation angles. Incorrect traceability of white standard materials leads to inaccurate values for the determination of luminance and chromatic coordinates. Figure 6.13 below shows photographs of common diffuse reflection standards, illuminated at grazing incidence, and the effects of surface texture can be clearly seen.

Considering the application of reflectance standards, they can be divided into two major categories:

i. Glossy surfaces—Glossy surfaces are usually easy to clean, but their application can be sophisticated, because the specularly reflected radiation must be specified to be included or excluded from the measurement. Reflection standard materials that are available with glossy surfaces are opal glasses and ceramics. A material that approximates the angular reflection behavior of the perfect reflecting diffuser (PRD) in the polar angle range of 0–40° for incidence and reflection are high-quality glossy opal glasses (like BCR 406), which also have a high opacity and uniformity.
ii. Matte surfaces—Generally, cleaning of matte surfaces is difficult and requires a carefully established cleaning procedure in order to yield repeatable results. If detergents are used, it is extremely important to use only high purity chemicals, in order to avoid further contamination due to the cleaning process itself.

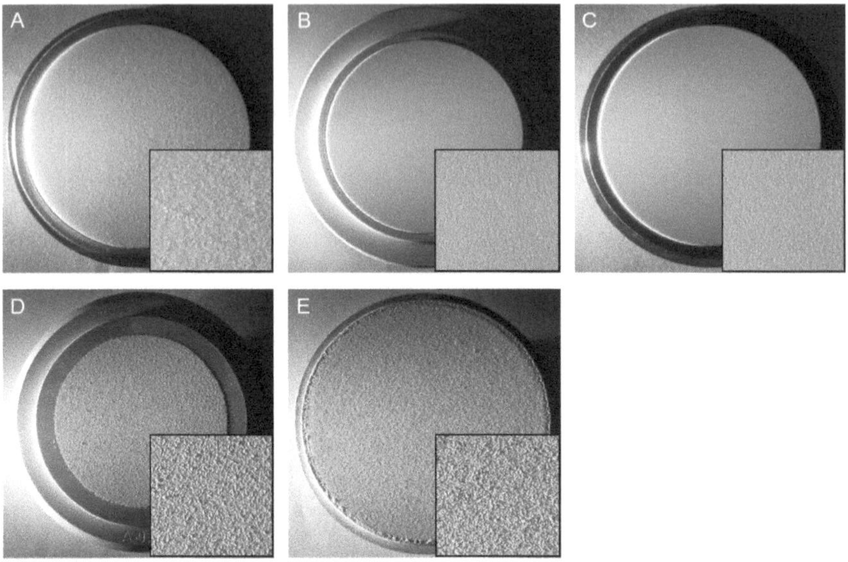

FIGURE 6.13 Standard reflection materials under grazing irradiation in order to show the surface structure. The inserts are magnified views. (A) Sintered PTFE, (B) opal glass, (C) ceramic tile with 88% white, (D) pressed $BaSO_4$ powder, and (E) primed $BaSO_4$ powder.

The spectral properties of four matte standard reflection materials are shown in Fig. 6.14. The most interesting information here is the dramatic decrease in reflection of opal glasses and ceramics below 400 nm. Also, all of the materials shown exhibit absorption bands in the infrared range above 1300 nm.

6.6.2.1 *Historical standards, Smoked MgO, and Vitrolite*

As mentioned before, smoked MgO layers were used for a long time as reference standards for the diffuse reflectance scale. These layers were produced by burning Mg ribbons or turnings in air and collecting the smoke on a substrate. Samples based on smoked MgO were prepared by first compressing a 5-mm layer of MgO powder in an aluminum well and then covering it with a 1-mm layer of smoked MgO [43]. Independent of wavelength, the reflection value of unity or 100% was assigned to them. In general, however, it was very difficult to reproduce high-quality MgO surfaces with constant reflectance. The obtained reflectance is dependent on the kind of carrier, the preparation technique, the environmental conditions like humidity, UV radiation, composition of the atmosphere during burning, the thickness of the layer, impurities of the ribbon used, and the grain size of the deposited particles [44,45]. These variables all lead to the fact that the reproducibility of the smoked MgO layers was never better than $\pm 1\%$, even if they were produced by well-trained

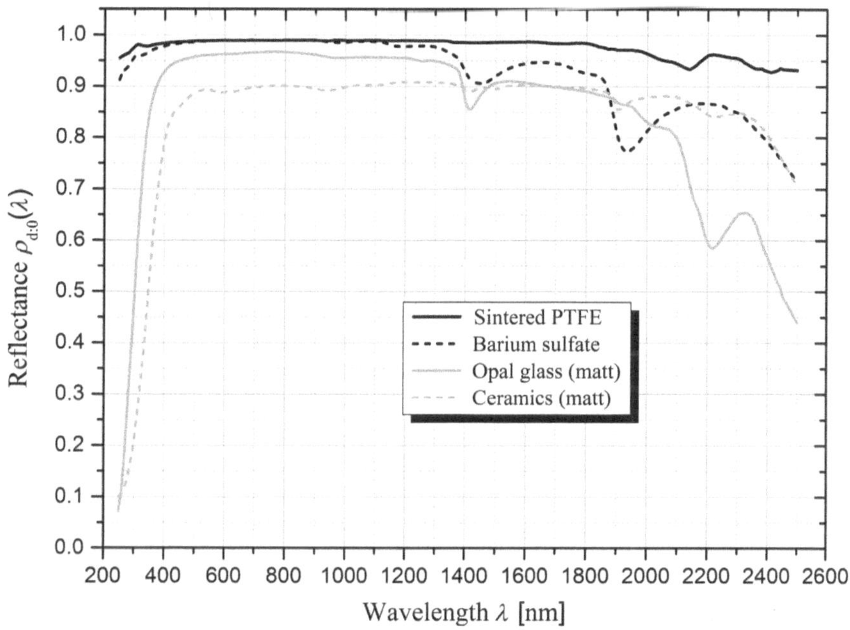

FIGURE 6.14 Spectral dependence of the hemispherical–directional reflectance $\rho_{d:0}(\lambda)$ of four standard reflection materials in the d:0° geometry.

personnel. A furthermore drawback is that smoked MgO surfaces are rather fragile and not stable with time. Therefore, replacement of MgO powder by more suitable materials was investigated.

Such a material was the Vitrolite white opal glass. Vitrolite was a product line of Pilkington Brothers in the UK and widely used in the 1930s and through the 1960s for large-scale cladding and for shop facias and interiors. The product was produced in a wide range of colors and careful temperature control was necessary at all stages in the production process to achieve the required degree of opacity. Vitrolite in its form of a white opal glass has been used as a diffuse spectral reflectance standard since 1944. Its uniformity and long-term durability made it useful as an everyday working standard for spectrophotometric measurements in the visible spectral range. The translucency of Vitrolite, however, can cause systematic errors in diffuse reflection measurements, if improperly used. The principal problem here is the loss of a portion of the incident flux due to diffusion of the light out of the back, the edges, or behind the integrating sphere sample port edge. Vitrolite reflectance standards had a fire-polished surface which gives them durability and stability, along with good cleanability. This polished surface reflects approximately 4% of the incident light as specular reflection.

Different kinds of standard reflection materials are commonly used now (see Fig. 6.13). These materials could be divided into four main groups that are highlighted in the following sections.

6.6.2.2 $BaSO_4$-Based Standards

Barium sulfate is a white crystalline powder. Because of its high electron density, it is also used as a contrast media in X-ray imaging. $BaSO_4$ has a long tradition as a standard material for diffuse reflectance measurements. In the past, it was the white standard reference for colorimetry and photometry in a series of German Institute for Standardization (DIN, Deutsches Institut für Normung) standards [46]. The original white standard, according to the DIN standard, was made of pressed $BaSO_4$ [47]. The pressing is done with a special powder press, originally introduced by the Zeiss company, where the powder is pressed against a matted glass plate. The structure of the glass plate is impressed into the $BaSO_4$ powder. This kind of reflection standard is very fragile and not suitable for shipping. Due to this fact, a version of the standard suitable for overseas shipping with a special internal structure was developed several years ago at PTB [48]. This primed $BaSO_4$ standard has an internal grid structure made of stainless steel, and the first one-third of the filling is an adhesive mixture of $BaSO_4$ with polyvinyl alcohol as a binder. The remaining two-thirds of the filling are made in layers by a composition with $BaSO_4$ diluted in distilled water. There are still a number of users from industry who use $BaSO_4$-based reflection standards, mostly for the calibration of Zeiss Elrepho® integrating sphere reflectometers. Over the years the Zeiss Elrepho® became the *de facto* reference instrument for the paper, pulp, and board industry worldwide.

6.6.2.3 Opal Glasses

Opal glasses have been utilized as diffuse reflection standards since the 1940s. The first samples available were the Vitrolite white opal glasses mentioned above [49]. Due to their relatively high translucency, they were replaced in the 1960s by Russian opals, which had much better spectral reflection properties [50]. The most prominent representative of this class is the MC-20 Russian opals, which became synonymous for the class of opal glasses.

Opal glass materials, which were developed by the European Commission in the 1980s and were purchasable from 1990 until 2012 as certified reference materials, were the BCR-406 opal glasses [51]. These opal glasses had two useable surfaces; one side was glossy, while the other side had a matte surface finish. It is possible to select the desired surface structure by turning the sample over in its protective container. Use of the BCR opal glasses has spread within Europe as a reliable reflection standard. They provide good cleanability and proven long-term stability [52]. Their steep drop in reflectance, however, starting around a wavelength of 400 nm (see Fig. 6.14), makes them unsuitable for UV-based applications.

6.6.2.4 Ceramic Standards

Ceramic reflection standards were also developed in the mid-1960s [53]. The intention was to have a set of permanent, uniform standards which would provide a check on the performance of color measuring instruments over time. The original development was based on the fact that ceramics are inherently very durable and stable over time, as can be seen in many examples from archeological artifacts. Ceramic reflection standards are commercially available with glossy and matte surface finishes. The major advantage of ceramic reflection standards is that they are easy to clean, and they are available in a wide range of colors. The main disadvantage is their relatively moderate maximum reflectance of about 0.9, even for white ceramics, and the massive drop in reflectance starting around 400 nm with values down to 0.1 around 250 nm (see Fig. 6.14). One drawback of ceramic reflection standards is their tendency to exhibit thermochromism, an effect that is most prominent for specific colored samples and can be neglected for white ceramic standards.

6.6.2.5 PTFE-Based Standards

Pressed PTFE powder as a reflection standard was introduced in 1975 by NBS [16]. The PTFE powder material commonly used throughout the optics industry has been Halon PTFE, type G-80, which was abbreviated to Halon. The reflectance of the PTFE powder is influenced by the density and the thickness of the powder layer. When the PTFE powder is used for the production of a reflection standard, a density of about 1 g/cm^3 and a 10-mm thickness is necessary in order to achieve maximum reflectance values of more than 0.96 for the wavelength range 250–2500 nm.

When the PTFE powder is sintered one obtains a solid material with a density of about 2.15 g/cm^3, a crystallinity of about 50–70%, and a porosity of less than 1.0%. Sintered PTFE standards have reflection values greater than 0.93 in the range 250–2500 nm and values of greater than 0.99 in the narrower range between 400 and 1500 nm (see Fig. 6.14). The most prominent representative of sintered PTFE is Spectralon®. It was developed by Labsphere, Inc., and has been commercially available since 1986. There are now many other companies with sintered PTFE materials on the market, including Fluorilon® (Avian Technologies), OP.DI.MA® (Gigahertz-Optik), Zenith Polymer® (SphereOptics), and OptoPolymer (OptoPolymer). Sintered PTFE standards are a well-established reference material for diffuse reflection measurements [54]. Besides having good diffuse scattering behavior, the main advantage over other comparable materials, such as those mentioned above, is its superior reflectance even in the UV spectral region (see Fig. 6.14). The disadvantage of the material is long-term degradation, even when stored in the dark, due to volatile hydrocarbon impurities in the resin from the production process [53]. A second drawback is rapid dust contamination due to electrostatic charging.

6.6.3 Fluorescence of Diffuse Reflection Standards

Fluorescence in the context of diffuse reflection measurements is always a serious problem. The standard approaches for absolute and relative measurements are based on the assumption that the targets (both samples under test and reference standards) are nonfluorescent. This assumption is not always the case. Relying on this assumption, there is generally only spectral selection in one of either the incident irradiation or the detection parts of the instrument. This is an insufficient approach in the case of fluorescent samples, where it is necessary to use spectral selection on both the incident irradiation and the detection. In this case, the irradiation of the sample is carried out at a selectable wavelength, while the reflected/emitted radiation is separately analyzed with a spectrometer or spectrograph. The full reflection description of the fluorescent sample can be reported using the Donaldson-matrix approach [55]. Such measurements are complex and time consuming.

All of the standard methods for measuring diffuse reflection, however, assume nonfluorescent samples. This assumption is more or less fulfilled for the four types of commercially available reference standards mentioned above [37,53,56,57].

Opal glasses can show fluorescence if irradiated below 370 nm. This fluorescence appears as a faint (or if the material is of poor quality, strong) orange emission, which can also have some spatial distribution across the standard (see Fig. 6.15). Often, the nonuniformity of the fluorescence can be also seen

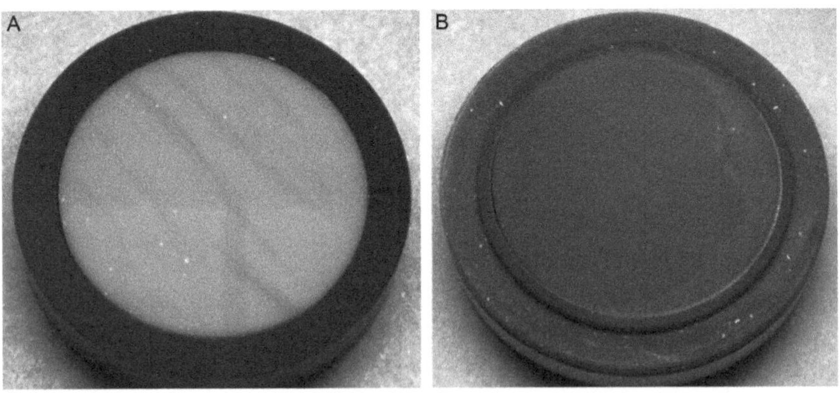

FIGURE 6.15 Black-and-white shot of two different opal glass standards under UV irradiation at 365 nm, (A) is a poor-quality standard showing an intense inhomogeneous orange fluorescence, and (B) is a high-quality standard appearing with a black surface indicating the absence of fluorescence.

as a very faint difference in the surface texture, observable under grazing incidence on a polished surface. Opal glass reflection standards should be inspected for inhomogeneities under a switchable fluorescent lamp and not be used if conspicuous features are noticed or if the fluorescent emission is high.

There is no clear evidence that intrinsic barium sulfate or PTFE powder show fluorescence. However, coatings made from PTFE powder can show very weak fluorescence. This fluorescence may be associated with slight contamination from the plastic bag used to contain the PTFE powder in its shipping drum [16] or from contaminants in the air during fabrication.

6.6.4 Handling Recommendations

Reflectance standards should be handled with great care, the same way as other optical elements, like filters, mirrors, gratings, etc. Although a dedicated material might be durable, care should be taken to prevent contaminants such as fingerprints, oil, or dust from the surface. To a high degree, reflectance is a surface characteristic, even if the reflected radiation is primarily attributed to volume scattering from inside of the material. Objects, such as tweezers can create smooth indentations in the surface and change the reflectance properties dramatically. In handling reflectance standards, it is also recommended that one wear protective gloves made of cotton.

To clean a standard, regardless of its material composition, it should first be cautiously blown clear with dry oil-less air or nitrogen. A good hand device for this activity is an air blower used for photographic or video cleaning applications. If the standard is made of pressed or primed powder, air cleaning is the only appropriate cleaning procedure, because any further treatment would serve to damage the fragile surface of these standards. For all other standards, a soft brush for removing dust can be used in a second step.

Further surface cleaning of a standard depends on the material. Standards made of ceramics or opal glass can be cleaned by using lens cleaning paper held by tweezers and soaked with pure solvents [58], like ethanol or acetone, both in the highest, synthesis-grade quality. In doing so, it is important to work carefully in order to obtain a clean surface, without streaks or scratches.

Occasionally, one will encounter information about reprocessing the surface of sintered PTFE samples using light, wet sanding. From the author's point of view, this is not an appropriate way to clean a standard. Sanding is a massive treatment for a surface and the former surface texture can never be restored. Furthermore, it is more probable that the procedure will produce scratches and a predominant lay to the surface. Given that reflection standards generally have uncertainties near 0.001, any sanding treatment should void any certified value assigned to them.

Finally, standards should be stored in a clean laboratory environment. It is recommended to keep them always in a protective, nontransparent box. The best way is also to control the environmental conditions of temperature and humidity, by storing them in the dark in an desiccator.

6.7 RELATIVE SPHERE METHODS FOR MEASURING DIFFUSE REFLECTION

Integrating sphere setups for relative reflectance measurements measure the reflectance relative to an absolute calibrated reflection standard. These methods can be divided into two categories: substitution methods and comparison methods.

An integrating sphere of the substitution type is by default equipped with a single port that is used successively for both measurements, attaching the unknown sample, followed by the calibrated standard. Such a sphere instrument is simple in design. It can be built very compact and requires neither moving parts nor relative motion or switching between different beams in the integrating sphere. The major drawback of the substitution method is that the sphere throughput changes between the measurement of the sample under test and the reflection standard, especially if the samples are quite different. This mismatch can be corrected by detailed knowledge on the sphere parameters, such as the wall reflectance and the specific sizes and locations of the different ports [33].

An integrating sphere of the comparison mode type is one that uses two ports in order to attach sample and standard simultaneously, having a mechanism for moving either the sphere or the input beam to switch between sample and reference measurement. Since the sphere does not change between measurements, the throughput of the sphere does not change, and no correction is necessary. Most commercial spectrophotometers now operate as double-beam devices, comparing the signal between two light paths, one path containing the reference standard and the other one equipped with the sample under test. Figure 6.16 shows a typical design for measuring relative reflectance in comparison mode.

Commercial spectrophotometers are designed as multifunction instruments used in physics, chemistry, or biology for measuring reflection or transmission properties of materials as a function of wavelength. These instruments typically cover the ultraviolet, visible, and near-infrared (UV–vis-NIR) wavelength range from 200 nm to 3.2 μm. Because these instruments are mostly developed for the measurement of regular transmittance, it is necessary to use attachments in the sample beam to accommodate measurements of diffuse reflectance and transmittance, as well as specular reflectance. For these purposes, a variety of accessories are available [59].

In Section 6.3.1, it was stated that the CIE recommends six different integrating sphere geometries for diffuse reflectance measurements [7]: di:8°, de:8°, 8°:di, 8°:de, d:0°, and d:d. The first five of these geometries are of

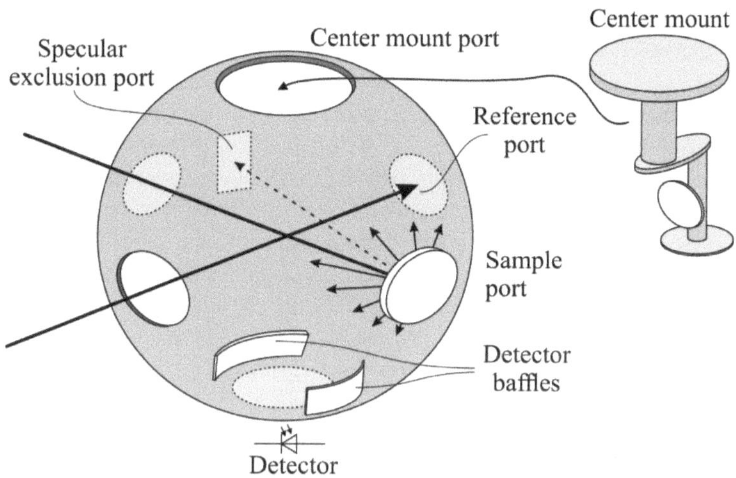

FIGURE 6.16 Typical layout of a commercial integrating sphere accessory for a double-beam spectrophotometer.

the most practical importance and are realized in commercial devices. The differences between specular-included and -excluded geometries are bridged by specular exclusion ports, also referred to as gloss traps, that can be operated open (for 8°:de) or closed up by a piece of sphere wall material (for 8°:di), as can be seen in Fig. 6.16.

Another class of accessory for diffuse reflectance measurements with integrating spheres are center mount sample holders (see Fig. 6.16). This type of sphere design could be either used in substitution or comparison mode with the ability to hold cuvettes, thin films, or reflection standards in the middle of the sphere and where the angle of incidence can be varied due to rotation of the attached holder [60,61].

The geometries d:8° (diffuse irradiation selectable as di and de) and the d:0° geometry are realized in special bench-top spectrophotometers that can come in horizontally or vertically aligned devices. This kind of equipment is mainly used for reflectance and transmittance color measurements. The d:8° geometry offers particular advantages in the measurement of glossy or structured samples. Observation of the diffusely illuminated sample at an angle of 8° allows the influence of the mirror-reflected component of glossy samples to be eliminated by inserting a beam trap at an angle of −8° in the integrating sphere. An alternative measurement of gloss is possible. Here, an additional light source illuminates the sample under test with a directional light beam at −8° to the sample normal. The relation between directional and diffuse reflection allows calculation of the gloss component.

There are two main advantages of the hemispherical–directional (e.g., d:0°) geometry, in comparison to the directional–hemispherical (e.g., 0°:d) geometry.

The incident flux is significantly greater, since the integrating sphere provides total light collection, thus increasing the signal-to-noise ratio for the instrument. Polychromatic irradiation will stimulate any inelastic scattering, such as fluorescence, which often needs to be quantified in color and appearance measurements. The main disadvantage of hemispherical–directional (e.g., d:0°) instruments is that potential sample heating can induce thermochromic effects in some samples. Many commercial color measurement instruments utilize flash lamps in order to reduce such thermal measurement problems.

Due to historical reasons, based on ISO-standards (ISO: International Organization for Standardization) [62,63], the so-called ISO-brightness is linked to the Zeiss Elrepho® [64–66] which is a d:0° instrument. The concept of the ISO-brightness was developed at the beginning of the 1970s as a way of monitoring the bleaching process in paper production. At the same time, paper manufacturers became interested in measuring the brightness, not only of the pulp they purchase, but also of the paper and board they produce. This is the reason that the d:0° geometry is now commonly used in the pulp and paper industry, and the Zeiss Elrepho® reached a status as the *de facto* reference instrument there, as well as in the sugar industry.

At the end of this section, we make a concluding remark. It has been found that for viewing angles θ_r in the range $0° \leq \theta_r \leq 10°$, the spectral radiance factor β for diffuse illumination rarely changes within the uncertainty budget of any measurement [66].

6.8 DIFFUSE TRANSMITTANCE MEASUREMENTS

When a beam of radiation passes through a transparent material along its surface normal, its intensity will decrease due to absorption and reflection. If the direction of propagation does not change, the transmittance is referred to as regular transmittance. When incident radiation passes through a hazy or translucent material along its surface normal, some of the light will scatter into other directions. This transmittance is referred to as diffuse transmittance.

One obvious difference between diffuse transmittance and diffuse reflectance measurements with integrating spheres is the opposing directions of the scattered radiation. This situation makes diffuse transmittance measurements more complicated than diffuse reflectance measurements. After diffuse transmission through the sample, the cross section of the incident radiation will increase, making it necessary to construct the entrance port of an integrating sphere used for diffuse transmission measurements to be about twice as big as the incident beam cross section.

Measurements of diffuse transmittance are made using an integrating sphere, where the sample is placed in front of the sphere entrance port through which the irradiating beam is entering. The sphere collects all of the radiation that is transmitted through the sample. Figure 6.17 illustrates how a sphere can be configured to distinguish between the total diffuse transmittance τ_{total}

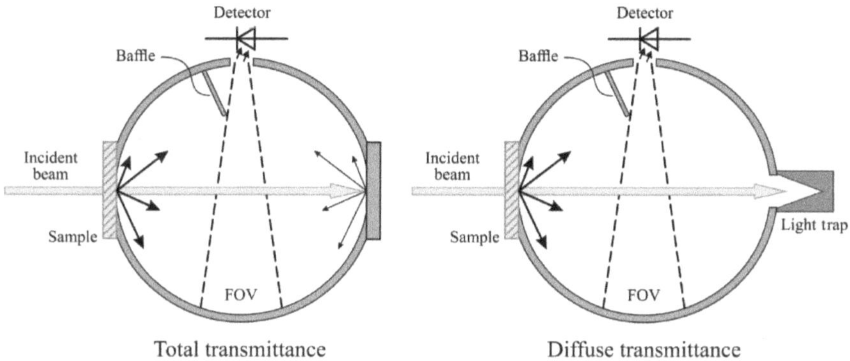

FIGURE 6.17 Typical setup of an integrating sphere system for transmittance measurements.

and diffuse transmittance τ_{diff}, in a manner similar to specular inclusion or exclusion in the reflection measurement. For measurements of total transmittance, the back wall of the sphere is covered by a material identical to the sphere wall, so that all of the transmitted radiation is detected. For measurements of diffuse transmittance, excluding the regular component, the back of the sphere is opened up and replaced with a light trap (see Fig. 6.17).

The CIE recommends six irradiating geometries for transmittance measurements [8]. Most transmittance measurements are made with commercial spectrophotometers designed for analytical chemistry applications configured to measure in the normal/normal (0°:0°) geometry. The incident radiation is directed along the surface normal, and the detector on the other side of the sample is also viewing along the surface normal. With this arrangement only regular transmittance can be measured.

Total diffuse transmittance can be measured using the 0°:d geometry or the equivalent inverse d:0° geometry using an integrating sphere accessory. Regular transmittance is measured by placing the sample as far as possible from the entrance port of the integrating sphere. Diffuse transmittance is measured by placing the sample in contact with the sphere port. Instruments designed especially for transmittance measurements have only a sample and a detector port. Instruments designed for reflectance measurements can also be used by blocking the sample port with a material similar to the inner wall coating and using the specular-included mode of the instrument. The entrance port would then serve as the sample port for the transmittance measurement.

The last geometry recommended by the CIE for transmittance measurements is the less common diffuse–diffuse (e.g., d:d) geometry. In this case, one integrating sphere is placed in front of the sample and used to provide diffuse irradiation, while a second sphere behind the sample is used for the collection of the diffusely transmitted radiation. The diffuse–diffuse configuration is used in life science applications for the measurement of diffuse transmission of skin or thin slices of biological tissue [67].

An interesting application of the measurement of total diffuse transmittance is the determination and certification of textiles that claim protection from UV sunlight radiation. Measurements are made according to appropriate international standards. These require the measurement of the total diffuse transmittance of textile samples at UV wavelengths from 280 to 400 nm. The integrating sphere enables collection of all the light that passes through the textile sample. The resulting spectral diffuse transmittance values are then used in conjunction with solar spectra data and biological activation data to calculate the required UV and solar protection factors [68].

REFERENCES

[1] X.D. He, K.E. Torrance, F.X. Sillion, D.P. Greenberg, A comprehensive physical model for light reflection, Comput. Graph. 25 (1991) 175–186.

[2] H.B. Westlund, G.W. Meyer, Applying appearance standards to light reflection models, SIGGRAPH '01, in: Proceedings of the 28th Annual Conference on Computer Graphics and Interactive Techniques, 2001, pp. 501–551.

[3] Practical Methods for the Measurement of Reflectance and Transmittance, International Commission on Illumination, 1998, CIE 130:1998.

[4] G. Kortüm, Reflexionsspektroskopie, Grundlagen, Methodik, Anwendungen, Springer-Verlag, Berlin, 1969.

[5] B. Judd, CIE Committee W-1.3.1 "Colorimetry" in: Proceedings of the CIE Session Bruxelles, 1959, pp. 91–107.

[6] A. Höpe, K.-O. Hauer, Three-dimensional appearance characterization of diffuse standard reflection materials, Metrologia 47 (2010) 295–304.

[7] F.E. Nicodemus, J.C. Richmond, J.J. Hsia, I.W. Ginsberg, T. Limperis, Geometrical Considerations and Nomenclature for Reflectance, 1977, NBS Monograph 160.

[8] Colorimetry, International Commission on Illumination, 2004, CIE 15:2004.

[9] L.M. Hanssen, K.A. Snail, Integrating spheres for mid- and near-infrared reflection spectroscopy, in: J.M. Chalmers, P.R. Griffiths (Eds.), Handbook of Vibrational Spectroscopy, John Wiley & Sons Ltd, Chichester, 2002.

[10] W.E. Sumpner, The diffusion of light, Phys. Soc. Proc. 12 (1892) 10–29.

[11] R. Ulbricht, Die Bestimmung der mittleren räumlichen Lichtintensität durch nur eine Messung, Elektrotechnol. Zeit. 21 (1900) 595–597.

[12] R. Ulbricht, Das Kugelphotometer (Ulbricht'sche Kugel): Darstellung seiner Theorie, Ausbildung und Anwendung, unter besonderer Berücksichtigung der Fehlerquellen, R. Oldenbourg, 1920, p. 110.

[13] E.B. Rosa, A.H. Taylor, Theory, Construction and Use of the Photometric Integrating Sphere, in: Scientific Papers of the Bureau of Standards, vol. 75, 2013.

[14] F. Grum, G.W. Luckey, Optical sphere paint and a working standard of reflectance, Appl. Opt. 11 (1968) 2289–2294.

[15] F. Grum, T.E. Wightman, Absolute reflectance of Eastman white reflectance standard, Appl. Opt. 16 (1977) 2775–2776.

[16] V.R. Weidner, J.J. Shaw, Reflection properties of pressed polytetrafluoroethylene powder, J. Opt. Soc. Am. 71 (1981) 856–861.

[17] J.A. van den Akker, L.R. Dearth, W.M. Shillcox, Evaluation of absolute reflectance for standardization purposes, J. Opt. Soc. Am. 56 (1966) 250–252.

[18] Y. Ohno, Detector-based luminous-flux calibration using the absolute integrating-sphere method, Metrologia 35 (1998) 473–478.
[19] L.M. Hanssen, S. Kaplan, Infrared diffuse reflectance instrumentation and standards at NIST, Anal. Chim. Acta. 380 (1999) 289–302.
[20] A. Höpe, Homogeneity measurements in a Korte-Schmidt-type integrating gold sphere, Appl. Opt. 43 (2004) 4467–4473.
[21] S. Winter, M. Lindemann, W. Jordan, U. Binder, M. Anokhin, Convenient integrating sphere scanner for accurate luminous flux measurements, Metrologia 46 (2009) S248–S251.
[22] A.H. Taylor, The measurement of diffuse reflection factors and a new absolute reflectometer, J. Opt. Soc. Am. 4 (1920) 9–23.
[23] F.A. Benford, An absolute method for determining coefficients of diffuse reflection, General Electr. Rev. 23 (1920) 72–75.
[24] C.H. Sharp, F.W. Little, Measurement of reflection factors, Trans. Illum. Eng. Soc. 15 (1920) 802–810.
[25] E. Karrer, Use of the Ulbricht sphere in measuring reflection and transmission factors, in: Scientific Papers of the Bureau of Standards, vol. 17, 1921, Scientific Paper 415.
[26] D.G. Goebel, Generalized integrating-sphere theory, Appl. Opt. 6 (1967) 125–128.
[27] L.M. Hanssen, Effects of restricting the detector field of view when using integrating spheres, Appl. Opt. 28 (1989) 2097–2103.
[28] B.J. Hisdal, Reflectance of perfect diffuse and specular samples in the integrating sphere, J. Opt. Soc. Am. 55 (1965) 1122–1125.
[29] H.L. Tardy, Matrix method for integrating-sphere calculations, J. Opt. Soc. Am. A 8 (1991) 1411–1418.
[30] J.A. Jacques, H.F. Kuppenheim, Theory of the integrating sphere, J. Opt. Soc. Am. 45 (1955) 460–470.
[31] J.A. Jacques, W. McKeehan, J. Huss, J.M. Dimitroff, H.F. Kuppenheim, An integrating sphere for measuring diffuse reflectance in the near infrared, J. Opt. Soc. Am. 45 (1955) 781–784.
[32] H.L. Tardy, Flat-sample and limited-field effects in integrating sphere measurements, J. Opt. Soc. Am. A 5 (1988) 241–245.
[33] F.J.J. Clarke, J.A. Compton, Correction methods for integrating-sphere measurement of hemispherical reflectance, Color. Res. Appl. 11 (1986) 253–262.
[34] J.A. van den Akker, L.R. Dearth, W.M. Shillcox, Evaluation of absolute reflectance for standardization purposes, J. Opt. Soc. Am. 46 (1956) 378–379.
[35] W.H. Venable, J.J. Hsia, V.R. Weidner, Establishing a scale of directional-hemispherical reflectance factor I: the van den Akker method, J. Res. Natl. Bur. Stand. 82 (1977) 29–55.
[36] H. Korte, M. Schmidt, Über Messungen des Leuchtdichtefaktors an beliebig reflektierenden Proben, Lichttechnik 19 (1967) 135A–137A.
[37] W. Erb, Änderungen der strahlungsphysikalischen Eigenschaften von Reflexionsnormalen nach UV-Bestrahlung, Lichttechnik 7 (1973) 345–349.
[38] W. Erb, Eine Anordnung zur Messung des spektralen Reflexionsgrades $\rho(\lambda)$ für die Geometrie 0/d und die erforderlichen Korrekturen, PTB Mitteilungen 87 (1977) 283–288.
[39] W.W. Coblentz, The diffuse reflecting power of various substances, Bull. Bur. Stand. 9 (1913) 283–325.
[40] SEMI MF 1048-1111, Test method for measuring reflective total integrated scatter. Semiconductor Equipment and Materials International, San Jose, CA, USA. 2011.
[41] ISO 13696:2002, Optics and optical instruments—Test methods for radiation scattered by optical components. International Organization for Standardization, Geneva, Switzerland.

[42] J. Schanda (Ed.), Colorimetry: Understanding the CIE System, Wiley-Interscience, New Jersey, 2007.
[43] P.A. Tellex, J.R. Waldron, Reflectance of magnesium oxide, J. Opt. Soc. Am. 45 (1955) 1922.
[44] W. Erb, Requirements for reflection standards and the measurement of their reflection values, Appl. Opt. 14 (1975) 493–499.
[45] W. Budde, Calibration of reflectance standards, J. Res. Natl. Bur. Stand. A 80 (1976) 585–595.
[46] Colorimetry-Part 9: Reflectance Standard for Calibration in Colorimetry and Photometry, Beuth Verlag GmbH, Berlin, 2005, DIN 5033-9 (Issues 1970–2005).
[47] P. Nikolaus, Die Abhängigkeit der Reflexionswerte von Weißstandards nach DIN 5033 von der Art ihrer Herstellung, LICHT-Forschung 5 (1983) 93–96.
[48] W. Erb, Ein Reflexionsnormal für den spektralen Strahldichtefaktor, Lichttechnik 27 (1975) 250–253.
[49] V.R. Weidner, White Opal Glass Diffuse Spectral Reflectance Standards for the Visible Spectrum (SRM's 2015 and 2016), NBS Special Publication, 1983, pp. 260–282.
[50] F.J.J. Clarke, F.A. Garforth, D.J. Parry, Goniophotometric and polarization properties of white reflection standard materials Light, Res. Technol. 15 (1983) 133–149.
[51] W. Erb, Zertifizierte BCR-Opalgläser als Reflexionsnormale, PTB-Mitteilungen 104 (1994) 20–22, 1/94.
[52] J.C. Zwinkels, W. Erb, Comparison of absolute d/0 diffuse reflectance factor scales of the NRC and the PTB, Metrologia 34 (1997) 357–363.
[53] F.J.J. Clarke, P.R. Samways, The spectrophotometric properties of a selection of ceramic tiles: NPL Report MC2, 1968.
[54] A.W. Springsteen, A novel class of Lambertian reflectance materials for remote sensing application, in: Optical Radiation Measurements II, Proc. SPIE, Vol. 1109, 1989, pp. 133–141.
[55] R. Donaldson, Spectrophotometry of fluorescent pigments, Br. J. Appl. Phys. 5 (1954) 210–214.
[56] W. Möller, K.-P. Nikolaus, A. Höpe, Degradation of the diffuse reflectance of spectralon under low-level irradiation, Metrologia 40 (2003) 212–215.
[57] P. Kruppa, H. Marchandise (Eds.), The Certification of Opal Glass, Commission of the European Communities, Community Bureau of Reference, CEC, Brussels, Belgium, CRM 406, 1990.
[58] C.J. Williamson, Durability and cleaning of matt surface ceramic colour standards, Anal. Chim. Act. 380 (1999) 413–416.
[59] S.L. Storm, A. Springsteen, T.M. Ricker, The use of center mount sample holders in reflectance spectroscopy: Labsphere Appl. Note No. 02, Labsphere, North Sutton, NH, 1998.
[60] H.H. Safwat, Effect of centrally located samples in the integrating sphere, J. Opt. Soc. Am. 60 (1970) 534–541.
[61] S.L. Storm, A. Springsteen, The integrating sphere reflectance accessory, Labsphere Appl. Note No. 03, Labsphere, Inc., North Sutton, NH 03260, USA, 2000.
[62] ISO 2469:1977, Paper, board and pulps—Measurement of diffuse radiance factor.
[63] ISO 2470:1973, Paper and board—Measurement of diffuse blue reflectance factor (ISO brightness).
[64] H.J. Höfert, Ein Filterphotometer zur Remissionsmessung, Z. Instr. 67 (1959) 118–124.
[65] W. Budde, S.M. Chapman, The calibration of standards for "Absolute Brightness" measurements with the Elrepho, Pulp Pap. Mag. Canada 69 (1968) 153–156.

[66] W. Erb, K.-P. Nikolaus, Untersuchung der Abhängigkeit des Strahldichtefaktors diffus angestrahlter Reflexionsnormale vom Abstrahlwinkel, PTB-Mitteilungen 107 (1997) 311–315.

[67] J.W. Pickering, S.A. Prahl, N. van Wieringen, J.F. Beek, H.J.C.M. Sterenborg, M.J.C. van Gemert, Double-integrating-sphere system for measuring the optical properties of tissue, Appl. Opt. 32 (1993) 399–410.

[68] P.G. Parsons, R. Neale, P. Wolski, A. Green, The shady side of solar protection, Med. J. Aust. 168 (1998) 327–330.

Chapter 7

Spectral Fluorescence Measurements

Joanne C. Zwinkels*, Paul C. DeRose† and James E. Leland‡
*National Research Council Canada, NRC, Measurement Science and Standards, Ottawa, Ontario, Canada
†National Institute of Standards and Technology (NIST), NIST, Material Measurement Laboratory, Gaithersburg, Maryland, USA
‡Copia LLC, Goshen, New Hampshire, USA

Chapter Outline

7.1 Introduction 222
7.1.1 Fluorescence Measurement Applications 223
7.2 Fundamental Concepts and Terminology 224
7.2.1 Quantities Common to Analytical and Colorimetric Applications 225
7.2.2 Specific to Analytical Applications 230
7.2.3 Specific to Colorimetric Applications 232
7.3 Measurement of Fluorescence Characteristics 238
7.3.1 Bispectral Measurements 238
7.3.2 Abridged Measurements 239
7.4 General Instrument Design and Measurement Considerations 240
7.4.1 Instrument Components 240
7.4.2 Spectral Measurement Methods 241
7.4.3 General Considerations and Good Laboratory Practices 243
7.5 Specialized Instrument Designs and Measurement Methods 244
7.5.1 Analytical Applications 244
7.5.2 Colorimetric Applications 247
7.5.3 Quantum Yield Measurements 256
7.6 Instrument Characteristics That Impact Spectral Fluorescence Measurements 260
7.6.1 Spectral Characteristics 260

Experimental Methods in the Physical Sciences, Vol. 46. http://dx.doi.org/10.1016/B978-0-12-386022-4.00007-8

7.6.2 Detector Nonlinearity 264
7.6.3 Polarization Effects 266
7.6.4 Stray Light 269
7.6.5 Instrument Lineshape 270
7.6.6 Sampling Effects 271
7.7 Sample Characteristics That Impact Fluorescence Measurements 274
7.7.1 Optical Density 274
7.7.2 Solvent/Matrix Properties 275
7.7.3 Refractive Index 275
7.7.4 Impurities 276
7.7.5 Luminescence Lifetime 277
7.7.6 Scattering 277
7.8 Standards for Spectral Fluorescence Measurements 278
7.8.1 Correcting Relative Excitation and Emission Spectra 278
7.8.2 Intensity Verification Standards 283
7.8.3 Fluorescent Color Standards 284
7.8.4 Quantum Yield Standards 286
References 287

7.1 INTRODUCTION

Photoluminescence is a general term for the phenomenon caused by absorption of optical radiation in one band of wavelengths, which we refer to as the *excitation range*, and emission of some of this radiation in another band of wavelengths, which we refer to as the *emission range*. When a fluorescent material absorbs light in its excitation band, it causes a *fluorophore* to be excited from its ground state to a higher electronic energy state of the same multiplicity. This fluorophore then loses its excess vibrational energy by collision or some other radiationless process so that further processes occur from the lowest vibrational level of the excited energy state. *Fluorescence* occurs if the fluorophore returns directly to any of the vibrational levels in the ground state, where the emitted radiation generally appears within 10 ns after the excitation [1]. This emission time is short because it involves an allowed transition of an electron from an excited state to its ground state, typically via a singlet–singlet transition.

Phosphorescence is another species of photoluminescence and is distinguished from fluorescence by the type of electron state transfer involved. Specifically, phosphorescence is light emission caused by a forbidden transition of an electron between states of different multiplicity, typically via a triplet–singlet transition. In phenomenological terms, the primary difference between these two species of photoluminescence is the average delay time between excitation and emission of the excited molecules. For fluorescent materials, the average delay time between excitation and emission, or *lifetime*, is very short, on the order of 10^{-10}–10^{-8} s, whereas for phosphorescent materials, the lifetime can be much longer, on the order of 10^{-3}–10 s. In both cases, since energy is conserved, the emitted energy of the given fluorophore is lower than the absorbed energy, and the emission wavelength is longer than the excitation wavelength. This characteristic is called *Stokes'-law* behavior and the separation between the peak of the excitation and the emission band is called the *Stokes shift*. However, excitation of multiple fluorophores or absorption from a high vibrational level of the ground state into the low energy tail of the first excited state [2] can produce an apparent exception to this law giving *anti-Stokes* scattering where the emitted radiation is at a shorter wavelength than the excitation wavelength; this phenomenon is outside the scope of this chapter.

For the purpose of measuring the photoluminescence properties of a material under steady-state conditions of illumination, it is not necessary, in principle, to distinguish between fluorescence and phosphorescence. In practice, it will be necessary to make this distinction only when a material exhibits an emission lifetime on the order of the time constants of the measurement apparatus, or which is significant in the context of the intended application. In the following discussion, whenever the term *fluorescence* is used, it may be understood to include both fluorescent and phosphorescent phenomena with time constants short enough to be ignored for the purpose of measurement.

The measurement of a fluorescent signal can be performed at fixed wavelength settings or as a function of wavelength. The general term *fluorometer* applies to any instrument that can make such measurements, with or without spectral collection capabilities. This chapter, however, will focus on spectral measurements, in keeping with the overall scope of this book.

7.1.1 Fluorescence Measurement Applications

Applications for fluorescence measurements may be divided into two broad categories. In *analytical applications*, the purpose is generally to identify and/or quantify the presence of fluorescent compounds in the specimen. In *colorimetric applications*, the purpose is generally to measure the radiation transfer properties of the specimen under test, in order to quantify its color appearance. Another important colorimetric application is for color formulation using Kubelka–Munk analysis [3]. In this application, it is necessary to separate the reflected and luminescent components to determine the intrinsic color quantities of the nonfluorescent substrate and the fluorophores so that their concentrations can be adjusted to obtain the desired color quantities.

7.1.1.1 Analytical Applications

Fluorescence spectrophotometry is an important analytical tool in chemistry, biochemistry, and material science. It plays a particularly important role in the pharmaceutical and biotechnology industries, in which high-throughput fluorescence measurements of analytes at low concentrations (e.g., <ppm) with small sample volumes (e.g., μl) are routinely required. The purpose of

measurement may be to identify one or more unknown chemical components within the specimen by its fluorescence characteristics, to characterize a known component or to determine the concentration of a particular fluorescent analyte. One of the important analytical applications of fluorescence measurements is for the pharmaceutical industry, which is described in detail in Chapter 11.

7.1.1.2 Colorimetric Applications

Accurate measurement of the color of fluorescent materials is important in various industries and applications. In the paper industry, fluorescent additives are widely used to enhance the appearance, and thus the commercial value, of white paper products; the industry has developed standardized metrics of "whiteness" and "brightness," which are based on specialized spectrophotometric measurements. Similarly, in the textile industry, fluorescent whitening agents (FWAs) are ubiquitous, and other fluorescent colorants, though less common, are regularly used. Fluorescent colorants are heavily used in the manufacture of plastics and are critical components in many ink systems— desktop, consumer printers, and commercial presses. In all of these industries, color control—and therefore, color measurement—is essential. Fluorescent materials are also used for visual signaling, in applications including signage for highways, waterways, and industrial safety; markings for slow and heavy vehicles; and personal safety markings for roadway workers.

7.2 FUNDAMENTAL CONCEPTS AND TERMINOLOGY

It is evident that fluorescence measurements, because of their extremely high sensitivity and selectivity, are used in a wide range of applications. Depending upon the application, different optical properties of fluorescence measurements are important. For colorimetric applications, what is usually desired is the total radiation leaving the sample surface which is responsible for its color appearance. For analytical applications, on the other hand, the desired optical properties are derived from the photoluminescent component which is considered to be an intrinsic property of the fluorescent molecular species in a particular environment. Thus, this field of fluorescence measurements encompasses interest from the molecular level to macroscopic phenomenological processes. To further complicate this description, both radiometric and photon quantities and units are used. This diversity of interest and practice leads to confusion in establishing metrological procedures. In this chapter, we separate the description of terminology to terms that are commonly used for all these fluorescence measurement applications and then present separately the specialized terminology that is used for analytical applications and for colorimetric applications. The main distinction here is that *molecular optical properties* are generally desired in analytical applications, whereas *sample optical properties* are generally desired in colorimetric applications.

For quick reference, these fluorescence terms, notations, and their use are summarized in Tables 7.1 (common), 7.2 (analytical), and 7.3 (colorimetric).

7.2.1 Quantities Common to Analytical and Colorimetric Applications

The quantitative description of the optical properties of any fluorescent material essentially involves the comparison of the quantity of radiation incident on the material, which may be absorbed, to that of the corresponding emission, under specified instrumental and experimental conditions. Since fluorescence measurements do not involve a simple ratiometric measurement at the same wavelength, the spectral conditions of the measurement are important quantities that need to be specified. The *excitation* wavelengths are denoted by the symbol μ and the *emission* wavelengths are denoted by the symbol λ.

TABLE 7.1 Quantities and Symbols Common to any Application of Fluorescence Measurements

Quantity	Notation	Reference for Quantity	Notes
Excitation spectrum	$X(\mu)$	CIE ILV [1]	
Emission spectrum	$M(\lambda)$	CIE ILV	
Excitation wavelength	μ	CIE ILV	For any geometry
Emission wavelength	λ	CIE ILV	For any geometry
Instrument spectral bandwidth	$\Delta\lambda$		Also known as spectral bandpass
Spectral radiance	L_λ	CIE ILV	For bidirectional geometry
Spectral radiant flux or radiant power	Φ_λ	CIE ILV	
Number of photons	N_p	CIE ILV	Also known as photon number
Absorptance	α	CIE ILV	Also known as absorption coefficient
Fluorescence or quantum yield	ϕ_f	ASTM [4]	Intrinsic property of single fluorophore
Fluorescence quantum efficiency	QE	ASTM [4]	Extrinsic property; concentration dependent
Fluorescence anisotropy	r	ASTM [4]	

TABLE 7.2 Quantities and Symbols Specific to Analytical Applications of Fluorescence Measurements

Quantity	Notation	Reference for Quantity	Notes
Excitation-emission matrix	EEM	ASTM	
Intensity Fluorescent intensity	I I_f	Common use	Used for any geometry
Absorbance	A	ASTM	
Absorption coefficient	α	ASTM [4]	Also known as absorptivity
Specific absorbance	ε	ASTM	Also known as molar extinction coefficient and molar absorptivity

TABLE 7.3 Quantities and Symbols Specific to Colorimetric Applications of Fluorescence Measurements

Quantity	Notation	Source of Term	Comment
Bispectral luminescent radiance factor	$\beta_{L,\lambda}(\mu,\lambda)$	CIE ILV	Limited to bidirectional geometry
Total spectral radiance factor	$\beta_T(\lambda)$	CIE ILV	Limited to bidirectional geometry
Reflected spectral radiance factor	$\beta_R(\lambda)$	CIE ILV	Limited to bidirectional geometry
Luminescent spectral radiance factor	$\beta_L(\lambda)$	CIE ILV	Limited to bidirectional geometry
Spectral efficiency factor	$b(\mu)$	ASTM [5]	Limited to bidirectional geometry
Photoluminescent radiant yield	RY	CIE ILV	Generalized for any geometry
Photoluminescent quantum yield	QY	CIE ILV	Generalized for any geometry
Spectral quantum efficiency of the fluorescent process	$\eta_\mu(\mu)$	CIE ILV	Generalized for any geometry
Donaldson radiance factor matrix	$D(\mu,\lambda)$	ASTM [5]	Can be generalized for any geometry

The radiometric quantities that are usually involved in these fluorescence measurements are *spectral radiance*, L_λ, for bidirectional measurement geometries, for example, 0°/90°, 45°:0°, and spectral radiant flux (or *spectral radiant power*), Φ_λ, for integrating sphere geometries. See Chapter 2 for the definition of these radiometric terms. It can be seen that these quantities are spectral concentrations, and thus, these measured fluorescence optical properties depend critically upon the *instrument spectral bandwidth* (or *spectral bandpass*), $\Delta\lambda$.

7.2.1.1 The Bispectral Nature of Fluorescence

In the foregoing section, we have discussed the fact that the spectral distribution of the radiation emitted by the fluorescence process depends upon two discrete wavelengths, the *excitation wavelength*, μ, and the *emission wavelength*, λ. Thus, in order to completely describe a fluorescent material, it is necessary to characterize its behavior *bispectrally*, that is, as a function of these two independent spectral variables.

The theoretical description of this fluorescent process in terms of the integration of a bispectral quantity, the *bispectral bidirectional reflectance distribution function* (BBRDF) over the incident illumination, was introduced in Chapter 2. The term that is commonly given in the standards literature to describe this bispectral nature of luminescence is the *bispectral luminescent radiance factor*, $\beta_{L,\lambda}$, defined as the ratio of the radiance per unit emission bandpass, at wavelength λ, due to photoluminescence from the specimen when irradiated at wavelength μ, to the radiance of the perfect reflecting diffuser (PRD), identically irradiated and viewed [1]. The functional bispectral dependence on μ and λ is sometimes explicitly shown as $\beta_{L,\lambda}(\mu, \lambda)$, while other times it is implicitly assumed.

Though fluorescence is essentially a bispectral phenomenon, fluorescence characteristics are frequently reduced to a simpler, unispectral representation. Various types of unispectral representations are possible; these may be distinguished by which spectral dimension (excitation or emission) is eliminated and which is retained. Spectral representations of fluorescence may also be distinguished by whether the reduction from a bispectral to a unispectral function is accomplished by simply fixing the value of one wavelength parameter, or by effectively integrating the bispectral function over one dimension.

7.2.1.2 Excitation and Emission Spectra

When the emission wavelength parameter, λ, of the bispectral function that describes the fluorescence characteristics of a material is fixed at a specified constant wavelength, λ_0, leaving the excitation wavelength, μ, as the sole spectral variable, the resulting function is known as an *excitation spectrum*. For example, if the applicable bispectral function is the bispectral luminescent radiance factor, $\beta_{L,\lambda}(\mu, \lambda)$, an excitation spectrum $X(\mu)$ may be derived as

$$X(\mu) = \beta_{L,\lambda}(\mu, \lambda_0). \tag{7.1}$$

When the excitation wavelength is fixed at μ_0, leaving the emission wavelength λ as the sole spectral variable, the resulting function is known as an *emission spectrum*, denoted $M(\lambda)$. For example, the emission spectrum $M(\lambda)$ may be derived from $\beta_{L,\lambda}(\mu, \lambda)$ as

$$M(\lambda) = \beta_{L,\lambda}(\mu_0, \lambda). \tag{7.2}$$

In either case, the fixed wavelength parameter is typically set to a value corresponding to a peak in the bispectral function.

These excitation and emission spectra can be referred to in various ways depending upon the level of spectral correction and adjustment to an absolute radiometric scale. They are most commonly referred to as (1) raw or uncorrected; (2) technically corrected; (3) normalized; or (4) true spectra. The "raw" data are exactly the measured raw signal without any correction, and these are displayed in arbitrary units (A.U.). The "technically corrected" spectra have been corrected for various instrument parameters described in detail in Section 7.6, where independent excitation and emission correction factors are determined, which is the usual requirement for most analytical applications. Technically corrected data are also usually displayed in relative intensity units. The "normalized" spectra have been corrected also for various instrument parameters described in Section 7.4, but where the interdependence of the excitation and emission correction factors has been determined, and both these corrections are applied at all wavelengths, including a required *normalization* of the bispectral fluorescence data so that it is dimensionless. This dimensionless scale is a *quasi-absolute intensity scale* for analytical applications [4] or an *absolute reflectance scale* for colorimetric applications; in the latter case, these data are displayed on an absolute radiance or reflectance factor scale from zero to unity. Finally, "true" spectra are technically corrected absorption and fluorescence spectra for analytical applications that have been further corrected for various sample-related effects, for example, self-absorption due to concentration, and sampling effects, such as excitation depth in the sampling volume; these effects are described in detail in Section 7.5; these data are displayed on an absolute intensity scale.

An example of technically corrected excitation and emission spectra for a fluorescent paper sample containing FWAs is shown in Fig. 7.1. These data are displayed on a corrected intensity scale (A.U.)

7.2.1.3 Quantum Optical Properties

The basic equation to convert from measured radiometric quantities, for example, radiant flux and radiance, to quantum or photon quantities, for example, photon flux and photon radiance, is given in Chapter 2, Section 4.6. The specific example given there is for the conversion from radiant flux to photon flux. The optical properties of a fluorescent material may

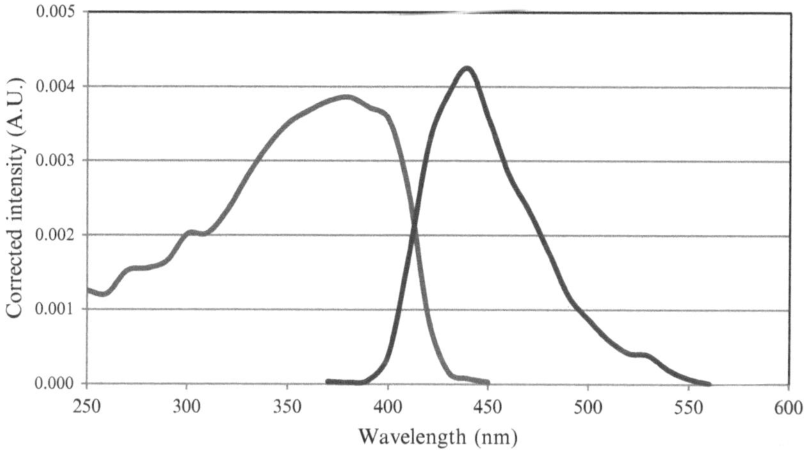

FIGURE 7.1 Technically corrected (left, red curve; light gray in print) excitation and (right, blue curve; dark gray in print) emission spectra for a fluorescent paper sample containing fluorescent-whitening agents (FWAs).

also be described in quantum or photon terms. The excitation spectrum, $X(\mu)$, and the emission spectrum, $M(\lambda)$, can be converted to a relative photon unit scale by multiplying these spectra by $1/\mu$ and λ, respectively.

Two other important quantum optical properties of fluorescent materials are the fluorescence quantum efficiency and fluorescence quantum yield. Unfortunately, these terms are often used interchangeably which can cause confusion. The American Society for Testing and Materials (ASTM) [4] defines *fluorescence quantum yield*, ϕ, as the probability that a molecule or species will fluoresce once it has absorbed a photon, whereas *fluorescence quantum efficiency* (no symbol) is defined in terms of number of fluorescence photons leaving the emitter, N_{em}, to the number of the photons absorbed, N_{abs}. In this chapter, we denote these two quantities by the following notation, ϕ_{f} and QE, respectively. Thus, the fluorescence quantum efficiency, QE, is given by

$$QE = \frac{N_{\mathrm{em}}}{N_{\mathrm{abs}}}. \tag{7.3}$$

It can be seen that ϕ_{f} is an *intrinsic molecular property*, that is, a property of the fluorophore itself that depends only on molecular structure and is independent of sample form or sample concentration. On the other hand, the QE, which represents the fraction of excited molecular species that return to the ground state with emission of fluorescent photons, is an *extrinsic molecular property*, that is, a property that depends upon these external factors. Thus, the QE can depend on scattering losses (reflectance and transmittance) from the surface structure (e.g., depend on grain sizes in powdered samples) and on the sample concentration (e.g., self-absorption and self-quenching effects in high concentrations of liquid samples).

If a fluorophore has simple excited-state dynamics, where the various external and sample-related effects which can impact the fluorescence yield are negligible (e.g., excitation wavelength and sample concentration), then the quantum efficiency, QE, is the product of its absorptance, α, and the fluorescence quantum yield, ϕ_f:

$$QE = \alpha\phi_f. \tag{7.4}$$

It can be seen that in the case of sufficiently dilute solutions (e.g., $A < 0.05$), where $\alpha = 1.0$, the QE is equivalent to ϕ_f.

To further complicate matters, the quantum efficiency can also be distinguished as a measure of the *internal quantum efficiency*, ϕ_f, defined above which depends only on the radiation absorbed by the sample, that is, internal scattering, or it can be a measure of the *external quantum efficiency*, EQE, which depends upon both *internal* and *external* scattering losses. The EQE compares the number of fluorescence photons leaving the material to the number of incident photons, N_{in}:

$$EQE = \frac{N_{em}}{N_{in}}. \tag{7.5}$$

It can be seen that EQE and QE are analogous to the measured "apparent" transmittance and the internal transmittance of a transmitting sample, where the former includes reflection losses. The EQE is an important device property, for example, of an LED.

7.2.2 Specific to Analytical Applications

In practice, however, such radiometrically quantitative data may not be necessary to meet the demands of a given analytical application. Some compounds may be identified on the basis of their excitation and/or emission peak locations alone; others may be identified by the relative magnitude of certain peaks. The concentration of a specific analyte might be determined by comparison of the test spectrum with that of a reference solution. For these reasons, analytical bispectrometers may be uncompensated, or only partially compensated, and the units of fluorescence signal that they produce may be more or less arbitrary.

7.2.2.1 *Fluorescent Intensity*

For this reason, in analytical applications of fluorescence measurements, the measurand is described as an *intensity*, I. However, this is not strictly an intensity in a radiometric sense where the quantity "intensity" has a specific measurement geometry. This intensity is simply a measure of the signal that is detected by the fluorescence measuring instrument and the precise measurement quantity might be spectral radiance, spectral radiant flux, photon flux, or number of photons, depending upon the measurement geometry and the measuring detector. Also, the units of this intensity are generally given in A.U.

These intensity values need to be "background corrected." This involves measuring the signal from a blank, I_0, for example, a cuvette with the solvent only, and subtracting this "blank" determination from the measured signals from the analyte, I_s. Thus, the relative fluorescent intensity due to the analyte, I_f, is given by

$$I_f = I_s - I_0. \tag{7.6}$$

For absolute fluorescence measurements, for example, for quantum yield determinations, the quasi-absolute fluorescent intensity due to the sample, I_f, is measured with, I_{in}, and without, I_{out}, the sample in the beam, as

$$I_f = I_{in} - I_{out}. \tag{7.7}$$

7.2.2.2 Molecular Optical Properties

In analytical fluorescence spectrometry, the greatest interest is in the intrinsic properties of the analyte or molecular species. The two molecular optical properties of the fluorescent analyte that are of greatest interest are the *fluorescent quantum yield*, ϕ_f, defined above in Section 7.2.1.3 and its *specific absorbance*, ε (also known as the *molar absorption coefficient* and *molar extinction coefficient*; see Chapter 2, Section 4.5). Differences in the fluorescence quantum yields of similar analytes with nearly identical fluorescence spectra are often used to distinguish between them. The product of these two intrinsic properties, $\phi_f\varepsilon$, is sometimes termed the "brightness," particularly for comparing the performance of fluorescent labels and dyes. In order to obtain these intrinsic molecular optical properties, it is necessary to first obtain the "true" excitation and emission spectra. As discussed in Chapter 12 on The Use of Spectrophotometry in the Pharmaceutical Industry, the most common type of fluorescent sample in analytical applications is a dilute solution (μM/L) of the fluorescent species or *fluorophore* that obeys the Beer–Lambert law.

The *absorption coefficient* of the analyte, α (also known as the *absorptance*), can then be calculated from the measured *absorbance*, A, of a sample of known analyte concentration, c, using the Beer–Lambert law, $A = \varepsilon bc$, such that

$$a = 1 - 10^{-A} = 1 - 10^{-\varepsilon bc}, \tag{7.8}$$

where b is the path length of the excitation light through the sample.

The integrated fluorescence intensity, I_f, at a given excitation wavelength and emission range, (μ_0, λ), is proportional to the product of the incident intensity, $I_0(\mu_0)$, the fluorescent analyte concentration, c, its fluorescence quantum yield, ϕ_f, and its specific absorbance, ε, according to

$$\int_\lambda I_f(\mu_0, \lambda) = K I_0(\mu_0)\varepsilon\phi_f c. \tag{7.9}$$

The proportionality constant K depends upon the spectral and geometric characteristics of the fluorometer used for these measurements. The use of this simple relationship for determining the fluorescent quantum yield is described in detail in Section 7.5.3.

7.2.2.3 Excitation–Emission Matrix

In most analytical applications, the fluorescence measurements are performed unispectrally, that is, as a function of one spectral variable, either excitation wavelength or emission wavelength. Unknown chemical components may be identified and quantified by examination of the spectral location and magnitude of one or more peaks in the specimen's excitation or emission spectra. Thorough characterization of a specimen may involve the acquisition of a complete bispectral array of data, that is, a three-dimensional spectrum, or *excitation–emission matrix* (EEM) which can be used to express the radiometric measurements as a function of both wavelength parameters, simultaneously. In principle, the values in this matrix might be those of the bispectral luminescent radiance factor, $\beta_{L,\lambda}(\mu,\lambda)$, or an equivalent quantity.

7.2.3 Specific to Colorimetric Applications

For colorimetric applications, the interest is in macroscopic optical properties, such as color, brightness, and whiteness, under given conditions of broadband illumination [6]. The fluorescent materials of interest are typically opaque materials that incorporate fluorescent dyes or pigments. In the measurement of the color of these fluorescent materials, surface geometries are most commonly employed. As we have seen above, the approach is generally to compare the spectral radiance of the sample under test to that of the PRD.

7.2.3.1 Spectral Efficiency Factor

The bispectral function that describes the fluorescence characteristics of a given material may also be reduced to a unispectral function by integration over one of the two wavelength variables. When a material's bispectral radiance factor, $\beta_\lambda(\mu,\lambda)$, is simply integrated over the complete range of emission wavelengths, from $\lambda_{\min}$ to $\lambda_{\max}$, the resulting function is the *spectral efficiency factor*, $b(\mu)$ as [5]

$$b(\mu) = \int_{\lambda_{\min}}^{\lambda_{\max}} \beta_\lambda(\mu, \lambda) \mathrm{d}\lambda. \tag{7.10}$$

Other functions of excitation wavelength, derived by weighted integration over emission wavelength, may also be meaningful, as discussed below (see Section 7.2.3.2). It should be noted that the use of this term is limited to measurements of spectral radiance, that is, to fluorescence measurements in a bidirectional geometry only. Currently, there is no equivalent term recommended for other measurement geometries, for example, for measurements of spectral radiant flux, which is important for a number of practical colorimetric applications that are based on integrating sphere methods.

7.2.3.2 Total Spectral Radiance Factor

From the foregoing, we have seen that the fundamental bispectral quantity that is important to colorimetric fluorescence applications is the bispectral luminescent radiance factor. In the case of fluorescent reflecting materials, it is necessary to measure not only the luminescent radiance but also the total radiance leaving the sample which is responsible for the material's appearance. Therefore, the fundamental measurand contains two components, a luminescent and a reflected component. The ILV [1] describes this quantity as the *total spectral radiance factor*, $\beta_T(\lambda)$, which is given by the sum of the *reflected radiance factor* $\beta_R(\lambda)$ and the *luminescent radiance factor* $\beta_L(\lambda)$ according to

$$\beta_T(\lambda) = \beta_R((\lambda) + \beta_L(\lambda)). \tag{7.11}$$

For fluorescence colorimetric applications, it is implicit that the quantity used in the colorimetric calculations is the total radiance factor, and it is common to omit the subscript T.

Thus, we can expand the formulation of the bispectral luminescent radiance factor given in Chapter 2, Section 4.6, by considering a sample of fluorescent material which is uniformly irradiated by a source with the given spectral irradiance, $E_\mu(\mu)$. The corresponding spectral radiance, $L_\lambda(\lambda)$, of the sample can be calculated by integrating the product of the spectral irradiance and the bispectral radiance factor of the material, $\beta_\lambda(\mu, \lambda)$, over the complete range of excitation, $\mu_{\min}$ to $\mu_{\max}$, and dividing by a geometric constant, Ω_{proj}, representing the projected solid angle into which the radiant flux from the sample is directed:

$$L_\lambda(\lambda) = \frac{1}{\Omega_{\text{proj}}} \int_{\mu_{\min}}^{\mu_{\max}} E_\mu \beta_\lambda(\mu, \lambda) \mathrm{d}\mu. \tag{7.12}$$

The *spectral radiance factor*, $\beta(\lambda)$, of the sample can then be calculated by comparison of its spectral radiance, $L_\lambda(\lambda)$, with the corresponding spectral radiance of the PRD, $L_{\lambda,\text{PRD}}(\lambda)$, where

$$\beta(\lambda) = \frac{L_\lambda(\lambda)}{L_{\lambda,\text{PRD}}(\lambda)} = \frac{\int_{\mu_{\min}}^{\mu_{\max}} E_\mu(\mu) \beta_\lambda(\mu, \lambda) \mathrm{d}\mu}{E_\lambda(\lambda)}. \tag{7.13}$$

For the reflected component, this comparison is made at the same wavelength, $\mu = \lambda$, and thus the properties of the source cancel from the above equation. This is not the case for the luminescent component. To underline the importance of the spectral quality of the illuminating source, $E_\lambda(\lambda)$, on the luminescent component and, thus, the total radiance factor, Eq. (7.13), can be rewritten to show this functional dependence more explicitly

$$[\beta_T(\lambda)]_{E(\lambda)} = \beta_R(\lambda) + [\beta_L(\lambda)]_{E(\lambda)}. \tag{7.14}$$

If a complete, illuminant-independent description is required, the material must be characterized bispectrally and scaled by the instrument-specific scaling factor, in order to determine the normalized bispectral luminescent radiance factor, $\beta_{L,\lambda}(\mu,\lambda)$.

The luminescent radiance factor data for the desired illuminant is obtained at each wavelength in the emission range λ, by weighting the normalized bispectral radiance factor data by the relative spectral power distribution of the desired illuminant, $E_{st}(\mu)$ and integrating this product for all wavelengths μ in the excitation band according to

$$\beta_L(\lambda) = \int_{\mu} E_{st}(\mu)\cdot\beta_{L,\lambda}(\mu,\lambda)\mathrm{d}\mu / E_{st}(\lambda). \tag{7.15}$$

The sum of the luminescent radiance factor with the reflected radiance factor then gives the total spectral radiance factor corresponding to the desired illuminant condition, as in Eq. (7.14). This total spectral radiance factor is then multiplied by the relative spectral power distribution of the desired illuminant condition to give the relative color stimulus function to be used in the calculation of the CIE tristimulus values and chromaticity coordinates of this fluorescent surface color. The details of these colorimetric calculations are given in Chapter 10.

Note that in this approach, the spectral radiance factor $\beta_T(\lambda)$ is treated as if it were simply the more familiar spectral reflectance factor $R(\lambda)$ (see Chapter 6). This approach is necessary for the one-monochromator method, in which the sample's total spectral radiance factor—for the illuminant simulated by the instrument's source—is measured directly, and the user does not have access to illuminant-independent bispectral data. An example of the implementation of this colorimetric calculation procedure is given in Section 7.8.3.

7.2.3.3 Spectral Quantum Efficiency of the Fluorescence Process

A key control characteristic of fluorescent materials used for lighting applications is its quantum efficiency. This differs from the quantum efficiency defined above for analytical applications since the interest here is in the macroscopic (*sample*) optical property, and the measurement comparison is made on the photon flux rather than on number of photons. The term that is relevant to these colorimetric applications is described by the ILV [1] as the *photoluminescent quantum yield*, described in Section 7.2.1.3. The corresponding radiant quantity is the *photoluminescent radiant yield*, which is the ratio of the radiant flux of the radiation emitted by the photoluminescent material to the radiant flux of the radiation absorbed by that material. In this chapter, we denote these two photoluminescent yields by the symbols, *QY* and *RY*, respectively, where *QY* is analogous to ϕ_f.

Thus, *RY* and *QY* defined here are *intrinsic sample properties*. It can be seen that the determination of the *RY* or *QY* requires absolute measurement techniques, for example, angular integration of goniometric measurements of the BBRDF over the hemisphere for an opaque reflecting sample and over the full sphere for a transparent or liquid-phase sample, or by using an integrating sphere to measure the total radiant flux. These absolute methods are described in more detail in Section 7.5.3.2.

For some applications of photoluminescent materials, such as improving lighting efficiency, the more meaningful optical quantity is the *extrinsic sample property* which compares the output radiant power to the incident radiant power. The ILV [1] distinguishes these optical quantities as being *external* when referring to incident radiation or photons (or radiant power) and *internal* when referring to absorbed radiation or photons (or radiant power). The ILV term given to this external quantum efficiency is the *spectral quantum efficiency of the fluorescent process*. This is defined as the ratio of the total number of photons of all wavelengths emitted from the specimen by the photoluminescent process for an excitation at wavelength, μ, to the number of photons of wavelength μ reflected from the PRD identically irradiated and viewed. The ILV gives this quantity the symbol, $\eta_\mu(\mu)$. This is perhaps an unfortunate notation, since the subscript, μ, suggests that this is a spectral concentration with respect to excitation wavelength, which it is not. This quantity may be derived from the integral of the emission peak to the excitation peak. This is calculated from the normalized bispectral luminescent radiance factors, $\beta_{L,\lambda}(\mu,\lambda)$, by converting this quantity from a wavelength to an energy scale and integrating over the complete emission range, from $\lambda_{\min}$ to $\lambda_{\max}$, as

$$\mu_\mu(\mu) = \frac{\int_{\lambda_{\min}}^{\lambda_{\max}} \lambda \beta_{L,\lambda}(\mu,\lambda)\mathrm{d}\mu}{\overline{\mu}}, \tag{7.16}$$

where the $\overline{\mu}$ in the denominator of Eq. (7.16) is the wavelength of the incident photon of average energy, within the excitation bandpass, $\Delta\mu$. For a narrow excitation bandpass, for example, 1–5 nm, this can be substituted simply by the incident wavelength, μ.

7.2.3.4 Donaldson Radiance Factor Matrix

The bispectral data obtained in colorimetric applications are generally presented graphically as a three-dimensional plot, as shown in Fig. 7.2, where the three axes are excitation wavelength, emission wavelength, and intensity. The complete bispectral data can also be expressed in tabular form. This is

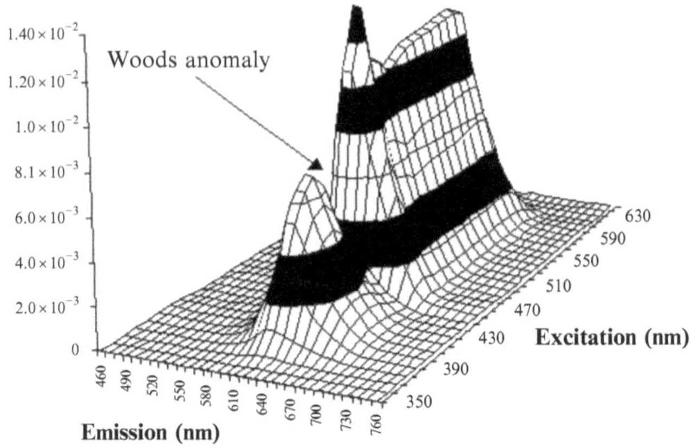

FIGURE 7.2 Raw measured bispectral luminescent radiance factor data on a bidirectional instrument, 45:0 geometry, s-polarized conditions, for a fluorescent red paint sample.

referred to as a *Donaldson* matrix representation, $D(\mu, \lambda)$, after the work of Donaldson at the National Physical Laboratory, who introduced this matrix concept for describing the reflected and luminescent components of the radiance factor for a fluorescent pigment sample [7]. In this approach, the bispectral data are presented as a two-dimensional array of values, or as a matrix, with one dimension corresponding to excitation wavelength and the other to emission wavelength, that is, as a function of two discrete spectral variables (μ_j, λ_i). In order to present both the reflectance and bispectral fluorescence data, which have different dimensions, in the same table, it is necessary to first transform the data into a dimensionless, discrete function, the *Donaldson Radiance Factor*, $D(\mu_j, \lambda_i)$ [5]. This quantity is approximately equal to the ratio of the sample's radiance within a rectangular waveband of width, $\Delta\lambda$, equal to the sampling interval, centered at λ, compared to that of the PRD when each is irradiated with a rectangular waveband of width, $\Delta\lambda$, centered at μ. Thus, when the bispectral measurand is the total spectral radiance factor, the $D(\mu_j, \lambda_i)$ is

$$D(\mu_j, \lambda_i) = \beta_{T,\lambda}(\mu_j, \lambda_i)\Delta\lambda. \tag{7.17}$$

Note that while the Donaldson radiance factor matrix is a radiometric quantity which can, in principle, be applied to any geometry, it is most directly relevant to colorimetric applications and surface geometries.

An illustration of this Donaldson matrix representation is shown in Table 7.4. The excitation wavelengths, μ, are tabulated as columns, the emission wavelengths, λ, are tabulated as rows, and the reflected radiance factors, $\beta_R(\lambda)$, appear on the diagonal where the excitation and emission (detection) wavelengths are equal.

TABLE 7.4 Illustration of the Bispectral Radiance Factor (%) Data of a Fluorescent Red Paper Sample Presented in a Donaldson Matrix, $D(\mu, \lambda)$

		Excitation Wavelength, μ (nm)											
		400	*420*	*440*	*460*	*480*	*500*	*520*	*540*	*560*	*580*	*600*	*620*
Emission wavelength, λ (nm)	**400**	0.3	–	–	–	–	–	–	–	–	–	–	–
	420	–	0.1	–	–	–	–	–	–	–	–	–	–
	440	–	–	0.1	–	–	–	–	–	–	–	–	–
	460	–	–	–	0.1	–	–	–	–	–	–	–	–
	480	–	–	–	0.4	0.5	–	–	–	–	–	–	–
	500	0.6	1.6	2.0	2.6	2.8	1.9	–	–	–	–	–	–
	520	1.6	9.6	11.0	12.0	12.0	12.0	26	–	–	–	–	–
	540	2.0	13.0	14.0	15.0	16.0	16.0	5.2	40	–	–	–	–
	560	2.6	7.4	8.6	8.8	5.2	10.0	3.6	0.2	40	–	–	–
	580	2.0	4.4	5.4	5.2	6.0	6.2	2.2	–	–	50.3	–	–
	600	1.6	2.6	2.6	3.2	3.4	3.8	1.2	–	–	–	40.1	–
	620	0.6	3.0	1.4	1.4	1.4	1.8	0.6	–	–	–	–	39.3

7.3 MEASUREMENT OF FLUORESCENCE CHARACTERISTICS

7.3.1 Bispectral Measurements

The complete and accurate characterization of the optical properties of a fluorescent material requires monochromatic irradiation and monochromatic detection, using the so-called *two-monochromator method*, first introduced by Donaldson [8]. A fluorescence measuring instrument that is based on the two-monochromator method is called a *bispectrometer*—an optical instrument equipped with a source of irradiation, two monochromators, and a detection system, such that a material sample can be measured at independently controlled irradiation and viewing wavelengths. The bispectrometer's *excitation monochromator* separates radiation from the source into its spectral components before it reaches the sample, and its *emission monochromator* separates radiation leaving the sample surface into its spectral components once again before it reaches the detection system. A bispectrometer may be designed to allow for calibration to provide quantitative determination of the bispectral radiation transfer properties of the sample.

The basic measurement procedure involved in the *two-monochromator method* may be described as follows: With the bispectrometer's excitation monochromator fixed at a given wavelength position, μ_j, the instrument acquires from the emission spectrometer an array of readings, or emission spectrum $M(\lambda)$, covering a range of emission wavelengths, λ_i. By stepping through the range of excitation monochromator positions, μ, and recording an array of emission readings at each position, a complete bispectral matrix of readings is acquired. For instruments with a scanning emission monochromator, it is also possible to proceed conversely, by fixing at a series of emission wavelengths λ_i and acquiring an excitation spectrum $X(\mu)$ for each fixed emission position.

The readings obtained directly from the instrument are influenced by many instrument-specific factors. Among these are the spectral distribution of the source, the bandpass functions of the excitation and emission monochromators, the spectral transmittance of the instrument's transfer optics, and the spectral responsivity of the instrument's detection system. Figure 7.2 illustrates the raw bispectral data for a fluorescent red paper sample measured for s-polarized conditions on a bispectrometer with a 45°:0° geometry. It can be seen that the data show an anomalous dip near 460 nm. This large dip is due to the characteristics of this instrument's grating which has a Wood's anomaly at this particular wavelength and, thus, is not a function of the sample's optical properties.

In some bispectrometers, no provision is made to correct for these influences. These are described as *uncompensated* bispectrometers, designed not for accurate radiometric measurement of materials, but rather to provide instrument-specific *apparent spectra*. Though such apparent spectra are not radiometrically quantitative, having an arbitrary magnitude scale, they may nevertheless be useful for analytical purposes. A *compensated* bispectrometer, on the other hand, may be designed to allow for complete or partial radiometric calibration using standards for spectral fluorescence measurements, as described in Section 7.8, depending on the demands of the particular application.

7.3.2 Abridged Measurements

Complete characterization of the fluorescence properties of a material requires the acquisition of a bispectral array of data, spanning the material's excitation and emission wavelength ranges. For many applications, however, some form of abridged measurement may be sufficient. For example, a bispectrometer can be used to provide a single excitation spectrum by selecting a fixed emission wavelength and scanning with the excitation monochromator. Conversely, a single emission spectrum can be provided by fixing the excitation monochromator and acquiring a complete range of emission readings. A bispectrometer may also be used to perform a *synchronous scan*, in which a scan is performed with a fixed wavelength offset established between excitation and emission wavelengths. For a zero offset between the wavelength scales of the excitation and emission monochromators, only the elastic scattering properties (reflectance or transmittance) of the fluorescent sample are obtained.

A single-monochromator instrument is generally designed to have one of two modes of operation, depending upon whether the monochromator precedes the sample, *monochromatic mode*, or follows the sample, *polychromatic mode*. Single-monochromator instruments may also be designed specifically to perform certain types of abridged fluorescence measurements, tailored to particular applications. If emission spectra are needed only for a single excitation wavelength, the system's excitation monochromator may be replaced by a narrow bandpass filter or the broadband lamp and excitation monochromator may be replaced by a laser source. Conversely, if excitation spectra are needed only for a single emission wavelength, the system's emission spectrometer may be replaced by a detection system equipped with a narrow bandpass filter.

More generally, various types of broadband filters may be introduced to provide appropriate specialized measurements. For example, if an instrument with only an excitation monochromator is equipped with a spectral balancing filter, which makes the spectral response of its detection system approximately neutral, or nonselective, then (with appropriate calibration), the instrument could be used to measure directly spectral efficiency factor $b(\mu)$ or equivalent quantities. Photopic or colorimetric detectors could be used instead, to simulate the response of the human visual system. Likewise, if an instrument with only an emission monochromator is equipped with a filtered source, which simulates the relative spectral power distribution of a desired standard illuminant, then (again, with appropriate calibration) the instrument could be used to measure directly the total spectral radiance factor, $\beta_T(\lambda)$, for that particular illuminant condition. Similar measurements could be supported for nonstandard illuminants, such as the spectrum of a particular light-emitting diode (LED), which is intended to excite an LED phosphor material under test. A single-monochromator instrument may also be equipped with multiple, interchangeable bandpass filters, to support a variety of related abridged bispectral measurements. Finally, both the excitation and the emission monochromator may be eliminated, and replaced by simple, non-spectrally selective illumination and detection systems, in order to provide a single-valued measurement result.

7.4 GENERAL INSTRUMENT DESIGN AND MEASUREMENT CONSIDERATIONS

Selecting an instrument for measuring spectral fluorescence depends on several considerations. First, it depends upon whether the application is colorimetric or analytical, where different measurement geometries are used. Other important options are the following: the instrument can be single-beam or double-beam operation, one-monochromator or two-monochromator measurement method, scanning or array detection, continuous or pulsed source, and monochromatic or polychromatic illumination. Here, we discuss the major components of an instrument for measuring spectral fluorescence and the general considerations for selecting the optimal instrument design and operation for different measurement applications.

7.4.1 Instrument Components

Figure 7.3 shows a schematic optical layout for a common instrument setup for measuring spectral fluorescence. The main components are a light source, an optical system for wavelength separation to define the excitation wavelength(s), a means of diverting part of the incident beam to a monitor detector that records a reference signal R that can be used to correct for any changes in the output of the light source during the fluorescence scan, a sample holder, an optical system for collecting the radiation leaving the sample, and an analyzing detector for recording the sample signal S due to emitted and/or reflected radiation. The term *optical head* is sometimes used to describe the combination of the sample holder and the system coupling the source and detection subsystems to the sample with the desired measurement geometry.

The emission spectrum is measured either wavelength by wavelength in a scanning system or simultaneously using a multichannel detector, such as a photodiode array and a CCD spectrograph.

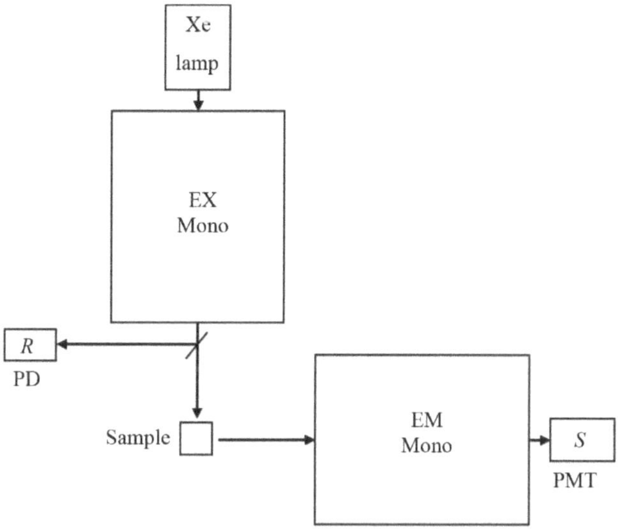

FIGURE 7.3 Schematic of a common instrumental setup used for measuring the fluorescence signal S from a sample by a photomultiplier tube (PMT). The reference signal R is monitored by a photodiode (PD). EX Mono and EM Mono stand for excitation monochromator and emission monochromator, respectively.

7.4.2 Spectral Measurement Methods

The fluorescence spectrophotometer, also known as a fluorescence spectrometer or spectrofluorimeter, has been commercially available for more than 50 years. Conventional fluorescence spectrometers have typically used a broadband light source, such as a xenon (Xe) lamp, radiating over a wavelength range from the UV to the NIR, monochromators for wavelength selection of both excitation and emission light, and a photomultiplier tube (PMT) for the detection of emission. This arrangement allows for the collection of a single photocurrent at any given pair of excitation and emission wavelength values, $\Phi_{ij}(\mu_j, \lambda_i)$. This value is expressed here as a radiant power, Φ, since the measured photocurrent should be directly proportional to the radiant power entering the detection system in Watts, assuming that the detector response is linear with Φ and the dark current is zero. An excitation spectrum is measured at a fixed λ, while μ_j is scanned from $j = 1,\ldots, m$. Similarly, an emission spectrum is measured at a fixed μ, while λ_i is scanned from $i = 1, \ldots, n$.

The fluorescence spectrometer most commonly used for the collection of spectra is a *steady-state* instrument, meaning that the excitation beam is continuous, not pulsed, and the integration time of the detector is on the order of seconds, resulting in a signal that is time averaged. This "steady-state" signal does not give fluorescence lifetimes or other time-resolved information related to the excited-state kinetics of the fluorophore. The use of the steady-state fluorescence spectrometer is the main focus of this chapter. The power of the excitation beam may change during the collection of a spectrum, either due to time-dependent fluctuations at a fixed wavelength or due to wavelength-dependent differences due to the spectral shape of the lamp radiation. A *compensated* fluorescence spectrometer uses a reference detector to monitor these power changes and to compensate for them. Since the sample's measured fluorescence signal (S) is directly proportional to the excitation reference signal (R), changes in the sample signal due to power changes in the excitation beam can easily be corrected by taking the ratio S/R to give a corrected signal S_C.

In the past 10 years, multichannel detectors have replaced PMTs, and polychromators or spectrographs have replaced monochromators in some instruments, enabling the collection time of spectra to be shortened by collecting intensity values at multiple wavelengths, simultaneously. It should be recalled that the actual radiometric quantity being measured can be a radiant flux, spectral irradiance, or spectral radiance, depending upon the geometric configuration, although it is common practice to simply use the term "intensity" (see Section 7.2.2).

The use of these multichannel detectors has opened up new fluorescence applications in the areas of *in situ* monitoring, chemical fingerprinting, and spectral imaging. The use of EEMs has also increased significantly due to the improved speed of collection that these components make possible, a factor of 10 faster being common. The ability to collect an entire spectrum at once has also made multichannel detectors desirable in portable instruments, where a fixed grating makes the instrument more rugged in the field and much cheaper to produce. Photodiode arrays (PDAs) and charge-coupled devices (CCDs) are the multichannel detectors most commonly used. Each photodiode or pixel in these detectors gives an independent signal, measuring the amount of light incident on it for a given sampling geometry. The polychromator is used to image the fluorescence spectrum onto the multichannel detector, such that each diode or pixel is measuring the intensity of light at a different wavelength.

Each of these fluorescence detection systems has its advantages and disadvantages. Multichannel detectors, in addition to having faster collect times for spectra, are less likely to be damaged by bright light, such as room light and sunlight, than PMTs. Si photodiodes and CCDs typically used for visible light detection are also more sensitive in the NIR out to 1100 nm, whereas corresponding PMTs are only sensitive out to about 800 nm.

The main advantage of a PMT is a larger dynamic range, as much as six orders of magnitude without the use of attenuators, which is one to two orders of magnitude greater than that of silicon-based multichannel detectors. A PMT also has a larger active area per channel, making it more sensitive under typical light-collection conditions. In addition, the most commonly used

PMTs for visible light detection are more sensitive in the UV down to about 250 nm, whereas the corresponding Si-based multichannel detectors are only effective down to about 400 nm. PDA and CCD detectors have similar performance in most areas, but the CCD is typically more sensitive than the PDA, making the CCD more effective for low-light intensity applications.

7.4.3 General Considerations and Good Laboratory Practices

Good laboratory practices should be used to minimize errors in fluorescence measurements. Some of the most important of these are reviewed here.

7.4.3.1 Handling and Cleaning

Impurities and contaminants can cause errors in fluorescence measurements by absorbing or scattering light, quenching fluorescence, or fluorescing themselves (see Section 7.7.4). The proper handling and cleaning of fluorophores, reagents, and contact surfaces are a critical part of preventing fluorophores from being contaminated. Since fingerprints are known to fluoresce, disposable, powder-free gloves should always be worn when handling fluorophores and contact surfaces.

Detergents used for cleaning glassware may also fluoresce, so thorough rinsing is needed. Soaking glassware in a solution of 50% concentrated nitric acid for several hours is one preferred cleaning method. This method is particularly useful for surfaces that cannot be scrubbed, such as the inside of cuvettes. A final rinse should always be done with a solvent that has been confirmed to be nonfluorescent.

7.4.3.2 Types of Sample Holders

Since the excitation beam is often incident on the sample holder, in addition to the sample, it is important to know its absorption and fluorescence spectra and, if possible, minimize its fluorescence intensity. Unwanted fluorescence can be minimized by choosing the sample holder with the least fluorescence. When the excitation beam and emission have to go through the sample holder, for instance, in the case of a cuvette, the holder must transmit both the excitation and emission wavelengths. The effective transmission range of a cuvette varies, depending upon its material type [4]. UV-grade fused silica (quartz) cuvettes have the widest transmission range from about 170 to 2700 nm. Infrared-grade quartz can extend out to 3800 nm. Glass and plastic are also used to make cuvettes for narrower spectral ranges. Optical glass stops transmitting below 350 nm, and the transmission of plastics varies, but are generally only recommended for visible light transmission.

7.5 SPECIALIZED INSTRUMENT DESIGNS AND MEASUREMENT METHODS

7.5.1 Analytical Applications

7.5.1.1 0°/90° Geometry

Analytical samples in chemical analysis are most commonly prepared in liquid form and measured in a standard cuvette. The most widely used optical geometry for fluorescence detection is a 0°/90° geometry, in which the sample is irradiated along an axis normal to the front surface of the cuvette (0°), and viewed in a direction perpendicular to the axis of illumination (90°) (see Fig. 7.4). Conventional fluorescence spectrometers have L-shaped excitation/emission configurations, where the detection system is perpendicular to the excitation beam. The conventional sample format is a cuvette with four transparent long sides for analyzing liquids and solutions. Sample sizes for this format are typically between 0.3 and 3.5 mL, with the latter volume being the amount needed to fill the traditional, standard cuvette with a 1-cm path length and internal dimensions of 1 cm × 1 cm × 4 cm. More recently, single-drop sample holders have been developed for fluorometers to handle sample sizes as small as 1 μL. These are particularly useful for biological applications

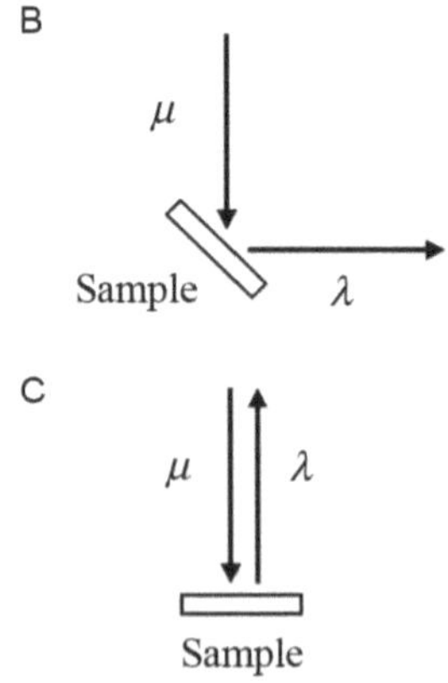

FIGURE 7.4 Schematic of a (a) 0°/90° transmitting detection geometry for an instrument with an L-shaped configuration, (b) front-face detection geometry for an instrument with an L-shaped configuration, (c) epifluorescence or front-face detection geometry for an instrument with fluorescence being detected at 0°/0° (parallel to) or at a small angle relative to the excitation beam.

where only a small amount of sample is available, such as when quantifying nucleic acids or proteins.

7.5.1.2 Front-Face Geometry

Analytical samples in chemical analysis can also be prepared as powder samples and other solid specimens. Solid samples are typically measured using a front-surface geometry. Solid sample holders are available for instruments with L-shaped configurations and are typically designed to hold a flat-surfaced sample, such as a microscope slide, optical flat, or pressed tablet, for excitation and fluorescence collection from its front face. The angle of the sample relative to the excitation beam varies for front-face measurements, commonly ranging from 0° to 45°. For an L-shaped instrument configuration, a sample angle of less than 45° is often used to avoid specular reflection of the excitation beam from the surface of the sample into the detection system (see Fig. 7.4b).

In contrast, an *epifluorescence* geometry (see Fig. 7.4c), in which the sample is viewed from a direction very close to the direction of the source of irradiation, is most commonly used in fluorescence microscopy with either solid or liquid samples [4]. More recently, portable and high-throughput fluorometers are using epifluorescence and transmitting geometries similar to those used in microscopy. For high-throughput applications, this can be an array of miniature wells combined in a single microwell plate (MP), a portable instrument using a bifurcated fiber, or an even more densely packed array of sample droplets deposited onto a microarray (MA). These arrays of specimens are then measured by specialized fluorometers or readers [9]. In these cases, filters are typically placed between the sample and detector to prevent stray light from reaching the detector.

7.5.1.3 One-Monochromator Instruments

The concentrations of fluorescent analytes being detected by common assays are often low, making high-power excitation sources desirable. Monochromators are designed to give high wavelength resolution, but are often not the best choice for excitation wavelength selection when high-power excitation is required. Monochromators are also one of the most expensive components used in fluorometers. For both of these reasons, the elimination of the excitation monochromator is advantageous in many applications. A spectral band-pass filter is often used to select a broader bandwidth of excitation light from the lamp, yielding a larger radiant power. Another option is to replace the lamp and excitation monochromator with a narrow bandwidth LED or laser. In this case, the light source determines the wavelength and bandwidth of excitation. Many instrument manufacturers offer sets of lasers, LEDs, or filters with accessories that make it easy to move these in and out of the instrument's optical path, such that the appropriate excitation wavelength needed for analyte detection can be selected.

7.5.1.4 Steady-State Versus Microsecond-Pulsed Instruments

Some fluorescence spectrometers can be used as steady-state instruments, even though their excitation light is pulsed on the order of microseconds and their emission signal is time gated, typically on the order of tens of microseconds to milliseconds. The term "time gate" refers to the time period, set by the user, during which the instrument measures luminescence, also referred to as the "detection gate." The period of time between the start of the excitation pulse and the start of the detection gate, called the "delay time," can also be set on the instrument by the user. These instruments give spectra with the same spectral shape, that is, relative intensity as a function of wavelength, as a true steady-state instrument, if the luminescence being detected has a much shorter lifetime than microseconds, which is true for fluorescence. Even if the luminescence lifetime is on the order of microseconds or greater, which is the case for phosphorescence, the spectra will not differ from a steady-state instrument, as long as the luminescence spectrum does not change over the length of its lifetime. But there are cases where the excited-state kinetics of a fluorophore causes its spectrum to change with time. For example, an inorganic phosphor may have multiple phosphorescent decay pathways, with various luminescence lifetimes, ranging from microseconds to milliseconds or longer, resulting in different emission spectra at different times after the excitation pulse.

Despite this caveat, microsecond-pulsed instruments offer several advantages over conventional steady-state instruments. First, the same signal-to-noise ratio can be achieved using less radiant power for sample excitation, due to the combination of pulsed excitation and time-gated detection. This is particularly advantageous for detecting fluorophores that are prone to photobleaching or photodegradation [9], such as organic dyes, because they can be detected over longer periods of time, that is, a greater number of spectra can be collected, without a significant decrease in fluorescence intensity due to these effects. A lower radiant power for sample excitation is also less damaging to living cells, which often die when exposed to too much light. This is particularly true when higher-energy radiation, such as ultraviolet and blue light, needs to be used for excitation.

Microsecond-pulsed instruments also have clear advantages over steady-state instruments when phosphorescence is being detected because the length of the detection gate and the delay time between the excitation pulse and the detection gate can both be set. This allows phosphorescence to be detected after the excitation pulse in time, which greatly improves the signal-to-noise ratio by preventing the excitation light from being detected. These instruments also have the capability to measure phosphorescence lifetime in the microsecond to millisecond time range.

7.5.1.5 Microwell Plate and Microarray Readers

MP and MA readers are two examples of specialized fluorometers that have emerged in the last two decades. They are widely used for high-throughput applications in the pharmaceutical, biotechnology, material science communities, and others. Most MP and MA readers use bandpass filters, LEDs, or lasers for selection of μ and bandpass filters for selection of λ. The design of most MP and MA readers is closer to that of a microscope with a raster-scanning stage or a spatial imaging camera than a typical fluorescence spectrometer, but as the trend for decreasing sample sizes continues, it is likely that more and more fluorometers will look like this. The main difference between MP and MA readers is sample size and number. An MP typically contains hundreds of millimeter-sized samples, whereas an MA contains thousands of micrometer-sized samples. Smaller sample sizes require smaller excitation beams and detection regions, using focusing optics to obtain irradiation and viewing areas that are comparable to the size of the samples, that is, on the order of millimeters and micrometers for MP and MA, respectively. We mention these instruments here, even though we have confined our scope to spectral instruments, because some of these instruments have more recently incorporated spectral capabilities into their design. The ultimate goal of these new instruments is to achieve effective high-throughput spectral analysis and real-time spectral imaging. We would be remiss to not mention these instruments, since they are likely to become more common and many of the issues discussed here that are related to fluorescence spectrometers are also applicable to these instruments, including calibration, uncertainties, and standardization.

7.5.2 Colorimetric Applications

As we have seen for colorimetric applications, the desired measurement quantity is generally the overall visual effect of the emitted (fluorescent) and reflected radiation, which is responsible for the material's color appearance, from an opaque reflecting material (e.g., a paint, textile, or plastic), under a specified broadband illumination condition. Thus, the instrument must be able to measure the sample's reflectance properties in conformance with one of the standard measurement geometries recommended by the International Commission on Illumination [10], as well as the sample's fluorescence properties under a specified source distribution.

As discussed in Section 7.2, colorimetric measurements must be referenced to the spectral reflectance scale. This is the important difference from the fluorescence measurements carried out for analytical applications, which are in A.U. given by the measured signal intensities. As in the case of reflectance measurements of nonfluorescent materials using conventional spectrophotometry discussed in Chapter 6, radiance factor measurements of fluorescent materials are referenced to the primary standard for reflectance measurements—the PRD.

Since the PRD cannot be realized in practice, the comparison of the specimen to the PRD is accomplished indirectly. Secondary white reflectance standards are used, whose properties are known in terms of the PRD via absolute reflectance methods (see Chapter 6). Examples of these secondary white reflectance standards are pressed tablets of polytetrafluoroethylene (PTFE) or barium sulfate powder. Thus, for fluorescence measurements using spectrophotometry, the total flux (reflected plus emitted) leaving the sample is compared to that reflected by a nonfluorescent white reflectance standard, identically irradiated and viewed. It is important to note that the reflectance standard that is used to establish the radiance factor scale must be nonfluorescent. If this reflectance standard exhibits fluorescence, it can impact the calibration of fluorescent standards, under test, and cause colorimetric errors. This is particularly problematic if this fluorescent test standard is then used as a transfer standard to calibrate secondary fluorescence measuring instruments of different design, which have different spectral responsivity in the region where the reference standard exhibited fluorescence effects [11].

For a nonfluorescent opaque material, the color appearance depends upon the reflected radiation only in the visible portion of the spectrum, where this radiation is of the same wavelength as the incident radiation. This quantity can be measured on a conventional spectrophotometer in either monochromatic or polychromatic illumination mode. However, for a fluorescent sample, a simple ratio measurement of the total flux leaving the specimen to the incident flux as a function of the incident wavelength can provide erroneous results. This is illustrated in Fig. 7.5, which shows the measured apparent reflectance of a fluorescent red paper sample recorded on a conventional one-monochromator scanning system with monochromatic irradiation and polychromatic (broadband) detection. It can be seen that the detector has incorrectly recorded the enhanced reflectance in the blue portion of the

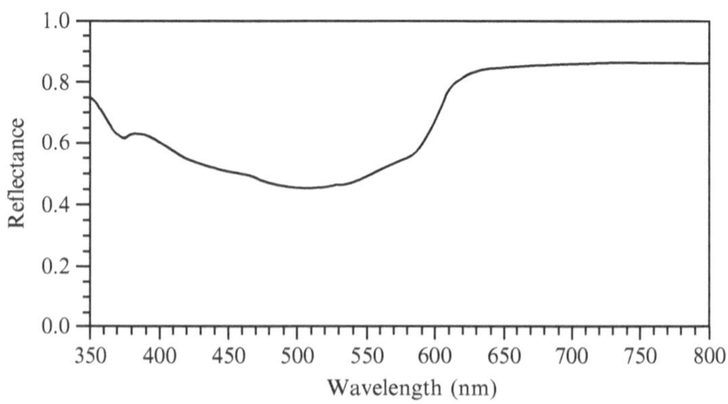

FIGURE 7.5 Measured reflectance of a fluorescent red paper sample recorded on a conventional one-monochromator spectrophotometer with monochromatic irradiation and broadband detection.

spectrum, at the wavelengths of excitation, instead of the red portion of the spectrum, where the absorbed light has been emitted (see also Fig. 7.8 which shows the spectral region of emission for this fluorescent red and its true reflectance spectrum). The measured spectrum in Fig. 7.5 is a meaningless result for colorimetric evaluation purposes. Thus, a one-monochromator system with monochromatic irradiation is unsuitable for the color measurement of fluorescent materials.

Due to these limitations of the monochromatic mode, if a one-monochromator system is used for the measurement of fluorescent materials, it must have polychromatic (undispersed) illumination and either monochromatic (scanning system) or spectrally dispersed (array system) detection of the reflected and emitted radiation. The measured total spectral radiance factors are then representative of the color appearance under the particular spectral distribution of the source used in the measuring instrument. Ideally, the polychromatic source of the spectrophotometer is designed to be a close match to the standard illuminant conditions required for the colorimetric application. Procedures for specifying the quality of the sources intended to simulate standard illuminants are given in CIE Publication 51 [12], and the influence of the instrument source spectral distribution on the measurement of fluorescent materials using the one-monochromator method is discussed in Section 7.5.2.4. In practice, it is difficult to achieve an accurate simulation, and various methods are used to improve the accuracy of this colorimetric result—these specialized one-monochromator methods are described below.

An instrument for color measurement of fluorescent materials must also conform to one of the standard illuminating and viewing geometries that are recommended for colorimetric applications of opaque reflecting materials [10]. These CIE color measurement geometries are described in detail in Chapter 10. They can be broadly classified as either a *hemispherical geometry* using an integrating sphere (d:0° or 0°:d, or de:8°, di:8°, 8°/de, 8°/di) or a *bidirectional geometry* (45°:0° or 0°:45°, or 45°a:0° or 0°:45°a), where the first term gives the direction of illumination and the second term gives the direction of viewing. The term "de" denotes the diffuse condition with the specular component excluded, and "di" denotes the diffuse condition with specular component included. The influence of these instrument geometries on the measurement of fluorescent materials is discussed in detail in the following sections.

7.5.2.1 Sphere Geometry

A hemispherical geometry was favored in some of the earlier spectrofluorimeters used for the measurement of fluorescent materials because the effects of surface structure and/or directional characteristics are minimized in this type of design. However, it was shown by Alman and Billmeyer [13] that this geometry is not ideal for the measurement of fluorescent materials, since the source irradiance on the specimen is altered by the sphere properties and the fluorescence emitted by the sample contributes to the diffuse illumination. Thus, the resulting source–sphere irradiance is dependent not only on the spectral reflecting nature of the sphere but also on the fluorescent emission from the sample itself so that it must be measured for each different fluorescent specimen.

When an integrating sphere instrument is used for measuring fluorescent samples, it is recommended that the spectral-selective properties of the sphere be minimized, for example, by reducing the size of the sample port in relation to the sphere diameter and by using a specular excluded geometry, since a light trap is more spectrally neutral than the white plugs that are used in the specular included geometry.

7.5.2.2 Bidirectional Geometry (45°:0° or 0°:45°)

Because of these complications, the recommended reference geometry for the color measurement of fluorescent materials is bidirectional. To reduce the measurement sensitivity to small changes in geometry or to small changes in specimen surface irregularities, an annular geometry is preferred, 45°a:0° or 0°:45°a. For this measurement geometry, the sample and nonfluorescent reflecting standard are either both mounted in a sample wheel and alternately positioned in the measurement beam or are alternately substituted at a fixed sample reflecting plane. The advantage of the first design is that it can be automated to improve measurement speed and to compensate for linear system drift. However, the reproducible and accurate positioning of the sample and standard in their respective mounts are critical to ensure that the same measurement geometry is used. Errors of several percent in the measured reflectance can occur if the sample and standard are not positioned at exactly the same reflecting plane. Direct substitution at the same fixed sample plane reduces this instrument geometry reproducibility error but is much more tedious and susceptible to system drift errors. In contrast, the fluorescent component is much less sensitive to the measurement geometry, due to the fact that it tends to be largely Lambertian in nature.

7.5.2.3 Bispectrometers and Two-Monochromator Measurement Method

For highest accuracy colorimetric applications and where it is desired to obtain the color specifications of fluorescent materials under different standard illuminating conditions, it is necessary to use the *two-monochromator method*, involving both monochromatic illumination and monochromatic detection.

Although two-monochromator instruments for analytical applications of fluorescence measurements have been available for many years, they do not conform to the geometric requirements of colorimetric applications. It is only within the past 15 years that instruments have become commercially available for bispectral color measurements of fluorescent materials; these instruments are sometimes referred to as *bispectral fluorescence color analyzers* or *bispectral fluorescence colorimeters* and are available in both bidirectional and sphere geometries. The basic instrument designs, components, and measurement methods are very similar to that described in Section 7.4.

The two-monochromator method, described in Section 7.3.1, is considered the reference method for the determination of the colorimetric properties of fluorescent materials. This method allows for the complete and accurate separation of the reflected and fluorescent components. This type of fluorescence instrument that is based on the two-monochromator method and is used for high-accuracy surface fluorescence measurements is referred to as a *reference spectrofluorimeter.*

The reflected radiance factor is determined by setting both monochromators to the same wavelength setting and scanning synchronously through the entire spectrum. The luminescent radiance factor is determined by the bispectral measurement procedure described in Section 7.3.1, repeating this process for all wavelengths μ in the excitation band of the fluorescent material and for all wavelengths λ in the emission spectrum. At the wavelengths $\mu=\lambda$, both the reflected and luminescent radiant flux are measured and special procedures are required to separate the two components which are described in detail in Section 7.6.6. The measured data are uncorrected, that is, they are distorted by the spectral characteristics of the instrument.

To obtain bispectral luminescent radiance factor data, defined in Section 7.2, it is necessary to carry out several calibration steps. To obtain *technically corrected* bispectral data, it is necessary to first carry out standard spectrophotometric calibration procedures, for example, wavelength and photometric scale accuracy, stray light, and polarization, whose effects are described in Section 7.6. It is then necessary to carry out several calibration procedures that are required for spectroradiometric measurements. These involve determining (1) the relative spectral distribution of the source illuminating the sample and (2) the relative spectral sensitivity of the detection system. To convert this technically corrected data to *normalized* data on an *absolute* reflectance scale, it is also necessary to determine (3) the spectral efficiency factor of the instrument. This can be considered to be an instrument-specific scaling factor which depends upon the geometric parameters of the instrument that determine its optical throughput, including the excitation and emission monochromator slit settings that are used in the fluorescence measurement. These calibrations are carried out using physical type standards, for example, calibrated sources (CSs), detectors, or white diffuse reflectors, which are known as *physical transfer standards* (PTSs) that link the fluorescence measurements to radiometric and physical scales [14]. The use of these PTS methods is described in detail in Section 7.8.1.1.

7.5.2.4 One-Monochromator Instruments and Methods

We have seen that if a one-monochromator instrument is to be used for color measurement of fluorescent materials, it must be used in polychromatic mode (undispersed, broadband illumination and monochromatic detection). Ideally, the source spectral power distribution illuminating the sample matches closely the desired standard illuminant condition. In this case, the total spectral radiance factor for this illumination condition is measured directly. In practice, this is often not the case, and the instrument must be calibrated to compensate for deviation of the source's spectral characteristics from the standard illuminant. This is accomplished using suitable fluorescent reference materials (RMs) that have been calibrated for the desired standard illuminant condition using a two-monochromator method. Depending upon the particular colorimetric application, and the desired colorimetric accuracy, different one-monochromator calibration methods are used. These are described in detail below.

7.5.2.4.1 One-Monochromator Instruments with UV-Control

One of the most important optical properties of white paper is its apparent reflectance in the blue region of the optical spectrum, which may be quantified as *ISO brightness* [15] or *TAPPI brightness* [16], depending upon measurement geometry. To improve the appearance and commercial value of white paper, papermakers typically use FWAs, which are excited in the near-UV and emit in the visible blue to give a preferred bluish-white appearance [3]. The objective evaluation of the optical properties of these materials requires the use of standardized methods, where the measuring instrument must conform to specific geometric and spectral characteristics. The linearity of the photometric scale of these instruments is calibrated using a nonfluorescent reference standard, which has also been calibrated in accordance with the provisions of the relevant standard.

To calibrate the colorimetric scale for the measurement of various color quantities, the spectral distribution of the instrument source must produce the same colorimetric result as would be obtained with an instrument whose source spectral distribution exactly matched the desired illuminant condition for the desired colorimetric specification. For this purpose, the measuring instrument must incorporate a UV control option. This option typically uses a 365-nm cutoff filter that allows the user to control (in a limited fashion) the spectral distribution of the instrument illumination in the UV region where excitation of the FWA occurs compared to the visible region where emission occurs. The position of the filter, and thus the relative UV to visible content of the illumination, is adjusted using a calibrated fluorescent standard. Since this is a simple ratiometric calibration method, the standard needs to have similar excitation and emission spectral properties to the sample under test. Thus, it is common to use fluorescent standards of the same type, that is, in this case, fluorescent paper standards are used for this UV adjustment.

This adjustment is also carried out for a specific colorimetric value (e.g., ISO brightness, defined for CIE illuminant C). Similar methods are used to calibrate the instrument using other colorimetric values, such as CIE whiteness, defined for CIE standard illuminant D65 conditions. It is important to note that an instrument calibrated to CIE D65 whiteness will give slightly different settings of the UV-adjustment filter than a calibration of the instrument to ISO brightness. This is a limitation of a one-point adjustment. Nevertheless, for the paper industry, the simplicity of such one-point UV adjustment outweighs the increased utility of a more complicated calibration procedure.

This type of instrument is also widely used in the textile and detergent industries for measuring fluorescent whitened fabrics using the *Ganz–Griesser method* of whiteness evaluation [17]. This method, unlike the CIE whiteness method, is not recognized as an international standard, but its applicability is not limited to samples that are measured on the same instrument at nearly the same time. The Ganz–Griesser formula and calibration method uses a set of four fluorescent textile samples containing different concentrations of FWAs that have been calibrated for CIE tristimulus values; these are used with the instrument's UV control to determine the instrument-specific parameters for the whiteness and tint deviation formulae [17]. An example of a four-step set of textile cotton reference standards commonly used with this Ganz–Griesser UV calibration procedure is shown in Fig. 7.6.

7.5.2.4.2 One-Monochromator Instruments Using Predictor Methods

One-monochromator instruments generally measure the total spectral radiance factor and do not give separate information on the reflection and fluorescence components. If the instrument source cannot be filtered or adjusted to have a relative spectral distribution matching that of the desired illuminant condition, then a one-monochromator predictor method can be used to improve the

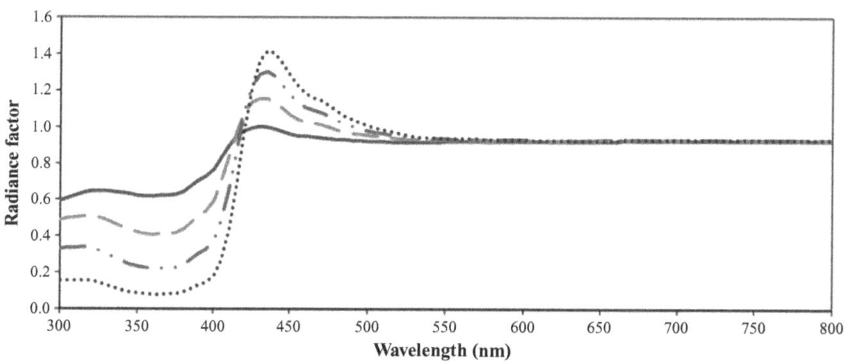

FIGURE 7.6 Spectral total radiance factors for CIE standard illuminant D65 of a set of four-step cotton fluorescence scale standards, available from the Hohenstein Institute.

colorimetric accuracy. The predictor method uses a mathematical model to estimate, or predict, the spectral radiance factors for the desired illuminant from various measured values obtained under controlled spectral filtering of the instrument source.

The most widely used predictor method is the *fluorescence-weakening method* first introduced by Allen [18]. This method involves measuring the specimen under three different spectral illuminating conditions: (a) bare source with no filter; (b) with *fluorescence-weakening* neutral density filter, of known transmittance $T_1(\lambda)$; and (c) with *fluorescence-killing* cutoff filter, of known transmittance, $T_2(\lambda)$. The fluorescence-weakening filter is a neutral density filter that does not alter the relative spectral distribution in the excitation range of the sample. The fluorescence-killing filter is selected so that its cutoff wavelength is no more than 10 nm below the wavelength of peak sample emission. The apparent total spectral radiance factors β_T recorded for each of these three conditions are denoted $\beta(\lambda)$, $\beta_{T_1}(\lambda)$, and $\beta_{T_2}(\lambda)$, respectively. The reflected radiance factor of the sample, $\beta_R(\lambda)$, is determined in a systematic fashion by analyzing the effect of the filters of known transmittance on the measured curve. For those wavelengths where a neutral density filter in the beam has no impact on the shape of the measured radiance factor data, that is, below the wavelengths of excitation and above the wavelengths of emission, the reflected component can be calculated in a straightforward manner from the measured radiance factor divided by the spectral transmittance of the attenuating filter, as

$$\beta_R(\lambda) = \beta(l), \text{ where } T_1(\lambda) < 0.30, \tag{7.18}$$

$$\beta_R(\lambda) = [\beta_{T2}(\lambda)/T_2(\lambda)], \text{ where } T_2(\lambda) \geq 0.30. \tag{7.19}$$

In the examples given in Figs. 7.7 and 7.8, this fluorescence-weakening filter is not neutral but has its cut-on wavelength well outside the spectral range of excitation. So, for those wavelengths below its cut-on, the reflected component is obtained from the measured radiance factor with no filters in the beam. In the intermediate spectral range, known as the *crossover region*, where there is simultaneous absorption of exciting radiation and emission of fluorescent radiation, the calculation is more complicated and requires analysis of the impact of the fluorescence-killing filter on the shape of the measured radiance factor and the average attenuation by the fluorescence-weakening filter. The spectral radiance factor used in this intermediate spectral range is

$$\beta_R(\lambda) = \frac{\beta_{T_1}(\lambda)\cdot T_1(\lambda) - \beta(\lambda)\cdot k}{T_1(\lambda) - k} \tag{7.20}$$

for wavelengths, where $T_1(\lambda) \geq 0.3$ and $T_2(\lambda) < 0.3$. The constant k is wavelength independent and is evaluated from the lowest wavelength reading, λ_e, obtained with the fluorescence-killing filter for which $T_2(\lambda) \geq 0.3$ from

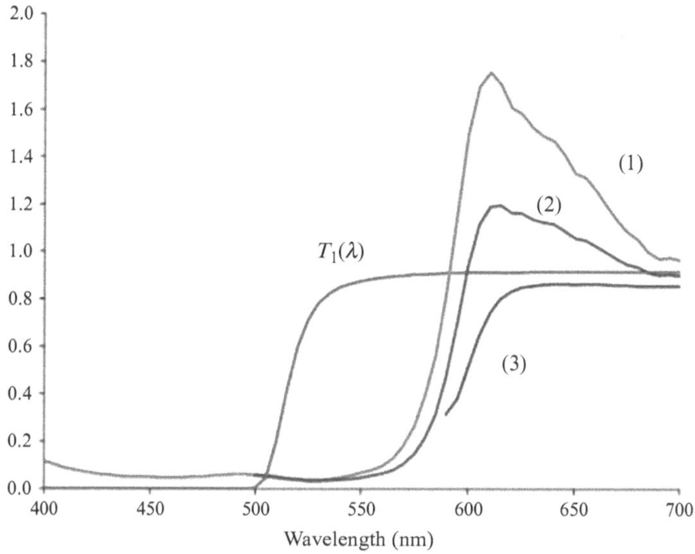

FIGURE 7.7 An illustration of the Allen [18] filter method applied to the measurement of a fluorescent red paper sample. The curve (1) was measured with no filter; curve (2) was measured with a fluorescent-weakening filter with a cutoff at 515 nm; and curve (3) was measured with a fluorescent killing filter with a cutoff at 590 nm. The transmittance curve, $T_1(\lambda)$, is the measured transmittance of the 515-nm filter; the transmittance of the fluorescence-killing filter is not shown.

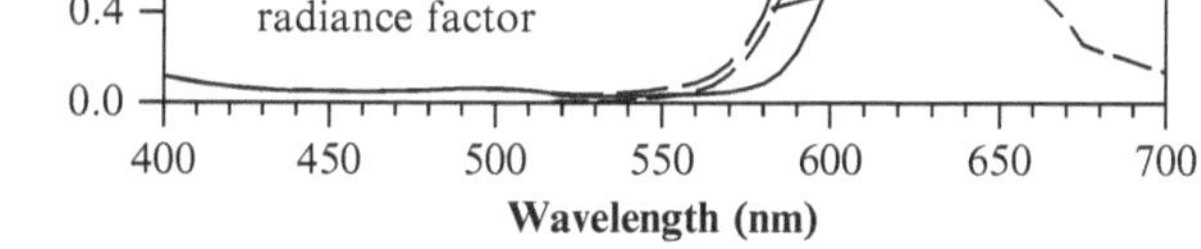

FIGURE 7.8 Total, fluorescent, and reflected radiance factor curves for the fluorescent red paper sample also measured in Fig. 7.5. Data shown here were calculated using the fluorescence-weakening method from one-monochromator measurements shown in Fig. 7.7 using a xenon source corrected to CIE standard illuminant D65.

$$k = T_1(\lambda_e)\frac{[\beta_{T_1}(\lambda_e) - \beta_R(\lambda_e)]}{[\beta(\lambda_e) - \beta_R(\lambda_e)]}. \tag{7.21}$$

Once $\beta_R(\lambda)$ has been calculated, the fluorescent radiance factor component can be calculated by the difference between the measured radiance factor without filters, $\beta(\lambda)$, and the calculated reflected-only radiance factor, $\beta_R(\lambda)$,

$$\beta_L(\lambda) = \beta(\lambda) - \beta_R(\lambda). \tag{7.22}$$

It should be noted that the luminescent radiance factor spectrum, so-determined, is strictly valid only for the light source conditions of the instrument. This predictor method is primarily used where knowledge of the reflected radiance factor of the sample is of great interest.

An application of this fluorescence-weakening predictor method is shown in Figs. 7.7 and 7.8. Figure 7.7 shows the measurements of the red fluorescent paper sample (also shown in Fig. 7.5, measured on a one-monochromator instrument in polychromatic mode with a xenon source) with both fluorescence-weakening and killing filters, along with the spectral transmittance of the fluorescence-weakening filters; Fig. 7.8 shows the corresponding predicted results for the total radiance factor, fluorescent radiance factor, and reflected radiance factor curves. These latter data were determined for CIE standard illuminant D65 conditions by also correcting the spectral distribution of the bare sphere-source to that of D65 by using a telespectroradiometer to measure the radiance from the sphere wall when the fluorescent sample was measured with no filter in the beam. The advantage of this procedure is its experimental simplicity. However, *a priori* knowledge of the spectral range of the excitation and emission bands of the fluorophore(s) in the fluorescent sample is required to select appropriate filters. This information is readily available for fluorophores of commercial interest.

7.5.3 Quantum Yield Measurements

Fluorescence quantum yields of both liquid and solid samples can be measured using either relative or absolute optical methods. The principles of these optical methods are described here.

7.5.3.1 Relative Quantum Yield Methods

For both analytical and colorimetric applications, it is most common to perform these measurements on a relative basis, using quantum yield standards that meet the requirements outlined later in Section 7.8.4. These relative measurements involve the direct comparison, by substitution, of the total luminescent flux or fluorescent intensity of the specimen when irradiated at the same excitation wavelength μ_0 to that of a reference fluorescent standard of known quantum yield.

In the case where the incident radiation is completely absorbed, as in the case of a *solid sample*, this determination is given simply by taking the ratio of the integrated areas of the corrected emission spectra from the specimen, $M_s(\lambda)$, to that of the reference standard, $M_{std}(\lambda)$, and multiplying by the known quantum yield of the standard, ϕ_{std}, to give the quantum yield of the specimen, ϕ_f, as

$$\phi_f = \frac{\int_{\lambda min}^{\lambda max} M_s(\lambda) d\lambda}{\int_{\lambda min}^{\lambda max} M_{std}(\lambda) d\lambda} \cdot \phi_{std}. \tag{7.23}$$

It should be noted that if both the sample and the standard have similar emission spectra and their integrated intensity is calculated over the same wavelength range, then it is not necessary to convert the fluorescence spectrum to energy units. However, this is generally not the case, so it is necessary to first convert these emission spectra to an energy scale before calculating the integrated value to be used in Eq. (7.23).

In the case of a *liquid* or *partly transmitting sample*, this determination requires additional measurements. These are the blank and dark count corrected absorbance values of the sample, $A(\mu_0)$, and standard, $A_{std}(\mu_0)$, at the excitation wavelength μ_0. These absorbance values can be measured on a conventional absorbance spectrophotometer, subtracting the contributions from the blank solvent and dark current reading. These absorbance measurements for both sample and standard should be performed with an instrument spectral bandpass $\Delta\mu$, similar to that used for the fluorescence measurement, and the measured absorbance should be below 0.05 so that it is in the linear absorbance range of the spectrophotometer. If it exceeds this value, the sample and standard should be diluted with additional solvent.

Furthermore, if the test sample and the standard are used in different solvents or in different matrices, it is also necessary to apply a correction for the difference in refractive indices, n(sample) and n(standard), of these two solvents. If we define a *shorthand notation for the integrated area of the fluorescence spectrum* given in Eq. (7.23) as $S_s(\lambda)$ and $S_{std}(\lambda)$ for the sample and standard, respectively, then the determination of the quantum yield for the molecular fluorophore of interest is given by

$$\phi_f = \frac{S_s(\lambda)}{S_{std}(\lambda)} \cdot \frac{A_{std}(\mu_0)}{A_s(\mu_0)} \cdot \frac{n^2(\text{sample})}{n^2(\text{std})} \cdot \phi_{std}. \tag{7.24}$$

For many fluorescent samples, the excitation and emission spectra overlap, and it is necessary to select an excitation wavelength that minimizes reabsorption or inner-filter effects (IFEs) that are discussed later in Section 7.7.2. Thus, for higher accuracy determinations of relative fluorescent yields, rather than carry out a single-point determination of the absorbance of the test sample and standards at μ_0, it is recommended to prepare calibration curves of the integrated fluorescence intensity versus absorbance at μ_0 for both the test sample and the standard, for a series of dilutions with a maximum concentration giving an absorbance of about 0.05. A *linear-regression analysis* is performed on these calibration curves for both sample and standard and the ratio of these slopes, $S_i(\lambda)$ versus A_i, where $i=$ s or $i=$ std are substituted for the first two terms given in Eq. (7.24).

An illustration of this procedure is given in Figs. 7.9 and 7.10 where the fluorescence reference standard is Rhodamine 101 solution in ethanol which has a quantum yield of 100% at room temperature over a wide temperature range and a wide range of excitation wavelengths from 450 to 550 nm. Figure 7.9 shows the technically corrected emission spectra of this dye when excited at 540 nm for several dilutions with a maximum absorbance of 0.05. These data have been plotted on an energy scale so that its quantum yield can be compared directly with test fluorescent samples that have slightly different emission shapes. Figure 7.10 shows a linear-regression analysis for a calibration curve of corrected fluorescent intensity versus absorbance for a test fluorescent sample, cresyl violet, which has been measured under exactly the same conditions (instrument and sampling parameters) as the reference standard. The slope of this plot of integrated fluorescent intensity to absorbance (2.43×10^6) is compared to that recorded for the reference standard (4.76×10^6) to determine the relative quantum yield of this dye. In this case, the quantum yield calculation is simply the ratio of these two slopes multiplied by 100% since $\phi_{std} = 100\%$ and both sample and reference are measured in the same solvent. The value of the relative quantum yield determined here is 51% which is in excellent agreement with the literature value for cresyl violet in ethanol [19].

In the case of a front-surface geometry, the external quantum efficiency can be measured using a nonfluorescent, diffusely reflecting standard of known spectral radiance factor for the same measurement geometry using Vavilov's substitution method [20]. This method essentially assumes that the ratio of the bispectral radiance factors can be interpreted in terms of a ratio of fluxes to obtain absolute fluorescent yields relative to a diffuse reflectance

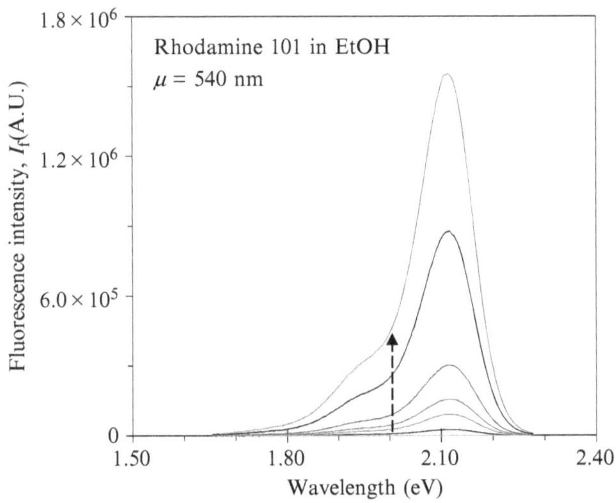

FIGURE 7.9 Corrected fluorescent intensity spectra of Rhodamine 101 solutions in ethanol for excitation wavelength of 540 nm, 2.5-nm bandpass.

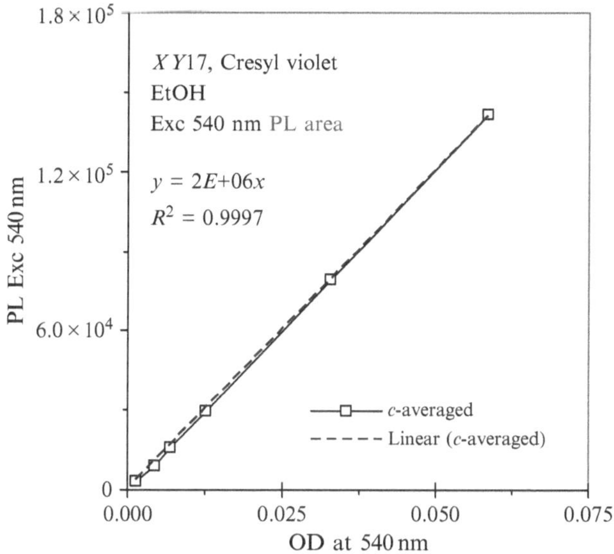

FIGURE 7.10 Linear-regression plot for integrated fluorescent intensity for excitation at 540 nm versus absorbance (OD) at 540 nm of cresyl violet dye solutions in ethanol, 2.5-nm bandpass.

standard. The accuracy of this method depends on both the sample and the standard having fluorescence properties that are isotropic, that is, insensitive to geometry, so that the quantum yield measured in a particular direction is representative of the value that would be measured in a hemispherical (integrating sphere) geometry. Examples of suitable nonfluorescent diffusing standards are freshly pressed PTFE or barium sulfate powder.

Under this assumption that the luminescent radiance is independent of measurement geometry, then for excitation at a specified wavelength μ_0, the integrated area under the corresponding normalized bispectral luminescent radiance factor $\beta_{L,\lambda}(\mu_0, \lambda)$ is calculated to give the photoluminescent radiant yield, *RY*. If the normalized bispectral luminescent radiance factors are first converted to photon units, $\beta_{Lp,\lambda}(\mu_0, \lambda)$, as shown in Chapter 2, Section 4.6, then the measured quantum yield is the radiometric equivalent to the *spectral quantum efficiency of the fluorescent process*, $\eta_\mu(\mu)$, defined in Section 7.2.3, that is,

$$\eta_\mu(\mu_0) = \int_\lambda \beta_{Lp,\lambda}(\mu_0, \lambda) d\mu. \quad (7.25)$$

7.5.3.2 Absolute Quantum Yield Methods

The determination of absolute fluorescent quantum yields typically requires at least two separate experiments, the first to obtain the number of photons absorbed by the sample at the given excitation wavelength and the second to obtain the corresponding number of photons emitted by the sample, by integrating the photoluminescent emission peak after correcting for background emission. The intrinsic properties of the fluorophore(s) within the material, at the molecular level, may be derived by correcting for various sample measurement effects described in Sections 7.7 and 7.8.1.3. Such intrinsic properties are useful for analytical purposes as well as for modeling used in color formulation.

For many years, the optical method that was used to determine absolute quantum yields of many of the available liquid fluorescent standards for both 0°/90° and front-surface geometries was based on the use of a nonselective quantum detector, such as a thermal detector (e.g., thermopile) and a quantum counter (e.g., Rhodamine 101 solution). It is now more common to carry out absolute measurements of fluorescent quantum efficiency or quantum yield for both liquid and solid samples using an *integrating sphere method*, with either *collimated incident light* or *diffuse incident light* using a fluorescence bis-pectrometer that has been calibrated for its spectral characteristics as described in Section 7.6 to provide technically corrected spectral radiant flux data. The equipment and experimental setup for carrying out these absolute quantum yield measurements for both integrating sphere methods is very similar and is described in a new draft IEC Standard, 62607-3-1 [21]. Commercial sphere accessories are available for this purpose for use with conventional fluorometers. A key requirement for the success of this method is that the sphere wall coating exhibits both high uniformity and high reflectance so that there is no asymmetry in the integrating properties and a high sphere efficiency, respectively. Also, the sphere coating should not fluoresce because this will compromise the accuracy of the sample fluorescence quantum yield measurements. It has been found that sphere accessories made from sintered PTFE are particularly susceptible to UV-induced fluorescence. This largely originates from organic contaminants adsorbed by the sphere wall coating either during the manufacturing process or from subsequent environmental pollutants [22,23]. This sphere fluorescence, like instrumental stray light, can cause errors in absolute quantum yield determinations if not characterized and corrected. Zong at NIST has developed a matrix procedure for characterizing spectral stray light, and this methodology has also been successfully applied to quantifying and correcting for sphere fluorescence properties [24].

7.6 INSTRUMENT CHARACTERISTICS THAT IMPACT SPECTRAL FLUORESCENCE MEASUREMENTS

7.6.1 Spectral Characteristics

7.6.1.1 Source Spectral Distribution

For a two-monochromator instrument, or bispectrometer, the precise details of the source spectral distribution are not important since the instrument spectral characteristics are calibrated using either PTS or RM-based methods.

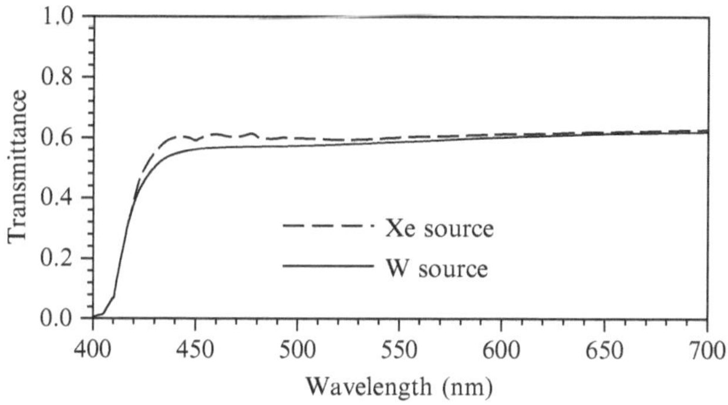

FIGURE 7.11 Spectral transmittance factor curves for a fluorescent white fabric measured on a sphere instrument using one-monochromator method and polychromatic illumination using two different instrument sources: xenon arc and tungsten lamp.

However, the source should have sufficient spectral output over the wavelength region of interest to give adequate measurement sensitivity. For a one-monochromator instrument with polychromatic illumination, the instrument source should ideally be a close match spectrally to the desired illuminant condition. But for most purposes, it is adequate that the illumination can be adjusted to have similar spectral balance (ratio) of the total radiant output in the two wavelength ranges that correspond to the fluorescent sample's emission and excitation spectra.

The influence of the instrument source in a one-monochromator instrument is illustrated in Fig. 7.11, which is the measured transmittance of a fluorescently whitened fabric recorded using the one-monochromator method, with different sources: a xenon arc source and an incandescent tungsten lamp. As can be seen from Fig. 7.1, for this type of fluorescent sample which uses FWAs, the fluorescence is predominantly caused by UV excitation. Since the xenon source has relatively more output in the UV region compared with the tungsten lamp, it is capable of exciting more fluorescence and this produces an enhancement to the measured transmittance in the blue region of a few percent.

7.6.1.2 Other Instrument Components

There are several other instrument components, besides the light source, that contribute to the spectral signature of the instrument. These include the spectral sensitivity of the analyzing detector; the spectral efficiency of the monochromator; and the spectral reflectance/transmittance of the mirrors, lenses, filters, polarizers, and other optical components in the measurement beams. The spectral calibration of the instrument usually involves a separate spectral characterization and correction of the excitation and emission units, where the excitation unit encompasses the light source, excitation monochromator, and all other instrument components before the sample, and the emission unit encompasses the emission monochromator, analyzing detector, and all other instrument components following the sample. If any of these components is changed, it is necessary to carry out a new spectral calibration for that part of the system. For example, if the monochromator grating is changed, a new spectral calibration is required.

In practice, it is found that the spectral sensitivity of PMTs is quite stable with time and a recalibration period of 1 or 2 years is adequate, whereas the spectral irradiance due to the excitation unit changes significantly, primarily due to aging of the light source. For this reason, the excitation unit typically incorporates a reference or a monitor detector to dynamically compensate/correct its calibration for changes in the source output between recalibration periods.

7.6.1.3 Instrument Aging

As we have seen, instruments that are used for fluorescence measurements need to have broad spectral coverage in the UV and visible range, since many of the fluorescent samples of interest are excited by UV radiation. Unfortunately, prolonged UV exposure of the various optical components of the measuring instrument, such as mirrors, filters, and diffusers, can cause their gradual degradation. This is characterized by a general decline in the instrument throughput and measurement sensitivity, particularly at shorter wavelengths. The net effect is that as the instrument ages, it becomes less capable of exciting fluorescence in many fluorescent white samples, which are excited by radiation in the UV and violet region. The aging of the xenon arc can also cause the optical beam to shift its position, which makes it critical to recalibrate the monitor detector frequently enough to ensure that it is sampling the same geometric distribution of the optical beam that is incident at the sampling position.

7.6.1.4 Wavelength and Spectral Slit-Width Accuracy

As mentioned above, the spectral accuracy of the fluorescence measurements is important for obtaining reliable comparable data that are instrument independent. This spectral accuracy is achieved by accurately calibrating and correcting for the influences of the instrument's spectral characteristics. This requires, in turn, the accurate calibration of the instrument's wavelength scales and slit settings.

The wavelength and spectral slit-width accuracy of the excitation and emission wavelength scales of a fluorescence spectrophotometer can be calibrated as described in Chapter 3 using the spectral lines of a wavelength standard [4]. The most commonly used wavelength standard is a set of spectral line lamps, for example, Hg, Cd, Cs, Ar, Xe, Ne, Kr, and He. Low-pressure pen lamps can be placed at the entrance slit of the excitation monochromator for calibrating the excitation unit and at the sample position for calibrating the emission unit. As discussed in Chapter 3, it is important that the spectral line lamp is well aligned so that the entrance slit of the monochromator is uniformly illuminated, or this can introduce a spectral shift of the emission line and an asymmetric lineshape. For applications where wavelength accuracies of about 0.5 nm are needed, it is sufficient to determine a wavelength calibration function from a linear least-squares fit of the measured versus the known line positions. This level of wavelength uncertainty is readily attainable with research-grade dispersive instruments using a PMT-based detection system. For higher accuracy requirements, it is necessary to use a higher order polynomial least-squares fit (typically fourth or fifth order) wavelength calibration function, as illustrated in Chapter 3 to give typical wavelength accuracies of 0.2 nm throughout the UV–visible spectral range; for many grating instruments, the residuals show a sinusoidal variation due to the sine drives used with stepper motors [25]. Thus, it is possible to further improve the wavelength accuracy to about 0.05 nm by further fitting the residuals from the least-squares polynomial fit with a sinusoidal function. However, for CCD-based instruments, the wavelength-scale errors can be much larger because of their sensitivity to detector alignment and focus.

For fluorescence instruments, where it is not convenient to use spectral lamps, calibrated material wavelength standards with narrow well-separated spectral features in the UV/visible range can be used, for example, a solution of holmium oxide in perchloric acid, doped glass with a diffuse reflector, or laser crystals such as Dy-YAG. For these material wavelength standards, it is important to note that their peak positions are dependent upon the instrument spectral bandpass conditions, particularly for adjacent asymmetric bands [26]. For example, an international evaluation of the holmium oxide solution wavelength standard showed that a spectral peak certified at 640.48 nm for a 1-nm instrument bandpass shifted by ~0.6 to 641.09 nm for a 3-nm instrument bandpass. Thus, it is recommended that these wavelength standards be calibrated for the same conditions of spectral bandpass that will be typically used with the fluorometer (see Ref. [4]).

For analytical applications, the water Raman band provides a useful relative wavelength check using deionized water. This method requires that the wavelength accuracy of either the excitation or emission monochromator is known; it can then be used to check the relative accuracy of the other scale. For example, if the accuracy of the excitation monochromator is known, then the excitation wavelength is set to a given wavelength, and the emission monochromator is scanned to longer (red-shifted) wavelengths. The Raman peak position is known to appear at a fixed lower energy value (3382 wavenumbers in air) relative to the excitation wavelength [27]. This wavelength check is generally carried out in the UV–blue wavelength range; for example, for an excitation wavelength of 350 nm, the water Raman peak occurs at 397 nm.

The accuracy of the spectral slit width of the excitation or emission monochromator can be determined using the experimental setup described above for wavelength calibration using spectrum lamps. In this case, a single line is selected and its spectral bandwidth, measured as the full-width at half-maximum (FWHM), is recorded as a function of slit-width setting. Alternatively, a diffuse reflector can be placed in the sample compartment in order to direct a portion of the radiation from the excitation monochromator to the emission monochromator. Then, for a fixed slit and wavelength setting for one monochromator, its lineshape can be scanned with the second monochromator, when the second is set at a much smaller spectral slit width than the first.

For the *normalization* of the bispectral fluorescence data, the accuracy of the spectral slit width is less important than its reproducibility since the instrument spectral efficiency factor depends critically upon this setting. Also, it is generally assumed that this normalization can be performed at a single arbitrary wavelength in the excitation range of the fluorescent sample. However, for fluorescence spectrophotometers, as with other types of spectrophotometer, the spectral bandpass for a fixed slit width can vary with wavelength, so it is important to check the accuracy of the slit setting for a few wavelengths across the instrument's wavelength range of operation. Provided that the variation in spectral bandpass with wavelength remains constant with slit setting, this effect compensates, and normalization at a single wavelength is valid. It is a good practice to verify this assumption by carrying out the normalization at a few wavelengths and checking that this gives near-identical normalized bispectral fluorescence data.

7.6.2 Detector Nonlinearity

The photometric linearity of the detection system should be verified throughout the dynamic range of the instrument operation and sampling conditions. Most fluorescence spectrophotometers use PMTs as their analyzing detector; these are highly nonlinear at high light levels, so linearity verification is critical to achieving accurate results. This verification involves determining the upper limits for linear response, where the detection system is generally considered to be linear if the deviations are within 1% of the signal's magnitude, which is the present state-of-the-art accuracy for these fluorescence measurements.

The dynamic range of measurements that needs to be checked for linearity depends upon the spectral measuring instrument and application. For example, for analytical applications, the dynamic range of the signal measurements is typically of the order of 1000:1 since only information on the excitation or emission spectrum is needed, that is, one type of measurand with units of power per spectral concentration. However, for colorimetric applications, using a two-monochromator method, there are two different types of measurand with different units. As we have seen, the reflected component is a simple function of wavelength, with units of power, so changing the emission bandpass has little effect on its measured value. However, the bispectral luminescent component has units of power per spectral concentration and its value changes proportionally with the emission bandpass setting. For colorimetric applications where the emission bandpass is typically 5 nm, the dynamic range between the reflected and bispectral fluorescent signals is on the order of 10^6:1.

For both analytical and colorimetric applications using PTS-based calibration methods, it is important that the signal levels measured by the analyzing detector using the CS are similar to those typically encountered when the excitation beam is incident on the fluorescent samples under test. In some cases, it is necessary to use attenuating filters in the optical path or to increase the distance of the CS from the sample position. This will ensure that the system nonlinearity effects are similar for the instrument calibration and the sample measurement and will be largely compensated.

For analytical applications, the linearity of the fluorescence measurement system is usually evaluated by using an organic dye solution that fluoresces in the emission region of interest. A standard test method developed by ASTM (E578), using quinine sulfate dihydrate in sulfuric acid as a standard test solution, is commonly used [28]. The fluorescence intensity of a series of standard solutions is measured, and the resultant readings are recorded on a log–log plot. The point (concentration) at which the upper end of the curve deviates by more than 5% of the signal from the straight line (defined by the center region of the curve) is taken as the upper limit of linearity. The subsequent fluorescence measurements should then be carried out for solutions below this upper limit concentration. For typical fluorescent dye standards, this upper limit for linear instrument response is an absorbance of about $A=0.05$ for a 1-cm path length cell.

For most colorimetric applications using a one-monochromator spectrophotometer, the linearity is checked with a set of calibrated nonfluorescent and fluorescent RMs over the dynamic range of use. Suitable nonfluorescent reference standards are high-quality neutral reflecting or transmitting materials, such as the Ceram Research set of neutral gray ceramic tiles and the NIST 930 set of neutral absorbing glass filters. A variety of fluorescent standards are available for applications in the plastics, textile, and paper industries. A survey of the materials that are available is given in CIE Publication 182 [29].

For high-accuracy colorimetric and analytical applications using a two-monochromator reference spectrofluorimeter, the linearity is checked on a more fundamental basis—applying the light addition principle by using either multiple sources or multiple apertures [30] to precisely control the flux level incident on the detection system. Alternatively, this flux level can be varied using calibrated attenuators in the beam, such as optical filters with known transmittance (see Fig. 7.12).

For high-accuracy analytical applications, where it is not convenient to install a multiple source or an aperture device, or to incur the cost of

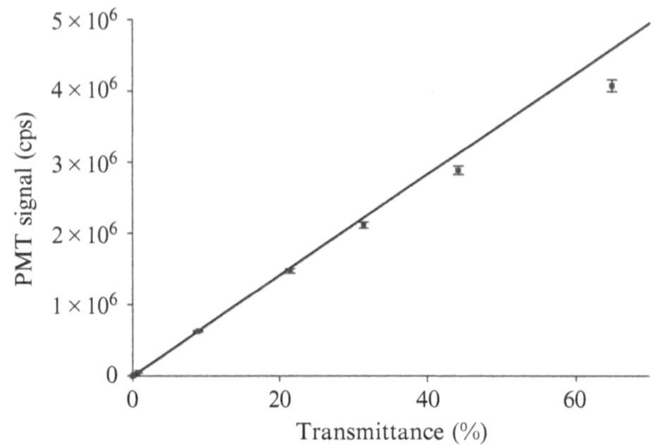

FIGURE 7.12 Plot of PMT signal at 440 nm excitation as a function of the transmittance of a series of calibrated filters used to attenuate the light reaching the detector. A scattering solution was placed at the sample position to scatter the excitation light into the detection system, and calibrated attenuation filters were placed before the sample. The line is a linear least-squares fit of the points in the linear range of the PMT. The associated uncertainties in the transmittance values are smaller than the point size [25].

calibrated filters, the linearity can be checked by using a method of signal ratioing that was developed by Federal Institute for Materials Research and Testing (BAM) [31]. In this method, the ratio of these signals should be invariant with light level, for example, the transmission ratio of two settings of a polarizer or the transmission ratio of an attenuator at two wavelengths. This ratio is plotted as a function of increasing spectral irradiance at the sample position, and where it deviates from a horizontal line within its measurement uncertainty determines the upper limit of the detector linear response.

Once the limits of the linear range of the detection system have been determined for typical sample fluorescence intensities, the fluorescence instrument should only be assumed to be linear within this dynamic range. However, it is also possible to determine a nonlinearity correction curve and apply this to the measured fluorescence signals, but the accuracy of this procedure is limited by the stability of the instrument's measurement beam under specified conditions of operation and the reliability of the nonlinearity correction function [30].

7.6.3 Polarization Effects

A fluorescence instrument can have a significant degree of polarization due to the polarization properties of the source, monochromator, beamsplitter, mirrors, and other optical components. Furthermore, the analyzing detector can have a responsivity that is sensitive to the polarization of the incident radiation, and this can vary spectrally. This polarization sensitivity of the instrument needs to be quantified in order to determine the polarization properties of the fluorescent samples under test. A detailed description of the nomenclature and basic methodology for analyzing the polarization properties in a radiometric measurement is described in Chapter 2, Section 5 and should be consulted for further details.

The instrument polarization bias for the excitation and emission systems, as a function of wavelength, is quantified by factors denoted by $F(\mu)$ and $G(\lambda)$, respectively. These factors are calculated, at each wavelength, as the ratio of the intensity measured at the output of the excitation and emission system for incident radiation that is polarized perpendicular (s) and parallel (p) to the plane of incidence, that is, I_s/I_p. For the measurement of the polarization bias that is due to the emission system, an isotropic emitting sample, for example, a white diffuser, is placed at the sample position.

The polarization bias curves for the excitation and emission systems of a state-of-the-art reference spectrofluorimeter are shown in Fig. 7.13. This instrument has a geometry of 45° annular illumination and normal detection (45°a:0°), so it is not surprising that the F factors are larger than the G factors, since the polarization sensitivity of the optical components will be greater for light incident at an oblique angle than at normal incidence. Also, the F factor curve shows a large jump at 450 nm, which, for the grating used with this instrument, is near the Wood's anomaly, where the grating efficiency changes dramatically with a change in polarization of the incident radiation. It can be seen that the values of the F and G factors are largely due to the intrinsic properties of the instrument's gratings and other optical components. Provided that these components are not changed, this instrument polarization bias is very stable with time. Because of the differences in the spectral shape of the polarization bias for the excitation and emission system, which is instrument specific, this will change the shape and intensity of the measured fluorescence spectra even for samples that are isotropic, that is, have no polarization sensitivity.

A term that is used to describe the polarization sensitivity of an instrument or a sample is the *degree of polarization*, *P*, which is defined as a function of

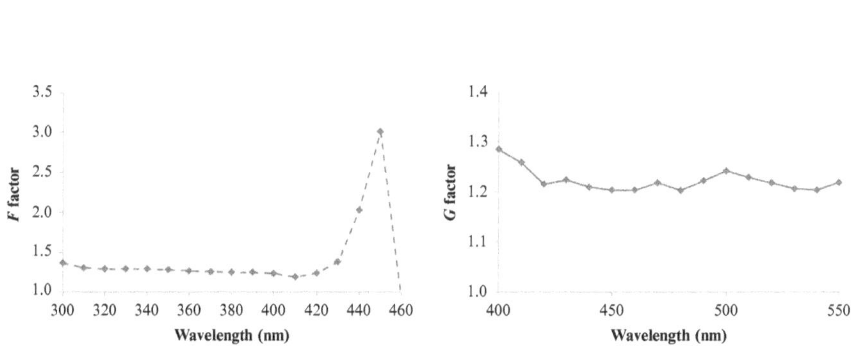

FIGURE 7.13 The range of F and G factors for the excitation and emission systems, respectively, of a state-of-the-art reference spectrofluorimeter (45°a:0° geometry).

the observed parallel ($I_{\|}$) and perpendicular intensities ($I_{\perp}$) (or p- and s-components, respectively, as described in Chapter 5),

$$P = \frac{I_{\perp} - I_{\|}}{I_{\perp} + I_{\|}}. \tag{7.26}$$

If the emission is completely polarized in the parallel direction, that is, the orientation of the electric vector of the exciting light is totally maintained, then the degree of polarization is +1. If the emitted light is totally polarized in the perpendicular direction, then the degree of polarization is −1. Thus, the degree of polarization that can be exhibited by a fluorescent instrument or a fluorescent sample ranges from +1.0 to −1.0.

Another term that is more commonly used to describe the polarization sensitivity of a fluorescent sample is *polarization anisotropy* r which is also defined in terms of the measured fluorescence intensity in the directions parallel, $I_{\|}$, and perpendicular, $I_{\perp}$, to the plane of incidence, but compared to the total fluorescent intensity, I_{T}, where

$$r = \frac{I_{\|} - I_{\perp}}{I_{\mathrm{T}}} = \frac{I_{\|} - I_{\perp}}{I_{\|} + 2I_{\perp}}. \tag{7.27}$$

The total intensity is given by the sum of $I_{\|}$ and two times $I_{\perp}$ because of symmetry, where the components polarized in the two orthogonal directions, $I_{\perp}$, have the same intensity, but in a given measurement geometry, only one of these polarized components is detected. Thus, if the emission is completely polarized in the parallel direction, it has a maximum value of +1.0; and if the emission is completely polarized in the perpendicular direction, it has a maximum value of −0.5. Thus, the polarization anisotropy that can be exhibited by fluorescent samples ranges from +1.0 to −0.5.

The measurement of fluorescence polarization is very important in biomedical analytical applications, such as measuring the binding of fluorophores to different target molecules, for example, proteins and antibodies [32]. For fluorescent samples that exhibit a significant fluorescence anisotropy ($r > 0.05$), the measured fluorescence signals can be corrected for the influence of instrument polarization if both the sample's polarization anisotropy, r, and instrument F and G factors are known. The measured fluorescence can then be considered a function of both the instrument degree of polarization and the sample's orientation and polarization anisotropy. This interaction can be described by a matrix equation where we solve for the true fluorescence vector of the sample for s- and p-polarized incident radiation and s- and p-polarized detection. The average of these calculated fluorescence vectors for pure s- and p-polarized conditions provides a fluorescent signal that is independent of instrument polarization effects [33].

For fluorescent samples that exhibit a small fluorescence anisotropy ($r < 0.05$), the influence of instrument polarization has very little effect on the measured fluorescence. Thus, for most analytical applications, no special precautions need to be taken. However, for colorimetric applications in a bidirectional geometry, the reflected component of many samples is polarization sensitive. Thus, where there is a large change in the instrument degree of polarization, for example, near the Wood's anomalies, there can be a significant effect on the measured sample reflectance. In this case, it is necessary to perform the measurements under well-defined polarization conditions with the use of polarizers in the excitation and/or emission channels, as needed. For an instrument with a 45°:0° geometry, it is only necessary to place a polarizer in the excitation path and make independent measurements using s- and p-polarized conditions, and then average the result to obtain the reflected and bispectral luminescent radiance factor corresponding to unpolarized conditions.

On the other hand, Minato [34] has extensively studied the influence of instrument polarization on surface fluorescence measurements and found that, while it affects the accurate shape of the measured reflected or luminescent radiance factor, it has negligible influence on the accuracy of the derived colorimetric results. This is largely because the colorimetric results depend on the spectrally integrated effect of these changes rather than on the absolute spectral changes at each wavelength.

7.6.4 Stray Light

Stray light, which is any undesired or spurious signal, is a significant source of error in fluorescence measurements due to the fact that the fluorescence intensities are generally very low. As discussed in Chapter 3, stray radiant energy (SRE) is defined as the ratio of this measured spurious radiation to the detected primary radiation. Its effects are generally greatest when there is significant scattering in the instrument or at wavelengths of decreased primary signal, for example, near the source energy and detector response limits, as well as when the grating efficiency is low. It is therefore expected that the influence of stray radiation energy will be significant in the excitation range of many fluorescence spectrophotometers, where the primary radiation is relatively low.

Stray light can be particularly problematic with CCD-based instruments because they detect all wavelengths simultaneously; a small percentage of stray light originating in the red portion of the spectrum, where the detector sensitivity is usually high, will have a very large impact at the shorter wavelengths in the UV and blue region, where the primary radiation level is low. As discussed in Chapter 3, methods have been developed by Zong *et al.* to obtain a spectral stray light correlation matrix for a given CCD-based instrument [24]. These matrix methods have since been incorporated in several commercial array-based spectroradiometers to reduce their spectral stray light error [24].

SRE can be described as either isochromatic or heterochromatic in origin. *Isochromatic* SRE is radiation of the same wavelengths as incident on the sample, that is, within the instrument bandpass, but which reaches the detector without actually interacting with the sample. *Heterochromatic* SRE is radiation of wavelengths outside the instrument bandpass setting. For a bispectrometer, it is more complicated to separate the stray light into isochromatic and heterochromatic contributions because there are effectively two different instrument bandpass conditions determined by the settings of the excitation monochromator and the emission monochromator, respectively.

The heterochromatic stray radiant energy (HSRE) in a dispersive instrument is primarily caused by scattering from inside the monochromator and higher order light diffraction by the grating (see also Chapter 3). This error can be minimized in a well-designed fluorescence instrument with optimized gratings, order-sorting filters, and appropriately placed baffles. For high-accuracy requirements, a double-grating monochromator is frequently used in the excitation path but at the sacrifice of measurement sensitivity. A second source of HSRE is reentry spectra which can be particularly problematic at shorter wavelengths in the excitation range of a fluorescence instrument. As discussed in Chapter 3, this phenomenon is sometimes called double refraction and occurs when light diffracts back to the collimator, and returns to the grating and diffracts again, eventually emerging from the exit slit at a constant offset with respect to the set wavelength of the monochromator. If the precise wavelength difference of this double refracted light is known, it is possible to reduce this directional stray light by inserting additional bandpass filters in the excitation path of the instrument.

To characterize and correct for the effects of isochromatic SRE, it is common to use a highly absorbing material. For example, in regular transmittance measurements discussed in Chapter 5, this requires blocking the beam, for example, with an opaque metal shutter. For analytical fluorescence measurement applications, this is referred to as a *blank* or *background* correction and is generally carried out using a cuvette with the solvent only under identical conditions to the fluorescent sample under test. In colorimetric applications, for correcting the reflected component, this can be a glossy black standard or a light trap; and for correcting the fluorescent component, the apparent bispectral fluorescence signal from a nonfluorescent white reflectance standard can be used [35]. In order to correct for this effect, the measured signal for the highly absorbing sample is subtracted from all subsequent readings. In the latter case, using the nonfluorescent reflectance standard, it is necessary to first scale the measured SRE signal to the reflectance level of the fluorescent sample under test, before subtracting it from the apparent fluorescence signal. This procedure is described in detail in Section 7.6.6.

7.6.5 Instrument Lineshape

The instrument lineshape or line-spread function (ILS) is effectively a transmission function. For a single scanning monochromator instrument, the ILS is the convolution of the rectangular function that is the image of the entrance slit produced by the grating and the rectangular function that is the opening of the exit slit. In the case of rectangular entrance and exit slits that are of equal width and uniformly illuminated, this ILS is a triangular function (see Chapter 3). However, for a two-monochromator instrument operated under similar rectangular slit conditions for both monochromators, the ILS is very nearly Gaussian in shape. This is because it is effectively the convolution of two such triangular functions from the individual monochromators. The spectral extent of the instrument bandpass function, that is, the range of wavelengths passed by the instrument, is also much wider than for a one-monochromator system with the same nominal spectral bandpass. For example, for a one-monochromator instrument with rectangular slits set for 5-nm nominal bandpass, the spectral extent of the triangular lineshape is 10 nm (full width at base of triangular lineshape), whereas for a two-monochromator instrument operated under the same 5-nm nominal bandpass, the spectral extent can be up to 20 nm and needs to be checked for accurate normalization of the bispectral fluorescence data.

It is also important to take these differences in ILS into account when transferring the calibration of a fluorescent standard from a two-monochromator reference instrument to a one-monochromator instrument since the different instrument transmission functions produce spectral differences in the measured fluorescence data, even for the same nominal bandpass. For example, if fluorescence calibration data have been obtained for a transfer standard on a two-monochromator instrument with a nominal 5-nm bandpass and are to be used to calibrate a one-monochromator instrument with a nominal 10-nm bandpass, it is necessary to first convolve the fluorescence calibration data with a Gaussian function to obtain calculated fluorescence data at nominal 10-nm bandpass and 10-nm measurement interval. It has been found that this procedure gives good inter-instrument agreement in the colorimetric values that are calculated from the original calibration data and from the convolved data (using the ASTM E308 Tables 6 for instruments without bandpass correction) [36]. By comparison, in the example above, if the fluorescence calibration data recorded at nominal 5-nm bandpass is convolved with a triangular function to provide calculated data at 10-nm bandpass, the differences in the calculated colorimetric results can be a few percent (e.g., a change in CIE whiteness of 1 unit for a fluorescent white paper sample).

7.6.6 Sampling Effects

Sampling parameters are especially critical in bispectral measurements and must be carefully considered. An instrument's sampling characteristics can be described in term of its spectral bandpass functions and the wavelength intervals at which measurements are performed. An instrument's bandpass function is typically specified in terms of its FWHM, but the spectral shape of the function is also important. In bispectral measurements, these parameters may be defined independently in the excitation and emission wavelength domains, so two bandwidth values ($\Delta\mu_{\text{bp}}$, $\Delta\lambda_{\text{bp}}$) and two sampling intervals ($\Delta\mu_{\text{int}}$, $\Delta\lambda_{\text{int}}$) must be specified.

In the colorimetry of fluorescent materials, certain standard sampling guidelines apply. For colorimetry in general, it is recommended that the instrument bandpass function be approximately triangular or trapezoidal, with bandwidth equal to, or an integer multiple of, the measuring interval [10]. For bispectral measurements with colorimetric applications, this recommendation is applied to both the excitation and emission wavelength domains, with the added requirement that the excitation and emission wavelength intervals be equal. When these recommendations are followed, two complications may arise, which have been described as *reflection overspill* and *diagonal fluorescence* [5].

7.6.6.1 Reflection Overspill and Diagonal Fluorescence

When considering a continuous bispectral function such as the bispectral radiance factor $\beta_\lambda(\mu,\lambda)$, the separation of the reflected and luminescent components is conceptually straightforward. By definition, the bispectral reflected radiance factor $\beta_{R,\lambda}(\mu,\lambda)$ equals zero, except on the diagonal, where $\mu=\lambda$. On the diagonal, it is the bispectral luminescent radiance factor $\beta_{L,\lambda}(\mu,\lambda)$ that equals zero. Thus, when considering the surface $\beta_\lambda(\mu,\lambda)$, reflection-related values are limited to the diagonal plane, while all other values are due solely to luminescence. For any real bispectral measurement, however, the specimen's bispectral radiance factor is convolved with the instrument's illumination and viewing bandpass functions so that there may be significant overlap between the instrument's illumination and viewing bands for measurement positions near the diagonal, as well as on the diagonal itself. Therefore, light reflected from the specimen will not be confined to the diagonal, but will be distributed across the near-diagonal region. The term *reflection overspill* is used to describe the contribution of reflection to off-diagonal values of a bispectral matrix, due to the partial overlap of illumination and viewing bandpass functions when the nominal excitation and emission wavelengths are not equal ($\mu\neq\lambda$). Conversely, the term *diagonal fluorescence* is used to describe the contribution of fluorescence to diagonal values of a bispectral matrix, due to the finite extent of actual illumination and viewing bands when the nominal excitation and emission wavelengths are equal ($\mu=\lambda$). For most practical purposes, the effect of diagonal fluorescence can be considered negligible, but if the colorimetric sampling guidelines described above are followed, the effect of reflection overspill may be significant and must be taken into account in order to obtain useful measurement results.

It is possible to reduce or eliminate the effect of reflection overspill by selecting excitation and emission bandwidths that are significantly smaller than the corresponding sampling intervals so that the instrument's illumination and viewing bandpass functions overlap only for diagonal readings ($\mu = \lambda$). Such an undersampling approach followed by mathematical interpolation to give data at the same interval as the instrument spectral bandpass, has been adopted for reference-level colorimetric measurements at some national metrology institutes [35]. However, for more routine measurements, it is necessary to consider the reduction in optical signal associated with decreased bandwidths. If the instrument's spectral bandpass functions are sufficiently well characterized, it is possible to employ conventional colorimetric sampling and then to correct for the effects of reflection overspill, in order to obtain results that approximate those that would be obtained using an idealized instrument with discrete rectangular bandpass functions. This is the industry standard approach for commercial bispectral instrumentation used for colorimetric applications [37].

7.6.6.2 Reflection Overspill—Characterization and Correction

Reflection overspill arises due to the convolution of the specimen's bispectral radiance factor with an instrument's illumination and viewing bandpass functions. It follows, then, that this effect can be corrected by bispectral deconvolution or an equivalent procedure. Deconvolution, however, is highly susceptible to errors in the measured data. One approach that has been taken is to estimate the spillover of the reflection component into the fluorescent, off-diagonal elements, subtracting that contribution from the off-diagonal elements and adding it back into the reflective, on-diagonal elements [5]. We start with the *discrete bispectral radiance factor*, $B(\mu, \lambda)$, which is the uncorrected Donaldson radiance factor, defined similar to Eq. (7.17):

$$B(\mu, \lambda) \equiv \overline{\beta}_{\lambda}(\mu, \lambda) \cdot \Delta\lambda_{\mathrm{bp}}, \tag{7.28}$$

where $\overline{\beta}_{\lambda}(\mu, \lambda)$ is the average bispectral radiance factor over the instrument bandpass $\Delta\lambda_{\mathrm{bp}}$. Equation (7.28) is essentially the result that is obtained when the instrument readings have been corrected by all of the other instrument-specific factors described in Section 7.6, leaving only bandpass effects uncorrected. A similar matrix, $B_{\mathrm{d}}(\mu, \lambda)$, is obtained for a nonfluorescent reflectance standard. Using $B_{\mathrm{d}}(\mu, \lambda)$, we can determine an overspill function

$$\kappa(\mu, \lambda) \equiv \frac{B_{\mathrm{d}}(\mu, \lambda)}{B_{\mathrm{d}}(\lambda, \lambda)}. \tag{7.29}$$

In practice, $\kappa(\mu, \lambda)$ is assigned a nonzero value only in the near-diagonal region, where overspill is determined to be significant. Generally, this should be limited to within three bands of the diagonal [5]. Outside this region, values may be set equal to zero.

The sample's discrete bispectral radiance factor, $B(\mu, \lambda)$, is separated into its reflected and fluorescent components. The reflected component is estimated by normalizing the overspill distribution function to the diagonal,

$$B_R(\mu, \lambda) \approx \kappa(\mu, \lambda) \cdot B(\lambda, \lambda), \tag{7.30}$$

while the fluorescent component of $B(\mu, \lambda)$ is estimated from the remainder,

$$B_F(\mu, \lambda) \approx B(\mu, \lambda) - B_R(\mu, \lambda). \tag{7.31}$$

The Donaldson radiance factor for the off-diagonal, fluorescent component is taken to be $B_F(\mu, \lambda)$, while the reflected component is taken by "diagonalizing" $B_R(\mu, \lambda)$,

$$D(\mu, \lambda) \approx \begin{cases} \sum_{\forall \mu} B_R(\mu, \lambda), & \text{if } \mu = \lambda \\ B_F(\mu, \lambda), & \text{if } \mu \neq \lambda. \end{cases} \tag{7.32}$$

The effects of reflection are now confined to the matrix diagonal, and diagonal values accurately represent sample reflectance.

7.7 SAMPLE CHARACTERISTICS THAT IMPACT FLUORESCENCE MEASUREMENTS

7.7.1 Optical Density

One of the most important applications of fluorescence measurements in chemical and biochemical analysis is the determination of analyte concentration. Calibration curves of fluorescence signal versus analyte concentration are commonly used to determine analyte concentration from the fluorescence signal of a particular instrument. It is often assumed that fluorescence signal is linearly proportional to analyte concentration, provided the response of the detection system is linear. However, this assumption is not true at higher concentrations. The absorbance of a fluorescent sample in a standard cuvette using a 0°/90° geometry needs to be less than 0.05 at the excitation wavelength to achieve deviations from linearity of 6 % or less. Decreases in the measured fluorescence intensity of a sample due to absorption of the excitation beam or reabsorption by the sample of the emission are called *primary* or *secondary IFE*s, respectively. The Beer–Lambert law, which describes the proportionality of a sample's absorbance, or optical density, to its path length and concentration, has been used to calculate the fluorescence signal corrected for IFEs, S_{ic} , using a 0°/90° geometry, [37] such that

$$S_{\text{ic}} = S \cdot \frac{2.303D(X_2 - X_1)}{10^{-DX_1} - 10^{-DX_2}} \cdot \frac{2.303D(Y_2 - Y_1)}{10^{-DY_1} - 10^{-DY_2}}, \tag{7.33}$$

where D is the optical density per cm, X_1 and X_2 are the distances (in cm) of the detection region boundaries from the incident face of the sample cell, and

Y_1 and Y_2 are the distances (in cm) of the detection region boundaries from the exit face of the sample cell. The first and second fractional terms in the Eq. (7.33) correspond to the primary and secondary IFE corrections, respectively.

For dilute solutions measured with a transmitting geometry, such as 0°/90°, the primary IFE tends to be significantly larger than the secondary. In contrast, the secondary IFE often predominates for highly reflecting or highly scattering solids. The secondary IFE is often significant when there is considerable overlap between the excitation and emission bands of the fluorophore. IFEs are important parameters to consider when optimizing the measurement conditions to obtain reliable and accurate optical constant data. Decreasing the sample concentration can reduce IFEs. Also, judicious selection of the excitation wavelength can reduce these effects. It is generally recommended that one select an excitation wavelength that is on the short wavelength side of the sample's absorption maximum.

7.7.2 Solvent/Matrix Properties

Interactions between the fluorophore and the solvent in solutions, or the matrix in solids, can significantly affect the peak wavelength position, shape, and peak width of emission spectra. No single model exists to describe the effects of environment on fluorescence emission spectra because they are complex and often specific to the fluorophore–solvent/matrix pair. General solvent effects due to the interactions of the fluorophore's dipole with the solvent's polarity are the most commonly observed and some of the easiest to describe. The Lippert equation was derived to predict the dependence of the Stokes shift, ($\Delta\upsilon_s$), of a fluorophore on the solvent's static dielectric constant, ε, the optical refractive index, n, and the difference between the fluorophore's excited and ground state dipole moments. Note that this model does not consider any pair-specific interactions [38].

In addition to understanding how a fluorophore and solvent/matrix interact, it is also important to be aware of the absorbance, Raman, and fluorescence spectra of the solvent/matrix alone. Absorbance by the solvent/matrix at the excitation and emission wavelengths can cause IFEs, requiring a correction. Furthermore, a fluorescence spectrum of the solvent/matrix alone, taken under the same conditions as that of the sample, can be used to subtract any background fluorescence or Raman lines due to the solvent/matrix. Reabsorption of emission from or by the matrix typically has an insignificant effect on the measured fluorescence (see Section 7.7.4).

7.7.3 Refractive Index

When the emission emanating from the detection region in a small cone emerges from the sample into the air, the angle of the emission cone enlarges, due to refraction according to Snell's law. The refractive index of the sample is larger than that of air, causing a decrease in emission reaching the detector. This decrease is emission wavelength dependent because the refractive index of the sample, n_s, is wavelength dependent. The multiplicative correction for this effect is directly proportional to n_s^2 for perpendicular viewing [39,40] (see Chapter 2, Section 6.1 for a detailed description).

A decrease in emission reaching the detector can also be caused by a decrease in the transmittance of light through the sample, due to a difference in the refractive indices of two media, that is, the sample and air, at their boundary. The transmittance, T, of light through such a boundary is given by the Fresnel equations, and when the incident light is normal to the boundary

$$T = \frac{4n_t n_i}{(n_t + n_i)^2}, \tag{7.34}$$

where the subscripts "t" and "i" identify the medium into which the light is transmitted and from which the light is incident, respectively [41]. The transmittance through the media boundary is also wavelength dependent due to the wavelength dependence of refractive indices (see Chapter 2, Section 6.2).

Both of these corrections must be applied in order to obtain true spectra of the highest accuracy. The refraction correction is significantly larger than the transmittance correction for most samples.

7.7.4 Impurities

Impurities that absorb or fluoresce in the same spectral region as the fluorophore of interest are often found either in the solvent or in the sample itself. Absorbance and fluorescence spectra of many solvents and fluorophores can be found in the literature and compared to the sample being measured, to insure that impurities are not adding fluorescence artifacts or IFEs to the analyte spectrum. To ensure solvent purity, it is advisable to use the highest grade available, for example, deionized/distilled water or spectrophotometric grade solvents. Other common sources of fluorescent impurities include those present in plastic containers, plastic gloves, detergents, and stopcock grease, which can be unknowingly introduced to a sample. Micro-organisms, which fluoresce, can also grow in sample reagents and aqueous solvents.

Impurities may also decrease the fluorescence intensity of a sample when they come into molecular contact (<0.2 nm apart) with fluorophores—a process known as *quenching* [42]. Another process that may be caused by other species in close proximity to the fluorophores is fluorescence resonance energy transfer (FRET). In this case, a donor molecule absorbs the excitation energy, but instead of being emitted, this energy is transferred (nonradiatively) to an acceptor molecule, which then fluoresces. The donor and acceptor typically have to be within a distance r of 10 nm or less of each other for this process to be significant, with the rate of FRET dropping off as $1/r^6$ [43].

7.7.5 Luminescence Lifetime

When a phosphor's luminescence spectrum changes over the course of its lifetime, differences in the luminescence spectrum may be measured on instruments with microsecond-pulsed versus continuous excitation, as discussed in Section 7.5.1.4. Spectral differences observed for such phosphors may also depend on the length of both the excitation pulse and the detection gate.

Some impurities have the ability to reduce the fluorescence quantum yield when they collide with fluorophores in the excited state, known as collisional *quenching*, which results in a decrease in the observed fluorescence intensity [40]. This decrease is directly proportional to the decrease in fluorescence quantum yield and is the most frequently encountered type of quenching. The most common collisional quencher is molecular oxygen, O_2, which is soluble in many solvents. O_2 is a particularly effective quencher due to the unusual property of its ground electronic state being a triplet, not a singlet, state. Removing the oxygen from solvents is often necessary to obtain accurate fluorescence quantum yields and lifetimes. Degassing solvents using a freeze–pump–thaw cycle is the most widely used method for removing unwanted gases from liquids.

7.7.6 Scattering

Highly scattering samples can cause significant amounts of excitation light to be detected, even when the emission monochromator is set at a wavelength that is far away from the excitation wavelength. This stray light reaches the detector because a monochromator has a stray light rejection that is small, but not zero; see Chapter 3 and Section 7.6.4 for a more thorough discussion on stray light. For instance, a single-grating monochromator may have a stray light rejection of 10^{-4}, meaning that the radiant power outside the set bandpass reaching the detector is 10^{-4} of that entering the monochromator. Many fluorescent samples can have high turbidity, biological samples in particular, and the combination of high scattering and low fluorescence can cause stray light to contribute significantly to the total detected signal. Because of this, double-grating monochromators, which have twice the exponential stray light rejection of single-grating monochromators, for example, 10^{-8} versus 10^{-4}, are often used for such applications.

Highly scattering or low fluorescence samples can also introduce peaks in the signal at higher order transmission wavelengths of the monochromator's grating at a particular grating position. For instance, the position at which a grating transmits 600 nm light at its first order will also transmit 300 nm light at its second order, simultaneously and coincident with each other.

The grating equation determines the wavelengths that will be transmitted (λ_m) coincidently at a particular grating position, such that $\lambda_m = \lambda_1/m$, $m = 1, 2, \ldots$, where m is the order of the grating [44]. When excitation light is scattered at $\mu = 300$ nm, an emission peak at $\lambda = 600$ nm is often observed due to second-order transmission of the emission grating. Wavelength-dependent transmission filters, such as bandpass filters before the sample and longpass filters after the sample, are often used to remove light at higher orders before it reaches the detector. This is called order sorting.

7.8 STANDARDS FOR SPECTRAL FLUORESCENCE MEASUREMENTS

7.8.1 Correcting Relative Excitation and Emission Spectra

7.8.1.1 Physical Transfer Standards

To achieve state-of-the-art accuracies in the correction of excitation and emission spectra measured on a fluorescence spectrometer, it is necessary to use calibration methods based on PTS that link the fluorescence measurements to radiometric scales that are traceable to the International System of Units (SI). For this purpose, various PTSs are used, for example, (1) a standard detector; (2) a standard source; and (3) a nonfluorescent reflectance standard. The standard detector provides traceability to the spectral responsivity scale. The standard source can be either a calibrated standard lamp used in combination with a calibrated reflector or a calibrated sphere source that provides traceability to the spectral radiance scale. The reflectance standard also provides traceability to the diffuse reflectance scale, which is required for colorimetric applications. For minimum uncertainty in the use of PTS-based calibrations, these PTSs need to be calibrated under measurement conditions identical to those used with the spectrofluorimeter for the measurement of fluorescent samples. These conditions include the same spectral radiance levels, geometric, and polarization conditions. This can be a particular challenge with typical calibrated light sources, which can have spectral radiance of two to four orders of magnitude higher than the fluorescence from dilute fluorescent dye solutions. In this case, the PTS should be used with additional calibrated optical components, for example, attenuating filters.

For the technical correction of excitation spectra, so that it has the correct spectral shape, it is necessary to determine the relative spectral irradiance incident at the sample position from the excitation system. For this purpose, a calibrated detector (CD) is typically used. A monitor detector is used to dynamically compensate for changes in the source output from the time of this PTS calibration and the subsequent measurement of the fluorescent samples. This monitor detector is permanently mounted in the instrument so that it intercepts a small fraction of the incident radiation (e.g., by use of a beamsplitter) and is calibrated at the same time of this PTS calibration of the excitation system. This is known as a CD method. Such CDs are typically UV-enhanced silicon photodiodes that are useful over the spectral range of 250–1000 nm.

For the technical correction of the emission spectra, it is necessary to determine the relative spectral response of the detection system. For this purpose, a CS of spectral radiance at the sample position is needed. This can be either a sphere source of spectral radiance or a combination of a calibrated spectral irradiance source, used to illuminate a calibrated diffuse reflectance standard (CR). In some cases, a CR is also used with a sphere source if the sphere is too large to be mounted in the sample compartment; in this case, it is used to steer the beam from the excitation beam to the emission system. These are known as CS methods. A schematic of a typical instrumental setup for the CS method is shown in Fig. 7.14. The sources used are typically tungsten lamps or quartz halogen lamps, which are useful over the spectral range above about 350 nm. An illustration of the measured PMT signal using a tungsten halogen lamp illuminating a CR is shown in Fig. 7.15 and compared with the true spectral irradiance distribution of this CS.

For a standard lamp with calibrated spectral irradiance $E_{\lambda\text{ref}}(\lambda)$, illuminating a reflectance standard with calibrated spectral radiance factor $\beta_{\text{std}}(\lambda)$, the spectral radiance of the illuminated reflectance standard, $L_{\lambda\text{ref}}(\lambda)$, can be derived from

$$L_{\lambda\text{ref}}(\lambda) = E_{\lambda\text{ref}}(\lambda)\beta_{\text{std}}(\lambda)\cdot g/\pi, \tag{7.35}$$

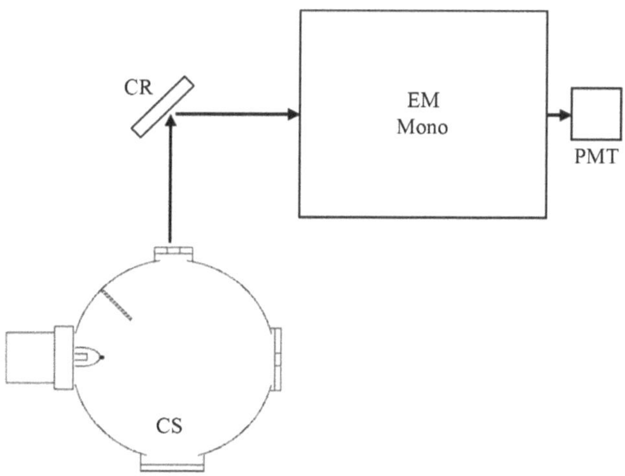

FIGURE 7.14 Schematic of a typical instrumental setup used for measuring the signal from a sphere-mounted lamp as a calibrated source CS that enters the detection system after diffusely reflecting off a calibrated reflector CR. Alternatively, a calibrated spectral irradiance lamp can be used to illuminate the calibrated reflector.

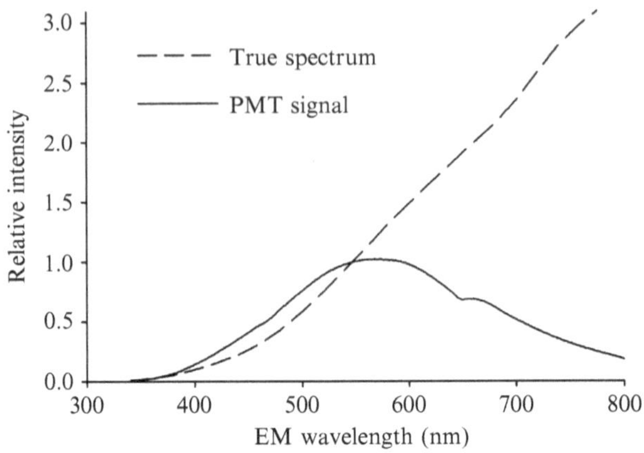

FIGURE 7.15 Comparison of the true spectrum and measured PMT signal of a tungsten halogen lamp in relative intensity units.

where g is a geometric constant or a view factor that represents the fraction of the radiant flux leaving the surface of the reflectance standard that enters the limiting aperture. For the purpose of relative spectral calibration, the value of this constant may be assumed to be equal to one. However, for the purposes of obtaining "normalized" data, it is necessary to measure and correct for this instrument geometric characteristic, as described below.

Given the spectral radiance of the illuminated reflectance standard, $L_{\lambda\mathrm{ref}}(\lambda)$, the signal of the analyzing detector, $i_{\mathrm{sw}}(\lambda)$, is measured across the spectral range of the emission unit. The spectral responsivity of the emission unit, $R_{\mathrm{sys}}(\lambda)$, can then be calculated from

$$R_{\mathrm{sys}}(\lambda) = \frac{i_{\mathrm{sw}}(\lambda)}{k_2 E_{\lambda\mathrm{ref}}(\lambda)\beta_{\mathrm{std}}(\lambda)}, \tag{7.36}$$

where k_2 is an instrument-specific constant dependent on geometric factors only, notably the slit settings of the excitation and emission monochromators that determine the instrument throughput. Thus, this instrument factor is essentially the quantum yield of the instrument under these specific measurement geometry conditions. To determine k_2, the instrument absolute yield is measured using a diffuse reflectance standard. The total integrated reflectance signal from this sample recorded over the emission bandpass is ratioed to the incident signal over the excitation bandpass (usually measured using a monitor detector), and this ratio is set equal to the mean spectral reflected radiance factor of this standard over the spectral extent of the instrument lineshape function (ILS). The choice of instrument wavelength to carry out this determination of k_2 is somewhat arbitrary since the quantum yield of the instrument due to geometric parameters only should be wavelength independent, but this should be verified for any particular fluorescence instrument.

However, the value of this instrument factor is not necessarily constant with time due to drifts or aging of the source and other optical components. Thus, for high-accuracy normalization of bispectral luminescent radiance factor data, it is recommended that the determination of this instrument factor be made at approximately the same time as the measurements on the fluorescent sample. Also, the selection of normalization wavelength here is not entirely arbitrary since it needs to be a wavelength within the excitation range of the sample and, for reliable results, it is best to choose a wavelength that is not close to a wavelength where the instrument's spectral or polarization characteristics change dramatically, for example, near a Wood's anomaly. Further details of this normalization procedure can be found in Ref. [30].

For the technical correction of both the emission and excitation spectra, the CD and CS methods described above can be combined. Alternatively, the CD method can be used to first calibrate the excitation unit, including the monitor detector, and then the known relative spectral irradiance from the excitation system illuminating a white reflectance standard can be used to calibrate the spectral response of the detection system. The latter approach has the advantage that a separate CS is not needed and that the measurement geometry is more representative of the actual fluorescence measurements. However, the excitation system typically uses a xenon source, which does not have as smooth a spectral distribution as standard tungsten lamps. Thus, it is much more sensitive to wavelength errors such as spectral mismatch of the wavelength scales of the excitation and detection systems [4].

7.8.1.2 Spectral Fluorescence Standards (RMs)

Fluorescence RMs are commonly used to perform instrument calibrations and qualifications of the responsivity of the detection system to emission, or spectral correction of emission, for fluorometers. In fact, for analytical applications, RMs are more commonly used than PTSs because they come in standard sample formats and can therefore be easily handled by nonexperts. They do not require any special attenuation or setup procedures. In addition, they can be measured using the same optical geometry and instrument settings as the intended sample. This last advantage has been shown to minimize instrument-dependent, systematic errors, allowing better agreement of corrected spectra between instruments in different laboratories and with different designs [45,46].

A general consideration for selecting fluorescence RMs for spectral correction is the need for the excitation or emission wavelength region to cover that of the test sample for excitation or emission spectral correction, respectively. Other instrument parameters, such as spectral bandwidths, integration time, detector gain, and wavelength increment, should also be the same for RM and sample signal collection. On the other hand, the fixed wavelength parameter used when collecting a relative emission or excitation spectrum (i.e., the fixed excitation wavelength used for an emission spectrum, or the fixed emission wavelength for an excitation spectrum) need not be the same for RM and sample, unless absolute optical properties are required, for example, absolute quantum yields.

Spectral correction factors can also be dependent on sample format due to the wavelength and incident angle dependencies of sample anisotropy and reflection. These effects may, for instance, cause the spectral correction factors for front-face and 0°/90° detection geometries to be different.

For most chemical and biological applications, a transparent (low absorption), fluorescence RM of high purity, high stability, high quantum yield, and broad wavelength range is ideal for spectral correction of emission. Dilute solutions of organic, fluorescent dyes were the first standards of this type to be widely used [47,48]. More recently, inorganic fluorophores in a glass matrix have been developed as fluorescence standards with much greater photostability than organic dyes [49–53].

Certified reference materials (CRMs) are RMs that are supplied with values and uncertainties by a certifying body, which has high confidence in their accuracy, such that all known or suspected sources of error have been investigated. CRMs represent the highest level of quality and quantitative accuracy in RMs. CRMs for spectral correction of emission are available from national metrology institutes (NMIs) as both solid inorganic glasses [54–58] and organic dye solutions [59]. Similar CRMs are also available from secondary standards manufacturers [60]. Note that the certified values for these CRMs are given as relative intensity of emission as a function of wavelength, which supplies the user with the correct spectral shape of the emission, but not the absolute intensity.

Similar corrected emission spectra of other materials have been published in the literature [61,62]. These published values are usually not given with uncertainties, and the values themselves have not been measured on multiple instruments [45,46]. In some cases. it is even unclear as to whether quantities are being given in radiant or quantum units. In addition, it is often difficult to obtain high-purity samples corresponding to the RMs found in the literature. For all of these reasons, CRMs are more reliable than literature-reported RMs and are typically preferred by users.

It is important to remember when using either RMs or CRMs that to obtain a high degree of comparability of fluorescence measurement data, it is critical to perform the instrument calibration using these materials under conditions identical to those employed for subsequent fluorescence measurements [45]. These include instrument components such as monochromator gratings and filters, instrument settings, and measurement geometries.

7.8.1.3 True Spectra

In the last two sections, transfer standards for spectral correction of the excitation beam and detection system of instruments were discussed. In addition, instrument, environment, and sample characteristics that impact spectral fluorescence measurements were reviewed in Sections 7.6 and 7.7. These topics cover the major factors for which a user must correct in order to obtain true spectra. Each of these potential sources of error can be relatively small or large, depending on the particular case. Therefore, it is important for a user to realize which sources of error are most significant for their sample and measurement conditions.

It is also important to note that many, if not most, of the corrected spectra given in the literature, as well as those from NMIs, are instrument independent, but not true spectra. As described in Section 7.2, a *true spectrum* is an intrinsic property of the fluorophore, excited under the measurement conditions, while the term *instrument independent* implies only that the spectrum has been corrected for factors that may vary between instruments, that is, been *technically corrected*. Solvent refractive index correction (see Sections 7.5.3 and 7.7.3) is a good example of an error correction that is required for a true spectrum, but not for an instrument-independent one. In this case, the correction is proportional to n^2, which is an intrinsic property of the solvent, and will not change between instruments. The "corrected" spectra of most RMs are not corrected for such sample-specific factors because the spectra are intended for spectral correction of an instrument, which should not include sample-specific corrections that are instrument independent.

7.8.2 Intensity Verification Standards

7.8.2.1 Day-to-Day Intensity Standards (Performance Verification)

The most common use for fluorescence standards is for performance verification of instruments. Such standards are used to verify the consistent performance of a single instrument over time or to compare different instruments. The easiest way to achieve performance verification is by using fluorescence intensity standards, such that the intensity of these standard samples can be compared over time and between instruments, using the same instrument parameters. An RM of this type must be either reuseable, that is, photostable with a long shelf-life, or single use with a reproducible intensity. They should be measurable with typical instrument settings, but do not have to reproduce the spectra of analytes. The measured intensities of the standard can be used to create an absolute intensity scale without collecting absolute fluorescence measurements.

The best-known intensity standard for fluorescence instruments is high-purity water, where the Raman line is used as a pseudo-fluorescence signal [63,64]. The effective wavelength range of this standard is limited to UV and blue light, due to decreasing scattering intensity at longer wavelengths. Inorganic solid standards are recommended as reusable RMs for covering the entire visible spectrum, due to their photostability, robustness, and ease of use. Organic dyes are preferred as single use RMs because their characteristics can closely resemble those of fluorescent probes.

7.8.3 Fluorescent Color Standards

Fluorescent material standards are used for calibrating one-monochromator instruments described in Section 7.5.2.4. These fluorescence standards must be specially selected to match the optical properties, the sampling format, and other properties of the sample under test. A key requirement is that it has similar excitation and emission spectral bands. For these reasons, fluorescent material standards have been developed for specific applications, for example, for the plastics, textile, and paper industries. An example of a set of fluorescent cotton standards is shown in Fig. 7.6.

A survey of these fluorescent color standards and their availability is given in CIE Publication 182 [27], which also provides information on their long-term stability. It is generally found that the fluorescent white cloth standards and paper standards are not particularly stable and require frequent replacement, with a recommended replacement interval every 3–6 months. However, the short-term reproducibility of these standards is very good; for example, the relative standard deviation of the mean of several independent calibration results over a 1-month period is typically within 0.2%. Plastic fluorescent standards in custom sizes and configurations are available as both white scale sets and chromatic standards in a variety of emission colors, including multiple peaks. These plastic standards have high thermal stability and excellent lightfastness which makes them well suited for use as quality control check standards of the fluorescence measurement system (performance verification). These fluorescence color standards are also finding use in bioanalytical applications with increasing demands for reliable traceable fluorescence data, for example, in medical diagnostics as biological/biomedical probes.

If these fluorescent color standards are to be used for unambiguous color specification for different illuminant conditions, it is necessary that their calibration be performed using the two-monochromator method to provide normalized bispectral luminescent radiance factor data [65]. An example of the normalized bispectral luminescent radiance factor data for a red fluorescent paper standard is shown in Fig. 7.16. These data can be compared with those in Fig. 7.2 which shows the corresponding raw measured bispectral luminescent radiance data. As discussed in Section 7.2.3, these data can then be used as input to the spectral stimulus function with the desired standard illuminant.

Two of the most commonly used CIE illuminants for specifying fluorescence surface color are CIE standard illuminant D65, which represents a phase of natural daylight with a color temperature of about 6500 K, and CIE illuminant C, which is intended to represent an average daylight with a correlated color temperature of approximately 6800 K and is often used as an approximation to an indoor daylight condition [9]. More details on the various CIE illuminants used for colorimetric applications can be found in Chapter 10.

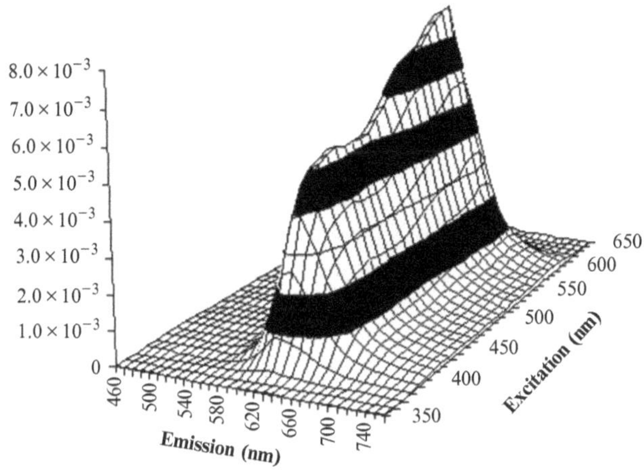

FIGURE 7.16 Calculated normalized bispectral luminescent radiance factor for a fluorescent red paint sample; 45:0 geometry, s-polarization; compare with Fig. 7.2 which shows the corresponding raw measured data on this instrument.

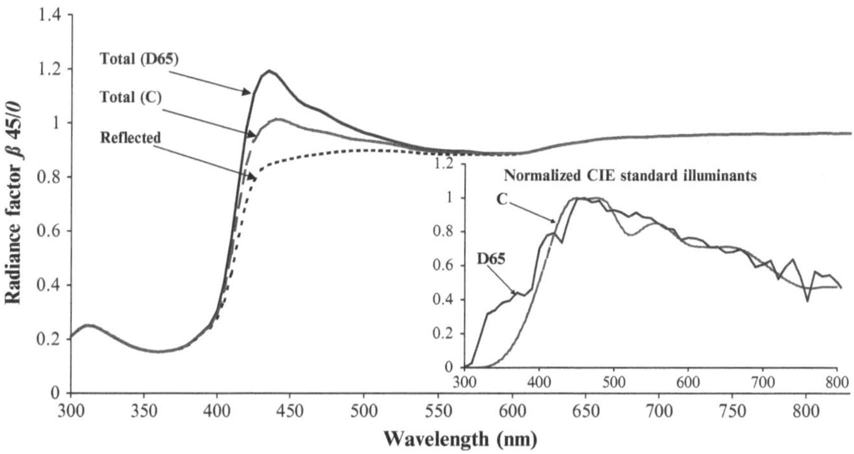

FIGURE 7.17 Comparison of reflected and total spectral radiance factors of a fluorescent paper standard calculated for CIE illuminants C and D65. Inset compares the relative spectral distributions of these two illuminants.

Figure 7.17 shows the calculated total spectral radiance factor $\beta_T(\lambda)$ for this fluorescent paper standard, calculated using these data, corresponding to two different illumination conditions. The colorimetric quantities (CIE tristimulus values: X,Y,Z, CIE chromaticity coordinates: x,y, CIELAB values, CIE whiteness, etc.) may then be calculated according to the expressions given in Chapter 10 where $\beta_T(\lambda)$ is substituted for $R(\lambda)$.

7.8.4 Quantum Yield Standards

To obtain reliable fluorescence quantum yields, with relative uncertainties less than 5%, it is important that the instrument be properly calibrated and that the fluorescence standard be properly chosen. In particular, the instrument needs to be checked for linearity over the dynamic range of spectral radiance/fluorescence intensity of the samples to be measured and spectrally characterized and corrected as described in Sections 7.6 and 7.8. To reduce errors due to imperfect instrument correction for nonlinearity and spectral characteristics, the fluorescent quantum yield standard should be selected according to the following criteria:

1. absorption/excitation spectrum similar in shape and wavelength range to the samples under test;
2. emission spectrum similar in shape and wavelength range to the samples under test;
3. a large Stokes shift, to minimize errors due to overlap of excitation and emission spectra;
4. a quantum yield that is slightly higher than that of the samples under test;
5. good photostability, having a quantum yield that is insensitive to environmental conditions (temperature, oxygen levels, time of exposure, impurities);
6. a quantum yield that is relatively insensitive to sampling conditions (e.g., excitation wavelength, pH, instrument polarization) [66].

For a 0°/90° geometry, quantum yield standards are typically fluorescent organic dyes of known quantum yield under specified experimental conditions (pH, solvent, and temperature) [66]. As discussed in Section 7.7.4, to avoid quenching effects, all solutions should be degassed, and solvents and cuvettes should be of spectral grade and checked for spurious emission. In the case of ideally dilute solutions ($A < 0.05$), the fluorescent quantum yield is independent of excitation intensity and wavelength, sample absorption, and concentration. Under these conditions, it is identical to the quantum yield of the molecular species. An extensive list of standards for photoluminescent quantum yield measurements can be found in Ref. [67] and recommendations on the selection of suitable quantum yield standards can be found in Ref. [68].

For a surface geometry, the radiant efficiency standards are white, diffusely reflecting standards of known spectral radiance factor with respect to the primary reflectance standard—the PRD, identically irradiated and viewed. In practice, secondary reflectance standards, such as pressed PTFE and barium sulfate powder, are used, where it is assumed that the reflected light from this comparison standard has the same angular distribution as the emitted light from the test sample.

REFERENCES

[1] CIE Standard 017:2011, ILV: International Lighting Vocabulary, Commission Internationale de l'Eclairage, CIE Central Bureau, Vienna, Austria, 2011.

[2] H.J. Eichler, D. Langhans, U. Klein, Quenching of two-photon fluorescence and stimulated anti-Stokes emission in rhodamine solutions, Appl. Opt. 18 (1979) 1383–1386.

[3] A. Brockes, Mechanism of whitening and quenching, in: R. Anliker, G. Müller (Eds.), Fluorescent Whitening Agents, in: EQS Environmental Quality and Safety, Suppl. vol. IV, George Thieme Publishers, Stuttgart, 1975, pp. 19–24.

[4] ASTM E 2719-09, Standard guide for fluorescence—instrument calibration and qualification, in: Annual Book of ASTM Standards, vol. 03.06, 2009.

[5] ASTM E2153-01, Standard practice for obtaining bispectral data for evaluation of fluorescent color, in: Annual Book of ASTM Standards, vol. 06.01, ASTM International, West Conshohocken, PA, 2011. www.astm.org.

[6] J. Schanda (Ed.), Colorimetry: Understanding the CIE System, John Wiley & Sons, New York, 2007.

[7] ASTM E 2152-12, Standard practice for computing the colors of fluorescent objects from bispectral photometric data, in: Annual Book of ASTM Standards, vol. 06.01, ASTM International, West Conshohocken, PA, 2013 www.astm.org.

[8] R. Donaldson, Spectrophotometry of fluorescent pigments, Br. J. Appl. Phys. 5 (1954) 210–214.

[9] R.Y. Tsien, A. Waggoner, Fluorophores for confocal microscopy, in: J.B. Pawley (Ed.), Handbook of Biological Confocal Microscopy, second ed., Plenum Press, New York, 1995, pp. 267–274.

[10] CIE Publication 15:2004, Colorimetry, third ed., Commission Internationale de l'Eclairage, CIE Central Bureau, Vienna, Austria, 2004.

[11] J.C. Zwinkels, M. Noël, S. Hillman, Near-infrared photoluminescence of orange color standards—then and now, J. Mod. Opt. 60 (2013) 1115–1123.

[12] CIE Publication 51, A method for Assessing the Quality of Daylight Simulators for Colorimetry, Commission Internationale de l'Eclairage, Paris, France, 1981.

[13] D.H. Alman, F.W. Billmeyer Jr., Integrating sphere errors in the colorimetry of fluorescent materials, Color Res. Appl. 1 (1976) 141–145.

[14] J.C. Zwinkels, Metrology of photoluminescent materials, Metrologia 47 (2010) S182–S193.

[15] ISO 2470-1, Paper, board and pulps—measurement of diffuse blue radiance factor—part 1: indoor daylight conditions (ISO brightness) 2009.

[16] TAPPI Test Method T452 om-08, Brightness of pulp, paper, and paperboard (directional reflectance at 457 nm) 2008.

[17] R. Griesser, Instrumental measurement of fluorescence and determination of whiteness: review and advances, Rev. Prog. Coloration 11 (1981) 25–36.

[18] E. Allen, Separation of the spectral radiance factor curve of fluorescent substances into reflected and fluoresced components, Appl. Opt. 12 (1973) 289–293.

[19] J. Olmsted, Calorimetric determinations of absolute fluorescence quantum yields, J. Phys. Chem. 83 (1979) 2581–2584.

[20] K. Rurack, Fluorescence quantum yields: methods of determination and standards, in: Springer Ser. Fluoresc., 5, 2008, pp. 101–145.

[21] IEC 62607-3-1, Nanomanufacturing—key control characteristics of luminescent nanomaterials, part 1: quantum efficiency.

[22] P.-S. Shaw, U. Arp, K.R. Lykke, Measurement of the ultraviolet-induced fluorescence yield from integrating spheres, Metrologia 46 (2009) S191–S196.

[23] P.-S. Shaw, L. Shigang, U. Arp, K.R. Lykke, Ultraviolet characterization of integrating spheres, Appl. Opt. 46 (2007) 5119–5128.

[24] Y. Zong, S.W. Brown, C.B. Johnson, K.R. Lykke, Y. Ohno, Simple spectral stray light correction method for array spectroradiometers, Appl. Opt. 45 (2006) 1111–1119.

[25] P.C. DeRose, E.A. Early, G.W. Kramer, Qualification of a fluorescence spectrometer for measuring true fluorescence spectra, Rev. Sci. Instrum. 78 (2007) 033107.

[26] J.C. Travis, J.C. Zwinkels, F. Mercader, A. Ruiz, E.A. Early, M.V. Smith, M. Noël, M. Maley, G.W. Kramer, D.L. Duewer, An International evaluation of holmium oxide solution reference materials for wavelength calibration in molecular absorption spectrophotometry, Anal. Chem. 74 (2002) 3408–3415.

[27] C.A. Parker, Raman spectra in spectrofluorimetry, Analyst 84 (1959) 446.

[28] ASTM E-578:07, Standard test method for linearity of fluorescence measuring systems, in: Annual Book of ASTM Standards, vol. 03.06, ASTM International, West Conshohocken, PA, 2013. www.astm.org.

[29] CIE Publication 182:2007, Calibration Methods and Photoluminescent Standards for Total Radiance Factor Measurements, Commission Internationale de l'Eclairage, CIE Central Burea, Vienna, Austria, 2007.

[30] J.C. Zwinkels, D.S. Gignac, Automated high precision variable aperture for spectrophotometer linearity testing, Appl. Opt. 30 (1991) 1678–1687.

[31] U. Resch-Genger, D. Pfeifer, C. Monte, W. Pilz, A. Hoffmann, M. Spieles, K. Rurack, J. Hollandt, D. Taubert, B. Schönenberger, P. Nording, Traceability in fluorometry: part II. Spectral fluorescence standards, J. Fluoresc. 15 (2005) 315–336.

[32] D.M. Jameson, J.A. Ross, Fluorescence polarization/anistropy in diagnostics and imaging, Chem. Rev. 110 (2010) 2685–2708.

[33] J.C. Zwinkels, C.X. Dodd, Determination of spectrophotometer polarization and its application to rapid, accurate polarized transmission measurements, Appl. Opt. 28 (1989) 2381–2388.

[34] H. Minato, M. Nanjo, Y. Nayatani, Errors in spectrophotometry and colorimetry of fluorescent samples caused by polarization of the measuring system, Color. Res. Appl. 8 (1983) 238–244.

[35] J.C. Zwinkels, F. Gauthier, Instrumentation, standards and procedures used at the National Research Council of Canada for high-accuracy fluorescence measurements, Anal. Chim. Acta 380 (1999) 193–209.

[36] ASTM E-308-13, Standard practice for computing the colors of objects by using the CIE system, in: Annual Book of ASTM Standards, vol. 06.01, ASTM International, West Conshohocken, PA, 2013. www.astm.org.

[37] J.F. Holland, R.E. Teets, P.M. Kelly, A. Timnick, Anal. Chem. 49 (1977) 706.

[38] J.R. Lakowicz, in: Principles of Fluorescence Spectroscopy, third ed., Springer-Science+ Business, New York, 2006, pp. 208–212.

[39] R.A. Velapoldi, K.D. Mielenz, A Fluorescence Standard Reference Material: Quinine Sulfate Dihydrate, U.S. Government Printing Office, Washington, DC, 1980, pp. 79–82, NBS Spec. Pub. 260-64.

[40] K.D. Mielenz, Photoluminescence Spectrometry, in: K.D. Mielenz (Ed.), in: Optical Radiation Measurements: Measurement of Photoluminescence, vol. 3, Academic Press, New York, 1982, pp. 79–82.

[41] E. Hecht, Optics, third ed., Addison Wesley Longman, Reading, 1998, pp. 113–121.

[42] J.R. Lakowicz, Principles of Fluorescence Spectroscopy, third ed., Springer-Science+Business, New York, 2006, pp. 278–317.

[43] J.R. Lakowicz, Energy Transfer, in: J.R. Lakowicz (Ed.), Principles of Fluorescence Spectroscopy, third ed., Springer-Science+Business, New York, 2006, pp. 443–472.
[44] C. Palmer, Diffraction Grating Handbook, sixth ed., Newport Corporation, Rochester, 2005, pp. 19–27.
[45] U. Resch-Genger, P.C. DeRose, W. Bremser, J.C. Zwinkels, B. Ebert, D. Pfeifer, J. Voigt, D. Taubert, C. Monte, R. Macdonald, J. Hollandt, F. Gauthier, M. Spieles, A. Hoffmann, State-of-the art comparability of corrected emission spectra—part 1: spectral correction with physical transfer standards and spectral fluorescence standards by expert laboratories, Anal. Chem. 84 (2012) 3889.
[46] U. Resch-Genger, P.C. DeRose, W. Bremser, J.C. Zwinkels, B. Ebert, D. Pfeifer, J. Voigt, D. Taubert, C. Monte, R. Macdonald, J. Hollandt, F. Gauthier, M. Spieles, A. Hoffmann, State-of-the-art comparability of corrected emission spectra—part II: field laboratory assessment of calibration performance using spectral fluorescence standards, Anal. Chem. 84 (2012) 3899.
[47] W.H. Melhuish, Absolute spectrofluorometry, in: Accuracy in Spectrophotometry and Luminescence Measurements, Natl. Bur. Stand. (U.S.) Spec. Publ. 378, U.S. GPO, Washington, DC, 1973, p. 137.
[48] R.A. Velapoldi, K.D. Mielenz, A Fluorescence Standard Reference Material: Quinine Sulfate Dihydrate, U.S. Government Printing Office, Washington, DC, 1980, NBS Spec. Pub. 260-64.
[49] P.C. DeRose, M.V. Smith, J.R. Anderson, G.W. Kramer, Characterization of standard reference material 2944, Bi-ion-doped glass, spectral correction standard for red fluorescence, J. Lumin. 141 (2013) 9–14.
[50] P.C. DeRose, M.V. Smith, K.D. Mielenz, J.R. Anderson, G.W. Kramer, Characterization of standard reference materials 2943, Cu-ion-doped glass, spectral correction standard for blue fluorescence, J. Lumin. 131 (2011) 2509.
[51] P.C. DeRose, M.V. Smith, K.D. Mielenz, J.R. Anderson, G.W. Kramer, Characterization of standard reference materials 2942, Ce-ion-doped glass, spectral correction standard for UV fluorescence, J. Lumin. 131 (2011) 1294.
[52] P.C. DeRose, M.V. Smith, K.D. Mielenz, D.H. Blackburn, G.W. Kramer, Characterization of standard reference materials 2940, Mn-ion-doped glass, spectral correction standard for fluorescence, J. Lumin. 129 (2009) 349.
[53] P.C. DeRose, M.V. Smith, K.D. Mielenz, D.H. Blackburn, G.W. Kramer, Characterization of standard reference materials 2941, uranyl-ion-doped glass, spectral correction standard for fluorescence, J. Lumin. 128 (2008) 257.
[54] Certificate of analysis, Standard Reference Material 2944, Relative Intensity Correction Standard for Fluorescence Spectroscopy: Red Emission, National Institute of Standards and Technology, Gaithersburg, MD, 2011. https://www-s.nist.gov/srmors/view_detail.cfm?srm=2944.
[55] Certificate of analysis, Standard Reference Material 2943, Relative Intensity Correction Standard for Fluorescence Spectroscopy: Blue Emission, National Institute of Standards and Technology, Gaithersburg, MD, 2009. https://www-s.nist.gov/srmors/view_detail.cfm?srm=2943.
[56] Certificate of analysis, Standard Reference Material 2942, Relative Intensity Correction Standard for Fluorescence Spectroscopy: Ultraviolet Emission, National Institute of Standards and Technology, Gaithersburg, MD, 2009. https://www-s.nist.gov/srmors/view_detail.cfm?srm=2942.
[57] Certificate of analysis, Standard Reference Material 2941, Relative Intensity Correction Standard for Fluorescence Spectroscopy: Green Emission, National Institute of Standards

and Technology, Gaithersburg, MD, 2007. https://www-s.nist.gov/srmors/view_detail.cfm?srm=2941.

[58] Certificate of analysis, Standard Reference Material 2940, Relative Intensity Correction Standard for Fluorescence Spectroscopy: Orange Emission, National Institute of Standards and Technology, Gaithersburg, MD, 2007. https://www-s.nist.gov/srmors/view_detail.cfm?srm=2940.

[59] Certificate of analysis, Certified Reference Materials BAM-F001-BAM-F005, Calibration Kit, Spectral Fluorescence Standards, Federal Institute for Materials Research and Testing, Berlin, 2006.

[60] Matech, Sigma-Aldrich and Starna. These are the only companies known to the authors at present.

[61] J.A. Gardecki, M. Maroncelli, Set of secondary emission standards for calibration of the spectral responsivity in emission spectroscopy, Appl. Spectrosc. 52 (1998) 1179.

[62] R.A. Velapoldi, H.H. Tonnesen, Corrected emission spectra and quantum yields for a series of fluorescent compounds in the visible spectral region, J. Fluoresc. 14 (2004) 465.

[63] R.J. Kovach, W.M. Peterson, The measurement of sensitivity in fluorescence spectroscopy, Am. Lab. 26 (1994), G32-K32.

[64] P. Froehlich, Under the sensitivity specification for a fluorescence spectrophotometer, Int. Lab. (1989) 42–44.

[65] J. Zwinkels, Surface fluorescence: the only standardized method of measuring luminescence, in: Springer Ser. Fluoresc., 5, 2008, pp. 163–192.

[66] D.F. Eaton, Reference materials for fluorescence measurements, Pure Appl. Chem. 60 (1988) 1107–1114.

[67] A.M. Brower, Standards for photoluminescent quantum yield measurements in solution (IUPAC Technical Report), Pure Appl. Chem. 83 (2011) 2213–2228.

[68] U. Resch-Genger, K. Rurack, Determination of the photoluminescence quantum yield of dilute dye solutions (IUPAC Technical Report), Pure Appl. Chem. 85 (2013) 2005–2025.

Chapter 8

Angle-Resolved Diffuse Reflectance and Transmittance

Thomas A. Germer*, John C. Stover[†] and Sven Schröder[‡]
*National Institute of Standards and Technology, Gaithersburg, Maryland, USA
[†]The Scatter Works, Inc., Tucson, Arizona, USA
[‡]Fraunhofer Institute for Applied Optics and Precision Engineering, Jena, Germany

Chapter Outline

8.1 Introduction 292
8.2 Reference-Free Measurement Methods 294
8.2.1 The Over-Illumination Method 295
8.2.2 The Under-Illumination Method 297
8.3 Instrument Characterization 301
8.3.1 The Instrument Signature 301
8.3.2 Noise-Equivalent BRDF 303
8.3.3 Profile of Illumination Spot 303
8.3.4 The Field of View 304
8.3.5 Detector Location Sensitivity 304
8.3.6 Linearity 304
8.4 Goniometer Designs 304
8.4.1 In-Plane Scatterometer Designs 304
8.4.2 Out-of-Plane Scatterometer Designs 305
8.4.3 Multichannel Systems 309
8.5 Uncertainty Analysis 311
8.6 Normalization Schemes 315
8.6.1 Relative Normalization 315
8.6.2 Specular Normalization 316
8.7 Special Conditions or Considerations 316
8.7.1 Ultraviolet 316
8.7.2 Infrared 316
8.7.3 Polarization 316
8.7.4 Transmission Measurements 317
8.8 Applications 317
8.8.1 Pressed PTFE 318
8.8.2 Integration of BRDF 318
8.8.3 Diffuse Black Paint 321
8.8.4 Characterization of Surface Roughness 322

Experimental Methods in the Physical Sciences, Vol. 46. http://dx.doi.org/10.1016/B978-0-12-386022-4.00008-X

8.8.5 Scattering Characterization of Anisotropic Surface Structures 325
8.8.6 Predicting Out-of-Plane BRDF from In-Plane BRDF 326
References 328

8.1 INTRODUCTION

Angle-resolved light scattering measurements have a wide variety of applications ranging from the realistic quantitative description of the visual appearance of materials, machine vision, the modeling of the regolith of extraterrestrial planets, to the inspection and roughness characterization of optical materials, surface structures, and interference coatings, and the characterization and subsequent modeling of stray light in optical systems [1]. The observed angular scattering characteristics span the properties from being highly diffuse and relatively featureless in the angle dependence of the scatter, to highly specular, where the scatter is sharply peaked in the regular reflective direction and decays by many orders of magnitude from there, yet still being of significant importance for specific applications. Multilayer-coated optics, diamond-turned surfaces, and diffraction gratings usually have additional pronounced nonspecular features. In this chapter, we describe the measurement of the angle dependence of diffuse light scatter, with an emphasis toward making accurate traceable measurements.

The angle-resolved structure in the diffuse scattering is quantified by the bidirectional reflectance distribution function (BRDF) [2,3]. Given in Chapter 2, the BRDF is the ratio of the scattered radiance L_r normalized by the incident irradiance E_i on the surface, when that surface is uniformly irradiated (see Eq. 2.21),

$$f_r = \frac{L_r}{E_i}. \tag{8.1}$$

As pointed out in Chapter 2, it is common to use a small illumination spot, and under this condition, the expression (see Eq. 2.24)

$$f_r = \frac{d\Phi_r}{\Phi_i \, d\Omega_r \cos\theta_r} \tag{8.2}$$

holds, where $d\Omega_r$ is the solid angle of collection. The BRDF has the property of Helmholtz reciprocity for most materials. That is, the BRDF is invariant upon interchange of incident and reflected directions. The BRDF is a function of wavelength λ and polarization. As discussed in Chapter 2, its polarization properties are best described by a Mueller matrix [4].

Other expressions of the scatter distribution are often used *in lieu* of the BRDF and are defined in the International Lighting Vocabulary [5], including the *radiance coefficient*, the *radiance factor*, and the *reflectance factor*. The radiance coefficient is defined as the radiance emitted in a specified direction normalized by the irradiance, but differs from the BRDF in that the incident irradiance does not have to be unidirectional. For a given incident irradiance distribution $L_i = L_i(\theta_i, \phi_i)$, the radiance coefficient q_e is related to the BRDF by

$$q_e = \frac{\int f_r L_i \cos\theta_i \, d\Omega_i}{\int L_i \cos\theta_i \, d\Omega_i}. \tag{8.3}$$

In the limit that the incident illumination is unidirectional, the radiance coefficient is the BRDF.

The radiance factor β is defined as the radiance in a given direction normalized by the radiance from a perfectly reflecting Lambertian surface identically illuminated. For a given incident irradiance distribution L_i, the radiance factor is given by

$$\beta = \frac{\pi \int f_r L_i \cos\theta_i \, d\Omega_i}{\int L_i \cos\theta_i \, d\Omega_i}. \tag{8.4}$$

Thus, for identical illumination conditions, the radiance factor is related to the radiance coefficient by $\beta = \pi q_e$. Similarly, if the incident illumination is unidirectional, the radiance factor is related to the BRDF by $\beta = \pi f_r$.

The reflectance factor R is defined as the ratio of the radiant flux reflected into a given set of directions to that reflected by a perfectly reflecting Lambertian diffuser identically irradiated. For a given radiance distribution L_i and set of reflected directions Ω_r, the reflectance factor is

$$R = \frac{\pi \int_{\Omega_r} \int f_r L_i \cos\theta_i \cos\theta_r \, d\Omega_i \, d\Omega_r}{\int_{\Omega_r} \int L_i \cos\theta_i \cos\theta_r \, d\Omega_i \, d\Omega_r}. \tag{8.5}$$

If the incident irradiation is unidirectional, the collection angle subtends a very small solid angle, and the BRDF is uniform, the reflectance factor $R = \pi f_r$. Notice that if the incident irradiance is uniform, the denominators in Eqs. (8.3)–(8.5) reduce to the projected solid angle, given in Eq. (2.6), and the L_i drops out from the expressions. Likewise, the integral over $d\Omega_r$ in the denominator of Eq. (8.5) is the reflected projected solid angle.

Other alternative terminology can be found in the literature. Angle-resolved scatter (ARS) is often used by those who are not interested so much in radiometry, appearance modeling, or image rendering, but rather in measuring the scatter distribution in low-scatter optics and optical systems. The ARS is measured in a small spot configuration and is given by

$$\mathrm{ARS} = \frac{\mathrm{d}\Phi_\mathrm{r}}{\Phi_\mathrm{i}\,\mathrm{d}\Omega_\mathrm{r}}. \tag{8.6}$$

ARS is thus related to the BRDF by the factor of $\cos\theta_\mathrm{r}$. It has the relatively simple interpretation of being the fraction of light scattered per solid angle. Because of the reciprocity conditions imposed on the BRDF and the presence of the factor of $\cos\theta_\mathrm{r}$, the ARS is not a reciprocal quantity. Yet, because of its simplicity, it is sometimes convenient to use ARS to represent scatter data. If necessary, the conversion to BRDF is then straightforward.

All the quantities introduced in this section have the common property that they describe angle-resolved scattering independent from the actual instrument or parameters used. This is basically achieved by using radiometric quantities (instead of scattered power) and careful calibration. Two standard guides have been published that describe measurements of the BRDF and its related quantities [6,7].

8.2 REFERENCE-FREE MEASUREMENT METHODS

Measurements that do not rely upon a reference sample are considered in this section. They are often referred to as "absolute" measurements, because they can assign the units of inverse steradians through measurements of solid angles, dimensional measurements, and ratios of comparable radiometric quantities, usually performed with the same device.

There are two basic reference-free methods used for measuring BRDF that can be categorized by whether they use Eq. (8.1) or (8.2). We will refer to these as the *over-illuminated* and the *under-illuminated* methods, respectively, because they differ in how they illuminate the sample. The methods could also be referred to as the forward and backward methods, but which direction is forward and which is backward would depend upon the perspective of which methodology one follows. Historically, Eq. (8.1) is the fundamental definition for the BRDF, but it is shown that Eq. (8.2) is equivalent, and for some applications, simpler to apply. In effect, however, they are equivalent, but have the propagation of light reversed, one with respect to the other. There are advantages to each of the two measurement schemes that will be discussed later in this chapter.

The unit associated with the BRDF is the inverse steradian (sr^{-1}). It is thus natural that we have to effectively realize this unit in our measurement. In Chapter 2, it was shown that the projected solid angle, Ω_proj, for a circular aperture of radius r (and area $A = \pi r^2$) at a distance R is given by [8]

$$\Omega_\mathrm{proj} \equiv \int \mathrm{d}\Omega \cos\theta = \frac{\pi r^2 \cos\theta_0}{R^2 + r^2}, \tag{8.7}$$

where θ_0 is the polar angle of the center of the aperture, which is either θ_i or θ_r, depending upon whether the aperture is on the illumination side or on the collection side. For small apertures, where $r \ll R$, the projected solid angle simply reduces to $\Omega_{proj} = A \cos \theta_0 / R^2$. Generally, this latter expression can be used with insignificant effects on the result for all but the highest accuracy measurements performed with instruments having low angular resolution. For example, for $r = 25$ mm and $R = 500$ mm, the true value of Ω_{proj} will be only 0.25% lower than that determined by the simpler expression. This correction is on par with typical $k = 2$ expanded uncertainties of white diffuse materials at moderate angles and visible wavelengths [9].

8.2.1 The Over-Illumination Method

One approach that is used for measuring the BRDF relies on the radiance–irradiance relationship in Eq. (8.1) [10,11]. This method is fairly common for high-scatter samples, such as diffuse reflectance standards and materials of interest for their visual appearance. The idea is to provide a uniform irradiance over the entire sample and then to measure the radiance scattered from it. The irradiance, however, is rarely measured directly. Instead, a source of uniform radiance L_i of area $A = \pi r^2$ is placed at a known distance R from the sample under test. The irradiance at the sample is then given by

$$E_i = L_i \Omega_{proj,i}, \tag{8.8}$$

where $\Omega_{proj,i}$ is the projected sold angle subtended by the source aperture. The BRDF is thus given by

$$f_r = \frac{L_r}{E_i} = \frac{L_r}{L_i} \frac{R^2 + r^2}{\pi r^2 \cos \theta_i}. \tag{8.9}$$

The radiances L_r and L_i are measured by a single radiometer, so that only the ratio L_r/L_i needs to be measured accurately. The other terms are all geometric.

Figure 8.1 shows a schematic of the over-illuminated sample method. In practice, the radiator is an integrating sphere illuminator, shown schematically in Fig. 8.1. It is often broadband illuminated, with a lamp covering the spectral range of interest. The radiometer is then spectrally selective. Some configurations have the radiometer be an optic collecting light onto the aperture of a spectrometer or a spectrograph. Using a spectrograph, one can disperse the light over a wide range of wavelengths and thus have a multiplexing advantage. Provided the depth of focus of the radiometer is such that the width of its point spread function is smaller than both the sample at the sample position and the source area at the source position, and that the detection is linear, the relative radiance measurement can be accurate. Some facilities move the radiometer or the source, so that the source–radiometer distance during the measurement of L_i is the same as the sample–radiometer distance during the measurement of L_r [8,12].

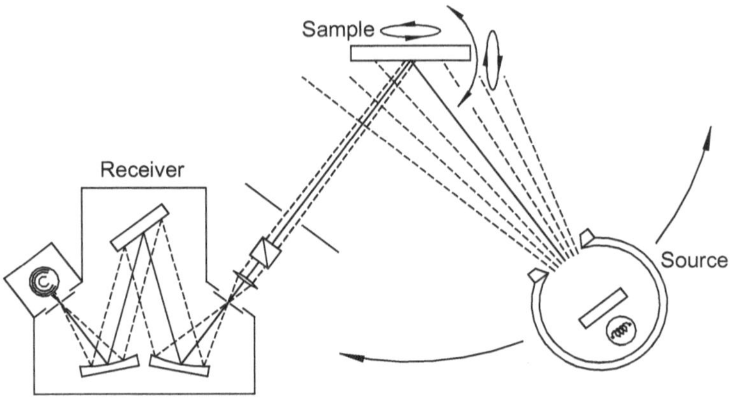

FIGURE 8.1 A schematic showing the over-illumination method of BRDF measurement. The source is shown as an integrating sphere, containing a light source and a baffle. The receiver is shown as a spectroradiometer, consisting of a polarizer, an imaging lens, a monochromator, and a detector.

There are several requirements that must be met in order for this method to achieve accurate results: (a) The field of view of the radiometer must be smaller than the sample at the sample distance and smaller than the sphere radiator aperture when measuring the incident radiance, and (b) the radiance emitted by the sphere radiator must be uniform over its entire aperture. These two conditions should be tested in order to validate the assumptions of the measurement.

To a lesser extent, other factors need to be considered as well. For example, the solid angle subtended by the illuminator Ω_i and the collection aperture of the radiometer should be sufficiently small that the angular resolution of the instrument captures the necessary structure in the BRDF. Stray light striking elsewhere in the room or enclosure and illuminating the sample needs to be minimized.

The over-illumination method is primarily used for highly diffuse materials. It is less ideal for measuring low-scatter materials, because it is relatively inefficient with its use of light. For example, only a small fraction of the light emitted by the illuminator reaches the part of the sample being measured by the radiometer. Furthermore, the radiometer field of view must fill only a fraction of the illuminator aperture, so that only a fraction of the area of the illuminator is used for measuring the incident radiance. This reduction in available light is further exacerbated if the angular resolution needs to be reduced, in order to capture the structure near the specular beam, because then the total illumination on the sample is reduced.

There are a number of advantages of this technique. It is relatively straightforward to guarantee that the illumination aperture provides a uniform radiance, and that the sample is uniformly irradiated. Requirements on the radiometer optics are also minimal. With stray light correction in the spectrograph radiometer [13], a wide range of wavelengths can be measured accurately at once.

8.2.2 The Under-Illumination Method

The scheme described in this section assumes that we illuminate a relatively small spot on the sample and then measure the amount of light scattered into a known solid angle. Thus, we rely on Eq. (8.2) for the measurement, which, using Eq. (8.7) for the projected solid angle of a finite circular aperture, is

$$f_{\mathrm{r}} = \frac{\Phi_{\mathrm{r}}}{\Phi_{\mathrm{i}}} \frac{R^2 + r^2}{\pi r^2 \cos\theta_{\mathrm{r}}}. \tag{8.10}$$

Often, the light incident upon the sample is monochromatic. In that case, it is very similar to the previous over-illumination method, except operated backward. The light being collected by the radiometer is replaced by a source and a monochromator, while the uniform source is replaced by a uniform detector. The light, however, is used more efficiently, because the receiver field of view includes the entire illuminated area. This feature makes it particularly useful for coherent sources, since the monochromator is not needed and multiplexing is of less interest. Figure 8.2 shows a schematic of the under-illuminated sample method.

In much of the remaining chapter, we assume that the under-illumination method is being used. However, most of the same design and characterization issues exist for the over-illumination method, except that the source and the receiver need to be interchanged, and the direction of propagation of the light reversed.

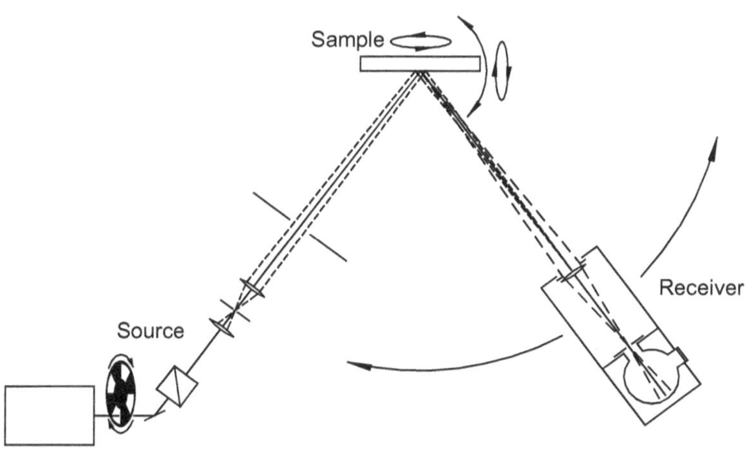

FIGURE 8.2 Schematic of the under-illumination method of BRDF measurement. The source is shown as a laser, chopping wheel, polarizer, and spatial filter. The receiver is shown as an aperture, lens, field stop, integrating sphere, and detector.

8.2.2.1 Illuminator

The illumination of the sample is intended to be as collimated as possible with as little stray light outside of the incident collimated beam. When the source is a laser, the beam can be spatially filtered, yielding a high-quality spatial profile. The output of the spatial filter can then be collected and collimated or slowly focused using a highly polished and focusing mirror. Care should be taken that either an aspheric mirror be used or that the mirror be as close to normally illuminated as possible, so that there is as little astigmatism as possible. Aspheric mirrors, however, tend to have higher scatter levels, and the optimum setup includes a super-polished spherical focusing mirror.

While a mirror is preferable to collimate or focus the light, a lens can be used. However, even the small amount of reflection from an antireflection coating may create a halo about the incident light beam. That halo not only has a detrimental effect on the ability to do small-angle scattering measurements but also increases the total size of the beam at the detector and can result in a loss of light during the measurement of incident radiant flux. Furthermore, scatter from the coating itself may contribute to stray light.

One can focus the source at different locations, depending upon the measurement interests. Focusing through the sample and onto the receiver aperture yields the highest angular resolution. With an *f*/100 beam and visible wavelengths as an example, the diffraction limited spot diameter is approximately 100 μm. If the detector arm is approximately 500 mm away from the sample, then the angular resolution, if the receiver aperture is approximately the same size, is about 0.01°, and the solid angle would be approximately $\Omega = 3 \times 10^{-8}$ sr. The maximum recorded BRDF will then be on the order of $1/\Omega = 3 \times 10^{7}\ \mathrm{sr}^{-1}$.

One of the disadvantages of focusing at the detector aperture is that this means that one is also focusing very close to the detector. This can potentially cause problems with performing an accurate measurement of the incident light intensity, since it will amplify any variations in the sensitivity of the receiver with position. Focusing the incident beam beyond the detector will further the collimation of the incident beam and will increase the size of the beam at the receiver aperture and reduce the effects of any detector sensitivity variations.

The third focusing configuration that would be of interest is that of focusing at the sample. This configuration gives the highest spatial resolution at the sample, which may be important for measurements of small targets or finding small surface features. However, the angular resolution will be degraded. For coherent light sources, speckle effects, discussed more in Section 8.5, will be increased.

It is useful, for low-scatter measurements or when using infrared sources and detectors, to chop the beam and use lock-in detection for the signal. When using laser sources, the chopper should be placed before the spatial filter, so that it does not contribute significantly to the scatter as the chopper blade is crossing the beam. When the edge of the blade is crossing the beam, there will be an increase in the scatter and diffraction. A partially blocked laser beam will distort the beam in a spatial filter. Thus, the beam should be small at the location of the blade so that the time the edge of the blade spends crossing the beam is minimized.

Measurements can also be performed using a shuttered system. This mode is effectively like performing a zero-frequency lock-in measurement. In this case, there will be no measurement made while the shutter is opening or closing. The disadvantage of this method is that any $1/f$ noise will be high, compared to that with a lock-in amplifier. Most infrared detectors cannot be operated in this fashion.

It is very useful to have a means for monitoring the incident radiant flux. A low-scatter pellicle beam splitter, placed in the beam at a small angle of incidence, can direct a fraction of the light toward a reference detector. In this way, fluctuations in the radiant flux can be compensated. If a lock-in amplifier is used, it is important that the time constants and filter setting be identical for the lock-in being used for measuring the reference as that being used to measure the radiant fluxes Φ_i or Φ_r, and that the reference and signal measurements be performed simultaneously.

The polarization state of the light can be selected with a polarizer, possibly with a waveplate, at any point in the optical path. For laser-based systems, and especially when small-angle scattering is of interest, it is best to place the polarization optics before the spatial filter. Since prism-style polarizers are thick optics, it is generally best to not rotate them, lest the beam walk on the spatial filter. In this case, a half-wave waveplate can effectively select the desired polarization. For spectroscopic systems with lower small-angle performance requirements, the location of the polarizer is less critical.

Besides single wavelength laser sources, the availability of widely tunable laser sources based on the effect of optical parametric oscillation enables the development of instruments for spectral ARS measurements with high sensitivity [14]. Also, the use of supercontinuum sources for this purpose is being explored [15].

8.2.2.2 *Receiver*

The receiver in the under-illumination method serves the purpose of defining the collection solid angle $d\Omega_r$. This is usually determined by having a well-defined circular aperture of radius r at a fixed distance to the sample, R, so that the projected solid angle is given by Eq. (8.7). The sensitivity of the detection over that area should be as uniform as possible. One design of a receiver is shown schematically in Fig. 8.3. The solid-angle defining aperture is in the front of the receiver, so that its area and distance can be determined and are well defined.

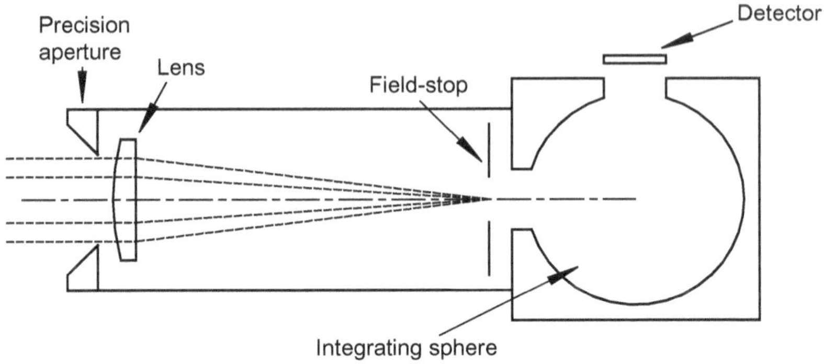

FIGURE 8.3 Schematic of a receiver.

Multiple apertures, perhaps on a computer-controlled translation stage, allow for different angular resolutions needed for measuring the BRDF of specular materials near the specular direction. The smallest aperture need only be approximately the size of the focused laser spot and is only used very close to the specular direction. As the receiver moves away from the specular direction, the signal can drop by orders of magnitude, and intermediate aperture sizes may be used, before the largest aperture is used for measurements where there is little structure in the BRDF.

In Fig. 8.3, the aperture is immediately followed by a focusing lens that images the sample plane onto a field stop. The field stop thus defines the field of view of the receiver. That field of view should be large enough that it encompasses all of the radiating area of the sample, keeping in mind that some materials diffuse light significantly away from the irradiated spot and that the maximum extent of the beam on the sample depends upon incident angle. However, too large a field of view can introduce unintended stray light into the system, especially when viewing very close to the specular direction on low-scatter materials, where the field of view might include optics upstream of the sample. Having a field of view that is adjustable for different measurement needs can increase the versatility of the instrument.

Behind the field stop, there is an integrating detector shown in Fig. 8.3. Direct illumination of a detector will enable the highest sensitivity, but at a cost of possible uniformity issues. For visible, ultraviolet, and near infrared light, direct illumination of a large area silicon photodiode can have sufficient spatial uniformity and a wide enough dynamic range for many applications. The best uniformity, however, is achieved when the detector is used in conjunction with an integrating sphere, as shown in Fig. 8.3. There is some loss of signal due to the efficiency of the integrating sphere, but this is minimal for all but the lowest scattering levels. Multiple detectors may also be used. For highly polished, low-scatter materials, the level of scatter may be low enough that a more sensitive detector, such as a photomultiplier tube

(PMT), is needed. A system having multiple detectors will need to account for the relative sensitivity of each detector, since the incident radiant flux Φ_i will be measured with a different detector than the scattered radiant flux Φ_r. A change in detector would add to the resulting uncertainty in a measurement, but the uncertainty requirement for low-scatter optics is usually significantly lower than that of high-scatter materials.

8.3 INSTRUMENT CHARACTERIZATION

There are a number of measurements that should be performed on a gonioreflectometer to characterize its operation. These may not directly enter into the measurement result, but they are essential to assess whether the measurement is performed correctly or to improve the instrument performance. Optimizing the instrument performance is usually a nontrivial task and requires careful analysis of all the critical factors.

8.3.1 The Instrument Signature

The instrument signature is an attempt to record the signal that one would achieve from a perfectly flat, perfectly reflecting sample free of any scattering. While it is generally impossible to find such a perfect sample, it is sufficient for many applications to remove the sample from the goniometer and to measure the scattering as a function of angle from the straight-through direction. This measurement will be indicative of what one would measure if there was a perfectly reflecting sample. If a measurement from a low-scatter sample is performed, the instrument signature should be scaled by the reflectance of the sample and displayed overlaid upon the data, to assess how much of the scatter is from the sample and how much is simply an instrument artifact. There will undoubtedly be a small angle, within which the measured signal will be indistinguishable from the instrument signature and where the data should be disregarded. How much larger than the instrument signature the data signal should be is somewhat up to the user of that data and will depend upon the application.

Figure 8.4 shows a representative instrument signature measured using an instrument at NIST. There are three apertures used in this measurement, so that aperture changes occur at 0.1° and 1.5°. At each of these angles, the noise level rises quite a bit, because the signal level is low, but switching aperture size too soon will yield excessive signal from the side of the next larger aperture. There are three regimes observed in the data. Initially, the signal drops off with an approximately Gaussian behavior, which is due to the propagation properties of the near-diffraction-limited laser beam. At some point, the signal levels off, because scatter from optics in the source (mostly the primary mirror) is collected. At these angles, the field of view of the receiver is large enough that the source optics is still visible to the receiver. Once the field of view of the receiver

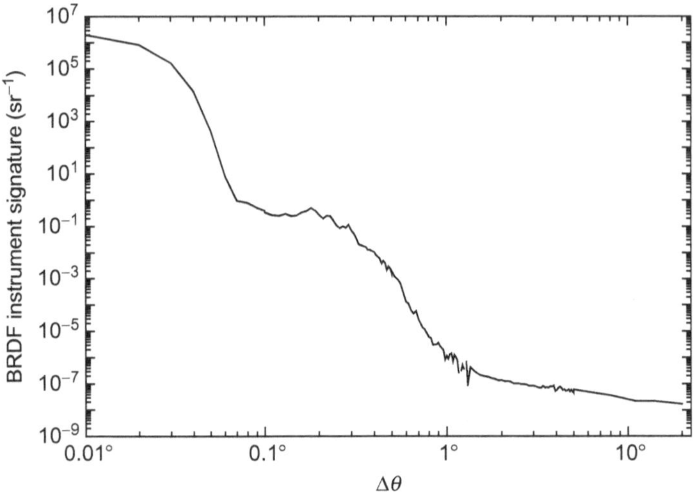

FIGURE 8.4 The BRDF instrument signature measured using a laser-based instrument at NIST.

no longer includes those optics, the signal drops again to another level. This lower level is limited by Rayleigh scatter in the air to an amount which is proportional to the length of beam subtended by the field of view. Reducing the field of view of the receiver will narrow the angle over which the upstream optics contribute to the signal and will reduce the amount of Rayleigh scatter observed. In the under-illuminated geometry, with the sample removed, the receiver will record a BRDF given approximately by [16,17]

$$f_{\mathrm{r}} = I_{\mathrm{Rayleigh}} \frac{l_{\mathrm{FOV}}}{\cos\theta_{\mathrm{r}} |\sin\theta_{\mathrm{ir}}|} \begin{cases} 1 & \text{for s polarization} \\ \cos^2\theta_{\mathrm{ir}} & \text{for p polarization} \end{cases}, \tag{8.11}$$

where

$$I_{\mathrm{Rayleigh}} = \frac{4\pi^2}{\lambda^4} \frac{(n-1)^2}{N}, \tag{8.12}$$

λ is the wavelength of the light, n is the refractive index of the air, N is the volume number density of molecules in the air, θ_{ir} is the angle between the incident and viewing directions, and l_{FOV} is the transverse length of the field of view of the receiver at the sample position. At 20 °C, a wavelength of 532 nm, and at standard atmospheric pressure, $I_{\mathrm{Rayleigh}} = 1.29 \times 10^{-1}\,\mathrm{m}^{-6}\,\mathrm{sr}^{-1}$. Thus, for $l_{\mathrm{FOV}} = 10\,\mathrm{mm}$, the contribution to the BRDF perpendicular to the incident direction will be in the range of $10^{-8}\,\mathrm{sr}^{-1}$. The exact contribution to the BRDF measurement will be affected by the presence of the sample, in that there will be contributions from the directly illuminating beam, the beam reflected from the surface, and the images in the surface of these two beams [16,17].

The instrument signature also serves to quantify the amount of light haloing the incident beam. When the incident radiant flux is measured, there is an assumption that all of the incident light is being captured by the receiver. However, there may be a small fraction of light which is outside that receiver. Integrating the amount of light inside of the solid angle of collection and comparing it to the integral of the outside provides a measure of this potential error.

8.3.2 Noise-Equivalent BRDF

The noise-equivalent BRDF is a measure of the lowest scatter level that an instrument can measure. It is generally measured by shutting off or blocking the light source and measuring the standard deviation of the resulting measured radiant flux Φ_r, then applying Eq. (8.1) or (8.2), as if one were performing a measurement. This value will be an indication of the repeatability when measuring the lowest level scatter signal. For higher level scatterers, the sampling uncertainty may be higher and can be assessed by repeated measurements. In practical applications, either the noise level or Rayleigh scattering from the ambient air surrounding the sample limits the sensitivity of the instrument.

8.3.3 Profile of Illumination Spot

It is important to know what the illumination spot on the sample looks like. There are a number of reasons for this. In the under-illuminated method, the receiver field of view must include all of the light scattered by the sample (including that which is diffused in the sample). Thus, it is important to know its size and to verify that it is smaller than the receiver field of view. More subtly, however, it is important to know that the illumination spot is symmetric and centered at the center of goniometer. At oblique viewing angles, any asymmetry in the illumination will cause an error in the solid angle calculation, because the intensity-weighted mean distance from the sample will not be at the center of the goniometer [15].

Measurement of the illumination spot can be performed with a beam profiler, either one based upon a scanning aperture, a scanning slit, or a camera. If the goniometer includes the automated ability to translate the sample, then it is natural to place a detector with a small aperture in the sample holder to measure the illumination spot. Another way to estimate the profile of the illumination spot is to map the scattering of a small defect on a smooth and generally clean surface at a fixed scatter angle. The measured scatter map is the result of a convolution of the illumination spot profile with the defect geometry. If the defect is much smaller than the spot, the scatter map directly represents the beam profile.

8.3.4 The Field of View

In the case of the under-illumination method, the field of view of the receiver needs to be known to ensure that the sample illumination spot does not extend beyond it and that the sensitivity over that spot is uniform. One way to measure this is to scan a small light source over the sample plane and to measure the signal.

8.3.5 Detector Location Sensitivity

If the detector is moving about the sample, there may be a slight variation in the sensitivity of the detector to its location or orientation. A PMT, for example, can be sensitive to magnetic fields, and if not properly shielded, there may be a variation in its response depending upon how it is oriented with respect to the Earth's magnetic field [18].

8.3.6 Linearity

BRDF measurements, even for white diffuse materials, require a relatively large dynamic range. For example, a modest-sized 25-mm aperture, located 500 mm from a perfectly reflecting Lambertian sample, viewed at 45° will have a scattered-to-incident signal ratio of $\Phi_r/\Phi_i = 4.4 \times 10^{-4}$. It is therefore necessary to ensure the detectors, along with all of their associated electronics, are linear. Often, the gain settings are changed between the measurements of these two signals, so that the relative gain settings should be checked. Furthermore, in the case of lock-in detection, the gains should be checked at the same frequency as the chopping frequency, since changing gain is often associated with changes in amplifier gain-bandwidth ratio.

8.4 GONIOMETER DESIGNS

Key to angle-resolved measurements is having the ability to orient the sample, the source, and the receiver with respect to one another. There are a number of ways that this can be accomplished, and this section reviews some of the more common ones, as well as some innovative methods.

8.4.1 In-Plane Scatterometer Designs

The simplest, and probably most common, method for sample orientation is shown schematically in Fig. 8.5. This method is used when only the scatter in a single plane, usually the plane of incidence, is desired. A simple rotation stage is used to hold the sample, while another one is used to rotate the detector. This particular scheme can even be used in a completely manual "protractor"-driven application (albeit with reduced accuracy and angular precision). Several considerations need to be considered in this arrangement. First, the two rotation stages should have axes that are coaligned with one

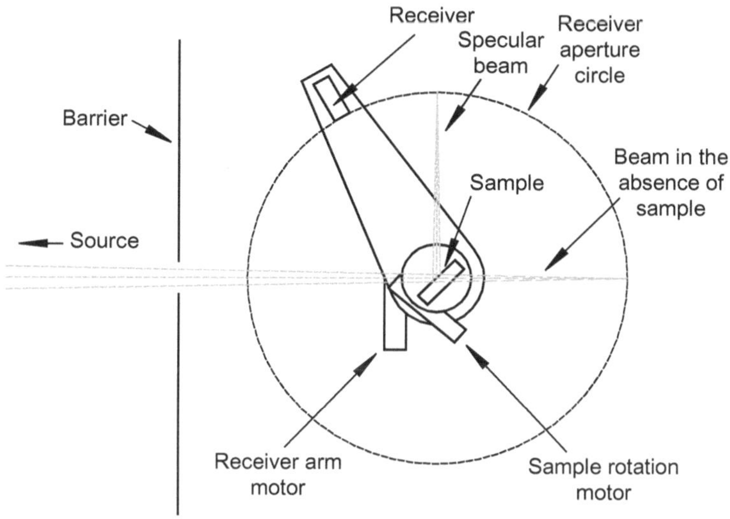

FIGURE 8.5 Schematic of an in-plane scatterometer.

another, so that the detector–sample distance does not vary as the detector moves about the sample and to ensure that the angles are accurate. Second, there should be arrangements made to place the front sample face coincident with these axes.

In-plane scatterometers have been developed for practically all wavelength ranges from the ultraviolet throughout the visible up to the infrared spectral regions [1,19]. Special tools are also available for measurements in the deep and vacuum ultraviolet [20] and even in the extreme ultraviolet (EUV) [21].

8.4.2 Out-of-Plane Scatterometer Designs

Measurements out of the plane of incidence require more complex motions for either the sample, the detector, or the source. In this section, we discuss a number of different schemes that have been used in the past.

In many respects, the simplest way to achieve out-of-plane measurement capability is to replace the sample rotation stage, shown in Fig. 8.5, with a device that rotates and moves the sample into any orientation. In this case, all of the necessary degrees of freedom exist to allow for arbitrary incident and scattering directions. Figure 8.6 shows a schematic of a basic Eulerian cradle [22], so named because the successive stages are stacked so that the rotations follow an Euler transformation. On top of the in-plane design described in Section 8.4.1 is a cradle allowing rotation of the sample about a horizontal axis. Inside of that is a rotation stage that rotates the sample about its surface normal. In addition to the rotation axes, linear translation stages can provide for sampling on different points along the sample surface.

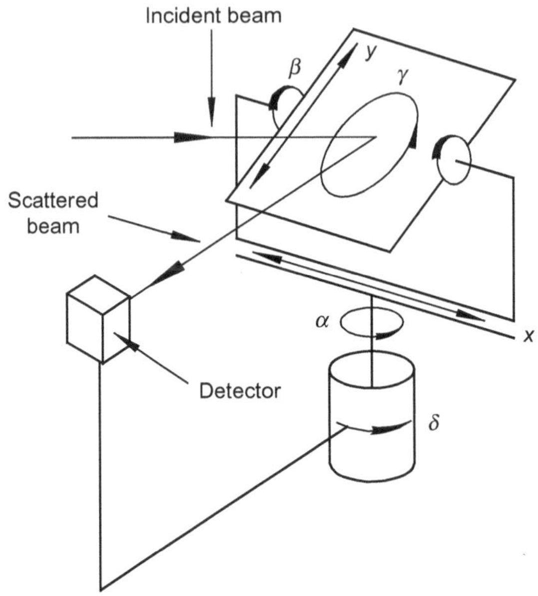

FIGURE 8.6 Schematic of an Eulerian cradle.

Alignment of such a system is straightforward, but more complicated than that of the simpler system described in Section 8.4.1. Ultimately, the vertical axes must both be coincident, the horizontal axis of rotation must pass through and be perpendicular to the vertical axes, and the surface normal axis must be perpendicular and pass through the intersection of the other axes. When aligned, the conversion from the axis angles to sample scatter angles is straightforward. The angles shown in Fig. 8.6 are related to the incident and scattering directions by [22]

$$\begin{aligned}
\delta &= \cos^{-1}[\cos\theta_i \cos\theta_r + (\cos\phi_i \cos\phi_r + \sin\phi_i \sin\phi_r)\sin\theta_i \sin\theta_r],\\
\alpha &= \tan^{-1}[(\cos\theta_r - \cos\theta_i \cos\delta)/(\cos\theta_i \cos\delta)],\\
\beta &= \sin^{-1}[\sin(\phi_r - \phi_i)\sin\theta_r \sin\theta_i/\sin\delta],\\
\gamma &= \phi_i - \tan^{-1}(-\sin\alpha, \cos\alpha\sin\beta),
\end{aligned} \tag{8.13}$$

where $\tan^{-1}(a,b)$ is the two-argument arctangent, which is $\tan^{-1}(b/a)$, taking into account in which quadrant the point (a,b) lies. These expressions lead to intermediate singularities in the case of retroreflection, for which

$$\begin{aligned}
\delta &= 0,\\
\alpha &= \theta_i,\\
\beta &= 0,\\
\gamma &= \phi_i.
\end{aligned} \tag{8.14}$$

There are, in general, two solutions for the sample pose for these goniometers. When one takes the arccosine in Eq. (8.13), one has a choice of sign. Choosing a specific sign (either positive or negative) will keep the detector on one side of the incident beam and will direct the specular direction toward one side of the room. Another option is to choose the sign of δ so that the sign of β is either positive or negative, that is, to choose it based upon the sign of

$$\sin(\phi_r - \phi_i)\sin\theta_r \sin\theta_i \tag{8.15}$$

This choice will keep the sample oriented upward or downward for measurements out of the plane of incidence.

The Eulerian cradle design can be simplified somewhat by not having rotation about the sample normal or by not having the linear translations. However, such a system cannot maintain a constant illumination direction on the sample for out-of-plane geometries.

One concern with the Eulerian cradle is that the specular direction is not kept in a horizontal plane. Specularly reflecting samples can thus create a safety hazard, if the area surrounding the measurement is not appropriately enclosed and interlocked.

A recent development in gonioreflectometer design is based upon the articulated robotic arm [8,11,23]. An example of one is shown in Fig.8.7. This design is very versatile and is particularly useful if large angles of

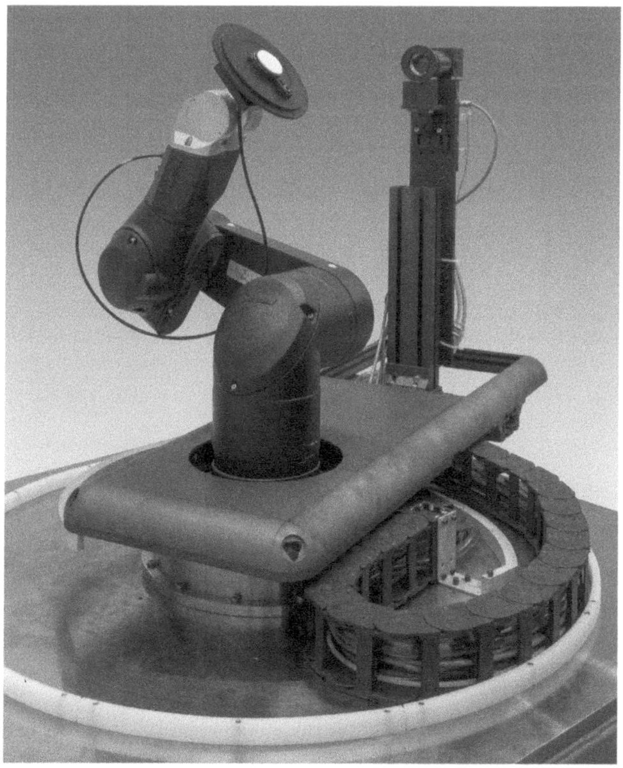

FIGURE 8.7 A goniometric BRDF instrument using a robotic arm.

incidence or viewing are required, because nothing obscures the front of the sample, or if large samples need to be measured. It also has the advantage that the sample can be easily moved out of the incident beam path to allow the receiver to directly view the source. Conceptually, however, the joint angle transformations are very complicated. Software embedded into the robot controllers generally make this problem trivial, since they provide various means for controlling the coordinates and orientation of not just the hand flange, but any stack of objects extending from the hand flange, such as the sample holder, and even the front face of the sample. Because the tools are built for industrial manufacturing applications, they are nearly a commodity item and can be quite cost effective.

Robotic arms suffer slightly in this application from having precision and accuracy that are difficult to assess or quantify. However, the vast majority of BRDF measurements are performed in an "in-plane" geometry, such that only the bottom, or shoulder, axis is the only one rotated, the other axes being used only to position the point of interest on the sample face at that axis.

Other options have been employed that move the detector along multiple axes and move the source. Figure 8.8 shows the True Angle Scatter

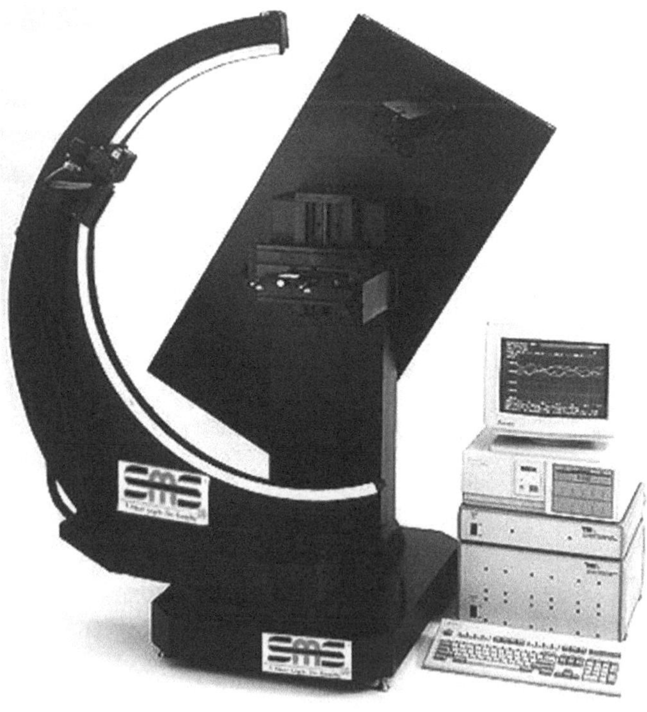

FIGURE 8.8 The TASC scatterometer is capable of reflective and transmissive measurements over the full hemisphere, while keeping the sample horizontal. *Used with permission.*

Coordinate (TASC) system [24], which can measure scatter over the reflective and transmissive hemispheres, while keeping the sample horizontal. The system places all of the source optics on a vertically mounted and rotating rectangular optical breadboard and brings the incident beam out on an arm (toward the viewer in Fig. 8.8) and then down onto the horizontal sample. The incident angle is controlled by rotating the breadboard. One advantage of this system is that the sample remains horizontal, allowing measurements from powders and liquids. The receiver moves along an arc, and that arc is on a rotation stage with a vertical axis, allowing out-of-plane measurements.

Another goniometer design, which holds the sample at a fixed angle while moving the source along one axis and the detector along another two, is shown in Fig. 8.9 [25,26]. This system is designed to be a tabletop system. The source is contained below the chassis, and the beam is directed up an arm, which rotates about the sample. The beam is then directed horizontally toward the sample. The detector assembly rides on a nested pair of arms, whose orthogonal axes both pass through the sample face. A mirror is used to direct the scattered light toward the detector, minimizing its size and allowing light close to the retroreflection direction to be collected. A similar system, where the sample is rotated and the incident laser beam is fixed, but designed to characterize even extremely large samples with diameters of up to 700 mm, can be found in [19].

8.4.3 Multichannel Systems

A serious issue associated with taking full hemispherical measurements with a single detector is speed. Measurements over the entire hemisphere often take several hours. An obvious way to improve the measurement speed is to employ multiple detectors. One of the early ones, called the detector array for laser light angular scatttering, was built at NIST and involved swinging an arc of optical fiber ends over the illuminated sample [27]. Measurement time was fast, but the small collection apertures did not allow measurement of low-scatter surfaces. Other designs using several discrete detectors might solve both the speed and sensitivity issues often associated with sampling the full reflective and transmissive hemispheres.

Another multichannel approach is to employ an array detector. A new generation of compact scatterometers has been developed, in which a camera images a screen illuminated by the scatter pattern. The basic components are shown in Fig. 8.10 in a configuration with a reflective sample and a flat transmissive screen. Scatter from the sample illuminates a screen that is viewed by a digital camera. The advantage to these systems is that there are no moving parts and a large part of the scattering hemisphere can be captured at once. Systems have been built with hemispherical screens that are reflective or transmissive [28,29]. Also, systems can directly image the scattering pattern onto an array detector [30–33]. These systems allow fast measurement

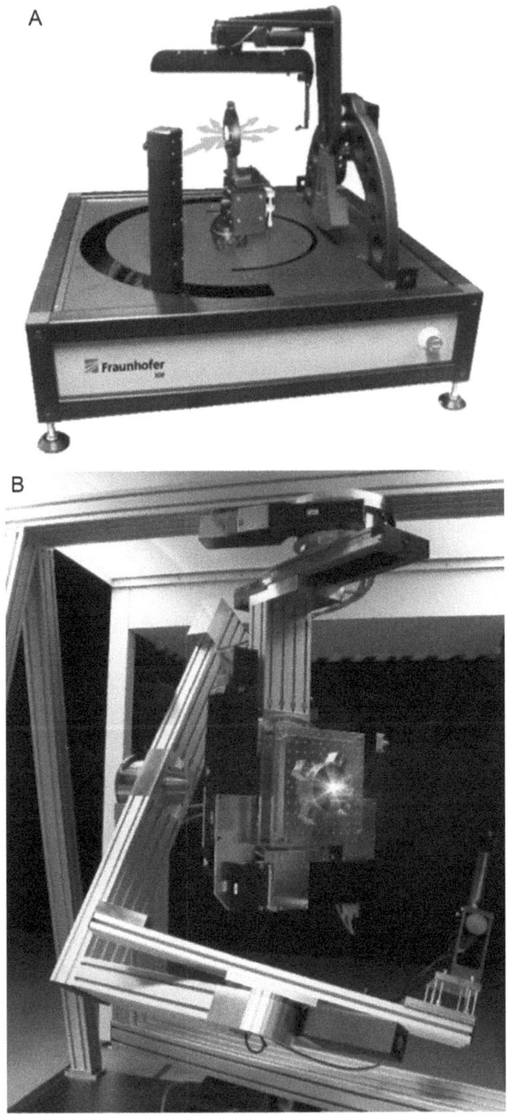

FIGURE 8.9 A tabletop scatterometer with out-of-plane capabilities (a) and a larger system with a similar cradle (b). *Panel (a) is adapted from [25] and panel (b) is adapted from [19].*

for applications where the BRDF levels are not too low and the required dynamic range is not too large. Calibration is often performed using a known white diffuse standard. Dynamic range can be increased to five or six orders of magnitude by taking several images at different exposure times, but the noise-equivalent BRDF is often above $10^{-3}\ \mathrm{sr}^{-1}$, which is well above the values needed for inspecting high-end optical components. Nevertheless, for

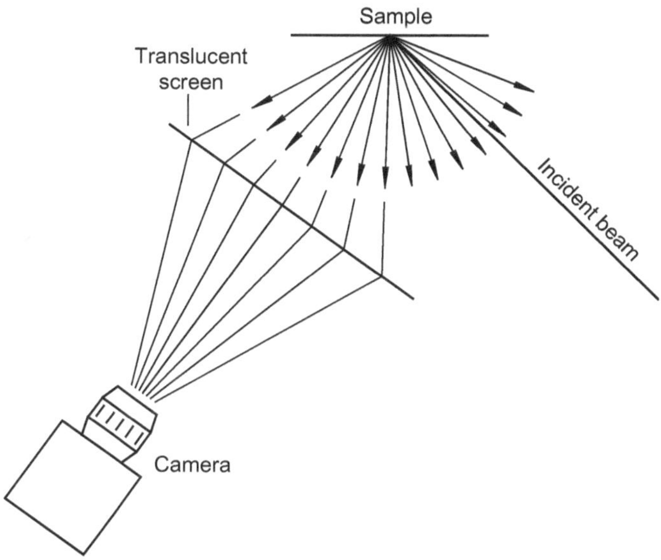

FIGURE 8.10 Schematic of a camera-based light scattering instrument.

diffuse samples, these systems can offer fast, reliable measurements and provide insights that are not apparent from incident plane data.

8.5 UNCERTAINTY ANALYSIS

Considering the measurement equation, Eq. (8.10) (or Eq. 8.9, if the overfilled sample geometry is being used), we can take the standard approach to determine the uncertainty in the measurement. A complete description of such an analysis is given in detail in the *Guide to the Expression of Uncertainty in Measurement* [34], but is also provided in summary in other texts [35,36] and in Chapter 1. In short, we must estimate variances of each contribution to the uncertainty and determine the resulting variance of the measured BRDF.

Particular care must be taken when considering errors in the angle θ_r. Considering the $\cos\theta_r$ term in the denominator of Eq. (8.2), and differentiating with respect to the θ_r to obtain the fraction uncertainty in f_r due to an error $\Delta\theta_r$, we would arrive at

$$\frac{\Delta f_r}{f_r} = \tan\theta_r\,\Delta\theta_r. \tag{8.16}$$

However, we can alternatively consider the effect of angle on the quantity we actually measure, the radiant flux Φ_r, which is given by

$$\Phi_r = f_r \Phi_i \Omega_r \cos\theta_r. \tag{8.17}$$

Differentiating Eq. (8.17) with respect to θ_r, we find that the fractional error in the radiant flux, due to an error $\Delta\theta_r$ is

$$\frac{\Delta\Phi_r}{\Phi_r} = \left(\frac{1}{f_r}\frac{\partial f_r}{\partial\theta_r} - \tan\theta_r\right)\Delta\theta_r. \tag{8.18}$$

Since f_r is proportional to Φ_r in Eq. (8.2), Eq. (8.18) is also the fractional error in f_r due to an error $\Delta\theta_r$. By comparing Eqs. (8.16) and (8.18), we see that in one case the dependence of the BRDF on angle is not included, while in the other, it is. We do not want to include both errors, because we will double count that which results from the cos θ_r dependence of the BRDF. This is, in effect, because including errors in the cos θ_r term in the denominator of Eq. (8.2) is akin to including errors in the abscissa rather than the measurement value itself. In what follows, we will use Eq. (8.18), so that the contribution to the uncertainty in the BRDF, $u(f_r)$, can be found from

$$\frac{u^2(f_r)}{f_r^{\,2}} = \frac{u^2(\Phi_r)}{\Phi_r^{\,2}} + \frac{u^2(\Phi_i)}{\Phi_i^{\,2}} + \frac{4u^2(r)}{r^2} + \frac{4u^2(R)}{R^2} + \left(\frac{1}{f_r}\frac{\partial f_r}{\partial\theta_r} - \tan\theta_r\right)^2 u^2(\theta_r), \tag{8.19}$$

where $u(\Phi_r)$, $u(\Phi_i)$, $u(r)$, $u(R)$, and $u(\theta_r)$ are the estimated standard uncertainties in Φ_r, Φ_i, r, R, and θ_r, respectively, and where $u(\Phi_r)$ encompasses remaining uncertainties that are *not* included in $u(\theta_r)$. Note that we have assumed in Eq. (8.19) that $r \ll R$, which has an inconsequential effect on the error budget for any angle-resolved instrument.

The uncertainty from errors in θ_r can be significant when θ_r is large or when the BRDF is varying rapidly. For diffuse materials ($df_r/d\theta_r \approx 0$) at $\theta_r = 80°$ and $u(\theta_r) = 0.1°$, the relative uncertainty to the BRDF is about 1%. However, for measurements very close to the specular direction, where $df_r/d\theta_r$ is large and the BRDF drops by an order of magnitude every 0.2°, an uncertainty $u(\theta_r) = 0.01°$ can still lead to a relative uncertainty in the BRDF of several percent.

Having properly accounted for errors in θ_r, we now turn our attention to estimating the contributions to uncertainty in the other terms on the right-hand side of Eq. (8.19). In general, each uncertainty component, $u(\Phi_r)$, $u(\Phi_i)$, etc., will include multiple contributions that should be added in quadrature to arrive at the total fractional uncertainty in that component. In the following discussion, we identify common sources of uncertainty and show one method for including them in the total fractional uncertainty in f_r, taking care not to double count any uncertainty source.

A precision aperture should be delivered with an estimate of the uncertainty $u(r)$ in its radius r. For the highest accuracy measurements and for traceability, the aperture should be specifically measured for its area or average radius, and that measurement should be traceable to International System of Units (SI) through national standards. National Metrology Institutes have facilities for this purpose, because of the importance apertures have on the ultimate accuracy of a radiometric measurement [37].

In a similar way, the uncertainty $u(R)$ in the distance between the sample and the aperture needs to be determined. This can be performed with an inside micrometer that has been appropriately calibrated to be traceable to the SI.

Care should be taken that this measurement and its uncertainty be performed over a number of different detector and sample orientations, so that runout error in the rotation stages is included in the error estimate. More subtly, a nonuniform illumination can result in a distance error, and such effects should be included in the error budget [15]. For this reason, the uncertainty in R can be the largest contribution to uncertainty in a high-accuracy scatterometer.

The uncertainty in the incident radiant flux, $u(\Phi_i)$, should contain contributions from linearity [which we could instead assign to $u(\Phi_r)$, but should not assign to both], uniformity of the detection, and sampling statistics. The incident radiant flux and the scattered radiant flux are at least a few orders of magnitude different, so linearity can be a big issue. The linearity can be difficult to assess, but should include not only the linearity of the detector, but all of the electronics, including gain changes that are made at different intensities. As an example, consider a system which has been determined to be linear over every factor of, say, δ to within some amount, say ϵ. This may, for example, be determined by checking the signal at different intensities, with two different gain settings. The total linearity may then scale proportional to the root-sum-of-squares of the number of factors of δ there are between Φ_r and Φ_i, with each factor of δ providing a variance of ε^2. An analytic expression that replicates this behavior may be given by

$$\frac{u(\Phi_i)}{\Phi_i} = \varepsilon\sqrt{\frac{\log(\Phi_i/\Phi_r)}{\log\delta}}. \tag{8.20}$$

Thus, a carefully calibrated system, which has all of its nominal $\times 10$ gain ratios ($\delta = 10$) known to 0.1% ($\varepsilon = 0.1\%$), will have a relative linearity uncertainty of 0.25% when $\Phi_r/\Phi_i = 10^{-6}$.

The uniformity of detection can be estimated by performing several measurements with the incident beam aimed into the receiver at slightly different angles. An instrument signature measurement with the largest aperture should appear flat when the incident beam is contained within the aperture, and any deviations from a constant value should be included as an uncertainty in this contribution. Contributions from sampling statistics (which can also be attributed to source stability) can be simply obtained by repeated measurements.

Having assigned the uncertainty from linearity to the incident power and that with direction to the uncertainty in θ_r, the contributions to the uncertainty in the scattered radiant flux, $u(\Phi_r)$, arise primarily from wavelength errors and sampling statistics. The sampling statistics, of course, can be measured by repeated measurements of the signal. However, if a laser source is used, and no effort is made to reduce its coherence, then speckle will be an issue. The contribution from laser speckle depends upon the projected area viewed by the receiver and the solid angle of collection. The characteristic diameter of a speckle at a distance R from a circular spot of diameter D is approximately [38]

$$d_{sp} \approx \frac{1.2\lambda R}{D}. \tag{8.21}$$

Given the obliquity of the illumination spot and the cross-sectional area of viewing, the area of a characteristic speckle is then approximately given by

$$a_{sp} = \frac{\pi d_{sp}{}^2}{4} \approx \frac{0.89 R^2 \lambda^2 \cos\theta_i}{A_i \cos\theta_r}, \tag{8.22}$$

where A_i is the cross-sectional area of the illumination beam. The collection aperture, having an area A_d, will have a random number of speckles, $N = A_d/a_{sp}$, and the standard deviation of an ensemble of such speckle patterns is

$$\frac{u(\text{speckle})}{\Phi_r} = N^{-1/2} = 0.94 R\lambda \sqrt{\frac{\cos\theta_i}{A_i A_d \cos\theta_r}}. \tag{8.23}$$

This expression assumes that only one measurement is performed. One can alternatively assess this contribution to the uncertainty by performing measurements at different locations on or rotations of the sample. As an example, a sample spot of diameter $D = 1$ mm, a collection aperture of diameter 7 mm, a detector distance of $R = 500$ mm, and a wavelength $\lambda = 632.8$ nm, the relative contribution to the uncertainty from laser speckle will be 5.4%.

When some laser sources are used, the contribution from the uncertainty in the wavelength is negligible. With the replacement of gas lasers with diode lasers, when using tunable lasers, or when using broadband sources with a monochromator, wavelength uncertainties can contribute to errors. Thus, we have a contribution to the uncertainty in the measured radiant power

$$\frac{u(\Phi_r)}{\Phi_r} = \frac{1}{f_r} \frac{\partial f_r}{\partial \lambda} u(\lambda). \tag{8.24}$$

This source of uncertainty can be important for colored materials. More importantly, it should be recognized that the light coming out of a monochromator, or measured by a spectrometer, may not be purely limited to the intended band, but have a significant amount of stray light coming from outside that band. Stray light can thus be a limiting factor to making accurate measurements of scatter from colored samples [9]. If we consider the incident light having broadband out-of-band power $\Phi_{i,\text{out-of-band}}$ and an in-band power $\Phi_{i,\text{in-band}}$, and we make assumptions about the uniformity of the detector spectral response, we can determine an uncertainty

$$\frac{u(f_r)}{f_r} = \frac{\langle f_r \rangle_\lambda}{f_r} \frac{\Phi_{i,\text{out-of-band}}}{\Phi_{i,\text{in-band}}}, \tag{8.25}$$

where $\langle f_r \rangle_\lambda$ is the BRDF averaged over the detector spectral response. For white samples, this contribution to the uncertainty is negligible, but for a typical single monochromator stray light level of $\Phi_{i,\text{out-of-band}}/\Phi_{i,\text{in-band}} = 10^{-3}$

and a colored sample, the relative uncertainty in the spectral region where the sample has a low BRDF may be a few percent.

8.6 NORMALIZATION SCHEMES

The considerations given above for measuring the BRDF are all based upon what is often referred to as the absolute normalization method. That is, a single receiver is used for both the incident power measurement and the scattered power measurement, and the collection solid angle is assumed to be known. In this way, no reference sample is needed. However, such measurements put relatively severe requirements on the mechanics of the instrument. The sample must be taken out of the path of the incident beam, and the detector must be able to swing all the way into the incident beam; the aperture area and distance need to be well known; and the sensor must be linear over a relatively wide dynamic range. For many applications, one can rely upon a reference standard to make the measurement. This can alleviate some or all of these requirements and allow for a much simpler instrument design. In the following sections, we discuss methods that can be used to obtain measurement results that can be nearly as accurate as the available reference standard.

8.6.1 Relative Normalization

Relative measurements can significantly increase the speed and reduce the need for wide dynamic range and linearity. A reference sample of a known BRDF, $f_{r,ref}$, calibrated elsewhere, is placed into the system and the radiant power $\Phi_{r,ref}$ measured in a geometry for which it is calibrated. Rearranging Eq. (8.2), we can infer

$$\Omega_i \Phi_i = \frac{\Phi_{r,ref}}{f_{r,ref} \cos\theta_{r,ref}}, \tag{8.26}$$

where $\theta_{r,ref}$ is the angle of scattering for the reference measurement. The BRDF of the unknown sample is then

$$f_r = f_{r,ref} \frac{\Phi_r}{\Phi_{r,ref}} \frac{\cos\theta_{r,ref}}{\cos\theta_r}. \tag{8.27}$$

Diffuse reflectance standards that are approximately Lambertian with reflectance near unity, based upon pressed or sintered powdered polytetrafluoroethylene (PTFE), are readily available for near-ultraviolet, visible, and short-wave infrared wavelengths [39–43]. Other materials are also available and may be more appropriate in certain applications [42,43]. Ultimately, one would like a reflectance standard that has a certified BRDF that is as close to the value being measured, so that detector linearity is not an issue. Unfortunately, standards in a low BRDF regime are difficult to obtain, in large part because they are easily damaged.

8.6.2 Specular Normalization

Specular normalization is useful in cases where the dynamic range of the system is sufficient to perform the measurement, but the geometry of the goniometer is such that the receiver cannot be placed behind the sample. In this case, a reflectance standard, having a reflectance ρ_{ref} measured separately can be placed in the sample holder and the incident radiant flux $\Phi_i{}'$ measured through reflection. In this case, the incident radiant flux $\Phi_i = \Phi_i{}'/\rho_{ref}$. The measurement should be performed at the same incident angle as the reference measurement was made.

8.7 SPECIAL CONDITIONS OR CONSIDERATIONS

8.7.1 Ultraviolet

Scatter measurements in the ultraviolet spectral range impose several additional challenges compared to measurements in the visible region [44,45]. One is that air is strongly absorbing at wavelengths below 190 nm, which makes operation under purge gas atmosphere or even high-vacuum conditions mandatory. Even if absorption is negligible, the Rayleigh scattering from air molecules substantially influences the instrument signature as becomes evident when looking at the strong wavelength dependence in Eq. (8.12). Therefore, even at 193 nm, high-vacuum is used to achieve highest sensitivity [44]. Another issue is that there are no lasers available for wavelengths shorter than 157 nm. Instead, lamps, plasma light sources, or synchrotron light sources have to be used [21,46]. Finally, since there are no diffuse reflectance standards available for wavelengths below 250 nm, only the calibration method described in Section 8.2 can be used. Another approach is to create an appropriate reference sample that allows determining its BRDF using the energy balance [44].

8.7.2 Infrared

Scatter measurements in the infrared spectral region are usually challenging because thermal noise in the available detector systems critically affects the instrument signature [47,48]. Therefore, thermoelectric cooling or liquid nitrogen cooling methods are necessary, leading to larger detector dimensions. Alignment and calibration are other issues that are more demanding in the infrared than in the visible region.

8.7.3 Polarization

The BRDF is dependent upon the incident polarization state and the scattered light will have a polarization state that depends upon the polarization state of the incident light. As discussed in Section 2.5, the BRDF is most completely described by a Mueller matrix [4], representing the relationship between the incident Stokes vector and the scattered Stokes vector. Measurements of the

Mueller matrix require modulation of the incident polarization state as well as modulation of the measured polarization state [49]. The properties of the Mueller matrix can be used to learn about the mechanisms of light scatter, to discriminate among materials, or to fully characterize a material's impact in an optical system.

While it is beyond the scope of this chapter to describe polarimetric measurements in detail, it is important to bear in mind that a measurement with a single polarization state will very likely not reflect the BRDF one would obtain with unpolarized light. Even diffuse reflectance standards, such as sintered PTFE, can have marked polarization dependence, especially at nonnormal incidence [50,51]. Sometimes, circularly polarized light or light linearly polarized at 45° to the plane of incidence can be used to effectively obtain the result for unpolarized light. However, such assumptions would not be expected to hold anywhere except in the plane of incidence from isotropic samples. Out of the plane of incidence, differences have been predicted and observed [52].

It is therefore considered as the best practice, if one wishes to report an unpolarized result, to perform the measurement with at least two orthogonal incident polarizations, and to average the results. Some attention needs to be taken to ensure that the receiver does not have significant polarization sensitivity. This, however, is usually alleviated by the use of an integrating sphere.

8.7.4 Transmission Measurements

Bidirectional transmittance distribution function (BTDF) measurements can be carried out in a manner very similar to that done with BRDF measurements. Since the sample can have a finite thickness, there arises a question about what distances one should use when evaluating the solid angle and where to place the rotation axes of the incident light and the receiver. Following the definition of the bidirectional scattering-surface reflectance distribution function in Chapter 2, from which we determined the BRDF, we can ascertain that one should evaluate the distances according to the side of the sample the solid angle is defined. For a system with a single axis, this makes it natural to place the viewing face of the sample at the axis. However, due to the thickness of the sample, incident angles may be limited to a small range.

The second concern with BTDF measurements is stray light. With samples that are transmitting, scattered light can become totally internally reflected within the sample, creating guided waves that escape only at the edges of the sample. Inclusion or rejection of these effects may depend upon the application. For these various reasons, the measurement of BTDF is not as well standardized as for BRDF.

8.8 APPLICATIONS

In the following sections, we give some examples of BRDF measurements.

8.8.1 Pressed PTFE

Despite its common use as a diffuse reflectance standard, pressed PTFE is hardly perfectly Lambertian. Figure 8.11 shows the results of measurements of the BRDF of pressed PTFE for three different incident angles, 0° (normal incidence), 45°, and 70°. For normal incidence, the BRDF is relatively flat, but about 10 % higher than the ideal Lambertian value of 0.318 sr^{-1} at normal viewing angles and up to 20 % lower at large angles. As the angle of incidence increases, the BRDF at large forward scattering angles increases, while that in backward scattering angles decreases. By 70° angle of incidence, the scattering in the forward (near-specular) direction is over twice that expected from a Lambertian scatterer. A plausible explanation for this result is that the scattering has contributions from single scattering events that are highly forward scattering. As the angle of incidence increases, the tail of the forward scattering lobe of the volume phase function begins to poke out into the scattering hemisphere, leading to the increase in the scattering [50].

8.8.2 Integration of BRDF

Integrated reflectance can be obtained by integrating the BRDF. For normal incidence and isotropic samples, integration is fairly simple, since measurements along different azimuths are not required. In that case

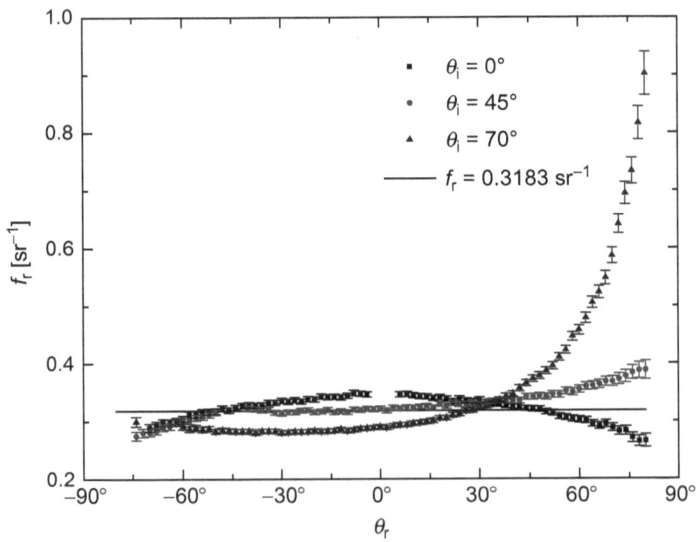

FIGURE 8.11 The unpolarized BRDF measured from pressed PTFE for three incident angles at a wavelength of 405 nm. *From [53].*

$$\rho(0;2\pi)=\int_0^{\pi/2} 2\pi f_r(0;\theta_r)\cos\theta_r\sin\theta_r\,d\theta_r. \tag{8.28}$$

This integral, in practice, must be replaced by a summation over a discrete number of data points. In this case, on could approximate Eq. (8.28) as

$$\rho(0;2\pi)\approx\sum_i 2\pi f_r(0;\theta_{ri})\cos\theta_{ri}\sin\theta_{ri}\,\Delta\theta_r. \tag{8.29}$$

Even for samples that are nearly Lambertian, this summation can have errors that result from the approximation. That is, even with perfect data, there will be errors. For example, the approximation

$$1=2\int_0^{\pi/2}\cos\theta\sin\theta\,d\theta\approx\sum_{i=0}^{10}\frac{\pi}{10}\cos\frac{i\pi}{20}\sin\frac{i\pi}{20}=0.9918. \tag{8.30}$$

It is thus useful to perform the integral in a manner that accounts for the limited accuracy of the numerical method [25]. For example,

$$\begin{aligned}\rho(0;2\pi)&=1-2\int_0^{\pi/2}[1-\pi f_r(0;\theta_r)]\cos\theta_r\sin\theta_r\,d\theta_r,\\&=1-\frac{2\int_0^{\pi/2}[1-\pi f_r(0;\theta_r)]\cos\theta_r\sin\theta_r\,d\theta_r}{2\int_0^{\pi/2}\cos\theta_r\sin\theta_r\,d\theta_r},\\&\approx1-\frac{\sum_i[1-\pi f_r(0;\theta_{ri})]\cos\theta_{ri}\sin\theta_{ri}\,\Delta\theta_r}{\sum_i\cos\theta_{ri}\sin\theta_{ri}\,\Delta\theta_r}.\end{aligned} \tag{8.31}$$

Expressed in this manner, two things are accomplished. For a sample, whose reflectance is close to unity, the contributions from the summation will be small. Second, the numerical errors in the summation are accounted for. Materials that are thought to be isotropic, however, often are not. Even sintered PTFE has been shown to exhibit small amounts of anisotropy [50]. Thus, for the highest accuracy, it is important that the sample be measured along several azimuths.

When the incident light is not directed along the surface normal, it is important that the BRDF be measured for directions both in and out of the plane of incidence. Such measurements become much more time consuming. Consider the scattering directions in cosine space,

$$\begin{aligned}\xi&=\sin\theta_r\cos\phi_r,\\\eta&=\sin\theta_r\sin\phi_r.\end{aligned} \tag{8.32}$$

These directions correspond to projections of the scattering direction onto the sample plane. We then have

$$d\theta_r\, d\phi_r \cos\theta_r \sin\theta_r \rightarrow d\xi\, d\eta, \tag{8.33}$$

and the integrated reflectance becomes

$$\rho(\theta_i; 2\pi) = \iint d\xi\, d\eta f_r(\theta_i; \theta_r, \phi_r). \tag{8.34}$$

The integral can be approximated by the sum

$$\rho(\theta_i; 2\pi) \approx \frac{\pi \sum_i \sum_j f_r\left(\theta_i; \theta_{rij}, \phi_{rij}\right) \Delta\xi\, \Delta\eta}{\sum_i \sum_j \Delta\xi\, \Delta\eta}. \tag{8.35}$$

This may seem like a relatively complicated formula, but its significance is that the reflectance is simply the average

$$\rho(\theta_i; 2\pi) \approx \left\langle \pi f_r\left(\theta_i; \theta_{rij}, \phi_{rij}\right) \right\rangle, \tag{8.36}$$

provided the sampling interval in ξ and η is uniform. Like the procedure above for normal incidence, the summation here serves to reduce errors in the reflectance measurement. Patrick *et al.* [53] used this method to measure the diffuse hemispherical reflectance from pressed PTFE and graphitic samples. Figure 8.12 shows the measurement results for pressed PFTE when the grid spacing was 0.1. The result for the pressed PTFE was very close (within 0.001) of the generally accepted reflectance of this material. One consideration is the fineness of the grid used. For the PTFE, this was not very important, because the BRDF had very little structure. However for a graphitic sample, a change in the grid spacing from 0.1 to 0.05, shown in Fig. 8.13, resulted in the measured reflectance at $\theta_i = 70°$ changing from $\rho(\theta_i; 2\pi) = 0.0100$ to $\rho(\theta_i; 2\pi) = 0.0106$. Given that there is some specularity to the

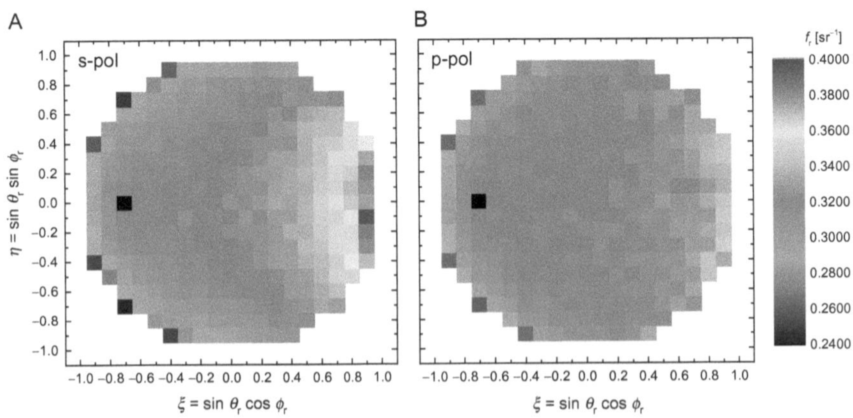

FIGURE 8.12 Hemispherically scanned BRDF from a pressed PTFE sample, $\lambda = 405$ nm and $\theta_i = 45°$ for (a) s-polarization and (b) p-polarization. The data were measured with spacing $\Delta\xi = \Delta\eta = 0.1$.

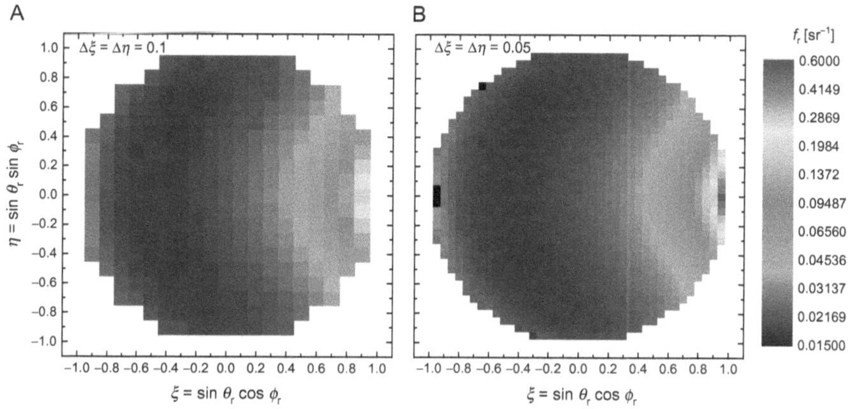

FIGURE 8.13 Hemispherically scanned, unpolarized BRDF measured for a graphitic sample. The results are shown for $\lambda = 658\,\text{nm}$, $\theta_i = 70°$, and measured for (a) 305 points with $\Delta\xi = \Delta\eta = 0.1$ (305 points) and (b) $\Delta\xi = \Delta\eta = 0.05$ (1245 points). The color scale is logarithmic.

graphitic sample, it is not surprising that such an error would exist. A secondary advantage of collecting data in this fashion is that data points known to have issues can be easily removed. In the data shown in Figs. 8.12 and 8.13, points exist where the incident beam is blocked by the receiver or by part of the sample holder assembly. Eliminating these points from the average or replacing their values by the average of adjacent points has little effect on the outcome of the reflectance.

8.8.3 Diffuse Black Paint

Diffuse black paints are often used to control stray light in optical systems. BRDF data are used by a number of stray light analysis packages during the development of commercial, military, and space optical systems. Good diffuse black paints tend to have BRDF levels in the 10^{-2} to 10^{-3} sr^{-1} range. Furthermore, measurements in the plane of incidence are often used to predict scatter over the entire hemisphere, and interpolation between the results obtained at different incident angles is used to predict scatter levels for geometries outside of those measured. Figure 8.14 shows the BRDF for one commercially available flat black paint, measured at 633 nm for a number of incident angles θ_i, spread approximately uniformly in equal increments of $\cos\theta_i$ to observe the behavior at high incident angles. In these measurements, the incident polarization was chosen to be linear at an angle of 45° with respect to the incident plane to simulate an unpolarized source, an assumption that is only valid for data measured in the plane of incidence and for isotropic samples, but useful nonetheless for this geometry. The data show how non-Lambertian this particular paint was. For larger incident angles, this paint exhibits scatter in the forward direction significantly higher than even the

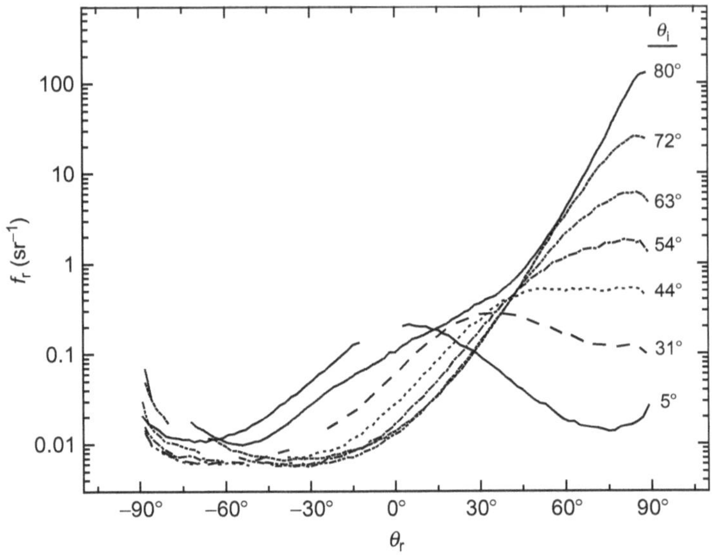

FIGURE 8.14 The BRDF of a commercially available flat black paint measured in the plane of incidence for a number of incident angles, a wavelength of 633 nm, and s-polarization.

pressed PTFE sample, shown in Fig. 8.11 for comparable geometries. These data also show how strong a function the scatter is of θ_i and θ_r, indicating how much data might need to be taken to interpolate or extrapolate the data to properly model the stray light.

8.8.4 Characterization of Surface Roughness

Roughness characterization is important for a wide variety of applications, ranging from precision machining to optical manufacturing. In many cases, stylus profilometry, interferometry, or atomic force microscopy (AFM) provide the necessary information. However, light scattering is particularly effective, since it gathers statistical information over a relatively large area, is noncontact, and is relatively fast. First-order vector perturbation (Rayleigh–Rice) theory, described in Chapter 2, provides a relationship between the power spectral density (PSD) of the roughness, $\langle |Z(\mathbf{f})|^2 \rangle$, and the BRDF [1],

$$f_r = \frac{16\pi^2}{\lambda^4} \cos\theta_i \cos\theta_r |r_{pq}|^2 \langle |Z(\mathbf{f})|^2 \rangle. \tag{8.37}$$

The two-dimensional spatial frequency of the roughness, **f**, is evaluated from the incident and scattering angles and has components

$$f_x = (\sin\theta_\mathrm{i} - \sin\theta_\mathrm{r}\cos\phi_\mathrm{r})/\lambda,$$
$$f_y = -\sin\theta_\mathrm{r}\sin\phi_\mathrm{r}/\lambda. \tag{8.38}$$

(Note that Chapter 2 uses angular spatial frequency difference, $\mathbf{k}_0 - \mathbf{k} = 2\pi\mathbf{f}$.) The reflection coefficients, r_{pq}, depend upon incident (q) and scattered (p) polarization states and are given in Eq. (2.84). The relationship Eq. (8.37) should hold, provided the scatter originates only from surface roughness (i.e., not from volume scattering or particulate contaminants), the amplitude of the surface modulations are much less than the wavelength of the light, and the surface slopes are small.

In our example here, we present work that assesses mirrors used for extreme ultraviolet (EUV) photolithography. Small-angle scattering leads to degradation in contrast and resolution in these imaging systems, needed to perform state-of-the-art microlithography. Therefore, scattering at the EUV wavelengths of 13.5 nm in particular in the near-specular range needs to be evaluated. Measurements at the design wavelength, however, are extremely challenging at small angles. According to the grating equation, this scattering range corresponds to spatial frequencies between 0.1 and 5 μm^{-1}. These spatial frequencies are partly accessible by a combination of white light interferometry (WLI) and AFM. However, these techniques are very time consuming, are extremely sensitive to vibrational noise, and cover only small areas and thus often underestimate the impact of localized defects.

Roughness measurements utilizing light scattering techniques are robust, noncontact, highly sensitive, and can provide information also over larger surface areas. By using longer wavelengths, the roughness components at spatial frequencies relevant for near-angle scatter at 13.5 nm will produce scattering at larger angles that are easier to access. The roughness components relevant for near-angle scattering at 13.5 nm also give rise to scattering at 193 nm, but into a wider range of scatter angles between 1° and 85°. This effectively zooms into the relevant spatial frequency range and relaxes of the near-angle scattering requirements. The technique can be used on reflective substrates before coating or on EUV coatings directly, which are reflective at 193 nm and longer wavelengths.

Angle-resolved light scattering measurements of EUV mirrors consisting of Bragg coatings (60 Mo/Si bilayers) designed for 13.5 nm and deposited onto super-polished fused silica substrates were performed at 193 nm using the instrument described in [44]. Figure 8.15 shows the resulting BRDF, along with the instrument signature. The BRDF is above the instrument signature over most of the angular range.

Figure 8.16 shows the surface PSD determined from the BRDF using Eqs. (8.37) and (8.38). Figure 8.16 includes an estimate of the PSD determined by combining WLI and AFM measurements. There is only one curve shown, which represents a weighted geometric mean of the different measurements [54]. The agreement between the PSD determined by ARS and that

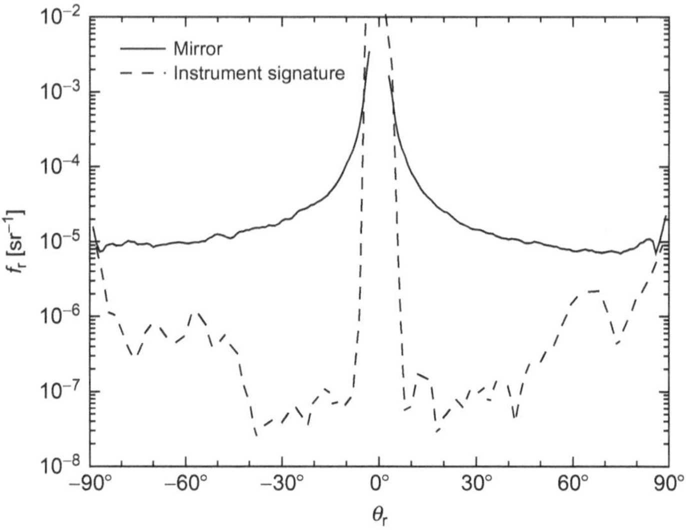

FIGURE 8.15 BRDF from an EUV mirror, measured at normal incidence and 193 nm. Also shown is the instrument signature.

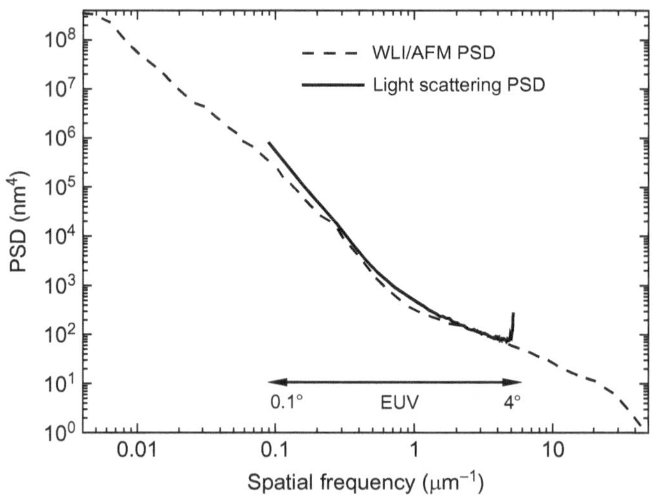

FIGURE 8.16 The surface PSD measured with (dash) profiling methods and (solid) 193-nm BRDF measurements. The region of interest for scattering at 193 nm is shown.

determined by the other roughness techniques is very good. The light scattering results, in fact, cover a regime that is poorly covered by the other two profilometry techniques. This significantly enhances the information in the spatial frequency range relevant for near-angle scattering at 13.5 nm. Moreover, statistical fluctuation effects are automatically suppressed, because the illumination spot diameter of 1 mm for the ARS measurements is equivalent to as many as 10^5 AFM measurement areas.

The rms roughness relevant for near-angle scatter calculated by integrating the PSD is (0.12 ± 0.02) nm. Using a single surface approximation, an integrated near-angle scattering at 13.5 nm of $(0.7 \pm 0.3)\%$ was predicted, which was later found to be virtually identical with the result of an at-wavelength scatter measurement at 13.5 nm using the instrumentation described elsewhere [21]. A further optimization of the procedure allows high spatial frequency roughness to be measured using light scattering even on very large and curved EUV mirror substrates [55].

8.8.5 Scattering Characterization of Anisotropic Surface Structures

New methods in optical manufacturing enable increasingly complex surfaces to be fabricated. Metrology faces increasingly challenging demands to keep up with these developments. For techniques like subaperture polishing, the homogeneity of roughness is an important issue, and mapping surface roughness using light scattering is a superior way to characterize it. Diamond turning is a particularly efficient technology to manufacture metallic mirrors even with large diameters and freeform shapes. Although the process parameters can be optimized to achieve considerably smooth surfaces with roughness in the nanometer range, it is well known that the relative motion of the tool on the surface results in pronounced turning marks leading to diffraction effects in the plane perpendicular to the direction of turning [56,57]. In the past, such surfaces were often considered as one-dimensional rough surfaces, and mechanical profilometry along the direction perpendicular to the marks was used for their characterization. However, such characterization does not provide information about the full impact of the surface on its optical performance.

To illustrate this fact, Fig. 8.17a shows a WLI image of a diamond-turned aluminum surface. Although the surface is optically smooth, turning marks can be clearly observed. Figure 8.17b shows the result of a scatter measurement performed over the entire hemisphere using the instrument described in [19]. The distinct diffraction peaks along the horizontal plane are associated with the periodicity of the turning marks, while the broadening at larger scatter angles is a result of their curvature. Even more interesting, however, is the fact that considerable scattering is observed in directions perpendicular to the diffraction plane. This enhanced scattering is produced by chatter—vibrations of the cutting tool on the surface within the turning grooves. Very often, this effect is underestimated or completely ignored. The scatter measurement, however, reveals that for the given example, the roughness contribution of chatter is actually higher than that of the turning marks.

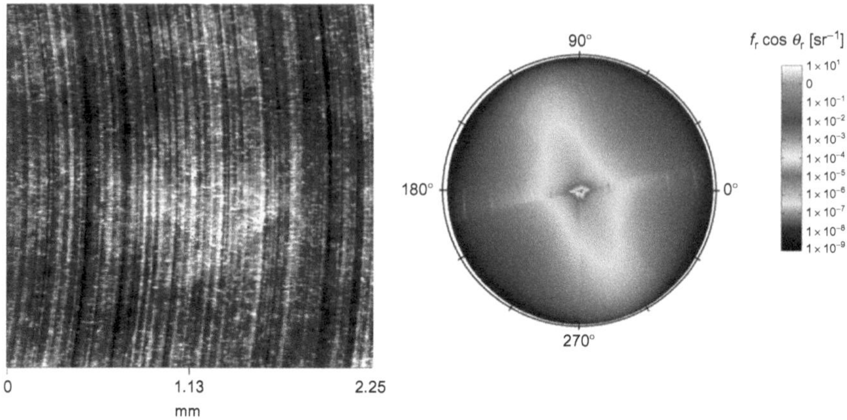

FIGURE 8.17 White light interferometry image (a) and ARS (b) of a diamond-turned surface. The image in panel (a) shows a field of view of 2.25 mm × 2.25 mm and a vertical intensity scale of 32 nm. The scattering results in panel (b) were obtained using 532-nm light, normal incidence, and vertical polarization. *Adapted from [58].*

8.8.6 Predicting Out-of-Plane BRDF from In-Plane BRDF

Three-dimensional ray-tracing simulations are often used to predict the stray light in high-performance optical systems. Besides knowing just the baffling materials, as described in Section 8.8.3, all of the different surfaces encountered by light, both in and out of the intended optical path, need to have BRDF data or models associated with them. These data or models need to cover the scattering hemisphere for a wide range of incident angles. However, as mentioned above, acquiring such data is usually prohibitively time consuming. Having robust methods for interpolating or extrapolating data is therefore useful for substantially reducing the volumes of data needed for stray light estimates.

Harvey first pointed out that the roughness-induced scattering intensity pattern obtained from a sufficiently smooth surface for different incident angles, when graphed as a function of $\sin\theta_r - \sin\theta_i$, tend to lie on a common curve, a process that he called *linear shift invariance* [59,60]. In fact, this invariance is a consequence of the diffractive nature of scattering, albeit there are usually other multiplicative factors that interfere with its universal use. Using the two-dimensional diffraction equations, linear shift invariance can be generalized to state that data, when plotted according to the two parameters

$$\begin{aligned}\beta_x &= \sin\theta_r \cos\phi_r - \sin\theta_i \cos\phi_i \\ \beta_y &= \sin\theta_r \sin\phi_r - \sin\theta_i \sin\phi_i\end{aligned} \tag{8.39}$$

will be invariant, or, if the material is isotropic, according to the magnitude of this vector,

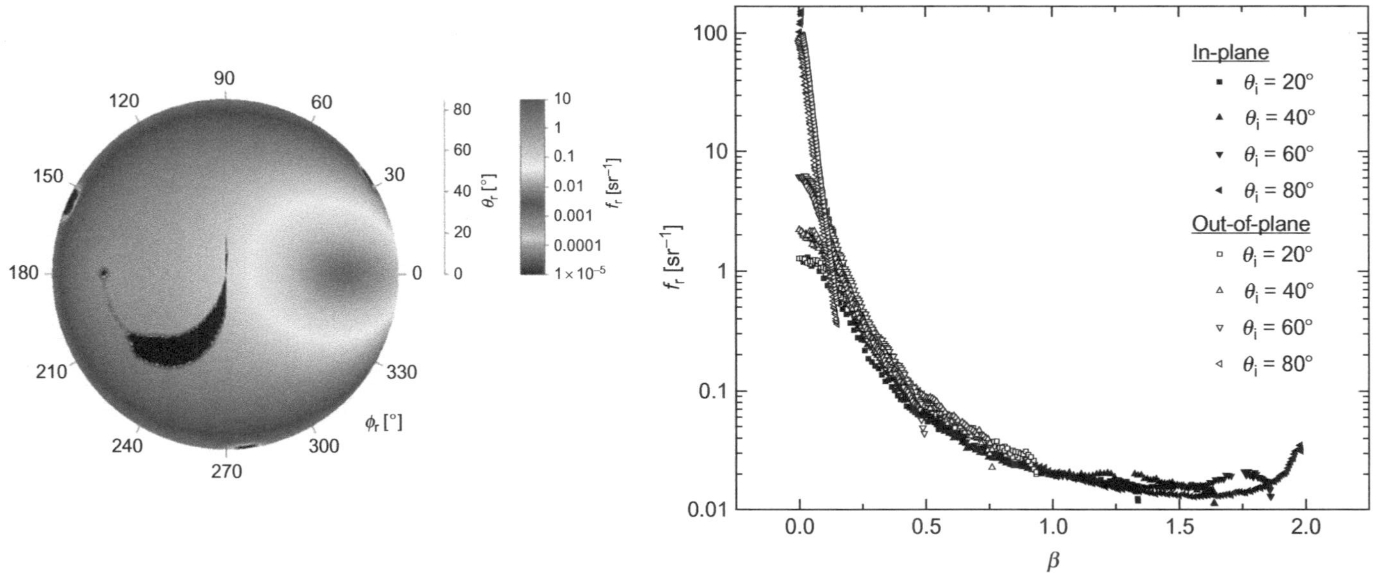

FIGURE 8.18 3D BRDF of rough stainless steel surface measured at 532 nm (a) and in-plane and out-of-plane BRDF data taken out of the 3D data and plotted versus β (b). The dark area in the graph in panel (b) are regions where the receiver or other parts of the apparatus block the incident light. *Adapted from [26].*

$$\beta = \sqrt{\sin^2\theta_i + \sin^2\theta_r - 2\sin\theta_i \sin\theta_r \cos(\phi_r - \phi_i)}. \tag{8.40}$$

Even if the BRDF is not entirely invariant with respect to these variables, if it is locally, then knowledge of that fact significantly reduces the amount of data needed to characterize materials.

More recently, Stover *et al.*, measured the BRDF from a rough (not optically smooth) stainless steel surface over most of the scattering hemisphere and for a number of incident angles at 532 nm [26]. Figure 8.18a shows the BRDF for $\theta_i = 60°$. Figure 8.18b shows horizontal and vertical slices through the data that pass through the specular direction, plotted as a function of β, given by Eq. (8.40). The data for each incident angle lie very close to a single curve, whether the data are taken horizontally (in-plane) or vertically (out-of-plane). Close to the specular direction, near $\beta = 0$, the results differ between incident angles, the sample becoming more specular at large angles of incidence. Because of the higher roughness, the scattering process can no longer be considered to be shift-invariant as discussed in [61,62]. As mentioned above, other aspects of the physics contribute to the full scattering behavior. For example, Eq. (8.37) has a number of factors, including the cosines of the angles and the reflection coefficients, which affect the linear shift invariance. Therefore, careful consideration and comparison with experiments is required when dealing with models and assumptions about the scattering behavior.

REFERENCES

[1] J.C. Stover, Optical Scattering: Measurement and Analysis, SPIE Optical Engineering Press, Bellingham, WA, 2012.

[2] F.E. Nicodemus, J.C. Richmond, J.J. Hsia, I.W. Ginsberg, T. Limperis, Geometrical Considerations and Nomenclature for Reflectance, National Bureau of Standards, Gaithersburg, MD, 1977.

[3] F.E. Nicodemus, Reflectance nomenclature and directional reflectance and emissivity, Appl. Opt. 9 (1970) 1474–1475.

[4] D.S. Flynn, C. Alexander, Polarized surface scattering expressed in terms of a bidirectional reflectance distribution function matrix, Opt. Eng. 34 (1995) 1646–1650.

[5] ILV: International Lighting Vocabulary, CIE, Vienna, Austria, 2011.

[6] E2387—Standard Practice for Goniometric Optical Scatter Measurements, ASTM International, West Conshohocken, PA, 2011.

[7] ME1392—Guide for Angle Resolved Optical Scatter Measurements on Specular or Diffuse Surfaces, SEMI International, San Jose, CA, 2011.

[8] R. Baribeau, W.S. Neil, É. Côté, Development of a robot-based gonioreflectometer for spectral BRDF measurement, J. Mod. Opt. 56 (2009) 1497–1503.

[9] P.Y. Barnes, E.A. Early, A.C. Parr, Spectral Reflectance, National Institute of Standards and Technology, Gaithersburg, MD, 1998.

[10] W. Erb, Computer-controlled gonioreflectometer for the measurement of spectral reflection characteristics, Appl. Opt. 19 (1980) 3789–3794.

[11] D. Hünerhoff, U. Grusemann, A. Höpe, New robot-based gonioreflectometer for measuring spectral diffuse reflection, Metrologia 43 (2006) S11–S16.

[12] R. Baribeau, J.C. Zwinkels, Comparison of NRC goniometric and integrating sphere methods for realizing an absolute diffuse reflectance scale, Proc. SPIE 8495 (2012) 84950A-1-10.

[13] Y. Zong, S.W. Brown, B.C. Johnson, K.R. Lykke, Y. Ohno, Simple spectral stray light correction method for array spectroradiometers, Appl. Opt. 45 (2006) 1111–1119.

[14] S. Schröder, D. Unglaub, M. Trost, A. Duparré, Spectral angle resolved scattering of thin film coatings, Appl. Opt. 53 (2014) A35–A41.

[15] H.J. Patrick, C.J. Zarobila, T.A. Germer, V.A. Ying, C.A. Cooksey, B.K. Tsai, Tunable supercontinuum fiber laser source for BRDF measurements in the STARR II gonioreflectometer, Proc. SPIE 8495 (2012) 84950K-1-13.

[16] C. Asmail, J. Hsia, A. Parr, J. Hoeft, Rayleigh scattering limits for low-level bidirectional reflectance distribution function measurements, Appl. Opt. 33 (1994) 6084–6091.

[17] T.A. Germer, C.C. Asmail, A goniometric optical scatter instrument for bidirectional reflectance distribution function measurements with out-of-plane and polarimetry capabilities, in: Z.-H. Gu, A.A. Maradudin (Eds.), Proceedings of SPIE, 1997, pp. 220–231.

[18] K. Saito, S. Hata, Effect of the Earth's magnetic field on photomultiplier photometry in astronomy, Publ. Astron. Soc. Jpn. 7 (1955) 1–8.

[19] S. Schröder, T. Herffurth, H. Blaschke, A. Duparré, Angle-resolved scattering: an effective method for characterizing thin-film coatings, Appl. Opt. 50 (2011) C164–C171.

[20] S. Schröder, A. Duparré, A. Tünnermann, Roughness evolution and scatter losses of multilayers for 193 nm optics, Appl. Opt. 47 (2008) C88–C97.

[21] S. Schröder, T. Herffurth, M. Trost, A. Duparré, Angle-resolved scattering and reflectance of extreme-ultraviolet multilayer coatings: measurement and analysis, Appl. Opt. 49 (2010) 1503–1512.

[22] T.A. Germer, C.C. Asmail, Goniometric optical scatter instrument for out-of-plane ellipsometry measurements, Rev. Sci. Instrum. 70 (1999) 3688–3695.

[23] A.M. Rabal, A. Rerrero, J.L. Fontecha, A. Pons, J. Campos, A. Corróns, Gonio-espectrofotómetro para medir la función de distribución bidireccional de dispersión (BSDF), Opt. Pura. Appl. 44 (2011) 127–137.

[24] F.M. Cady, J.C. Stover, D.R. Bjork, M.L. Bernt, M.W. Knighton, D.J. Wilson, D. R. Cheever, Design review of a multiwavelength, three-dimensional scatterometer, Proc. SPIE 1331 (1990) 201–208.

[25] A. von Finck, M. Hauptvogel, A. Duparré, Instrument for close-to-process light scatter measurements of thin film coatings and substrates, Appl. Opt. 50 (2011) C321–C328.

[26] J.C. Stover, S. Schröder, A. von Finck, A. Duparré, Estimating hemispherical scatter from incident plane measurements of isotropic samples, Proc. SPIE 8495 (2012), 84950F.

[27] T.V. Vorburger, E.C. Teague, F.E. Scire, M.J. McLay, D.E. Gilsinn, Surface roughness studies with DALLAS—detector array for laser light angular scatttering, J. Res. Natl. Inst. Stand. Technol. 89 (1984) 3–17.

[28] J.C. Stover, E.L. Hegstrom, Scatter metrology of photovoltaic textured surfaces, Proc. SPIE 7771 (2010), 777109.

[29] S. Wadman, S. Baumer, Appearance characterization by a scatterometer employing a hemispherical screen, Proc. SPIE 5189 (2003) 163–173.

[30] T. Herffurth, S. Schröder, M. Trost, A. Duparré, A. Tünnermann, Comprehensive nanostructure and defect analysis using a simple 3D light-scatter sensor, Appl. Opt. 52 (2013) 3279–3287.

[31] A. Kasper, H. Rothe, Evaluation of in-situ ARS sensors for characterizing smooth and rough surfaces, Proc. SPIE 3426 (1998) 252–261.

[32] E. Kawate, M. Hain, New scatterometer for spatial distribution measurements of light scattering from materials, Meas. Sci. Rev. 12 (2012) 56–61.

[33] P. Mattison, M. Dombrowski, J. Lorenz, K. Davis, H. Mann, P. Johnson, B. Foos, The handheld directional reflectometer: an angular imaging device to measure BRDF and HDR in real-time, Proc. SPIE 3426 (2013) 240–251.

[34] ISO Guide to the Expression of Uncertainty in Measurement, ISO, Geneva, Switzerland, 1995.

[35] P.R. Bevington, D.K. Robinson, Data Reduction and Error Analysis for the Physical Sciences, McGraw-Hill, New York, NY, 1992.

[36] B.N. Taylor, C.E. Kuyatt, Guidelines for Evaluating and Expressing the Uncertainty of NIST Measurement Results, National Institute of Standards and Technology, Gaithersburg, MD, 1994.

[37] J. Fowler, M. Litorja, Geometric area measurements of circular apertures for radiometry at NIST, Metrologia 40 (2003) S9–S12.

[38] A.E. Ennos, Speckle Interferometry, in: J.C. Dainty (Ed.), Laser Speckle and Related Phenomena, Springer-Verlag, Berlin, Germany, 1984, p. 207.

[39] V.R. Weidner, J.J. Hsia, Reflection properties of pressed polytetrafluoroethylene powder, J. Opt. Soc. Am. A 71 (1981) 856–861.

[40] V.R. Weidner, J.J. Hsia, B. Adams, Laboratory intercomparison study of pressed polytetrafluoroethylene powder reflectance standards, Appl. Opt. 24 (1985) 2225–2230.

[41] Certain commercial equipment, instruments, or materials are identified in this chapter in order to specify the experimental procedure adequately. Such identification is not intended to imply recommendation or endorsement by the National Institute of Standards and Technology, The Scatter Works, or by the Fraunhofer Institutes, nor is it intended to imply that the materials or equipment identified are necessarily the best available for the purpose.

[42] Spectralon®, from LabSphere, Inc., North Sutton, New Hampshire, USA.

[43] Fluorilon-99 W™, Avian Technologies, LLC, Sunapee, New Hampshire, USA.

[44] S. Schröder, S. Gliech, A. Duparré, Measurement system to determine the total and angle-resolved light scattering of optical components in the deep-ultraviolet and vacuum-ultraviolet spectral regions, Appl. Opt. 44 (2005) 6093–6107.

[45] S. Gliech, J. Steinert, A. Duparré, Light-scattering measurements of optical thin-film components at 157 and 193 nm, Appl. Opt. 41 (2002) 3224–3235.

[46] D.G. Stearns, D.P. Gaines, D.W. Sweeney, E.M. Gullickson, Nonspecular X-ray scattering in a multilayer-coated imaging system, J. Appl. Phys. 84 (1998) 1003–1028.

[47] L. Brooks, W. Wolfe, Microprocessor-based instrumentation for bidirectional reflectance distribution function (BRDF) measurements from visible to far infrared (FIR), SPIE Proc. 257 (1980) 182.

[48] J. Zeng, L. Hanssen, Development of an infrared optical scattering instrument from 1 μm to 5 μm, Proc. SPIE 7453 (2009), 74530Q.

[49] R.A. Chipman, Polarimetry, in: second ed., Handbook of Optics, vol. II, McGraw-Hill, New York, NY, 1995.

[50] T.A. Germer, H.J. Patrick, Mueller matrix bidirectional reflectance distribution function measurements and modeling of diffuse reflectance standards, Proc. SPIE 8160 (2011), 81600D-1-10.

[51] H.J. Patrick, L.M. Hanssen, J. Zeng, T.A. Germer, BRDF measurements of graphite used in high-temperature fixed point blackbody radiators: a multi-angle study at 405 nm and 658 nm, Metrologia 49 (2012) S81–S92.

[52] T.A. Germer, Large-angle in-plane light scattering from rough surfaces: comment, Appl. Opt. 40 (2001) 5708–5710.

[53] H.J. Patrick, L.M. Hanssen, J. Zeng, T.A. Germer, BRDF measurements of graphite used in high-temperature fixed point balckbody radiators: a multi-angle study at 405 nm and 658 nm, Metrologia 49 (2012) S81–S92.
[54] A. Duparré, J. Ferre-Borrull, S. Gliech, G. Notni, J. Steinert, J.M. Bennett, Surface characterization techniques for determining the root-mean-square roughness and power spectral densities of optical components, Appl. Opt. 41 (2002) 154–171.
[55] M. Trost, S. Schröder, T. Feigl, A. Duparré, A. Tünnermann, Influence of the substrate finish and thin film roughness on the optical performance of Mo/Si multilayers, Appl. Opt. 50 (2011) C148–C153.
[56] E.L. Church, H.A. Jenkinson, J.M. Zavada, Measurement of the finish of diamond-turned metal surfaces by differential light scattering, Opt. Eng. 16 (1977) 360.
[57] J.C. Stover, Light scattering metrology of diamond turned optics, Proc. SPIE 5878 (2005) 214–220.
[58] S. Schröder, A. Duparré, Finish assessment of complex surfaces by advanced light scattering techniques, Proc. SPIE 7102 (2008), 71020F.
[59] J.E. Harvey, Light-scattering characteristics of optical surfaces, SPIE Proc. 107 (1977) 41.
[60] J.E. Harvey, Surface scatter phenomena: a linear, shift-invariant process, Proc. SPIE 1165 (1989) 87–99.
[61] A. Krywonos, J.E. Harvey, N. Choi, Linear systems formulation of scattering theory for rough surfaces with arbitrary incident and scattering angles, J. Opt. Soc. Am. A 28 (2011) 1121–1138.
[62] S. Schröder, A. Duparré, L. Coriand, A. Tünnermann, D. Penalyer, J. Harvey, Modeling of light scattering in different regimes of surface roughness, Opt. Express 19 (2011) 9820–9835.

Chapter 9

Spectral Emissivity Measurements

Hiromichi Watanabe*, Juntaro Ishii*, Hidenobu Wakabayashi†, Tomoyuki Kumano‡ and Leonard Hanssen§
*National Metrology Institute of Japan, AIST, Tsukuba, Japan
†Kyoto University, Kyoto, Japan
‡Kobe City College of Technology, Kobe, Japan
§National Institute of Standards and Technology, Gaithersburg, Maryland, USA

Chapter Outline

9.1 Introduction 334
9.2 Measurement Methods 336
9.3 Spectral Emissivity Measurements Near-Ambient Temperature 338
9.3.1 Measurement Apparatus Based on the Blackbody Comparison Method 338
9.3.2 Calibration of the System Response 340
9.3.3 Correction for Reflected Radiance of the Surroundings 342
9.3.4 Traceability and Uncertainty of Spectral Emissivity Measurements 342
9.4 Spectral Emissivity and Reflectance Measurements of Oxidized Metals 345
9.4.1 A Broad-Spectral-Range High-Speed Spectrophotometer System 345
9.5 Spectral Emissivity Measurements of Molten Metals at High Temperatures 349
9.5.1 Importance of Molten Metal Emissivity Data 349
9.5.2 Spectral Emissivity Measurements Based on Containerless Techniques 351
9.6 Spectral Emissivity Measurements of Ceramics 357
References 362

Experimental Methods in the Physical Sciences, Vol. 46. http://dx.doi.org/10.1016/B978-0-12-386022-4.00009-1

9.1 INTRODUCTION

The thermal radiation properties of materials are important for modeling the radiative heat transfer between bodies and for radiation thermometry. In almost all cases, real materials cannot be treated as blackbody radiators. That is to say, the spectral radiance emitted by a real material is less than that emitted by an ideal blackbody radiator at the same temperature. The difference in thermal radiation from that of an ideal blackbody is expressed by the spectral emissivity, which is defined as the ratio of the spectral radiance emitted by the material to one emitted by an ideal blackbody radiator at the same temperature for the same optical and geometrical conditions. To distinguish between ideal and real blackbodies in this chapter, we use the terms "ideal blackbody" when we are discussing a Planckian radiator, and simply "blackbody" when discussing real, nonideal ones. As described in Chapter 2, by Kirchhoff's law the spectral emissivity of a material is equal to its spectral absorptance under thermal equilibrium conditions. Therefore, the emissivity can also be derived from two other optical properties, that is, reflectance and transmittance, because the sum of the three optical properties must be unity. Since the spectral emissivity is directly related to the optical properties of a material, it also depends on the direction of the emission from the surface and the temperature of the material as well as the physical and chemical conditions of the surface, that is, the degree of the surface roughness and contamination. The terms relating to the angular characteristics, "directional," "normal," and "hemispherical," apply to emissivity as well as to the other radiometric quantities. It is to be noted that the term, "total," referring to emissivity, represents "all wavelengths" in contrast with the term, "spectral."

The normal spectral emissivity of materials is of considerable interest for both practical and scientific reasons. In many industrial and scientific fields, radiation thermometry is applied to carry out noncontact temperature measurements of various materials. The basis of radiation thermometry is derived from Planck's law,

$$L_{\lambda,\mathrm{b}}(\lambda,T) = c_1\lambda^{-5}\left[\exp\left(\frac{c_2}{\lambda T}\right) - 1\right]^{-1}, \tag{9.1}$$

where $L_{\lambda,\mathrm{b}}(\lambda,T)$ is the spectral radiance, $c_1 = 2hc_0^2$ and $c_2 = hc_0/k_\mathrm{B}$ are the first and second radiation constants, respectively, λ is the wavelength, c_0 is the speed of light, and k_B is the Boltzmann constant. Equation (9.1) is introduced in Chapter 2, but in slightly different form. For real materials, the voltage signal measured by radiation thermometry is given by

$$V(\lambda,T_\mathrm{a}) = \Re(\lambda)\lambda^{-5}\left[\exp\left(\frac{c_2}{\lambda T_\mathrm{a}}\right) - 1\right]^{-1} = \Re(\lambda)\varepsilon(\lambda,T)\lambda^{-5}\left[\exp\left(\frac{c_2}{\lambda T}\right) - 1\right]^{-1}, \tag{9.2}$$

where $V(\lambda,T_\mathrm{a})$ is the voltage signal corresponding to the radiance temperature T_a, $\varepsilon(\lambda,T)$ is the normal spectral emissivity, a function of λ and T, and $\Re(\lambda)$ is the composite spectral sensitivity of the radiation thermometer times c_1. When $\Re(\lambda)$ is previously determined by the calibration process, the temperature of a material can be determined from Eq. (9.2) if $\varepsilon(\lambda,T)$ of the material is known. The accuracy of radiation thermometry is dependent on the accuracy of the input data of $\varepsilon(\lambda,T)$. This has led to numerous experimental and theoretical studies to estimate $\varepsilon(\lambda,T)$ of various materials, especially at high temperatures where normal contact thermometers cannot be used. To increase the efficiency of high-temperature industrial processes such as steel making and casting, strict process design, and control should be carried out based on mathematical modeling for the various steps. Heat transfer modeling is a crucial tool for optimizing the thermal design and process parameters. The radiation heat transfer is given by

$$\Phi_{\mathrm{rad}} = A\varepsilon_{\mathrm{ht}}\sigma_{\mathrm{SB}}\left(T^4 - T_0^4\right), \tag{9.3}$$

where Φ_{rad} is the radiant power of a real object, σ_{SB} is the Stefan-Boltzmann constant, $\varepsilon_{\mathrm{ht}}$ is the hemispherical total emissivity of the object, A is the radiating surface area of the object, and T_0 is the temperature of the object's surroundings. At high temperatures, heat is mainly transferred by two mechanisms, that is, lattice conduction and thermal radiation. Therefore, data of the $\varepsilon_{\mathrm{ht}}$ for various materials, as well as the thermal conductivity, are required for evaluating heat transfer based on computer simulations. The hemispherical total emissivity of solid metals at high temperatures can be generally derived by a direct heating calorimetric method [1,2]. However, it is difficult to measure $\varepsilon_{\mathrm{ht}}$ of insulating or ambient-temperature materials by the calorimetric method. $\varepsilon_{\mathrm{ht}}(T)$ can also be determined by

$$\varepsilon_{\mathrm{ht}}(T) = 1/\pi \int_0^{2\pi}\int_0^{\pi/2} \varepsilon_{\mathrm{t}}(\theta,\phi;T)\sin\theta\cos\theta\,\mathrm{d}\theta\,\mathrm{d}\phi, \tag{9.4}$$

where $\varepsilon_{\mathrm{t}}(\theta,\phi;T)$ is the directional total emissivity, θ is the polar angle with respect to the normal to the object surface, and ϕ is the azimuth angle. However, it is also difficult to measure $\varepsilon_{\mathrm{t}}(\theta,\phi;T)$ of materials at high temperatures, which are often surrounded by heat sources such as carbon heaters. A secondary and alternative way to estimate $\varepsilon_{\mathrm{ht}}(T)$ is to measure $\varepsilon(\lambda,T)$ over a broad wavelength range. The normal total emissivity $\varepsilon_{\mathrm{nt}}(T)$ is derived from the measured spectrum of $\varepsilon(\lambda,T)$ as follows:

$$\varepsilon_{\mathrm{nt}}(T) = \frac{\int_0^{\infty} \varepsilon(\lambda,T)\lambda^{-5}\left[\exp(c_2/\lambda T) - 1\right]^{-1}\mathrm{d}\lambda}{\int_0^{\infty} \lambda^{-5}\left[\exp(c_2/\lambda T) - 1\right]^{-1}\mathrm{d}\lambda}. \tag{9.5}$$

According to Wien's displacement law, the peak wavelength $\lambda_{\max}$ of the thermal radiation spectrum of the blackbody at temperature T is given by

$$\lambda_{\max} T = 2897.8\,\mu\text{m}\,\text{K}. \tag{9.6}$$

Therefore, $\varepsilon_{nt}(T)$ can be roughly evaluated from the spectrum of $\varepsilon(\lambda,T)$ measured over a wavelength range, which includes at least $\lambda_{\max}$. On the assumption that the directional dependence of the emissivity is weak, $\varepsilon_{ht}(T)$ can be calculated as the product of $\varepsilon_{nt}(T)$ and a conversion constant derived approximately [3].

The comparison of experimental results with theoretical estimations of emissivity is a robust way to validate the experimental results and vice versa. It is well recognized that the wavelength dependence of the emissivity of metals can be qualitatively explained by the simple Drude model, which assumes that all the electrons in metals are free. This classical model predicts a weak negative wavelength dependence of the emissivity, as shown by the Hagen-Rubens relation:

$$\varepsilon(\lambda, T) = 4\left(\pi c_0 \varepsilon_0 \frac{\rho_e}{\lambda}\right)^{1/2} = 0.365\left(\frac{\rho_e}{\lambda}\right)^{1/2}, \tag{9.7}$$

where ρ_e is the electrical resistivity, c_0 and ε_0 are the speed of light and the dielectric constant in vacuum, respectively. This relation corresponds to the simplest case of the simple Drude model of $\varepsilon(\lambda,T)$ under the condition that λ is sufficiently long (far-infrared wavelengths). The simple Drude model cannot explain the wavelength dependence of the emissivity of insulating materials, because bound electrons and phonons contribute to the thermal emission mechanism of insulators. In addition, the spectral emissivity of metals at ultra-violet wavelengths often deviates from the simple Drude model, because interband transitions affect the optical properties of metals at shorter wavelength range.

9.2 MEASUREMENT METHODS

The normal spectral emissivity of materials that are practically opaque in the wavelength range of interest is generally determined by one of two methods: (a) comparison between spectral radiation intensities measured from the sample and a blackbody by a radiation thermometer or spectrometer or (b) by measurements of spectral reflectance $\rho(\lambda,T)$, because the spectral emissivity $\varepsilon(\lambda,T)$ of an opaque material can be calculated as $1-\rho(\lambda,T)$.

In the first method using a radiation thermometer, the normal spectral emissivity of materials at the operating wavelength of the radiation thermometer is given by

$$\varepsilon(\lambda, T) = \left[\exp\left(\frac{c_2}{\lambda T}\right) - 1\right] \Big/ \left[\exp\left(\frac{c_2}{\lambda T_a}\right) - 1\right], \tag{9.8}$$

where λ is the operating wavelength of the radiation thermometer. This does not directly provide the full spectrum of $\varepsilon(\lambda,T)$. However, the additional use of a spectrometer can yield the wavelength dependence of the emissivity. In this case, the normal spectral emissivity is given by

$$\varepsilon(\lambda,T)=M(\lambda)V(\lambda,T)\left[\exp\left(\frac{c_2}{\lambda T}\right)-1\right], \tag{9.9}$$

where $V(\lambda,T)$ is the voltage signal corresponding to the spectral radiation from the sample measured by the spectrometer at wavelength λ and $M(\lambda)$ is the instrument coefficient for the spectrometer as a function of wavelength. In general, $M(\lambda)$ is experimentally determined by measuring blackbody radiation at various temperatures. The linearity of the signal as a function of the measured radiance, that is, the invariance of $M(\lambda)$ with the signal level, has to be confirmed to obtain accurate results. To achieve an accurate spectral emissivity measurement based on the first method, the following three requirements should be fulfilled: (a) the isothermal condition: a sample surface has the same temperature as the reference blackbody, (b) the freely radiating condition: thermal radiation emitted by the surface should be separated from background radiation, reflected radiation, and nonthermal emission, and (c) the blackbody condition: the effective emissivity of the reference blackbody radiator should be sufficiently close to unity for the wavelengths of interest.

At ambient or lower temperatures, it is more difficult to obtain the spectral radiance emitted with wavelengths below 3 μm due to low intensity levels, which can be several orders of magnitude less than radiation at higher temperature. In such circumstances, the second method becomes an attractive alternative, because the signal-to-noise ratio for the reflection measurement can be improved simply by increasing or modulating the probe light intensity. The spectral reflectance can be measured in either a relative mode, comparing the sample to a reference material whose spectral reflectance is known, or in an absolute mode (see Chapter 6 or [4]) independent of a reference. In either case, the reflectance quantity to be measured should be appropriate to the emissivity quantity desired. For example, the normal emittance ε_N is derived from the normal-incident hemispherical reflectance ρ_{NH}, as $\varepsilon_N = 1-\rho_{NH}$, where the subscripts N and H refer to normal and hemispherical, respectively. Since the normal spectral emissivity is often required, ρ_{NH} is one of the most important quantities to be measured. An alternative to the direct measurement of ρ_{NH} [5,6] is to perform the measurement of the hemispherical-normal reflectance factor, R_{HN}. Since the Helmholtz reciprocity law [7] holds that

$$\rho_{NH}=R_{HN}. \tag{9.10}$$

ε_N can be obtained by a measurement of R_{HN}. In addition, if both the specular reflectance ρ_{spec} at near-normal incidence, which is the specular reflection component of ρ_{NH}, and $\rho_{NH}(=R_{HN})$ are measured simultaneously, the diffuse-reflectance ρ_{diff} at normal incidence, which is the diffuse reflection component of the hemispherical reflectance ρ_{NH}, can be calculated by

$$\rho_{\mathrm{diff}} = \rho_{\mathrm{NH}} - \rho_{\mathrm{spec}}. \tag{9.11}$$

For practical purposes, the normal quantities are often actually measured at slightly off-normal or near-normal angles. In these cases, we continue to use N, with this understanding. An experimental variation of the second method is the use of a spectroscopic ellipsometer. The normal spectral reflectance can be deduced from the optical constants (the refractive index $n(\lambda)$ and the extinction coefficient $k(\lambda)$), which can be deduced from spectroscopic ellipsometry. Combining Fresnel's relation and Kirchhoff's law, we can obtain

$$\varepsilon(\lambda, T) = 4n(\lambda) \Big/ \left[(n(\lambda)+1)^2 + k(\lambda)^2\right]. \tag{9.12}$$

There are many experimental difficulties that can be encountered in spectral emissivity measurements. To emphasize the difficulties and present some solutions, we introduce and discuss experimental examples of spectral emissivity measurements where such difficulties were encountered and dealt with. In Section 9.3, an emissivity measurement system for objects around ambient temperature based on the first method, a comparison of the sample spectral radiance to that from a blackbody source, is described. In Section 9.4, an emissivity measurement system for metals being oxidized is described. Microstructure of the metal surface exposed to high-temperature air is far from that of the ideal smooth and clean surface, and the microstructure changes with time. In Section 9.5, various studies of the emissivity of high-temperature molten metals, heated by containerless techniques to minimize the sample contamination, are reviewed. In Section 9.6, an emissivity measurement method for polycrystalline alumina-based oxides is described. Since these ceramics are semitransparent materials in the visible and infrared regions, their emissivities depend on bulk parameters such as thickness, purity, grain size, and crystalline structure, which present a number of experimental difficulties.

9.3 SPECTRAL EMISSIVITY MEASUREMENTS NEAR-AMBIENT TEMPERATURE

9.3.1 Measurement Apparatus Based on the Blackbody Comparison Method

Figure 9.1 shows a design of an apparatus for measuring infrared spectral emissivity in the near-ambient temperature range based on the blackbody comparison method [8–10]. The apparatus consists of the sample unit, reference blackbodies, and a Fourier transform spectrometer (FTS). The measurable spectral range is mainly limited by the beam splitter and the infrared detector. For the thermal infrared spectral range, triglycine sulfate (TGS) or deuterated TGS infrared detectors are commonly used for the FTS. However, the sensitivity of such detectors is not sufficiently high for the room temperature thermal emission measurements. Therefore, the liquid nitrogen cooled photovoltaic

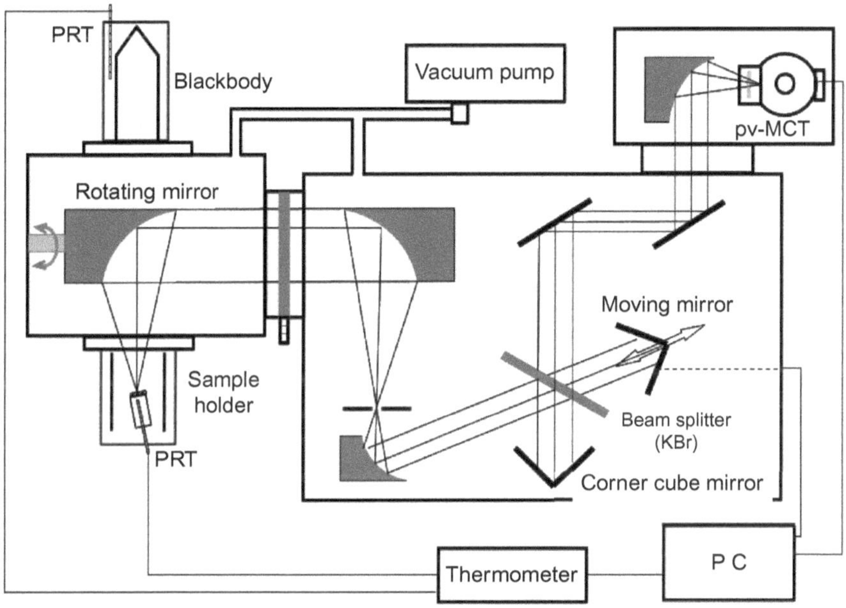

FIGURE 9.1 Schematic diagram of an FTS system for infrared spectral emissivity near ambient temperature.

(pv) mercury cadmium telluride detector and the KBr beam splitter, which covers the spectral range between 5 and 12 μm, are used in the apparatus. All of the optical components including the sample and the reference blackbodies are operated in vacuum to reduce atmospheric absorption effects.

Calibration of the system response and compensation for the offset drift caused by the ambient temperature change are performed using two reference blackbodies operated in vacuum. The first reference blackbody is a fixed temperature blackbody cooled by liquid nitrogen to just above its boiling point of 77 K, which can be used as a reference of the zero-level radiation. The second one is a variable temperature one operating from 253 up to 373 K (100 °C). The effective emissivity of these cavities is estimated to be greater than 0.999, based on an analytical routine.

Figure 9.2 illustrates the cross section of the sample holder for the emissivity measurement. The solid sample attached to the holder is heated or cooled by direct contact with a thermostatic fluid. The temperature of the fluid is measured with a calibrated platinum resistance thermometer (PRT) placed near the rear surface of the sample. Instead of measuring with a contact sensor attached to the sample surface, the temperature of the sample surface is derived indirectly by calculation assuming the equivalence of the heat flux between the fluid and the sample surface since the heat loss from the surface due to air convection is negligibly small because of vacuum operation. The sample surface is surrounded by a thermostatic radiation shield placed above

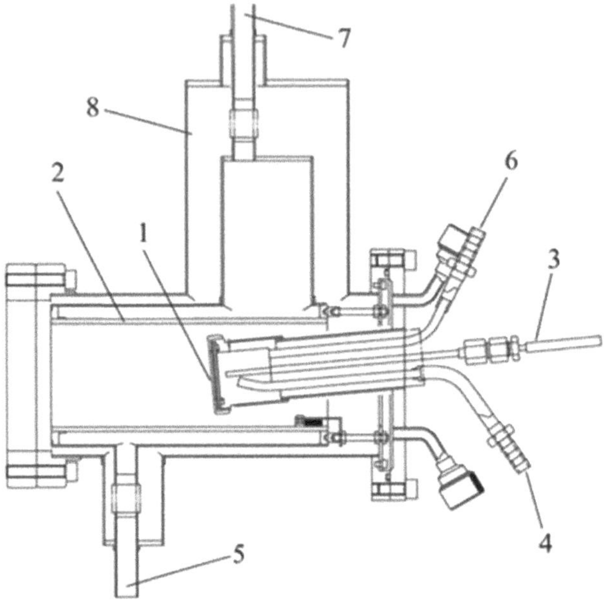

FIGURE 9.2 Cross-sectional view of the sample holder for emissivity measurement: (1) sample, (2) radiation shield, (3) PRT sensor, (4, 5) inlet of fluid, (6, 7) outlet of fluid, and (8) vacuum chamber.

the sample holder. The surface of the radiation shield is coated with a high emissive paint and the temperature of the surface is controlled with a thermostatic fluid. The radiation shield functions to control the irradiation by the surroundings, and the sample is tilted to reduce multiple reflections of radiation between the sample surface and surroundings.

9.3.2 Calibration of the System Response

One purpose of the calibration is to determine parameters of a mathematical model describing the spectrum as a function of the radiance incident into the FTS. The most conventional approach is to use two blackbody sources as references. On the assumption of linearity of the spectrometer response to the incident spectral radiance, the response of the FTS can be given by

$$C(\lambda) = \Re(\lambda)[L(\lambda) + L_0(\lambda)], \tag{9.13}$$

where $\Re(\lambda)$ is the spectral response, which includes the properties of the optical and electrical components, $L(\lambda)$ is the incident spectral radiance, and $L_0(\lambda)$ is the background spectral radiance emitted by the surroundings and the instrument.

If the measured spectra of the blackbodies are $C_c(\lambda)$ and $C_h(\lambda)$, where the subscripts c and h denote cold and hot, respectively, with different temperatures T_c and T_h ($T_c < T_h$), Eq. (9.13) gives

$$C_c(\lambda) = \Re(\lambda)[B(T_c, \lambda) + L_0(\lambda)] \tag{9.14}$$

and

$$C_h(\lambda) = \Re(\lambda)[B(T_h, \lambda) + L_0(\lambda)], \tag{9.15}$$

where $B(T, \lambda)$ is the spectral radiance of the blackbody. These equations can be solved yielding the instrument-related functions and calibration relationship:

$$\Re(\lambda) = \frac{C_h(\lambda) - C_c(\lambda)}{B(T_h, \lambda) - B(T_c, \lambda)} \tag{9.16}$$

and

$$L_0(\lambda) = \frac{C_h(\lambda)}{\Re(\lambda)} - B(T_h, \lambda). \tag{9.17}$$

Then, combining Eqs. (9.13), (9.16), and (9.17) the radiance is related to the FTS response by

$$L(\lambda) = \frac{C(\lambda) - C_c(\lambda)}{C_h(\lambda) - C_c(\lambda)}[B(T_h, \lambda) - B(T_c, \lambda)] + B(T_c, \lambda). \tag{9.18}$$

A more rigorous expression for Eq. (9.13) includes the effects of anomalous phase [11]. The complex Fourier-transformed spectrum of the measured interferogram can be expressed as

$$C(\lambda) = \Re(\lambda)\{L(\lambda) + L_0(\lambda)\exp[i\phi_0(\lambda)]\}\exp[i\phi(\lambda)], \tag{9.19}$$

where $C(\lambda)$ is a complex Fourier transform spectrum, $\Re(\lambda)$ is the spectral response of the FTS, $L(\lambda)$ is the incident spectral radiance originated from the object, $L_0(\lambda)$ is the background spectral radiance emitted by the surroundings and the instrument, $\phi(\lambda)$ is the normal phase of the instruments, and $\phi_0(\lambda)$ is the anomalous phase of the background radiation. The optical and electrical dispersion of the instrument and the sampling error cause the normal phase component. The anomalous phase arises from components due to the radiance from the surroundings and the instrument itself and can be especially pronounced for low temperature measurements.

In an ideal case, interferograms are symmetric with respect to the zero path difference (ZPD) position, and the spectrum of the source can be reconstructed by means of the simple cosine transform. However, if the background radiation is larger than the signal radiation at low temperatures, the apparent centerburst of the interferogram and the optical ZPD position do not agree, because of the dominant background radiation having anomalous phase [11]. In such a situation, because many conventional FTS's automatically identify the optical ZPD position of the interferogram with the apparent peak position of the center burst, the optical path difference cannot be assigned correctly to the interferogram. This misassignment of the optical ZPD position of the interferogram causes the anomalous phase error and deforms the spectrum. The spectrum derived from the relationship in Eq. (9.18) with the complex Fourier-transformed spectra affected with the anomalous phase error is given by

$$L(\lambda) = |L(\lambda)| \exp[i\phi_{\mathrm{L}}(\lambda)], \tag{9.20}$$

where $\phi_{\mathrm{L}}(\lambda)$ is the residual phase error. By employing the complex spectra from Eq. (9.19) in Eq. (9.18), the anomalous phase is cancelled out in the ratio. However, for FTS instruments that cannot maintain alignment of the interferograms through continuous fringe counting, an alternative iterative phase correction approach can be used. The residual phase error $\phi_{\mathrm{L}}(\lambda)$ of the spectrum in Eq. (9.20) must be zero if the correct ZPD positions are assigned to the measured interferograms. Therefore, this method corrects for the anomalous phase error by iterating the calculation of the complex spectral radiance in Eq. (9.18) by shifting the phase of the interferogram until the residual phase error in Eq. (9.20) becomes a minimum [12].

9.3.3 Correction for Reflected Radiance of the Surroundings

A fraction of the background radiation from the surroundings reflected at the sample surface can increase the apparent radiance from the sample and cause a misinterpretation of the emissivity value at near-ambient temperatures. If an isothermal condition of the surrounding can be assumed, the radiance of the surrounding is regarded as ideal blackbody radiation with ambient temperature. When we consider an opaque sample, the total spectral radiance of the surface can be written as

$$L(\lambda) = \varepsilon(\lambda, T_{\mathrm{sample}}) L_{\mathrm{b}}(\lambda, T_{\mathrm{sample}}) + [1 - \varepsilon(\lambda, T_{\mathrm{sample}})] L_{\mathrm{b}}(\lambda, T_{\mathrm{ambient}}), \tag{9.21}$$

where $\varepsilon(\lambda, T_{\mathrm{sample}})$ is the spectral emissivity of the sample at wavelength λ and temperature T_{sample}.

In the apparatus shown in Fig. 9.2, the sample surface is surrounded by a thermostatic radiation shield to control the emission from the surroundings. The radiances of the sample fixed at a given temperature can be measured while changing the temperature of the radiation shield, for example 0 and 50 °C, and the radiance emitted by the sample surface itself can be derived using Eq. (9.21).

9.3.4 Traceability and Uncertainty of Spectral Emissivity Measurements

For quality assessment or assurance of the spectral emissivity data, it is highly desirable that the emissivity measurements are traceable to SI units with evaluated uncertainty. As mentioned above, emissivity is defined as the ratio of the radiation emitted by the sample surface to the radiation emitted by an ideal blackbody radiator at the same temperature for the same optical and geometrical conditions. Take as an example the blackbody comparison method, which is the most commonly used for spectral emissivity measurements. The independent physical quantities to be measured are: (a) the radiance temperature scale of the reference blackbody, (b) the temperature of the specimen surface, (c) the wavelength scale over the range of the radiance spectrum, and (d) the geometrical factors of the radiance measurement.

Figure 9.3 shows the traceability scheme of the infrared spectral emissivity measurement described in this section. The radiance temperature scale is realized using variable temperature blackbodies with Planck's law, in which the effective emissivity of the blackbody cavity and the reference temperature of the blackbody cavity are constituent elements. The cavity emissivity can be measured based on reflectance methods [13,14]; however, reflectance measurements of cavities with very low reflectance can generally only be made with single wavelength laser sources with sufficiently small uncertainties. Another valid approach is to estimate the effective emissivity of the cavity based on theoretical calculations, such as the Monte Carlo method, using cavity surface reflectance data as input [15].

The reference temperature of the blackbody cavity can be measured using a contact thermometer, for example, PRT sensor, calibrated against a reference standard traceable to the SI unit. The temperature of the specimen surface can also be measured with a traceable contact thermometer.

Calibration of the wavelength scale of the FTS is not practical at the measurement site. The wavelength scale of the spectrum is given with satisfactory accuracy using the fringe patterns of He–Ne laser monitoring position of the moving mirror of the interferometer.

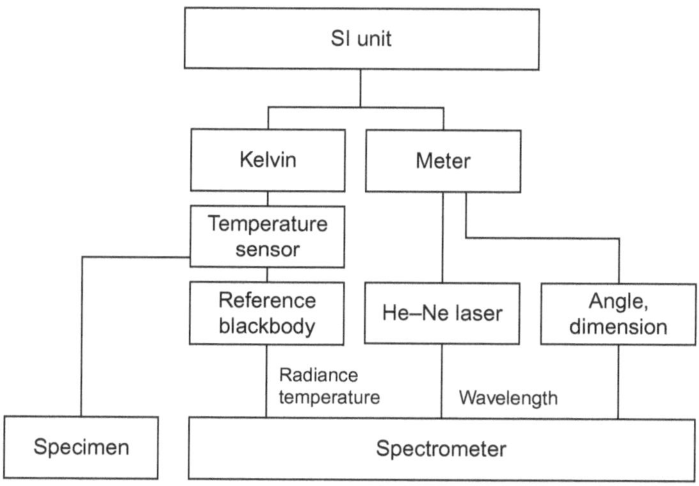

FIGURE 9.3 Traceability scheme of the spectral emissivity measurement based on the blackbody comparison method with an FTS.

High accuracy estimations of the geometrical factors, area, direction, and angle of emission, are also rarely needed for the emissivity measurement. The effect of the uncertainty of the geometrical factors can be reduced with a radiance comparison measurement between the specimen and the reference blackbody under the same conditions.

The main uncertainty components for the spectral emissivity measurement with the FTS based on the blackbody comparison method are described below. Uncertainty components related to the radiation measurements are as follows:

a. *Temperature scale*: uncertainty in the temperature scales include those of the reference PRT sensor and radiances of the reference blackbodies used.

b. *Reference blackbodies*: uncertainty of the reference blackbody radiances mainly originates from the difference of the effective emissivity of the cavity from unity. The calculated effective emissivities (with its standard uncertainties) for the system described are 0.9996 ± 0.0001 for the intrinsic emissivity of the cavity, and 0.96 ± 0.01 for the cavity wall surface, separately measured. Consequently, the discrepancy of the value of 0.9995 from unity is considered to be the standard uncertainty of the radiance of the high temperature reference blackbody. For the liquid nitrogen cooled blackbody used as a lower radiance reference, the apparent radiance temperature of the blackbody was determined by the reflection of the background emission, rather than the thermal radiation emitted by the liquid nitrogen cooled cavity itself.

c. *Nonlinearity of the spectrometer response*: an easily overlooked source of uncertainty for radiation measurements with the FTS is the nonlinearity of the spectrometer response. The nonlinearity between the incident and measured spectral radiance within the radiance levels of the reference blackbodies can be checked with measurements of the radiance from a third blackbody cavity placed at the sample port by changing the temperature after calibration against the reference blackbody cavities.

d. *Size-of-source effect*: effect of radiation from outside the nominal field of view (size-of-source effect) of the spectrometer is also one of the important characteristics for emissivity measurement, when the effective size of the sample surface is different from the aperture size of the reference blackbodies. For checking the effective target area of the spectrometer, a large area black target having removable plates of various aperture sizes can be used in place of the sample holder.

The uncertainty, $\Delta\varepsilon$, in the emissivity value produced by the uncertainty, ΔT, in the temperature measurement can be given by the expression

$$\frac{\Delta\varepsilon}{\varepsilon} = \left| \frac{c_2/\lambda T}{\exp(-c_2/\lambda T) - 1} \right| \frac{\Delta T}{T}. \tag{9.22}$$

Uncertainty components related to temperature measurements are as follows:

a. Uncertainty of the temperature scale, which is used for temperature measurement of the sample surface, is the most fundamental component of uncertainty in the blackbody comparison method.
b. Uncertainty of the calibrated PRT can be smaller than 0.01 K in practice. Instability of the surface temperature, which is affected by the short-term stability of the temperature of the sample-heating unit, should also be considered as an uncertainty component. Nonuniformity of the surface temperature inside of the FOV of the FTS and temperature differences between positions measured by the FTS and by the calibrated PRT can be major components of uncertainty of the sample surface temperature.

9.4 SPECTRAL EMISSIVITY AND REFLECTANCE MEASUREMENTS OF OXIDIZED METALS

9.4.1 A Broad-Spectral-Range High-Speed Spectrophotometer System

When most metals are heated in air from room temperature up to high temperature (around 1000 K), an oxide film is formed on the surface, and the film surface roughens with the inhomogeneous growth of the crystal grains of the film on a clean optically smooth specular-finished surface. In this case, the radiation reflected from the oxidized rough metal surface is prone to be diffusely scattered radiation. The normal emittance ε_N (normal incidence absorptance α_N), which is of importance as described above, can be derived by $(1-\rho_{NH})$ if Kirchhoff's law is valid. Therefore, the (normal-) hemispherical reflectance measurement can be important for investigating radiation properties of materials in thermal engineering. ρ_{NH} has been measured by various techniques [5,6,16–22], but primarily by integrating sphere techniques [16], because integrating sphere attachments for spectrophotometers are commonly available [23] (see Chapter 6). In addition, some integrating sphere devices have the capability to measure both ρ_{NH} and ρ_{spec} by removing the specular component from the measurement.

In order to measure ρ_{NH} of a time-varying real surface, Makino and Wakabayashi [24–26] have developed a broad-spectral-range high-speed spectrophotometer system having (a) a wavelength range from the visible to the infrared, (b) ability to measure the directional characteristics (that is, hemispherical, specular, and diffuse reflections) and (c) high-speed time scanning. Figure 9.4 shows the experimental system that can measure time evolutions of ρ_{NH}, ρ_{spec}, ρ_{diff}, α_N, and ε_N from the near-ultraviolet through the infrared spectral regions with wavelengths from 0.30 to 11 μm simultaneously and repeatedly with a cycle time of 2–6 s. The system is equipped with an incident optics subsystem for measuring hemispherical reflection by using two parabolic

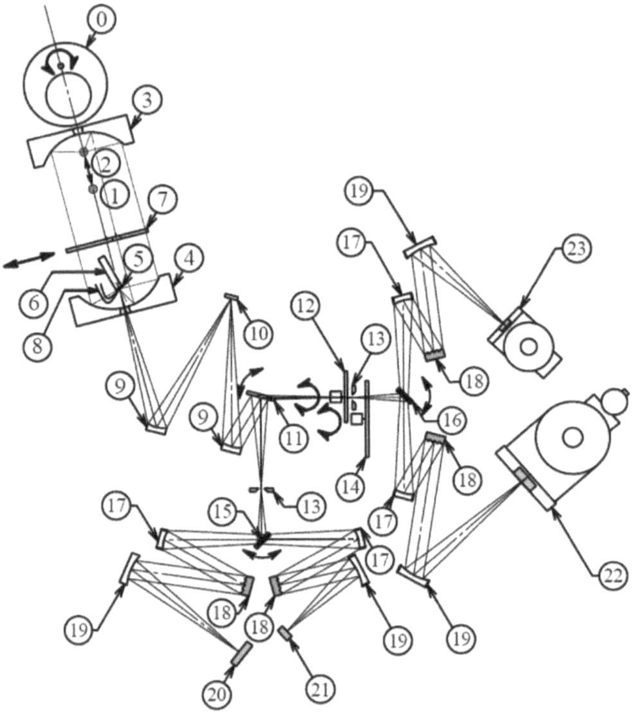

0. Cam mechanics
1. Tungsten-halogen lamp
2. Si_3N_4 light source
3. Parabolic mirror 1
4. Parabolic mirror 2
5. Specimen
6. Heater
7. Shutter disk
8. Thermocouple
9. Concave mirror
10. Plane mirror
11. Rotational plane mirror
12. Chopper
13. Entrance slit
14. Filter disk
15. Rotational plane mirror
16. Rotational plane mirror
17. Collimator
18. Diffraction grating
19. Camera mirror
20. 35-Si photodiode array
21. 16-Ge photodiode array
22. 32-InSb photovoltaic detector array
23. 16-HgCdTe photoconductive detector array

FIGURE 9.4 Schematic diagram of the broad-spectral-range high-speed spectrophotometer system.

mirrors. The system enables wide-spectral measurements by using four interlocking sets of Czerny–Turner-type diffraction grating spectrometers and realizes high-speed spectral measurement by using an array of four types of optical detectors. Figure 9.5 shows a sketch of the incident optics subsystem for measuring R_{HN} ($=\rho_{NH}$), ρ_{spec}, and ε_N. The two parabolic mirrors of 120 mm diameter are set to face each other. The two parabolic mirrors ((3) and (4) in Fig. 9.5) have holes of 12 mm in diameter in their centers.

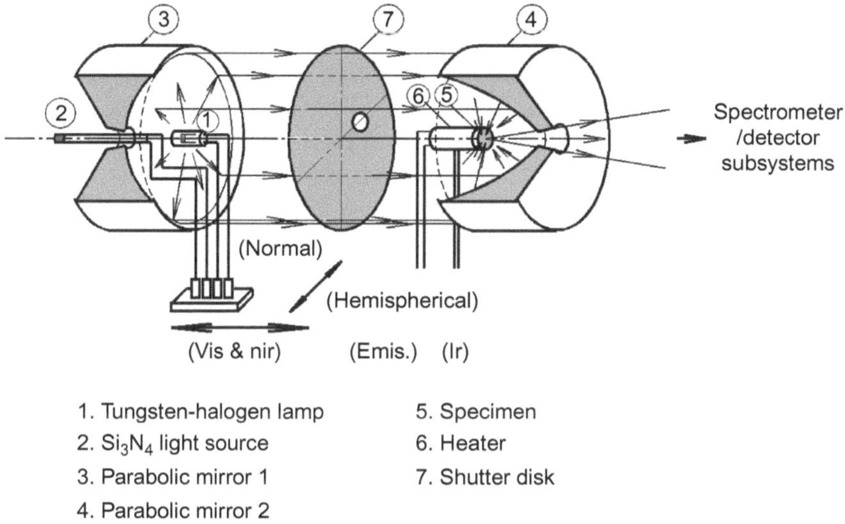

FIGURE 9.5 Sketch of the incident optics subsystem for measuring the hemispherical-normal reflectance factor R_{HN}, the near-normal specular reflectance ρ_{spec}, and the near-normal emittance ε_N.

A tungsten–halogen lamp (1) for the near-ultraviolet through near-infrared region or a silicon nitride (Si_3N_4) light source (2) for the infrared region is set at the focal point of the parabolic mirror (3). The sizes of halogen and Si_3N_4 light sources are 3.3 mm × 0.6 mm and 4.7 mm × 1.3 mm, respectively. The two light sources emit radiation whose intensity is approximately isotropic. The two light sources are alternatively set at the focal point of the parabolic mirror (3) or at an off-focal point by the help of cam mechanics (0) driven by a computer-controlled stepping motor. When one of the two light sources is set at the focal point, diffusely emitted radiation is transformed to parallel flux by the parabolic mirror (3) to proceed to another parabolic mirror (4). A specimen (5) is set at the focal point of the parabolic mirror (4). The surface of the specimen is inclined at an angle of 15° to the optical axis of the parabolic mirror (4). On the surface of the specimen radiation whose intensity is hemispherically homogeneous is incident from the parabolic mirror (4) and focuses. A component of the hemispherically reflected radiation which proceeds in the direction of 15° reflection is led through a hole on the parabolic mirror (4) to the spectrometer and detector subsystem and measured. The 15° is assumed to be normal.

In order to measure the 15° incidence specular reflection, a shutter disk (7) of 130 mm in diameter is inserted and removed in the optical path between the two parabolic mirrors (3) and (4). The disk has a hole of 15 mm in diameter at the position of 16 mm from the center of the disk. When the disk does not interrupt the parallel flux, the hemispherically homogeneous radiation is incident on the surface of the specimen. When the disk interrupts the parallel flux, a portion of the flux which comes through the hole on the disk is incident on the surface of the specimen from the direction of an incident angle of 15°. In the latter case, the component reflected on the surface of the specimen in the specular direction is led to the spectrometer and detector subsystem and measured.

The above-mentioned reflection components involve the radiation component emitted by the specimen surface itself in the direction of an emission angle of 15°, assumed to be normal emission, which is led through the hole of the parabolic mirror (4) to the spectrometer and detector subsystem. On the other hand, when both of the two light sources are set at off-focal points, the radiation is not incident on the specimen surface and only the radiation emitted from the specimen surface is measured. From these measured values, the normal reflection component for R_{HN}, ρ_{spec} at near-normal incidence and ε_N are individually determined.

In order to determine absolute values of R_{HN} and ρ_{spec}, a clean optically smooth nickel surface is used as the reference specimen whose spectra of the optical constants are published [27]. On the other hand, ε_N is determined based on the radiative method by which the specimen is replaced using a small blackbody cavity [25] when the blackbody radiation measurement is carried out. The blackbody is a cylinder type, having a 20 mm inner diameter with a 50 mm depth, which has a diaphragm aperture with a 10 mm inner diameter. The blackbody is made of stainless steel SUS304 whose inner surface is sufficiently air-oxidized. The apparent normal emittance of the blackbody is estimated to be 0.99 [28]. Table 9.1 lists the specifications of the blackbody.

The time evolutions of ρ_{spec} and ε_N of the specular-finished surface during high-temperature air-oxidation processes have been measured for nickel [24] and chromium [29], respectively. In Ref. [30], a diagnostic algorithm of electromagnetic wave theory and radiative heat transfer science including interference and diffraction (scattering) of radiation was proposed on the basis of the hardware performance of the broad-spectral-range high-speed spectrophotometer

TABLE 9.1 Specifications of Blackbody for Emittance Calibration

Diameter and length	ϕ 20 mm × 50 mm
Diaphragm	ϕ 10 mm
Material	SUS304
Sheath heater	ϕ 1 mm in diameter
	Three circuits of base, body, and neck
Thermocouple	K-type, ϕ 0.1 mm at base, body, and neck
Calibration temperatures	1000, 1050, and 1100 K
Apparent emittance	0.99

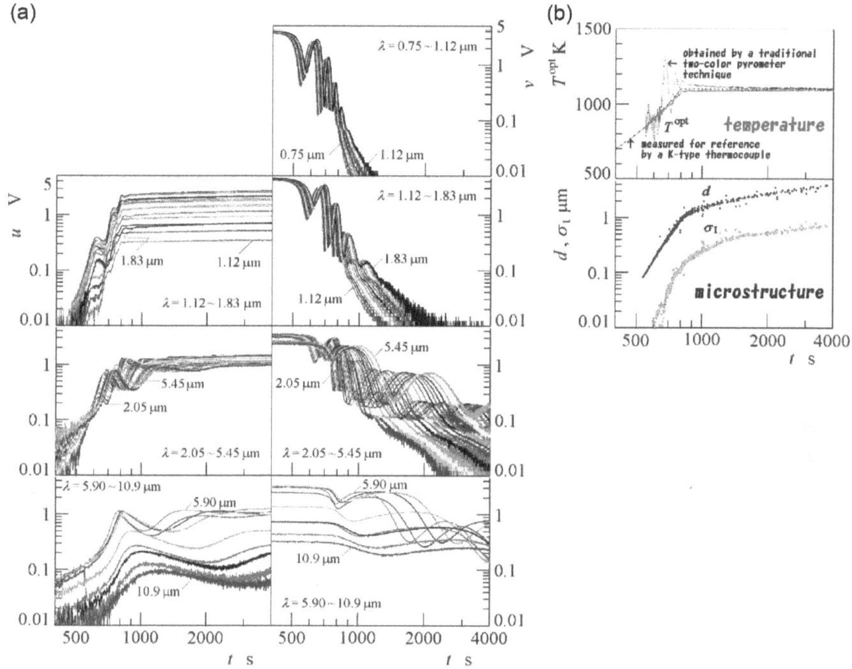

FIGURE 9.6 Measured quantities of (a) the normal emission energy u and the normal incidence specular reflection energy v (that is, input quantities to a proposed diagnostic system) and (b) diagnosis of surface temperature T^{opt}, surface film thickness d, and rms surface roughness σ_I of a nickel surface during a high-temperature air-oxidation process.

system. This diagnostic algorithm enables one to carry out noncontact and real time diagnosis of the surface temperature T^{opt}, the surface film thickness d, and the rms surface roughness σ_I at high temperature (1000 K level) and through change of states. The temperature diagnosis is used to accurately determine the momentary surface temperature without previous knowledge of the surface emittance, and the microstructure diagnosis is used to determine the surface film thickness of the order of 0.1–4 μm and the rms surface roughness σ_I of the order of 10–700 nm. The experimental results for a nickel surface in a high-temperature air-oxidation process are shown in Fig. 9.6.

9.5 SPECTRAL EMISSIVITY MEASUREMENTS OF MOLTEN METALS AT HIGH TEMPERATURES

9.5.1 Importance of Molten Metal Emissivity Data

Radiation thermometry is often used in a high-temperature process such as casting and crystal growth. Therefore, $\varepsilon(\lambda, T)$ of molten materials is indispensable. In addition, $\varepsilon(\lambda, T)$ of molten metals at their melting points is used as the constant input parameter in the radiation thermometry to determine true temperatures of the melts over a broad temperature range. This is because it is empirically recognized that $\varepsilon(\lambda, T)$ of molten metals exhibits insensitivity to temperature change or at most only a moderate temperature dependence. Information about the difference in the wavelength dependence between the emissivities of liquid and solid phases of a pure metal at the melting point is also important for understanding the behavior of electrons in molten metals. This is because the difference in emissivity, as well as other optical properties, is related to the differences between the electronic structures of the liquid and solid phases. Electrical resistivity of liquid metals has been theoretically explained by Ziman [31]. The theory of Ziman has been developed based on the accumulated theoretical attempts to explain the difference in the measured electrical resistivity between the solid and liquid phases of various metals at their melting points. Therefore, comparison of the spectral emissivities of a metal at the melting point in solid and liquid states could play a significant role in the development of a theoretical model of radiation properties of liquid metals. For example, the the sharp absorption edges in the visible spectral emissivity of copper and gold at the respective melting points is of strong interest in experimental research. This strong optical absorption (emission) is due to the interband transition between the upper *d*-band and the Fermi level. It can be considered that the band gap is drastically affected by the solid-to-liquid transition.

The behavior of the emissivity of semiconductors such as silicon at the solid-to-liquid transition is also of interest. Solid silicon is a semiconductor with a diamond cubic structure and its electrical resistivity is much higher than that of metals. On the other hand, silicon is metallic in the liquid state because its electrical resistivity is as low as mercury and decreases as the temperature increases. Therefore, the emissivity difference between liquid and solid silicon corresponds to the metal-to-semiconductor transition. It is another point of interest to investigate the effect of the band gap of solid silicon on the emissivity at the melting temperature, because silicon has a band gap of 1.12 eV at room temperature. If the band gap is preserved just below the melting temperature, we can expect that the emissivity spectrum around 1100 nm, which corresponds to the energy of 1.12 eV to drastically vary with the solid-to-liquid transition.

In general, the electronic properties of metals change discontinuously when they melt. However, some previous studies have reported that the spectral emissivities of some metals at a specific wavelength λ_x do not change on melting, while for shorter wavelengths, the spectral emissivities decrease on melting, and for longer wavelengths, the spectral emissivities increase on melting. The specific spectral emissivity ε_x, which is insensitive to the phase transition, has been observed for two noble metals (Cu and Au) [32–34] in the visible wavelength range and for nine refractory transition metals (Hf, Mo, Nb, Re, Rh, Ta, W, V, and Zr) [35] in the near-infrared region. Such a phenomenon has long been known as the so-called X-point [35,36] in the emissivity of metals, which generally refers to an intersection of spectra of the emissivity for a metal at various temperatures. For copper and gold, it is widely recognized that the X-points are related to opposing effects of the phase change on two optical emission mechanisms, viz., intra- and interband transitions. Namely, at the wavelength of λ_X, the expected increase in the intraband emission on melting compensates for the decrease in the interband emission caused by broadening of the absorption edge due to smearing out of the Fermi distribution. Therefore, the X-point phenomenon is of great interest to develop electronic models of optical properties of high-temperature molten metals. In addition, the phenomenon is of interest for the development of radiation thermometry, because the melting point of metals can provide a useful high temperature standard for the calibration of radiation thermometers.

9.5.2 Spectral Emissivity Measurements Based on Containerless Techniques

There are numerous measured emissivity data available on various liquid metals and alloys. However, significant discrepancies exist among the data even for a pure metal at its melting point. For example, a review on the normal spectral emissivity around 650 nm of liquid iron at the melting point indicates that the literature values range from 0.30 to 0.43 [37]. In many of the measurements of liquid metal emissivities, the metal samples were indirectly heated by conventional resistive heaters and the molten samples were contained in a conventional ceramic crucible. The discrepancy among the reported data is likely primarily caused by the use of the conventional sample-heating techniques. The emissivity is a physical property of the surface of a material. It is generally difficult to avoid the surface oxidation of metals at high temperatures. Since the density of oxides of a metal is usually smaller than that of the metal, the surface of the molten metal sample tends to be covered with an oxide film. The surface oxidation of a metal is prone to a larger emissivity value and a significant surface roughness, both of which result in an emissivity increase. In conventional emissivity measurements, thermal radiation is emitted from not only the sample but also the furnace heater and crucible. Furthermore, the temperature of the furnace heater and outer wall of the crucible is generally greater than that of the sample itself. Thus, stray radiation from the sample surroundings can cause considerable error in the emissivity measurement. To avoid surface oxidization, molten metal samples are usually contained in a furnace chamber. However, the window of the furnace chamber is often contaminated by evaporation from the sample and/or its surroundings at elevated temperatures. This can also lead to considerable error for radiation thermometry. Furthermore, it is experimentally difficult to distinguish between the liquid and solid emissivities when observing the thermal radiation from a mixture of liquid and solid phases at the melting point.

Heterogeneous nucleation may occur on the inner crucible wall in the solidification process of the molten metal contained in a ceramic crucible. Heterogeneous nucleation tends to make the solidification proceed rapidly and disorderly, which can increase the difficulty in determining the solid and liquid emissivity independently. To minimize such errors, containerless techniques have been attempted to melt the metal sample. This also avoids the problem of contamination from the containers in the high-temperature melts. In addition, many containerless techniques can quickly heat the sample up to the predetermined high temperatures above the melting point from room temperature, compared to more conventional heating processes. Such a fast heating process of the sample also can suppress sample contamination and evaporation. Some notable studies that adopted containerless techniques are reviewed below.

9.5.2.1 Rapid Resistive Self-Heating Techniques

An electrically conductive sample rod is rapidly heated up to the melting point by passing a large current through the sample itself. The spectral emissivity of the sample without any container, at and around the melting temperature, can be measured in a short period (less than 1 s) before the sample is completely melted. The shortness of exposure time of the sample to elevated temperatures can minimize the contamination of the sample surface by residual impurities in the vacuum or inert gas environment as well as minimize loss of the sample mass by evaporation.

The group led by Cezairliyan has determined the normal spectral emissivities (in the wavelength range of 500–1000 nm) of 10 pure metals (Ni, Fe, Pd, Ti, Zr, V, Nb, Mo, Ta, and W) at their melting temperatures based on the rapid resistive self-heating technique and on the radiance temperature measured with high-speed pyrometers [38,39]. The measurement results demonstrate excellent reproducibility of the normal spectral emissivity (generally within 1%) at the melting point of each metal. Because of the high degree of reproducibility, they suggested that radiance temperature or normal spectral emissivity of selected metals at their melting points can provide a valuable source of reliable reference values for applications that necessitate noncontact methods for determining temperature.

Normal spectral emissivities of four liquid metals (Ta, Nb, Ni, and W) at temperatures up to 5000 K have been determined by the group led by Pottlacher based on the rapid resistive self-heating of the wire-shaped samples by a large current from a capacitor-discharge circuit and on microsecond-resolution ellipsometry [40,41], though the measured emissivity is limited to that at the operating wavelength (684.5 nm) of the ellipsometer. In their experiments, which are of extremely short duration (less than 0.1 ms), the sample is heated at rates above 10^7 K/s and is ultimately exploded. Because of such a high heating rate, the sample wire keeps its shape, even in the liquid state at temperatures of several hundred degrees above its melting point.

9.5.2.2 Electromagnetic Levitation Techniques

An electrically conductive sample is suspended (levitated) in space within a high-frequency electromagnetic field generated by an induction coil at its center and is inductively heated at the same time. The resultant molten sample, having a spherical shape, is stirred by the eddy current occurring on the sample surface. This stirring effect makes the temperature distribution uniform within the sample melt. A notable feature of levitation melting is the possibility of realizing highly undercooled liquid states. Although the sample type is limited to electrically conductive materials, it is relatively easy to construct and operate the apparatus because it is a modified version of a conventional induction furnace. A shortcoming of this containerless technique is the difficulty of controlling both the sample temperature and position, because the variation of the electromagnetic field affects both the temperature and position and because the heating efficiency strongly depends on the sample position in the electromagnetic field. To solve these problems, several means to control either the sample temperature or position have been added to the electromagnetic levitator. In almost all cases with electromagnetic levitators, the sample is generally heated in a flow of inert gas, and the sample temperature is mainly controlled by changing the flow rate or the type of gas. In some cases, laser beam heating is used as a heating method to obtain temperatures above those achieved by electromagnetic heating alone. The best solution may be the use of an electromagnetic levitation technique under microgravity conditions. To measure the surface tension of molten metals, the group led by Egry has operated an electromagnetic levitator launched into space, where the electromagnetic field is used as the primary means of heating [42].

Krishnan *et al.* have determined the normal spectral emissivities in the visible region of 11 molten metals (Cu, Ag, Au, Ni, Pd, Pt, Zr, Ti, Al, Nb, and Si) by using a combination of an electromagnetic levitator and a high-speed spectroscopic ellipsometer [43–45]. Figure 9.7 shows a schematic diagram of the emissivity measurement system, which was used. Since molten metals generally emit high-intensity thermal radiation, the signal-to-noise ratio for the surface reflection of the probe polarized light has to be enhanced by the use of a lock-in amplifier or a high-intensity light source. The surface of molten metals is intrinsically smooth and flat if there is no significant contamination on the surface. This makes it easier to carry out spectroscopic ellipsometry on the molten metal surface, because the reflection of the probe light is not dispersed by the surface and because the incidence angle that is used to derive the optical constants can be accurately determined. The results have shown that nine of the metals, but not Nb and Al, exhibit only moderate temperature dependence in the liquid state over a large temperature range.

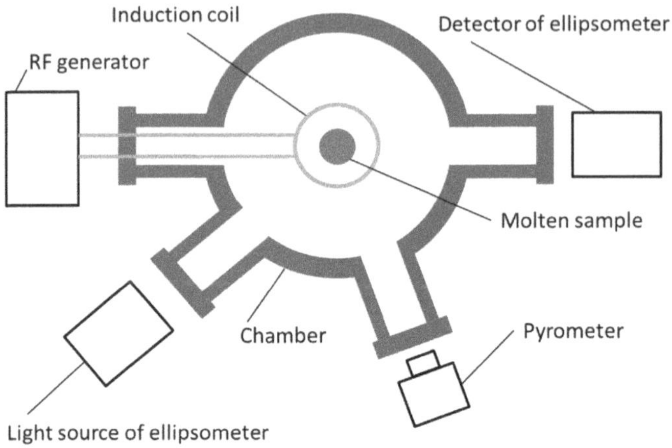

FIGURE 9.7 Schematic diagram of the spectral emissivity measurement system based on the electromagnetic levitator used by Krishnan *et al.* [44].

The group led by Fukuyama has measured the normal spectral emissivity at 807 and 900 nm of electromagnetically levitated molten copper droplets under a strong static magnetic field by the spectral radiance measurement method [46]. The additional use of a static magnetic field can suppress the translational motion and surface oscillation of the electromagnetically levitated droplet sample during the measurement, which increases the measurement accuracy.

9.5.2.3 *A Cold Crucible Technique*

An electrically conductive sample is placed in a crucible, consisting of water-cooled copper segments, which is surrounded by an induction coil. When a high-frequency alternating current passes through the coil, eddy currents occur on the sample surface and copper segments. As a result, the sample is launched by the repulsive electromagnetic force between the sample and the copper segments and is inductively heated at the same time. Therefore, this technique can be regarded as a modification of the electromagnetic levitation method. The difference is the capability of realizing a skull melting process. In many cases, the resultant molten sample is kept in contact with the copper segment located at the bottom of the sample melt. Water cooling of the segment keeps an outside layer of the otherwise molten sample cool enough to be solidified. Thus, the sample melt can be heated to at least 3000 °C, while maintaining the melt in the solid "skull" of the sample itself on the copper segment wall that isolates the segment from the melt and at the same time, and protecting the melt. The advantage of skull melting over electromagnetic levitation melting is the ability to produce a larger amount of molten metal, because the electromagnetic field is no longer used to launch the melt. In

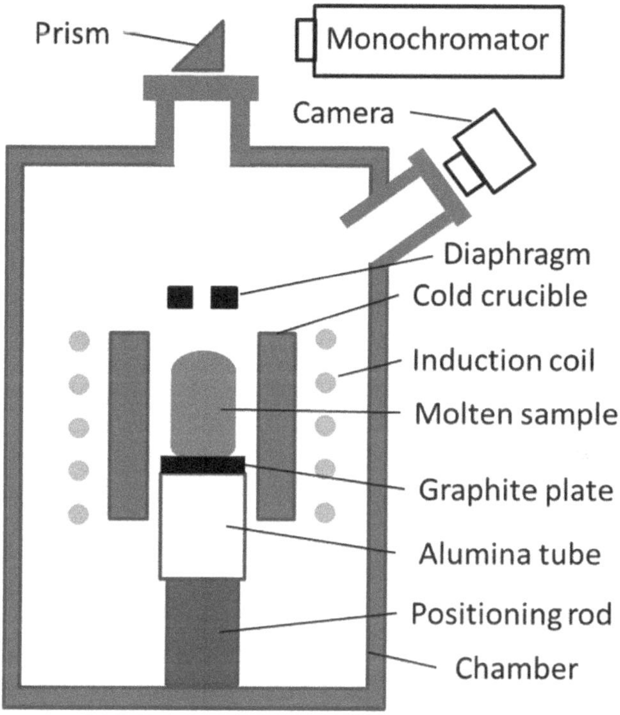

FIGURE 9.8 Schematic diagram of the spectral emissivity measurement system based on the cold crucible used by Watanabe *et al.* [47].

addition, it is relatively easy to control the temperature of a larger sample. Because of these advantages, Watanabe *et al.* have adopted the skull melting technique to measure normal spectral emissivities of Cu, Ag, Au, Fe, Co, Ni, and Si in the solid and liquid states at their melting points based on the blackbody comparison method [33,46–49]. Figure 9.8 shows a schematic diagram of the emissivity measurement system based on the cold crucible technique used by Watanabe *et al.* They have determined the visible and near-infrared spectral emissivities of the solid and liquid phases, respectively, at the melting point, by observing the slow-moving solidification front on the relatively large area of the sample melt surface. In addition, they have described the phase (liquid/solid) dependence of the normal spectral emissivity of metals and silicon.

9.5.2.4 *An Electrostatic Levitation Technique*

A charged sample is levitated between a pair of parallel disc electrodes where a large electric field is applied. The field is controlled through a feedback system to keep the levitated sample at a predetermined position. To melt the levitated sample, it is homogeneously irradiated by high-power multilaser beams

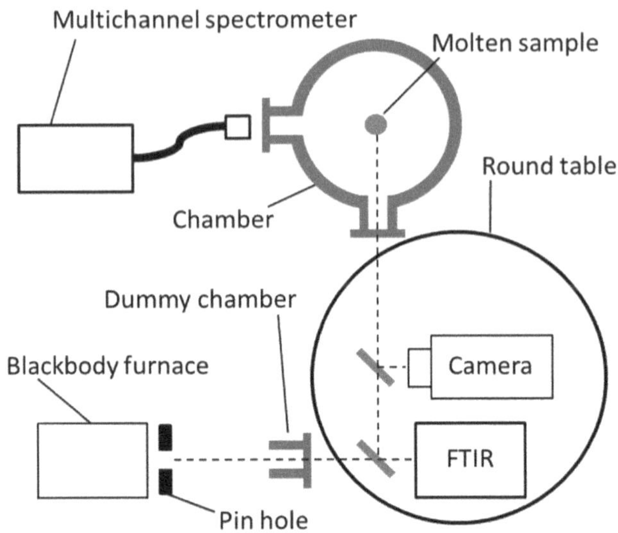

FIGURE 9.9 Schematic diagram of the spectral emissivity measurement system based on the electrostatic levitator used by Ishikawa *et al.* [51].

under an ultra-high vacuum atmosphere to avoid the discharge of the sample that results in the sample dropping down. Ishikawa *et al.* have measured hemispherical spectral emissivity of molten zirconium over a wavelength range of 700–6000 nm using spectrometers combined with an electrostatic levitator [51]. Figure 9.9 shows a schematic diagram of the apparatus based on the electrostatic levitator used by Ishikawa *et al.* Except for this work, however, electrostatic levitators have not been applied to measure spectral emissivity of molten materials. This may be because the electrostatically levitated sample, which has a typical diameter of approximately 2 mm, is smaller than that which can be measured with either the electromagnetic levitator or the cold crucible techniques.

9.5.2.5 Surface Heating by a Laser Technique

A partial area of the sample surface is melted by irradiation with a high-power laser, and thus, this technique conducts a kind of skull melting process for not only electrically conducting but also insulating materials. To carry out fast measurements of the normal spectral emissivity of nine refractory metals (Hf, Mo, Nb, Re, Rh, Ta, W, V, and Zr) in their liquid states, Hiernaut *et al.* have used a pulsed-multibeam laser heating system and a microsecond-resolution six-wavelength pyrometer based upon the assumption that the natural logarithm of spectral emissivity is a linear function of wavelength [52]. In their experiments, a small sphere of a sample material was supported on a needle of the same metal in a pure argon atmosphere. The sample sphere is heated by four pulsed laser beams in a tetrahedral configuration, which yields good temperature homogeneity over the irradiated sample surface area.

9.6 SPECTRAL EMISSIVITY MEASUREMENTS OF CERAMICS

Ceramics are versatile materials used in many high temperature applications due to their excellent chemical stability and heat resistant properties. Since radiation is the dominant form of heat transfer in the applications, it is important to investigate the radiative properties of refractory ceramics under high temperature conditions. There is exhaustive data of spectral emissivity of ceramics compiled by Touloukian and DeWitt in 1972 [53]. In addition, Chaney *et al.* compiled another comprehensive listing of emissivity literature in 1982 [54]. Recently, knowledge of the spectral emissivity of ceramics has gained interest due to spectral controlling technology. For example, specific ceramics that emit radiation selectively in limited near-infrared wavelength regions are necessary for the thermophotovoltaic (TPV) generation of electricity; this system can convert radiant energy, for wavelengths shorter than about 2 μm, into electricity using low-bandgap photovoltaic cells. In this section, some basic characteristics and experimental methods for spectral emissivity of white ceramics, that is, oxide ceramics such as alumina are described.

Aluminum-based oxides have dominant emission peaks on the basis of the natural frequency of lattice vibrations in the wavelength range from 7 to 10 μm. The spectral emissivity is close to unity in this range for all specimens, which means that the oxides are almost completely opaque. On the other hand, the oxides are also generally semitransparent media for visible light and near infrared radiation because the valence electrons are bound to the atoms through ionic or covalent bonds. With respect to a semitransparent medium, the spectral emissivity becomes a volumetric property, that is, it depends on bulk parameters such as thickness, purity, grain size and crystalline structure. Because of this, the term "emittance" will be used instead of emissivity later in this section. In fact, the compiled data of the values of normal spectral emittance of aluminum oxide vary widely particularly for wavelengths less than 4 μm, due to their semitransparency [53]; this implies that it may be possible to control the emittance using these parameters within this wavelength range. According to purpose of investigation, it may be desirable to obtain more fundamental properties such like absorption and scattering coefficient. In general, there is only a weak dependence on direction (zenith angle) and temperature of the spectral emittance for ceramics. Therefore, the hemispherical spectral emittance can be taken as equal to the normal spectral emittance and the spectral emittance measured at a certain temperature can be used within a wide temperature range. However, it must be noted that the total emittance of oxides depends strongly on temperature, due to the temperature dependence of the Planck function.

Typical experimental methods to measure the normal spectral emittance of alumina for temperatures above 1000 °C were reported by Nordine and Postlethwait [55,56]. In general, measurement of the emittance for high temperature semitransparent materials is a formidable task. This is because a temperature distribution exists more or less within the sample due to low thermal conduction and volumetric emission. Hence, it may be difficult to measure the surface and average temperatures accurately, which can lead to inaccurate comparisons of the sample surface and the reference blackbody. In order to avoid this problem, one must heat the sample as uniformly as possible. The use of a furnace can be an effective way to achieve quasi-isothermal conditions. Furthermore, one does not have to prepare a separate reference blackbody if the inside of the furnace can be considered as a blackbody cavity; one only needs to move the sample or the view point of the detector within the furnace. However, it is difficult to separately detect the radiation emitted from the sample itself when the sample is heated in the furnace. Laser heating can overcome this disadvantage, selecting a heating laser whose wavelength is outside of the measurement wavelength range, although it also tends to cause a temperature gradient within the sample. Although thermocouples are often used for the temperature measurement of the sample, the contact between the sample and thermocouple sometimes affects the measurement. If the temperature gradient within a sample is enough small, a small hole fabricated in the sample surface can be used as a blackbody, which can help to obtain an accurate temperature determination by radiation thermometers even if the sample emittance is unknown. To investigate the wavelength dependence of the emittance, as discussed in previous sections, spectral radiance from the sample is usually measured with a diffraction grating spectrometer or FTIR spectrometer.

With respect to the TPV application, several interesting studies for the development of a selective emitter based on ceramic oxides have been reported [57–60]. Rare-earth (RE) oxides are a good example because RE elements retain their somewhat atomic-like emissive characteristics in a variety of different binding environments; the specific emission can be attributed to radiative transitions between energy levels of internal 4f-orbitals of the RE cation. For instance, applying erbium (Er) with a sharp emissive peak near 1.5 μm, erbium aluminum garnet ($Er_3Al_5O_{12}$:ErAG) and a eutectic composite between the ErAG and alumina can be used as monolithic selective emitters, where the RE oxides have single crystalline structures exhibiting relatively high transmittance of the radiation emitted from the RE element as compared to their polycrystalline version. Namely, the optical attenuation by the grain boundary scattering can be avoided by using the single-crystal RE oxide. Researchers have managed to measure the normal spectral emittance of these RE oxide samples, which have a several hundred micrometer thickness and are heated up to a temperature above 1000 °C using a furnace and gas burner. In the case of a transparent sample, the three-layer configuration (sample/platinum foil/SiC plate) is usually fabricated because the platinum foil itself emits very little radiation and can improve the temperature uniformity within the sample. If the sample is very thin and transparent in the wavelength range of interest, it is quite difficult to accurately measure the sample temperature either by thermocouples or by radiation thermometers. Some researchers have reported that there is a severe temperature discrepancy (from 100 to 200 K) between the two surfaces of a plain sample even if it is very thin. The temperature gradient can cause a significant decrease of the spectral emittance as compared to that under isothermal conditions. On the other hand, RE oxide with single or polycrystalline structures are often coated onto a durable substrate such as SiC, $MoSi_2$, and cordierite to achieve good thermostability [61,62]. For example, a porous garnet coating with thickness sufficient to block the background emission from the substrate has been investigated for this purpose. When the spectral emittance of the substrate is known, the substrate can be used as an alternative to a reference blackbody for calibration.

Kumano *et al.* have measured the normal spectral emittance of various polycrystalline aluminum-based oxides in the wavelength range from about 1 to 5 μm, at temperatures above 1000 °C, by means of several heating methods [63–67]. Figure 9.10a shows the experimental setup used to measure the spectral radiant energy emitted from all the samples and a separate reference blackbody. The radiant flux is introduced into the grating spectrometer, which has a thermopile detector, using two concave mirrors under ambient conditions. In order to eliminate errors resulting from stray radiation entering the spectrometer, an aperture is set and the incident radiation is chopped. The results of the measured normal spectral emittance are summarized in Fig. 9.11. No emittance is shown for wavelengths around 2.7 and 4.3 μm due to the significant absorption by H_2O and CO_2 gas molecules in the atmosphere. To fill in such gaps in the data, it is necessary to measure this kind of spectral emission in an inert gas atmosphere or vacuum.

Figure 9.10b shows a schematic of the samples and sample holder for laser or gas burner heating. An alumina chip with a thickness of 0.7 mm is supported by three needles to minimize conductive heat loss. The alumina is 95% pure, with 3% of SiO_2 and trace components of MgO, Na_2O, and K_2O. The back surface of the sample is heated using a CW CO_2 laser up to 1393 and 1493 K; it is desirable to heat the sample on both surfaces using split laser beams to approach isothermal conditions. Since the wavelength of the laser beam (10.6 μm) is far from the wavelength range of interest, laser scattering on the sample surface does not cause any measurement error. However, the stability of the laser intensity is indispensable to minimize sample temperature variation. The sample temperature is measured by a thermocouple embedded in the alumina chip. Figure 9.11 shows that the difference in the normal spectral emittance between 1393 and 1493 K is slight, which indicates that the temperature dependence of the alumina chip emittance can be negligible. In addition, the emittance is relatively low for wavelengths below 4 μm. This is due to the small thickness of the sample. On the other hand, the normal spectral emittance of porous media made of aluminum oxide (foam ceramics)

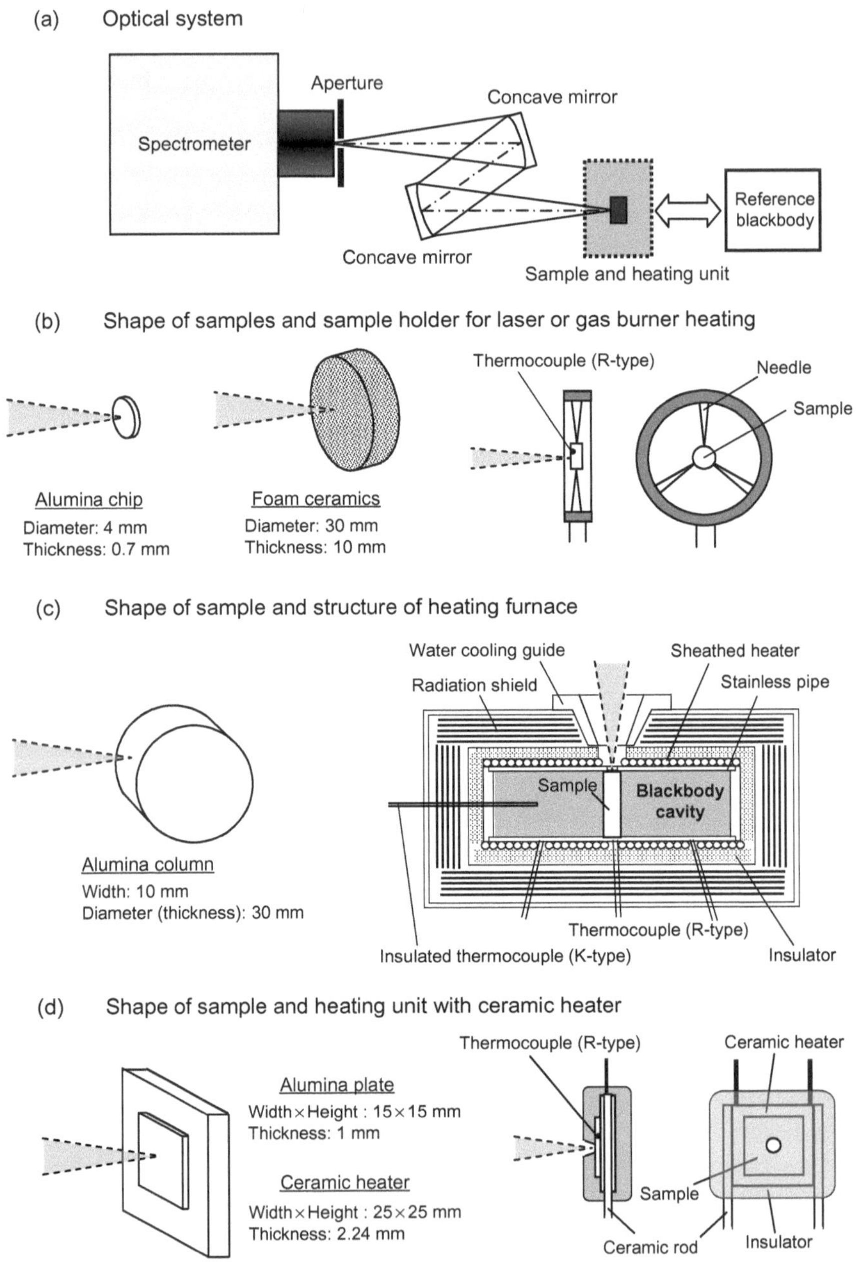

FIGURE 9.10 Schematic diagrams of the experimental setups used by Kumano *et al.* [63–67] for normal spectral emittance of various polycrystalline aluminum-based oxides by means of several heating methods.

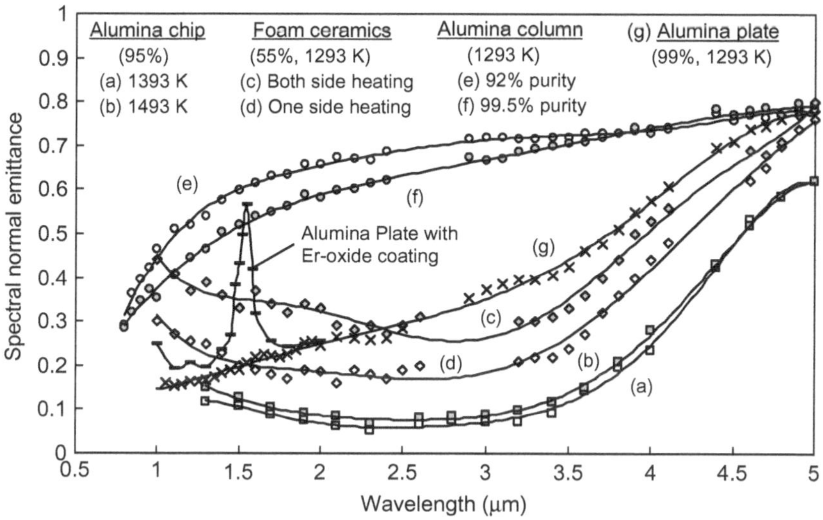

FIGURE 9.11 Normal spectral emittance of various alumina-based ceramics.

was measured at a temperature of about 1273 K. Foam ceramics are used as emitters in the TPV system proposed by Kumano *et al.*, which is based on a super-adiabatic combustion method with reciprocating flow in porous media [64]. The thickness and porosity of the foam ceramics is 10 mm and 87.5%, respectively. The sample is composed of 55% Al_2O_3, 37% SiO_2, 6% MgO, and others. Two liquefied petroleum gas burners are used to heat both surfaces of the sample and the temperature is measured by an embedded thermocouple. It is assumed that the radiation emitted by the burner flame does not affect the emittance measurement in the wavelength range of interest. However, the sample temperature variation can be considerable during burner heating, because the flame is in general unstable. Although foam ceramics have high porosity and the internal pores may be thought to behave like blackbodies, the measured spectrum of emittance is quite different from that of a gray body (a sample or source with a constant emissivity less than 1, independent of wavelength) as shown in Fig. 9.11. The emission spectrum measured of the sample when heated on only one side is also shown. Comparison between the two spectra (c) and (d) implies that the possible temperature gradient within the one-side heated sample decreases the measured emittance.

Figure 9.10c shows the structure of a heating furnace for column alumina. A stainless pipe with lids is heated by a coiled sheathed heater to provide thermal equilibrium for the internal cavity at a temperature of 1273 K. The cavity has a small hole for measurement. The hole can be regarded as a blackbody surface when the alumina sample is not set within the cavity. The column alumina sample has a diameter of 30 mm, which is the sample thickness in this measurement method. Two samples with purities of 99.5% and 92% (6% SiO_2 and infinitesimal MgO, CaO, Na_2O, and Fe_2O_3) were prepared. The furnace temperature is measured by a thermocouple welded onto the surface of the stainless pipe and it is assumed that the sample temperature is the same. The pipe is covered with an insulator and surrounded by a layered radiation shield. In Fig. 9.11, the normal spectral emittance is seen to be the highest for all wavelengths, because the sample thickness is much larger than the optical thickness of alumina. Especially, it is notable that spectral shape of the emittance for thick alumina is significantly different from that of thin alumina. The emittance of the 92% alumina sample is clearly higher than that of the 99.5% alumina in the wavelength range from 1 to 3.5 μm. This means that the spectral emittance depends on the type and quantity of the impurities in the alumina sample. An improvement to this experiment would be to develop a means to distinguish the radiation emitted from sample from that scattered through the sample.

The heating unit with a ceramic heater for the alumina plate sample is illustrated in Fig. 9.10d. The ceramic heater irradiates the back side of the alumina plate, whose purity and thickness are 99% and 1 mm, respectively. The ceramic heater consists of an electrical heating element and a polycrystalline alumina (2.24 mm thickness) matrix and is in contact with the alumina plate sample. Therefore, the maximum thickness of the alumina layer in the sample-heating configuration is 3.24 mm. The sample temperature is measured with an embedded thermocouple. The output of the ceramic heater is controlled to maintain the sample temperature at 1273 K. The sample-heating configuration is covered with an insulating medium, which has a hole for the emission measurement and is supported by two hollow ceramic rods. A comparison of the spectra (g) and (a) or (b) in Fig. 9.11 indicates that the normal spectral emittance of the alumina plate is greater than that of the alumina chip. This is because the plate sample is much thicker than the chip, whose thickness is only 0.7 mm. However, it should be noted that a significant temperature distribution, which can reduce the measured emittance, exists within the plate sample. A typical spectrum of the emittance for the alumina plate coated with a glaze-based erbium oxide film with thickness about 0.1 mm is also shown for the wavelength range from 1 to 2 μm. The spectrum indicates that the coating can enhance the emittance of the alumina substrate significantly around 1.5 μm. It is necessary to optimize the coating parameters such as the concentration of Er, film thickness, and crystal structure so that the emittance becomes an achievable maximum value and the film has also sufficient durability and thermostability for the TPV application.

REFERENCES

[1] A. Cezairliyan, M.S. Morse, H.A. Berman, C.W. Beckett, High-speed (subsecond) measurement of heat capacity, electrical resistivity, and thermal radiation properties of molybdenum in the range 1900 to 2800 K, J. Res. Nat. Bur. Stand. 74A (1970) 65–92.

[2] R. Taylor, W. Kimbrough, R. Powell, Thermophysical properties of tantalum, tungsten, and tantalum-10 wt. per cent tungsten at high temperatures, J. Less-Common Met. 24 (1971) 369–382.
[3] L.M. Jakob, Heat Transfer, vol. 1, Wiley, New York, 1967.
[4] L.M. Hanssen, K. Snail, Integrating spheres for mid- and near-infrared reflection in spectroscopy, in: J. Chalmers, P. Griffiths (Eds.), Handbook of Vibrational Spectroscopy, vol. 1, Wiley, New York, 2002, pp. 1175–1191.
[5] T. Yoshida, S. Tanaka, T. Makino, A new ellipsoidal mirror-type reflectometer for measuring normal-incident hemispherical reflectance spectrum, JSME Int. J. 36B (1993) 166–171.
[6] T. Makino, T. Yoshida, S. Tanaka, A new reflectometer for measuring the spectrum of hemispherical reflectance for perfect-diffuse hemispherical irradiation, Heat Transfer Jpn. Res. 23 (1994) 103–115.
[7] R. Siegel, J.R. Howell, Thermal Radiation Heat Transfer, third ed., Taylor & Francis, Bristol, 1992, pp. 47–91, pp. 189–364.
[8] J. Ishii, A. Ono, Uncertainty estimation for emissivity measurements near room temperature with a Fourier transform spectrometer, Meas. Sci. Technol. 12 (2001) 2103–2112.
[9] J. Ishii, A. Ono, Fourier transform spectrometer for thermal-infrared emissivity measurements near room temperatures, in: Proc. SPIE, vol. 4103-1, 2000, pp. 126–132.
[10] J. Ishii, A. Ono, A Fourier-transform spectrometer for accurate thermometric applications at low temperatures, in: D.C. Ripple (Ed.), Temperature: Its Measurement and Control in Science and Industry, vol. 7, AIP, New York, 2003, pp. 705–710.
[11] H.E. Revercomb, H. Buijs, H.B. Howell, D.D. LaPorte, W.L. Smith, L.A. Sromovsky, Radiometric calibration of IR Fourier transform spectrometers: solution to problem with the high-resolution interferometer sounder, Appl. Optics 27 (1988) 3210–3218.
[12] A. Shimota, H. Kobayashi, S. Kadokura, Radiometric calibration for the airborne interferometric monitor for greenhouse gases simulator, Appl. Optics 38 (1999) 571–576.
[13] S.N. Mekhontsev, A.V. Prokhorov, L.M. Hanssen, Experimental characterization of blackbody radiation sources, in: Z.M. Zhang, B.K. Tsai, G. Machin (Eds.), Radiometric Temperature Measurements: I. Fundamentals, Academic Press, Oxford, 2010, pp. 57–136.
[14] L.M. Hanssen, S.N. Mekhontsev, J. Zeng, A.V. Prokhorov, Evaluation of blackbody cavity emissivity in the infrared using total integrated scatter measurements, Int. J. Thermophys. 29 (2008) 352–369.
[15] A.V. Prokhorov, L.M. Hanssen, S.N. Mekhontsev, Calculation of the radiation characteristics of blackbody radiation sources, in: Radiometric Temperature Measurements: II. Applications, Academic Press, Oxford, 2010, pp. 181–240.
[16] D.K. Edwards, Thermal radiation measurements, in: E.R.G. Eckert, R.J. Goldstein (Eds.), Measurements in Heat Transfer, second ed., Hemisphere Publishing, Washington, DC, 1976, pp. 425–473.
[17] N. Terada, K. Ohnishi, M. Kobayashi, T. Kunitomo, Spectral radiative properties of a living human body, Int. J. Thermophys. 7 (1986) 1101–1113.
[18] J.R. Markham, K. Kinsella, R.M. Carangelo, C.R. Brouillette, M.D. Carangelo, P.E. Best, P.R. Solomon, Bench top Fourier transform infrared based instrument for simultaneously measuring surface spectral emittance and temperature, Rev. Sci. Instrum. 64 (1993) 2515–2522.
[19] K.A. Snail, L.M. Hanssen, Magnification of conic mirror reflectometers, Appl. Optics 37 (1998) 4143–4149.
[20] A. Seifter, K. Boboridis, A.W. Obst, Emissivity measurements on metallic surfaces with various degrees of roughness: a comparison of laser polarimetry and integrating sphere reflectometry, Int. J. Thermophys. 25 (2004) 547–560.

[21] L.M. Hanssen, C.P. Cagran, A.V. Prokhorov, S.N. Mekhontsev, V.B. Khromchenko, Use of a high-temperature integrating sphere reflectometer for surface-temperature measurements, Int. J. Thermophys. 28 (2007) 566–580.
[22] H.J. Lee, A.C. Bryson, Z.M. Zhang, Measurement and modeling of the emittance of silicon wafers with anisotropic roughness, Int. J. Thermophys. 28 (2007) 918–933.
[23] For example, Shimadzu Corporation, "Diffuse Reflectance Measurement," http://www.shimadzu.eu/diffuse-reflectance-measurement (2014).
[24] H. Wakabayashi, T. Makino, A new spectrophotometer system for measuring thermal radiation phenomena in a 0.30–11 μm wavelength region, Meas. Sci. Technol. 12 (2001) 2113–2120.
[25] T. Makino, H. Wakabayashi, A new spectrophotometer system for measuring hemispherical reflectance and normal emittance of real surfaces simultaneously, Int. J. Thermophys. 31 (2010) 2283–2294.
[26] T. Makino, H. Wakabayashi, New spectrophotometer system for measuring thermal radiation characteristics of real surfaces of thermal engineering entirely, J. Therm. Sci. Technol. 6 (2011) 80–92.
[27] T. Makino, K. Kawasaki, T. Kunitomo, Study of the radiative properties of heat resisting metals and alloys (1st report, optical constants and emissivities of nickel, cobalt and chromium), Bull. JSME 25 (1982) 804–811.
[28] G. Alfano, A. Sarnö, Normal and hemispherical thermal emittances of cylindrical cavities, Trans. ASME 97C (1975) 387–390.
[29] H. Wakabayashi, T. Makino, Thermal radiation phenomena of surfaces of chromium and palladium in a high-temperature environment, Int. J. Thermophys. 32 (2011) 2112–2126.
[30] T. Makino, H. Wakabayashi, Thermal radiation spectroscopy diagnosis for temperature and microstructure of surfaces, JSME Int. J. 46B (2003) 500–509.
[31] J.M. Ziman, A theory of the electrical properties of liquid metals. I: the monovalent metals, Philos. Mag. 6 (1961) 1013–1034.
[32] C. Stubbs, E. Prideaux, A spectro-photometric comparison of the emissivity of solid and liquid gold at high temperatures with that of a full radiator, Proc. R. Soc. Lond. A 87 (1912) 451–465.
[33] C. Stubbs, A spectro-photometric comparison of the emissivity of solid and liquid copper and of liquid silver at high temperatures with that of a full radiator, Proc. R. Soc. Lond. A 88 (1913) 195–205.
[34] H. Watanabe, M. Susa, H. Fukuyama, K. Nagata, Phase dependence (liquid/solid) of normal spectral emissivities of noble metals at melting points, Int. J. Thermophys. 24 (2003) 223–237.
[35] C. Ronchi, J. Hiernaut, G. Hyland, Emissivity X points in solid and liquid refractory transition metals, Metrologia 29 (1992) 261–271.
[36] D.J. Price, The temperature variation of the emissivity of metals in the near infra-red, Proc. Phys. Soc. 59 (1947) 131–138.
[37] H. Watanabe, M. Susa, K. Nagata, Spectral emissivities of metals and semiconductors at the melting point, Jpn. J. Thermophys. Prop. 13 (1999) 192–200.
[38] A. Cezairliyan, A. Miiller, F. Righini, A. Rosso, Radiance temperature of metals at their melting points as possible high temperature secondary reference points, in: J.F. Schooley (Ed.), Temperature: Its Measurement and Control in Science and Industry, vol. 5, AIP, New York, 1982, pp. 377–381.
[39] A. Cezairliyan, A.P. Miller, F. Righini, A. Rosso, Radiance temperature and normal spectral emissivity of metals at their melting-point as possible reference values, in: J.F. Schooley (Ed.), Temperature: Its Measurement and Control in Science and Industry, vol. 6, AIP, New York, 1992, pp. 377–382.

[40] G. Pottlacher, A. Seifter, Microsecond laser polarimetry for emissivity measurements on liquid metals at high temperatures—application to tantalum, Int. J. Thermophys. 23 (2002) 1281–1291.
[41] A. Seifter, F. Sachsenhofer, G. Pottlacher, A fast laser polarimeter improving a microsecond pulse heating system, Int. J. Thermophys. 23 (2002) 1267–1280.
[42] I. Egry, A. Diefenbach, W. Dreier, J. Piller, Containerless processing in space—thermophysical property measurements using electromagnetic levitation, Int. J. Thermophys. 22 (2001) 569–578.
[43] S. Krishnan, G. Hansen, R. Hauge, J. Margrave, Emissivities and optical constants of electromagnetically levitated liquid metals as functions of temperature and wavelength, High Temp. Sci. 26 (1990) 143–164.
[44] S. Krishnan, G. Hansen, R. Rauge, J. Margrave, Spectral emissivities and optical properties of electromagnetically levitated liquid metals as functions of temperature and wavelength, High Temp. Sci. 29 (1990) 17–52.
[45] S. Krishnan, J.R. Weber, P.C. Nordine, R.A. Schiffman, R.H. Hauge, J. Margrave, Spectral emissivities and optical properties of liquid silicon, aluminum, titanium, and niobium at 632.8 nm, High Temp. Sci. 30 (1990) 137–153.
[46] R. Kurosawa, T. Inoue, Y. Baba, K. Sugioka, M. Kubo, T. Tsukada, H. Fukuyama, Normal spectral emissivity measurement of molten copper using an electromagnetic levitator superimposed with a static magnetic field, Meas. Sci. Technol. 24 (2013) 015603.
[47] H. Watanabe, M. Susa, K. Nagata, Discontinuity in normal spectral emissivity of solid and liquid copper at the melting point, Metall. Mater. Trans. A 28 (1997) 2507–2513.
[48] H. Watanabe, M. Susa, H. Fukuyama, K. Nagata, Phase (liquid/solid) dependence of the normal spectral emissivity for iron, cobalt, and nickel at melting points, Int. J. Thermophys. 24 (2003) 473–488.
[49] H. Watanabe, M. Susa, H. Fukuyama, K. Nagata, Near-infrared spectral emissivity of Cu, Ag, and Au in the liquid and solid states at their melting points, Int. J. Thermophys. 24 (2003) 1105–1120.
[50] H. Watanabe, M. Susa, H. Fukuyama, K. Nagata, Emissivities of liquid and solid silicon at the melting point, High Temp. High Press. 31 (6) (1999) 587–594.
[51] T. Ishikawa, Y. Ito, J.T. Okada, P.-F. Paradis, Y. Watanabe, T. Masaki, Spectral emissivity measurements of liquid refractory metals by spectrometers combined with an electrostatic levitator, Meas. Sci. Technol. 23 (2012) 125602.
[52] J.P. Hiernaut, F. Sakuma, C. Ronchi, Determination of the melting point and the emissivity of refractory metals with a six-wavelength pyrometer, High Temp. High Press. 21 (1988) 139–148.
[53] Y.S. Touloukian, D.P. DeWitt, in: Thermophysical Properties of Matter, vol. 8, Plenum Press, New York, 1972.
[54] J.F. Chaney, V. Ramdas, C.R. Rodriquez, Thermophysical Properties Research Literature Retrieval Guide, Plenum, New York, 1982.
[55] C. Nordine, E.I. Hang Lee, J.L. Hurd, Spectral emittance of polycrystalline alumina, Rev. Int. Hautes Temp. Refract. 17 (1980) 165–171.
[56] M.A. Postlethwait, M.F. Modest, M.A. Botch, J.R. Hellmann, Normal spectral emittance of alumina based materials, in: Y. Bayaztoglu (Ed.), Radiative Heat Transfer, HTD-vol. 276, ASME, New York, 1994, pp. 73–77.
[57] A. Licciulli, D. Diso, G. Torsello, S. Tundo, A. Maffezzoli, M. Lomascolo, M. Mazzer, The challenge of high-performance selective emitters for thermophotovoltaic applications, Semicond. Sci. Technol. 18 (2003) 174–183.

[58] D.L. Chubb, A.T. Pal, M.O. Patton, P.P. Jenkins, Rare earth doped high temperature ceramic selective emitters, J. Eur. Ceram. Sci. 19 (1999) 2551–2562.

[59] H. Sai, H. Yugami, K. Nakamura, N. Nakagawa, H. Ohtsubo, S. Maruyama, Selective emission of $Al_2O_3/Er_3Al_5O_{12}$ eutectic composite for thermophotovoltaic generation of electricity, Jpn. J. Appl. Phys. 39 (2000) 1957–1961.

[60] D. Diso, A. Licciulli, A. Bianco, M. Lomascolo, G. Leo, M. Mazzer, S. Tundo, G. Torsello, A. Maffezzoli, Erbium containing ceramic emitters for thermophotovoltaic energy conversion, Mater. Sci. Eng. B98 (2003) 144–149.

[61] A. Licciulli, A. Maffezzoli, D. Diso, Porous garnet coatings tailoring the emissivity of thermostructural materials, J. Sol-Gel Sci. Technol. 32 (2004) 247–251.

[62] W.J. Tobler, W. Durisch, Plasma-spray coated rare-earth oxides on molybdenum disilicide—high temperature stable emitters for thermophotovoltaics, Appl. Energy 85 (2008) 371–383.

[63] T. Kumano, K. Hanamura, Spectral emission characteristics of a high-temperature alumina-based emitter for TPV, Therm. Sci. Eng. 13 (2005) 13–16.

[64] T. Kumano, K. Hanamura, Development of a prototype thermophotovoltaic power generation system using super-adiabatic combustion in porous quartz glass, J. Therm. Sci. Technol. 7 (2012) 549–562.

[65] T. Kumano, Direct measurement of spectral emittance of ceramics in high temperature, in: Proc. of 46th National Heat Transfer Symposium of Japan, vol. 1, 2009, pp. 75–76.

[66] T. Kumano, K. Hanamura, Development of ceramic glaze with specific near-infrared radiation by doping of rare-earth elements, in: Proc. 31st Jpn. Symp. Thermophys. Prop., 2010, pp. 203–205.

[67] T. Kumano, K. Hanamura, Study on coating conditions of rare-earth oxide film for control of spectral emittance of ceramics, in: Proc. 33rd Jpn. Symp. Thermophys. Prop., 2012, pp. 221–223.

Chapter 10

Color and Appearance

Maria E. Nadal*, Dave Wyble† and Clarence J. Zarobila*
*National Institute of Standards and Technology, Gaithersburg, Maryland, USA
†Avian Rochester, LLC, Webster, New York, USA

Chapter Outline

10.1 Introduction 367
10.2 Spectral Attributes—Color 368
10.2.1 Definition 368
10.2.2 CIE System 370
10.2.3 Metamerism 376
10.2.4 Chromaticity Coordinates and Color Spaces 377
10.2.5 Color Difference Metrics 380
10.3 Color-Measuring Instruments 383
10.3.1 Geometric Considerations 385
10.3.2 Source Considerations 389
10.3.3 Detection and Signal Processing Considerations 390
10.3.4 Uncertainties 392
10.4 Gonioapparent Materials 394
10.5 Geometrical Attributes 396
10.5.1 Specular Gloss 397
10.5.2 Haze 400
10.5.3 Distinctness of Image 403
References 404

10.1 INTRODUCTION

The quality and acceptability of manufactured products are often judged by their overall appearances. Although the need for accurate measurements of these attributes has been recognized for many decades, increasing globalization—particularly in manufacturing—and new materials, such as gonioapparent materials, have spurred efforts to develop further methods for the characterization of the appearance and color of objects. The measurable attributes of appearance are divided into color (spectral properties) and geometry (spatial properties). The appearance of an object cannot be measured; only the specific spectral and spatial reflectance or transmittance characteristics of the object are measured.

Experimental Methods in the Physical Sciences, Vol. 46. http://dx.doi.org/10.1016/B978-0-12-386022-4.00010-8

The appearance of an object is a complex interaction of the light field impinging on the object, the optical properties of the object, and the pyschophysiology of human perception [1]. The visual appearance of objects is determined by the parts of the visible light that are reflected and/or transmitted without being absorbed. The overall appearance of any object is a combination of its chromatic attributes and its geometric attributes. The chromatic attributes are those that vary the spectral distribution of the reflected or transmitted light and are those attributes associated with color. Color is a three-dimensional quantity usually divided into lightness, hue, and saturation. Lightness refers to the color being closer to black or white. Hue is the perceived color such as yellow or green, and saturation is the purity of a color. The scientific basis for the measurement of color attributes has been under active investigation for the better part of a century and has been codified by the Commission Internationale de l'Eclairage (CIE) as the CIE Standard Observer [2] which is a representation of normal human color vision. Since the appearance of an object is also a function of the spectral distribution of the light source, the CIE has also defined a series of standard spectral distributions which are known as the CIE illuminants [3]. The geometric attributes are based on the directional distribution of reflected or transmitted light, described by attributes like specular gloss, haze, orange peel, and so on. Color and gloss are the attributes most often use to characterize the appearance of objects.

10.2 SPECTRAL ATTRIBUTES—COLOR

10.2.1 Definition

Ultimately, the goal of color measurement is to model the perception of a human observer. The human visual system (HVS) has three overlapping broadband detectors: short (S)-, medium (M)-, and long (L)-wavelength cones. The average spectral sensitivities of these are shown in Fig. 10.1. In principle, an instrument that accurately quantifies the behavior of these three detectors should be able to predict color perception. Even with such a high spectral bandwidth, a three-channel system can do a very good job of color measurement. The limits of such a system will be shown below in the discussion on the relationship among the light source, the object, and the human observer.

To understand color measurement, we must explore the system under which a human observer experiences color. For color perception of an object to exist, there need to be three components: a light source, an object, and a human observer. Figure 10.2 shows the relationship between these components. Light from the source is incident on the object (arrow 1), and some of that light reflects off the object (arrow 2). Thus, the observer sees the original light after modification by the object. The third arrow indicates that the source also has a direct effect on the human observer, and hence the color perception. The perception cannot be described simply as the physical

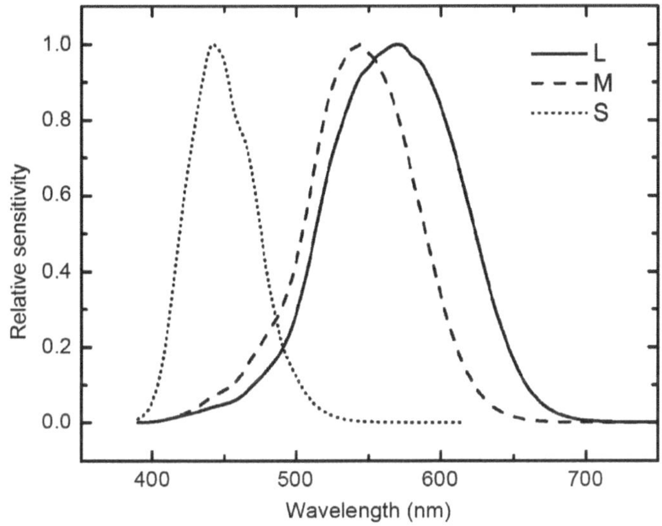

FIGURE 10.1 The average spectral sensitivity of the human visual system.

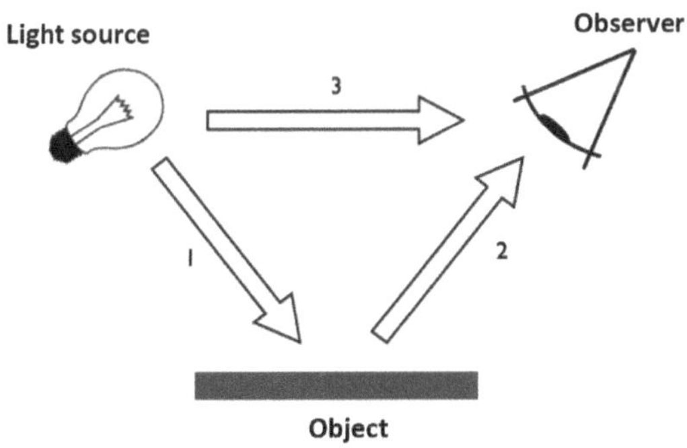

FIGURE 10.2 Requirements for the perception of color.

relationship between the spectral properties of the source and the object. Human observers adapt to the source, and color can be greatly affected by this adaptation. Adaptation—a physiological response—must be accounted for to properly estimate color from instrumental data. As will be explored in Section 10.2.2, the color specification of an object always includes information about the source and the observer. Without these data, a color specification is useless, akin to stating a distance without specifying the units.

Related to arrow 3 in Fig. 10.2, color appearance attempts to account for even more aspects of color than the properties of the light source and the object. Note that appearance has come to have two definitions: one as described above related to geometric properties such as haze, gloss, and others; and color appearance which implies the consideration of the complete state of the observer when determining the color perception. For example, color appearance terminology includes the background and surrounding fields beyond the actual color stimulus. Also, temporal aspects can account for the observer's recent experience. Consider coming indoors from a bright sunlight day. The room seems unreasonably dark until one's eyes adapt to the lower light level. Likewise, color appearance models can account for so-called illusions, such as simultaneous contrast, and edge effects.

We do not raise the concepts of color appearance to later connect to color measurement, but rather to aid in the understanding that color measurement does have some limitations. However, when accounted for, these limitations do not preclude the useful application of color measurement. For example, all useful color appearance models [4] also include detailed conditions under which they are valid. At a bare minimum, these models need to comprehend the spectral (or color) nature of the incident light. More in-depth models apply parameters for viewing surround (the visual field outside the actual sample); the adaptation state of the observer (dark adapted, or not); and other properties of the viewing geometry. Ultimately, to fully model the color seen by an observer, one must account for the properties of the entire field of view of the observer, as well as what that field of view has been over the last several minutes. Fortunately, although these complex models are of interest to the vision and appearance community, simplifications can be applied which predict color well enough to allow for reasonably accurate color communication required for most commercial applications.

The physical properties of reflectance, transmittance, and others, described in Chapter 2, are usually specified in spectral terms. That is, a given property is determined at each wavelength within a range of interest. Hence, an instrument measuring a spectral quantity can reasonably be considered to have a whole series of detectors, one at each wavelength of interest. The HVS, however, is not a spectral system, but a trichromatic one. This is due to the presence of three types of color "detectors" in the retina, which are roughly sensitive in the blue, green, and red areas of the spectrum, as shown in Fig. 10.1.

10.2.2 CIE System

Modern color specification systems are based on a system developed by CIE. In the 1920s, the CIE addressed a commercial need to accurately specify color by instrumental means. At that time, there were numeric methods to specify colors, the Munsell [5,6] system being one of note. Without instrumental measurement, a color specification could only be provided by visually comparing a sample to a set of physical Munsell colors of known coordinates. The system works well so long as the various sets of physical colors are identical and the people making the color judgments had similar color vision. In practice, physical standards will fade, and the color vision of humans can vary significantly. The CIE colorimetric system [7] advanced the ability to determine a set of color coordinates solely by instrumental means, thus removing the variability of the human observer and standardizing color measurement to a common basis of methods. To fulfill these requirements, a system must first account for the degree to which an average human observer reacts to light of different wavelengths. Extensive visual experiments were designed to determine the human response, resulting in a series of standardized color-matching functions.

10.2.2.1 Color-Matching Functions

The standardized color-matching functions were derived by visual experiment in which normal observers attempted to create mixtures of three colored lights that best matched a monochromatic stimulus [8]. The observer looked at a white screen through an aperture having a 2° field of view. At the center of the field, the white screen was separated into two halves by an opaque black partition. The bottom half of the screen was illuminated by the monochromatic light to be matched, and during the course of the experiment, the wavelength of this light was varied. The upper half of the screen was illuminated by an additive mixture of three "primary" colors (Red, Green, Blue or RGB). The observer adjusted the intensity of the three colored lights until both halves of the screen appeared identical in color. The amount of energy in each of the three lights was recorded, and the experiment progressed to the next stimulus wavelength. Three primary RGB colors were used, but in fact, any colors can be used so long as not one of them can be produced by an additive mixture of the other two.

By recording the amount of RGB light required for the best match at each wavelength, spectral curves can be derived that correlate to human perception at each wavelength. It was not actually possible for the observers to match each monochromatic test stimulus with the RGB primaries. The solution was to modify the apparatus to permit the addition of one or more of the RGB lights to the test side of the visual field. In essence, this means that a particular wavelength requires a negative amount of $\bar{r}$, $\bar{g}$, or $\bar{b}$ light to obtain a match. This is shown in Fig. 10.3 on the left. Note that the negative areas do not indicate any particular flaw or issue with the HVS; these are only based on the selection of the physical RGB lamps used for the visual experiment. However, the negative lobes presented difficulty for the technology available at the time, and for this and other reasons, the curves were transformed to be all-positive using techniques developed by Judd [9,10]. These transformed curves, known as the "$\bar{x}$," "$\bar{y}$," and "$\bar{z}$," primaries, are shown at the right in Fig. 10.3. They cannot be produced by any real lights and the observer that is defined by these color-matching functions is known as the CIE 1931 Standard Colorimetric Observer [11].

The transformation was constructed so that (a) negative coefficients in the transformation equations were eliminated; (b) to maintain consistency with

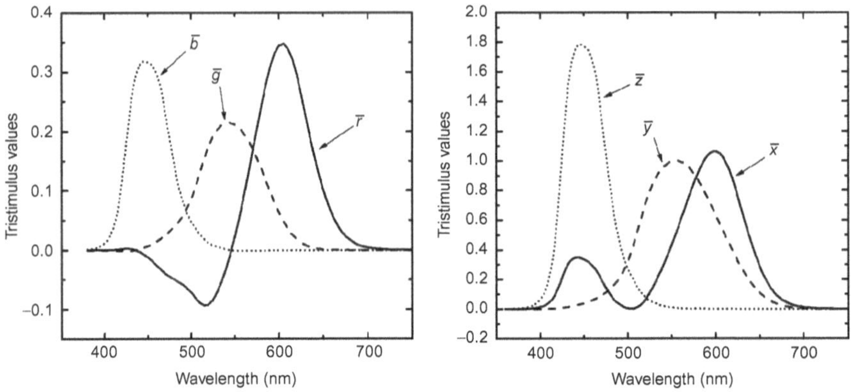

FIGURE 10.3 Basis of human perception (left) and the CIE 1931 standard observer color-matching functions (right).

existing systems, the "$\bar{y}$," curve was constrained to match the CIE 1924 spectral luminous efficiency $V(\lambda)$, which is introduced in Chapter 2 (see Eq. 2.26 and Fig. 2.4) and is a metric that describes how efficient the eye is at each wavelength in converting light power to luminous sensation; (c) one of the functions was selected to be close to zero as possible for as much of the spectrum as possible; and (d) the areas under the curves of each function were made equal for light of equal energy at all wavelengths. The discrete x_λ, y_λ, and z_λ values of X, Y, and Z at each wavelength are known as the respective CIE tristimulus values for that wavelength, λ.

It was subsequently pointed out that the CIE Z function does not give as much weight to the shortwave violet as actual human observers who graded industrial white panels with their eyes. Such grading involved a larger field of view than was used in the 2° color-matching experiments. Thus, new experiments were performed with a larger field of view and it was noted that observers do in fact have important differences in their shortwave response. Accordingly, in 1960, the CIE proposed additional color-matching functions in an effort to obtain better correlation with commercial judgments. The resulting $\bar{x}_{10}$, $\bar{y}_{10}$, and $\bar{z}_{10}$ functions are collectively known as the CIE 1964 color-matching functions [12,13]. Tables for the 1931 and 1964 CIE 1931 and 1964 color-matching functions can be found in the literature [14] and on the Internet.

10.2.2.2 Standard Illuminants

In addition to functions that represent the response of the average human observer, the CIE has defined "illuminants" that refer to specific energy distributions S_λ that are spectrally close to natural light sources and are realized by actual sources to close approximation in the laboratory. They are particularly useful as standard spectral distributions for color calculations. A brief description of some is given below.

CIE standard illuminant A represents light from a tungsten filament lamp. The relative power distribution is that of a Planckian radiator of approximately 2856 K according to the International Temperature Scale of 1990. The spectral distribution may be calculated using Planck's radiation law, which gives the spectral radiance of a blackbody radiator as a function of wavelength and temperature, given in Eq. (2.22). Illuminant A data are typically presented normalized to the result at 560 nm, namely, as 100 M/M_{560}. Usually, the results are rounded to two decimal places so that $M(560) = 100.00$.

An important property of any CIE illuminant or physical light source is its relationship to one particular expansion of Planck's equation. The *color temperature* of a blackbody source is simply the temperature T in Eq. (2.22). For the example of CIE standard illuminant A, the color temperature is 2856 K. The use of "color temperature" is limited to actual blackbody curves. For all others, the more appropriate metric is correlated color temperature (CCT). The CCT of a test light source is defined as the color temperature of the blackbody curve with the nearest visual match to the test source. Figure 10.4 plots the normalized blackbody curves for three different CCTs.

Illuminant C is an approximation of average daylight from the total sky having a CCT of approximately 6774 K and is also obtained using a combination of illuminant A and a special liquid filter. Historically, illuminant C was widely used in various industries, and many object color specifications written in terms of appearance under illuminant C are still in use. Illuminant C has

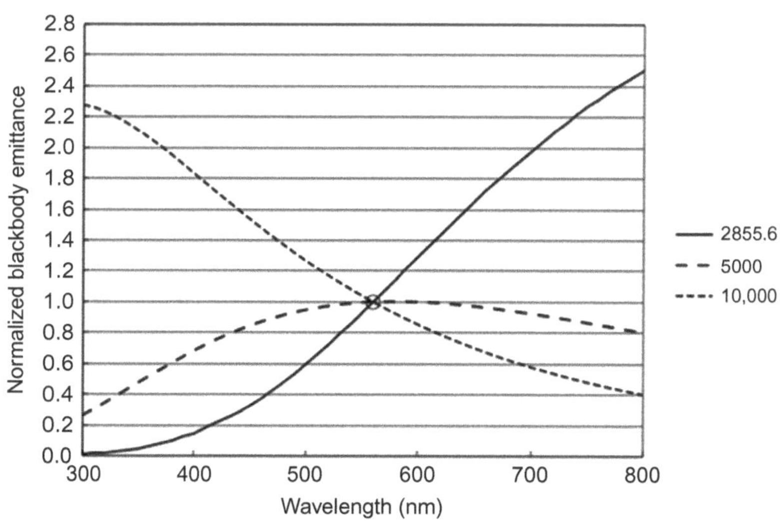

FIGURE 10.4 Normalized blackbody curves for three different color temperatures (K).

fallen out of favor as a standard illuminant (because of the introduction of the "D" series) despite its continuing use in colorimetric specifications.

In 1966, the "D" series of illuminants was proposed by the CIE and published in 1971 [6]. They are designed to more accurately represent daylight and are especially important in the evaluation of fluorescing or optically brightened materials where the absorption in the UV range of the spectrum contributes to color appearance. They are designated by the first two digits of the CCT, and thus D50, D55, D65, and D75 refer to illuminants with CCTs of about 5000, 5500, 6500, and 7500 K, respectively. If for some reason the color scientist desires a daylight simulator with some other CCT, methods are available to transform the daylight distribution to that having the desired CCT. In the interests of standardization however, the CIE recommends that D65 be used whenever possible. When D65 (known as "Average Daylight") cannot be used, the CIE recommends that either illuminant D55 (Noon Sky Daylight) or D75 (North Sky Daylight) be used whenever possible. Due to the specific spectral requirements of daylight simulators used to excite fluorescence, methods for assessing the quality of a test source meant to act as a practical reproduction of the D55, D65, or D75 illuminants have been developed [15]. See Fig. 10.5 for a plot of standard illuminants A and D65, and illuminant C.

As technology advanced and as artificial lighting from fluorescent lamps became more prevalent, as a matter of practical importance for color calculations, additional representative illuminants have been introduced by the CIE. Examples of such distributions include F2, F7, and F12. F2 represents the emission from a "standard" fluorescent lamp; F7 represents the broadband emission from a multiple-phosphor fluorescent lamp; and F12 represents emission from a narrowband fluorescent lamp weighted in the RGB regions of the spectrum.

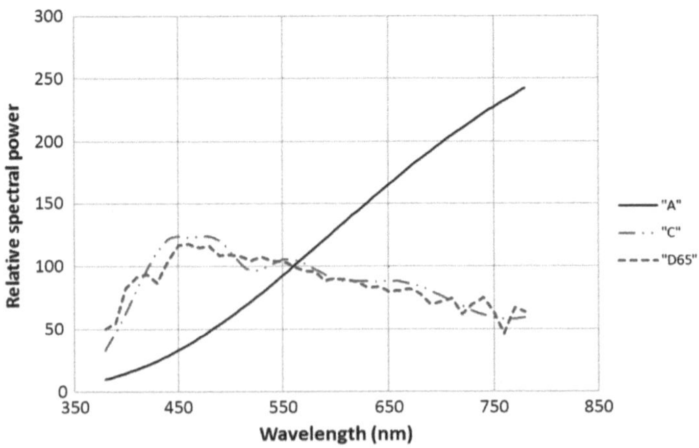

FIGURE 10.5 Relative spectral power of CIE standard illuminants A and D65 and illuminant C.

CIE standard illuminants A and D65 relative spectral power distributions are given in 1 nm increments in Table T.1 of Ref. [3]. Compilations of the spectral distributions of the CIE illuminants and the representative sources may be found in color handbooks and on the Internet. ASTM Standard Practice E308 also gives compilations for those illuminants of interest to the color scientist or industrial color professional [16].

10.2.2.3 Calculation of Integrated Tristimulus Values and Corresponding Weighting Functions

It was previously mentioned that the perceived color of an object depends upon the light source, the properties of the object that modify the light source upon reflection or transmission, and the human observer. Mathematically, the perceived object color can be found by integrating the product of the spectral source irradiance, the spectral reflectance (or radiance factor, or transmittance) of the object, and the CIE tristimulus values (1931 or 1964) over the entire wavelength range for which the eye responds. The results are the CIE tristimulus values (X, Y, Z or X_{10}, Y_{10}, Z_{10} of the color stimulus). In the absence of an object (where in this case the reflectance or transmittance equals unity), the perceived color of the light source is found, and in the case of light sources that are CIE illuminants, the calculations result in the *white point* of the illuminant.

In the case of a tristimulus colorimeter (see Section 10.3.3), the integrals are inherently carried out through the use of separate filters whose spectral transmittance curves closely match that of the standard observer functions or to some other values that transform to the desired illuminant/observer (e.g., $C/2°$) combination. Thus, the source stimulus must ultimately be carefully controlled to achieve reproducible results. With instruments that report spectral reflectance or transmittance data versus wavelength, the integrals are approximated by summations according to

$$\begin{aligned} X &= k\sum S(\lambda)R(\lambda)x(\lambda)\Delta\lambda, \quad Y = k\sum S(\lambda)R(\lambda)y(\lambda)\Delta\lambda, \\ Z &= k\sum S(\lambda)R(\lambda)z(\lambda)\Delta\lambda, \end{aligned} \tag{10.1}$$

where $S(\lambda)$ is the relative spectral irradiance distribution of the source, $R(\lambda)$ is the spectral reflectance (or spectral radiance factor, or spectral transmittance) of the object, $\Delta\lambda$ is the wavelength interval, and k is a normalization defined to be

$$k = \frac{100}{\sum S(\lambda)y(\lambda)\Delta\lambda}. \tag{10.2}$$

For the highest accuracy in color measurements, the CIE specifies that this tristimulus calculation be performed over the wavelength range of 360–830 nm at 1-nm intervals and a 1-nm spectral bandpass. However, for most practical applications, a spectral range of 380–780 nm with a 5-nm spectral bandpass and a 5 nm measurement interval gives adequate accuracy. Since the order of the factors in the product to be integrated does not matter, to facilitate calculations, the source, the observer functions, and the normalization can be combined to form weighting factors at each wavelength, λ, to give $WX(\lambda)$, $WY(\lambda)$, and $WZ(\lambda)$; thus Eq. (10.2) becomes

$$X=\sum W_x(\lambda)R(\lambda)\Delta\lambda,\quad Y=\sum W_y(\lambda)R(\lambda)\Delta\lambda,\quad Z=\sum W_z(\lambda)R(\lambda)\Delta\lambda. \quad (10.3)$$

The advantage to this approach is that the weighting factors need to be calculated only once for each desired illuminant–observer combination and thus may be carried as lookup tables in instrument memory. In most industrial spectrophotometers, and particularly those that are handheld and used outside the laboratory, the data are available over a more restricted wavelength range, for example, 400–700 nm and at a much larger spectral bandpass and measurement interval, for example, 10 or 20 nm. To reduce the error in the calculation of the colorimetric values using this abridged or truncated spectral data, CIE provides specific recommendations that include the use of the above weighting factors. Tables of these weighting factors are available in ASTM E308 for both 1931 and 1964 standard colorimetric observers and a number of illuminants and practical sources for increments of 10 and 20 nm [16]. More demanding color applications expand this wavelength range to include those wavelengths in the near UV and the near IR over which the observer functions have been defined. ASTM E308 provides the tables for increments of 1, 5, 10, and 20 nm.

It is important to understand that two colors of identical X, Y, Z coordinates will match. However, should two colors not match in X, Y, Z, it is not possible to estimate the degree of color difference perceived without further calculations. Thus, color difference metrics have been defined, and some are discussed in Section 10.2.5.

10.2.3 Metamerism

The dimensional reduction implied in the tristimulus equations has one critical effect: two different objects with different $R(\lambda)$ can result in the same X, Y, Z values. Effectively, this means that an observer viewing these two objects under identical viewing conditions will perceive the same color. Likewise, by viewing the same object under different light sources S_λ, a different color perception may result. Changing the observer can have the same effect. These phenomena are each different forms of *metamerism*. To be considered metameric, two objects must have the same X, Y, Z values under a given illuminant/observer combination, and different X, Y, Z values under another illuminant/observer combination.

Illuminant metamerism is the specific case of changing X, Y, Z values when the light source is changed. Practically, this implies that if two observers do not use the same light source when viewing a pair of objects, the perceived color might not be the same. Likewise, when calculating instrumental color, the same illuminant must be used in the tristimulus equations. *Observer metamerism* is the specific case of changing X, Y, Z values when a different observer is used. Again, the practical application implies that two observers with different sensitivities might not perceive a color match in the same way. More importantly, observer metamerism means that if a specific observer's sensitivity functions differ from those defined by the CIE, instrumental color matches using CIE functions might not match that observer's perception. The term *metameric failure* is often used to describe a failure of colors to retain a match under different light conditions. Metamerism would seem to be a fundamental flaw in the HVS, making color matching much more difficult than if people simply detected spectra as an instrument does. However, consider the very common application of color printing using a limited number of pigments (occasionally as few as three: cyan, magenta, and yellow). Using these three pigments alone, one could not possibly hope to create all of the spectra present in the natural world. However, metamerism allows the limited spectral variation present to produce a wide range of colors that, within some bounds, are sufficient for most commercial applications. The same argument applies to most computer and television displays, where only red, green, and blue lights are mixed to form the color of each individual pixel. ASTM 4086 describes the visual methods for determining observer and illuminant metamerism [17].

10.2.4 Chromaticity Coordinates and Color Spaces

Light sources are not perceived in the same fashion as object colors. Two-dimensional chromaticity spaces are used, which are projections of a three-dimensional color space along the luminance direction. Therefore, two lights of the same color but different luminance will map to the same point in the chromaticity space. This simplifies the measurement by removing the constraint for an absolute luminance calibration. The CIE has recommended a chromaticity system wherein "chromaticity coordinates" (also known as trichromatic coefficients) x, y, and z are related to the tristimulus values X, Y, and Z as follows:

$$x = \frac{X}{X+Y+Z}, \quad y = \frac{Y}{X+Y+Z}, \quad z = \frac{Z}{X+Y+Z}. \tag{10.4}$$

Only two of the three coordinates are necessary to specify chromaticity since $x+y+z=1$. The CIE recommends the use of x and y. The chromaticity coordinates x_{10}, y_{10}, and z_{10} may be similarly computed for tristimulus values resulting from the 10° observer [18].

The chromaticity coordinates x and y corresponding to the 1931 Standard Colorimetric Observer (and 1964 standard colorimetric observer) may be plotted on a diagram in Cartesian space, and the resulting spectrum locus represents the limits to chromaticity for which all perceived color must fall. Such a diagram is shown on the left in Fig. 10.6, which is the CIE 1931 chromaticity diagram.

The tristimulus value Y with its high correlation to luminance and object lightness can be combined with the chromaticity coordinates x and y to form a color space Yxy (alternately known as CIEXYZ) in an attempt to form a three-dimensional scale for object colors. This space has many useful properties, among which is color matching. If two similarly viewed colors produce identical tristimulus values, it can be concluded that the colors match. The shortfall of CIEXYZ is that should the colors mismatch, even slightly, the perceptual relevance of the color difference cannot be trivially estimated. Because of this and the fact that CIEXYZ space is not uniform, a variety of geometric transformations were developed to map CIEXYZ coordinates to Munsell space, the most perceptually uniform space available. There is a rich history to the development of such uniform color spaces, and in 1976, the CIE recommended a space known as CIELAB for use in the measurement of small color difference primarily when working with diffuse surfaces [19]. The CIELAB color space coordinates L^*, a^*, and b^* are defined by

$$\begin{aligned} L^* &= 116\left[f\left(\frac{Y}{Y_n}\right)-\frac{16}{116}\right], \quad a^* = 500\left[f\left(\frac{X}{X_n}\right)-f\left(\frac{Y}{Y_n}\right)\right], \\ b^* &= 200\left[f\left(\frac{Y}{Y_n}\right)-f\left(\frac{Z}{Z_n}\right)\right], \end{aligned} \tag{10.5}$$

where

$$f\left(\frac{W}{W_n}\right) = \begin{cases} \left(\frac{W}{W_n}\right)^{1/3}, & \frac{W}{W_n} > 0.008566 \\ 7.787\left(\frac{W}{W_n}\right)+\frac{16}{116}, & \frac{W}{W_n} \leq 0.008566 \end{cases}$$

and X_n, Y_n, and Z_n represent the white point of the illuminant mentioned in Section 10.2.2.3. L^* is considered to be a lightness dimension; a^* is a red–green dimension; and b^* is a yellow–blue dimension. The CIELAB chroma is given by

$$C^*_{ab} = \sqrt{a^{*2}+b^{*2}}, \tag{10.6}$$

and the hue is:

$$h_{ab} = \tan^{-1}\left(\frac{b^*}{a^*}\right). \tag{10.7}$$

In addition to this space, the CIE recommended a space for use in measurement of small color differences when working primarily with additive

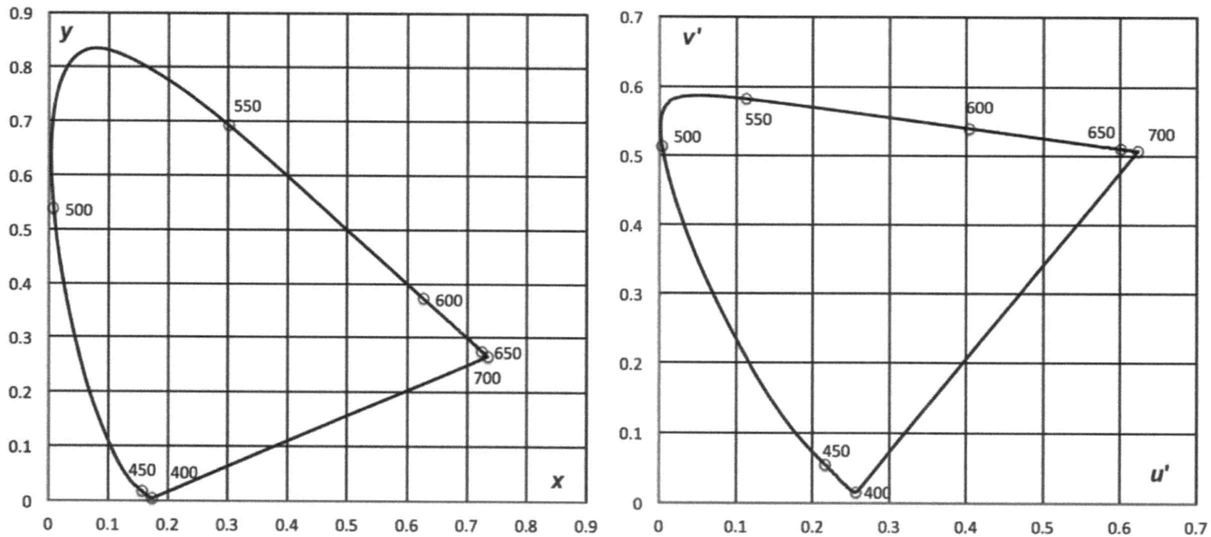

FIGURE 10.6 The CIE 1931 chromaticity diagram (left) and 1960 chromaticity diagram (right). The charts show the position of monochromatic colors in each respective color space. Note the nonuniformity of the wavelength spacing along the spectral locus. Due to this, the (x, y) diagram is not recommended for color difference calculations. The (u', v') space is somewhat better, as it was designed to have increased uniformity of color difference.

mixtures or illuminants. This space, known as the 1976 CIELUV and the 1960 $u'v'$ uniform chromaticity scale diagram, retains L^* as defined in Eq. (10.5) but modifies the red–green and yellow–blue dimensions by introducing the coordinates

$$u^* = 13L^*\left(u' - u'_n\right), \quad v^* = 13L^*\left(v' - v'_n\right), \tag{10.8}$$

where

$$u' = \frac{4X}{(X + 15Y + 3Z)}, \quad v' = \frac{9Y}{(X + 15Y + 3Z)}, \tag{10.9}$$

and the n subscript indicates corresponding u' and v' values calculated for the illuminant white point [20]. Figure 10.6 shows the CIELUV color space plotted on this 1960 chromaticity diagram.

CIELAB addresses the shortfall of X, Y, Z mentioned above, namely that the magnitude of perceived color difference cannot be estimated simply by the differences in X, Y, Z space. The benefits of Munsell space, and hence the Munsell-based CIELAB, are that small color differences correspond to approximately equal distance in the space. Commercially, for object color measurements, CIELAB represents the dominant color space used. The chief application is to create industrial tolerances by which product specifications can be maintained.

10.2.5 Color Difference Metrics

The traditional color difference metric based on CIELAB is the Euclidean distance between two colors denoted as (L_1^*, a_1^*, b_1^*) and (L_2^*, a_2^*, b_2^*) with component differences

$$\Delta L^* = L_1^* - L_2^*, \; \Delta a^* = a_1^* - a_2^*, \; \Delta b^* = b_1^* - b_2^*, \; \Delta E_{ab}^* = \sqrt{\Delta L^{*2} + \Delta a^{*2} + \Delta b^{*2}}, \tag{10.10}$$

or equivalently

$$\Delta E_{ab}^* = \sqrt{\Delta L^{*2} + \Delta C^{*2} + \Delta H^{*2}}, \tag{10.11}$$

where hue difference ΔH^* is defined circularly as the remainder of color difference after accounting for lightness and chroma differences:

$$\Delta H^* = \sqrt{\Delta E_{ab}^{*2} - \Delta L^{*2} - \Delta C_{ab}^{*2}}. \tag{10.12}$$

Often the comparison is made to a standard or reference color, against which individual product samples are compared. Depending on the application, the differences along individual axes should also be evaluated, that is, the differences $\Delta L^*, \Delta a^*, \Delta b^*$, and ΔC_{ab}^*. These component differences can uncover features in the distribution of color difference that are not accounted for by the implied spherical nature of ΔE^*_{ab}. That is, ΔE^*_{ab} compresses all the color differences into a single metric, since it defines a sphere around the reference color.

Over time, the use of the simple geometric distance implied by ΔE^*_{ab} has been shown to be insufficient for many applications. Subsequently, the CIE and other organizations have recommended several other color difference metrics. All of these are ultimately based on CIELAB measurements. Most apply the philosophy of separately weighting the lightness, chroma, and hue component differences. The general form of this family of equations is:

$$\Delta E = K\sqrt{\left(\frac{\Delta L}{k_L S_L}\right)^2 + \left(\frac{\Delta C}{k_C S_C}\right)^2 + \left(\frac{\Delta H}{k_H S_H}\right)^2}. \qquad (10.13)$$

With this form, the component differences ΔL, ΔC, and ΔH are weighted by the scalars S_L, S_C, and S_H to improve the uniformity in the particular region of color space in which the colors in comparison lie. This form effectively defines ellipsoids in color space around the target color. By default, the axes of these ellipsoids fall along the axes of constant L, C, and H. The scalars k_L, k_C, and k_H are parametric factors, applied to all comparisons identically, based on the application. In practice, only two of the three need to be specified (effectively choosing one to always set to unity); the resulting magnitude of color difference can be scaled with the global K parameter, if desired. Since the parametric factors are user defined, proper nomenclature for these color differences should include the values of two of the factors, typically in a form such as "ΔE_{12}." Care must be used, since the factor set to unity is dependent on the particular color difference equation. For example, if k_L is fixed at unity, ΔE_{12} implies $k_C = 1$ and $k_H = 2$.

The weights S_L, S_C, and S_H vary with the location in color space of the standard that is used for comparison. It is particularly important to address *chroma compression*, a visual aspect of color difference. Low chroma colors (neutrals or near-neutral pastels) will show a higher degree of visual difference than high chroma colors for a given ΔE^*_{ab}. To compensate for this, advanced formulae typically calculate S_C as a nonlinear proportion to C, effectively reducing the calculated color difference for high chroma comparisons. Similar adjustments are made for low versus high L depending on the location in the hue circle. Briefly, as the shortfalls of ΔE^*_{ab} became clear, several attempts at improvements have been made. For example, certain difference equations utilize the concepts of a standard or target color and a batch or test color. For some of these, the color difference will change when the standard and batch colors are interchanged.

There are many variations of color difference equations, and they continue to be adapted in both application-specific and general forms. These are too numerous to describe every form here. Some details are shown below for three popular equations: CMC *l*:*c*, CIE94, and CIEDE2000 [21]. The first widely accepted advanced color difference formula was developed by the

Colour Measurement Committee (CMC) of the Society of Dyers and Colourists and is denoted CMC l:c. It uses the general form for ΔE as specified in Eq. (10.12) with small changes in nomenclature:

$$\Delta E^*_{\mathrm{CMC}} = \sqrt{\left(\frac{\Delta L^*}{lS_L}\right)^2 + \left(\frac{\Delta C^*_{ab}}{cS_C}\right)^2 + \left(\frac{\Delta H^*}{S_H}\right)^2}. \tag{10.14}$$

CMC l:c is an example of an equation for which the color difference can change when standard and batch colors are interchanged. Users are free to select the parametric factors l and c. These are typically set to $l=2$ and $c=1$ for visual acceptability or process tolerance, and to $l=1$ and $c=1$ for the threshold of perception. These color difference values should be identified as $\Delta E^*_{\mathrm{CMC}(2:1)}$ and $\Delta E^*_{\mathrm{CMC}(1:1)}$, respectively. The most modern recommendation is CIEDE2000 [22]. The term "ΔE^*_{00}" adopts the general form shown in Eq. (10.13) with the addition of a rotation term:

$$\Delta E^*_{00} = \sqrt{\left(\frac{\Delta L'}{k_L S_L}\right)^2 + \left(\frac{\Delta C'}{k_C S_C}\right)^2 + \left(\frac{\Delta H'}{k_H S_H}\right)^2 + R_T\left(\frac{\Delta C'}{k_C S_C}\right)\left(\frac{\Delta H'}{k_H S_H}\right)}. \tag{10.15}$$

The notation is slightly different as the numerator terms are in turn functions of the color under comparison. The rightmost term under the radical addresses a serious shortfall of ΔE^*_{ab} where visual data show a large rotation of sensitivity in the blue region. The need for this correction is seen in Fig. 10.7, where

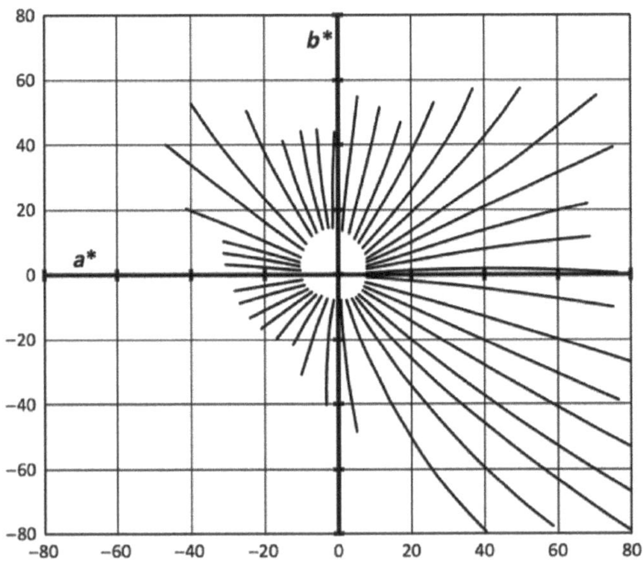

FIGURE 10.7 Munsell hue "iso-lines" plotted in CIELAB space, showing undesirable curvature.

constant Munsell hue lines are plotted in CIELAB. The blue region (fourth quadrant, lower right) shows the curvature addressed by the rotation term.

10.3 COLOR-MEASURING INSTRUMENTS

Chapter 2 provides an in-depth mathematical description of reflectance and transmittance. For the discussion related to color and appearance, the relevant aspect is that light eventually is incident upon the eye or upon an instrument detector that simulates the human eye responsivity. We can therefore consider that reflectance quantifies the light reflecting off an object, and transmittance quantifies the light passing through an object. If the end result of a measurement is to predict color, reflectance or transmittance is determined on a wavelength-by-wavelength basis, or at least captured using special filters to simulate color perception. An important attribute of a spectral measurement system is the bandwidth of light detected around each given wavelength. Again, the specific mathematical details of this are rigorously covered in Chapters 2 and 3. While it is often tempting to desire the smallest bandwidth, and therefore the best spectral resolution, the reality is that narrow bandwidth also implies lower signal and therefore lower signal-to-noise ratio. Color measurement instruments are designed to balance the physical needs of the detection system with the need to accurately quantify materials' physical properties. In general, more weight is placed on accurate color prediction and reproducibility of measurements than on accurate spectral properties. Instrument reproducibility is a key factor when evaluating consistency of manufactured product color from day to day, month to month, or even year to year. It is normally quantified by measuring a set of stable color standards during instrument production and comparing the results to the average of the readings obtained for a small and vetted "pilot" set of instruments. The color difference of the individual instrument's readings with respect to the average for the pilot set is computed, and steps are taken to minimize the difference if required.

Instruments that measure color are generally classified into two types: tristimulus colorimeters and spectrophotometers. Both normally measure values associated with the reflectance factor (see Chapter 6), which is the ratio of the light reflected by the sample being measured to that reflected by a reference specimen. For colorimetric measurements of opaque materials, the CIE recommended reference standard is the perfectly reflecting diffuser (PRD), an ideal reflector wherein 100% of the light is reflected equally in all directions when measured at a hemisphere above the specimen. Such an ideal diffuser cannot be realized and approximations have been in use for quite some time known as secondary white diffusers which have been calibrated using absolute methods to provide traceability to the PRD. ASTM International publishes a Standard Practice for the preparation of such a diffuser [23]. When the process is followed in strict accordance with this standard, the resulting reflectance factor values can be considered to be in good agreement with those provided by an absolute calibration as performed by a National Metrology Institute (NMI) such as the National Institute of Standards and Technology (NIST) in the United States, the National Research Council (NRC) in Canada, or the National Physics Laboratory (NPL) in the United Kingdom. Thus, a secondary laboratory or even the instrument manufacturer avoids the elaborate methods and expensive equipment associated with an absolute calibration. The results may then be transferred to the color-measuring instrument through an ancillary process.

Tristimulus colorimeters use multiple detectors windowed by colored glass filters or engineered, dielectric thin-film filters whose transmittances have been carefully selected to closely match the standard observer functions or to transform the instrument's broadband illumination to a specific illuminant/observer combination. The output is a direct report of X, Y, Z (refer to Eq. 10.1 with the summations replaced by integrals), and then the results are used to convert to the appropriate color scale (such as CIELAB). They generally use nominal 45°:0° or 0°:45° geometries (refer to Section 10.3.1) as these geometries are best for measuring color according to eye perception. The measurement of a set of stable color standards which span the gamut of color space should be performed as a matter of routine with tristimulus colorimeters since the spectral output of the optical source is generally not monitored and any degradation due to aging of the source or in the spectral transmittance of the filters will cause errors in the reported output. Because only tristimulus values X, Y, and Z are reported, metamerism (refer to Section 10.2.3) cannot be detected.

Spectrophotometers measure either reflectance (or reflectance factor) or transmittance directly on a wavelength-by-wavelength basis. A monochromator disperses the received spectrum into known wavelength bands, and the amplitude reported within each band is combined to report the reflectance. The reflectance information is subsequently converted to tristimulus values (again using Eq. 10.1). Spectrophotometers may employ 45°:0°, 0°:45°, or integrating sphere geometries depending on the surface characteristics of the object to be measured and the goal of the measurement.

Industrial spectrophotometers normally use broadband sources with diffraction grating spectrometers with diode array detectors. The primary design goal for handheld versions is to minimize instrument size (to within some of the constraints imposed by measurement practice and as dictated by customer needs), weight, and cost while still providing meaningful data. Benchtop versions tend to use designs that more closely follow the requirements of standard practice, where size, weight, and cost are of less concern. The purpose of all of these spectrophotometric instruments, as with the tristimulus colorimeter, is to identify, sort, quantify, and record color metrics and color differences primarily for product quality control.

Higher-cost (sometimes referred to as *analytical* or *research-grade*) instruments, designed for calibration and metrology, use broadband sources and perhaps multiple broadband sources. They generally employ scanning monochromator(s) as the spectral engine and a suite of detectors whose response is specific to the wavelength range of interest. The use of a scanning monochromator allows both the wavelength and the bandwidth of the selected light to be adjusted prior to reaching the object. Their size and design are such that most, if not all, of the geometric requirements imposed by colorimetric standard practices are met. The spectral ranges of these instruments is typically well beyond the range of human vision, and since this chapter deals with the measurement of color, these instrument types will not be discussed further.

Before proceeding, it is important to comment that object fluorescence can be a problem for any instrument that measures color. A fluorescent object absorbs light at one (or possibly more) wavelength band and reemits light at another, higher wavelength band (refer to Chapters 2 and 7 for detailed discussions). Consider a system that employs a monochromator before the object to be measured, and only one wavelength band is incident on the object at a time. The fluorescent signal of the longer wavelength will be improperly attributed to the lower excitation wavelength and therefore is a source of error. If the monochromator is instead placed between the object and the receiving detector, the fluorescent signal is correctly interpreted at the longer wavelength. Therefore, in the case of fluorescent objects, when using broadband sources, great care should be taken to faithfully reproduce the spectral characteristics of the desired source for the color measurement, particularly in the blue and near UV region of the spectrum where absorption most often occurs. Otherwise, different spectral reflectance curves will result giving variations in the reported color. Of course, this is only a problem if the color difference is outside the acceptable tolerance. This can occur if the object is thought to be free of fluorescence, but in fact exhibits the effect. In this case, if the same object is measured by different instruments of the same design wherein the excitation spectra is not adequately managed, the interpretation would be that there is a lack of color measurement reproducibility between instruments. This can be a significant problem leading to wasted time and effort, if the object serves as a color standard used to check reproducibility. Such an effect was recently discovered with certain ceramic color standards and the effect analyzed in detail [24]. Furthermore, many color instruments are used to report a Whiteness Index [25] and this index is calculated directly from the reflectance curve resulting from the fluorescence. Thus, once again, if the excitation spectrum is not adequately managed, the reported whiteness will vary due to the source, and not due to unintentional variations in the recipe for any fluorescent whitening agents that may have been added to the product by the manufacturer.

10.3.1 Geometric Considerations

In Section 10.1, it was mentioned that the color of objects is determined by the parts of the incident visible light that are reflected and/or transmitted without being absorbed. The theoretical description of this interaction was given in

Chapter 2 by considering a beam of light incident on a nonfluorescent object. If you refer to Fig. 2.1 of Chapter 2, this illustrates the notions of specular reflectance, the diffuse reflectance, the regular transmission, and the diffuse transmission. The light from the source is incident on the object from the upper left. A retroreflected component may exist and if so, it retraces the path of the incident beam but in the opposite direction. The regular or "specular" component of reflection traces a path that has the same angle with respect to the normal of the object as the incident beam, but on the opposite side of the normal. This is an illustration of the well-known law for mirror-like reflection. In addition to the specular component, there may be "diffuse" components where the light is reflected at many angles and not just the specular or retroreflected angle. Diffuse reflections are shown at the left hemisphere in the figure. The origin of the diffuse components may be thought to arise from roughness or scattering centers just below the surface where the light is incident. Light that propagates through the object is subject to Snell's Law (if it is mostly transparent), absorption, and scattering and then emerges from the object. The regularly transmitted beam takes the same angle with respect to the object normal as the incident beam assuming that the refractive index on the emerging side is the same as the incident side. There may or may not be diffuse components of transmission depending on scattering within and also at the exit surface of the object. Refer to Chapters 2 and 6 for a complete description of diffuse reflectance and transmittance.

An opaque object will not allow a beam to be transmitted through it, and if there is some surface gloss, the total components of reflection are due only to the specular component and the sum of all the diffuse components. Consider now what happens if the left hemisphere constitutes the surface of an integrating sphere. The radiance of the sphere wall is due to the integration of the specular ray and that of the diffuse components, and in this case, the sphere is operating in reflectance "*SpeCular Included*" mode, referred to as "SCI" (formerly, RSIN or SPIN). If the specular ray is omitted, perhaps either by adding a light trap at the point where the ray strikes the sphere wall or by adding a small hole to allow it to escape, the sphere operates in reflectance "*SpeCular Excluded*" mode, referred to as "SCE" (formerly, RSEX or SPEX). If the object primarily has a matte finish, there is no specular component, and the two modes are essentially indistinguishable. Furthermore, the specular and diffuse components can have significantly different color properties. Thus with instruments that employ an integrating sphere, the reported color of objects having surface gloss will be different in the SCI mode compared to the SCE mode. The user must be aware of this difference. Generally, the rule of thumb is the following. When measuring plane surfaces with high gloss, use SCE mode to eliminate the component of gloss, as this does not contribute to surface color. In the case of plane surfaces with intermediate gloss and possibly texture (e.g., textiles), measurements may be made in the SCI mode, but the correlation with visual color judgment may be poor. It is better to use

0°:45° or 45°:0° geometry discussed below. In the case of low gloss, matte specimens, either SCI or SCE geometries may be used. Bare metal surfaces should generally be measured with the specular component included. See Ref. [26] for general guidelines for measuring color using an instrument employing hemispherical geometry.

The CIE has established recommendations for hemispherical illumination using integrating spheres, and ASTM has established standard guidelines [27,28] for the measurement of reflection and transmission. If the sample to be measured is illuminated directly by an integrating sphere and the viewing angle does not exceed 10° from the normal to the surface of the sample, the reader may recognize this condition as the "di:0°" and "de:0°" geometries if the specular component is included or excluded, respectively. The integrating sphere may be of any diameter provided the total area of all sphere ports does not exceed 10% of the total internal reflecting surface area, and the angle between the axis and an arbitrary ray of the viewing beam should not exceed 5°. Helmholtz reciprocity (described in Chapter 2) applies so that the sample may be illuminated by a beam whose axis does not exceed 10° from the sample normal, and the reflected flux is collected by an integrating sphere. The geometry is then referred to as "0°:di" and "0°:de" if the specular component is included or excluded, respectively. Refer to CIE Publication 15.3 (see Ref. [29] herein) for a complete description of the geometric conditions for illumination and viewing in colorimetry, including the nomenclature to be used.

A schematic for a practical color-measuring instrument that uses hemispherical illumination by means of an integrating sphere is shown in Fig. 10.8. The sample is located at the measuring port of the sphere and illuminated diffusely by the light from the source reflecting from the sphere wall. The center of the view port is inclined at an angle θ with respect to the normal of the sample. The center of the SCE port is also inclined at θ with respect to the normal, thus providing a means to locate either a light trap or an opening at the port to allow the specular ray to be trapped or to exit. This configuration can be designated as "d:θ," where θ can take any value up to 10° if the CIE/ASTM guidelines are to be followed. Typically, θ is either 6° or 8°. A practical limit to the sphere size and θ is set by the dimensions of the SCE port. Referring to Fig. 10.8, the distance of the center of the SCE port to the axis of the sphere, which is the sample normal, is approximately given by the distance d. Suppose d is just a little larger than the radius of the SCE port and has a value of 15 mm (suppose the SCE port has a 25 mm diameter). Then

$$D \tan\theta = d, \tag{10.16}$$

where D is the diameter of the sphere. The product is therefore fixed at 15 mm. If $\theta = 10°$, the maximum permitted, then the diameter of the sphere can only be as small as 85 mm. Typically, θ is either 6° or 8° so that D becomes 143 or 107 mm, respectively. Assuming 6°, the notation would thus become "d:6°." The SCE port may lead to significant errors in comparisons of

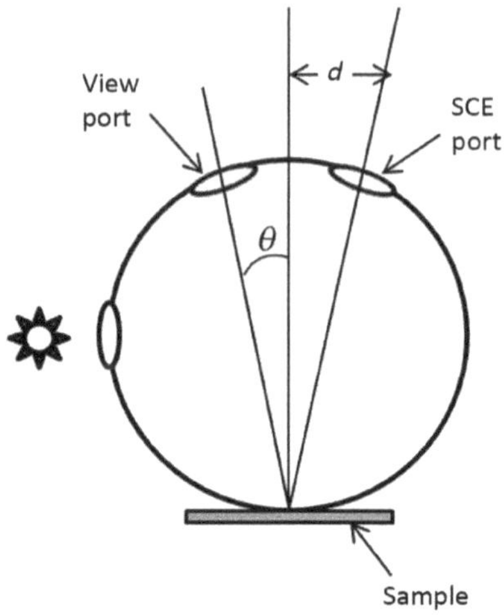

FIGURE 10.8 Schematic for sample measurement using an integrating sphere for illumination.

specular-excluded measurements of the same sample among instruments using spheres of different diameters and different size SCE ports.

If the sample is not opaque, light will propagate through the first surface, and transmission and scatter components will be generated, if they are not completely absorbed within the object. Some of the scattered components may be reflected back to the first surface where they will exit, thereby contributing to the diffuse components of the first surface. This added effect due to hazy, turbid, or translucent samples may be significant depending on the sample. The apparent surface color is affected by the size of the illuminated region of the sample, and interpretations of color may vary between instruments if the sample port sizes are not identical. Sample port sizes depend upon the area of the sample that is desired to be viewed. Typically, the sample port diameter is 25 mm. Other sizes may be used as dictated by the needs of the customer, with the caveat that at no time may the edge of the port be seen by the detector monitoring the light exiting the view port.

The surface color of objects is perceived because of the diffuse reflection (or diffuse transmission) only. Although sphere measurements can be used, it is generally recognized that 45°:0° and 0°:45° illumination/viewing geometries are best for measuring color since they produce results that correlate well with human perception, especially when surface texture or gloss is taken into account. ASTM International publishes a standard test method for the measurement of reflectance factor and color using a spectrophotometer in the 45°:0° geometry [30] and another similar test method when using a filter colorimeter [31].

Figure 10.9 shows a simplistic schematic of instruments that use 45° illumination and 0° viewing. On the left-hand side of Fig. 10.9, light is incident at a 45° angle with respect to the normal of the sample. The object is shown to

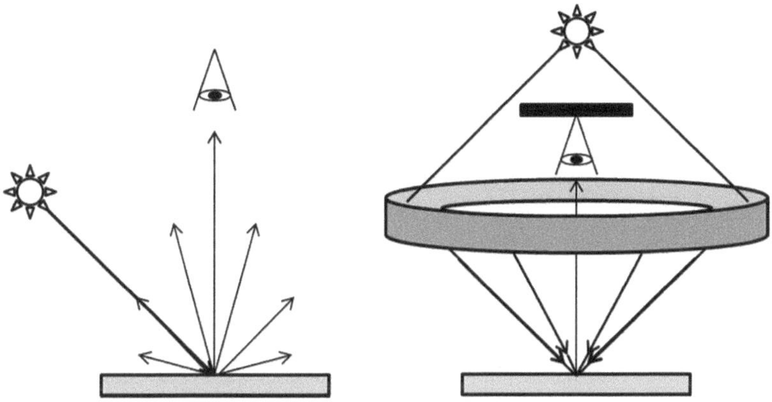

FIGURE 10.9 Schematic of a simplified 45°:0° configuration.

generate diffuse components, some of which are collected by the observer looking down the axis which is the normal to the object. Unidirectional illumination is shown; that is, the light is incident from only one azimuth and elevation (45°). For this illumination scheme, some objects (e.g., textiles, plastic pellets, and food) may exhibit directionality, meaning that the instrument readings will vary as the object is rotated about the normal, in the plane of the object. In this case, bidirectional, or even better, circumferential illumination is preferred, as shown on the right-hand side of Fig. 10.9, to reduce the sensitivity to object orientation. Direct rays from the source are blocked by an opaque screen, and an annular cone of illuminating rays is intercepted by a cylindrical mirror and directed to the object. Because of Helmholtz reciprocity [32], the illumination and viewing angles may be interchanged and should lead to identical results, provided the extent of the angular divergence of the illumination and viewing beams are the same. Geometric recommendations for the limits to the illumination and viewing fields can be found in the various ASTM guidelines and Standard Practices given in the References section.

10.3.2 Source Considerations

Optical sources for use in handheld and benchtop colorimeters and spectrophotometers are typically continuous wave (CW) tungsten, quartz–tungsten–halogen, or xenon flash. Xenon is particularly well suited as a daylight simulator because of its relatively high UV content and its spectral similarity to standard illuminant D65. Power supplies must be designed with consistency and stability in mind, especially if a separate instrument channel is not dedicated to monitor the light output.

In the case of tungsten lamps employed in tristimulus colorimeters, current stability and actual current value are both a concern, as variations will alter the relative spectral power distribution, thereby modifying the range of wavelengths intercepted by the observer filters during photodetection. This leads to a change in calibration and therefore errors in reported color. In the case of xenon flash, the spatial variations of the position of the arc can be particularly troublesome and will manifest themselves as reading repeatability failure if the light projection optics are not designed to mitigate the potential variations. In either case, cost is always a consideration, as is the potential for obsolescence. For example, high power, high brightness, and selectably addressable LEDs coupled to appropriate phosphors could replace both tungsten and xenon as a source of light in colorimetric instruments. Such LEDs are small, consume very little power, and have high stability with very long life. White organic LEDs [33] are also the subject of much research, particularly in solid-state lighting applications, and these may one day be utilized.

10.3.3 Detection and Signal Processing Considerations

Since the spectral range required for colorimetric measurements only spans the visible, silicon detectors, which have high sensitivity in this region, are commonly used. They may be packaged as discrete devices or in linear or square arrays. Discrete devices are typically employed for tristimulus colorimeters, and either linear or square arrays in the case of spectrophotometers.

Figure 10.10 shows a schematic of a generic tristimulus colorimeter. Light from the source is directed by the source optics to the sample under measurement. The source optics may be designed to produce 45° or 0° illumination. Light that is diffusely reflected by the sample is collected by the sample optics (at either 0° or 45°) and directed to a suite of filters and photodetectors. Three such filters and photodetectors are shown, but sometimes four are used where two are dedicated to replicating the "*X*" observer function. The output of the photodiode amplifiers may or may not be digitized.

The spectral transmittance of the individual filters can be designed to replicate the standard observer functions when the source has a relative spectral power distribution similar to that of standard illuminant A. However, since

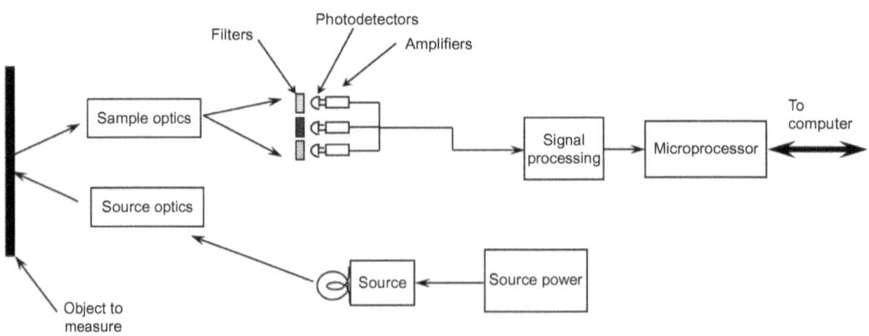

FIGURE 10.10 Schematic of a generic colorimeter.

many colorimetric specifications are written with the illuminant C in mind, this constraint can be taken into account and the filter transmittance may be appropriately modified.

A schematic of a dual-channel system that might represent a spectrophotometer is shown in Fig. 10.11. In this case, the colorimeter filter suite is replaced by grating spectrometer which disperses the collected light into its constituent wavelengths. The advantage of the second channel is that optical source variations can be divided out on a wavelength-by-wavelength basis.

Let's assume the output of the array detector and the processing electronics is a voltage V. There is a voltage associated with the sample signal, V_{samp}, and a voltage associated with the monitor signal, V_{mon}. Thus, the signal ratio $r_\lambda = V_{samp}/V_{mon}$ is determined. The subscript λ designates that there are as many such ratios as wavelengths that are reported.

The photometric scale of the instrument is established when the instrument is calibrated. Typically, a white standard is measured, thereby producing a reference signal ratio $r_{w\lambda}$. To allow for the measurement of fluorescent materials or materials whose reflectance factor is greater than unity, the sample scale is set so that the signal corresponding to the white sample is usually about half way to the maximum value allowed by the electronics. Next, a light trap is measured to set the dark signal, producing a dark signal ratio $r_{d\lambda}$. A linear photometric scale is assumed (which may or may not actually be the case but techniques described in Section 10.3.4 are later used to linearize the scale), and the ratio, $r_{s\lambda}$, obtained for an arbitrary sample presented to the measurement port is

$$r_{s\lambda} = \frac{(r_\lambda - r_{d\lambda})}{(r_{w\lambda} - r_{d\lambda})}. \tag{10.17}$$

If the reference sample is used for the arbitrary sample, $r_{s\lambda} = 1$, while if the light trap is used, $r_{s\lambda} = 0$. To complete the transfer to the reflectance scale, Eq. (10.17) is multiplied by the value of the reflectance factor $R_{w\lambda}$ for the reference standard, so that the reported reflectance factor of the sample of the sample is simply

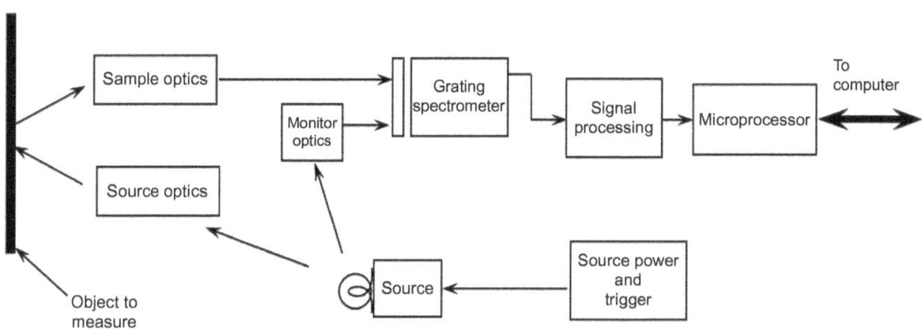

FIGURE 10.11 Schematic of a generic spectrophotometer.

$$R_\lambda = R_{w\lambda} \cdot r_{s\lambda}. \tag{10.18}$$

Instruments are usually delivered with a reference standard. Its reflectance factor $R_{w\lambda}$ is determined by transferring a calibration of a known standard of similar design, traceable to the PRD.

Since the individual pixels comprising the array detector usually have a physical extent that spans a much smaller wavelength than the wavelength band that is desired, typically 5 or 10 nm, their individual outputs are combined and weighted into constant triangular passbands centered on each reported wavelength, with a full width at half maximum (FWHM) commensurate with the reporting interval, as recommended by ASTM E308. Triangular passbands are normally employed, but their actual shape is not critical assuming that techniques (refer to Section 10.3.4 and Eq. 10.19) are ultimately used to minimize the effect of bandpass variations.

The wavelength scale for spectrophotometers employing array detectors is usually established by viewing monochromatic light of known wavelengths (e.g., lasers or the output of spectral calibration lamps; see Chapter 3) and then calculating the positions of the centroids of the monochromatic lines in pixel space, thus mapping wavelength to pixel number. For best results, the entrance aperture of the instrument should be illuminated in a manner identical to that used when actually measuring reflectance while the wavelength calibration is performed.

10.3.4 Uncertainties

The progression from measurement to surface color specification goes from measured spectral reflectance factor, to calculated tristimulus values, and chromaticity coordinates or color space values. Therefore, to arrive at the uncertainties in the final values describing the surface color of a specimen, the uncertainties in spectral reflectance factor, $R(\lambda)$, must be propagated through the calculations. The uncertainty associated with the quantitative description of the color of an object is often necessary for determining the acceptability of that object for its intended application. Uncertainties are also required for establishing the traceability of a measurement to an NMI. Determining the uncertainty in the measured color of a specimen has been a topic of interest for many years [34–43]. Chapter 1 outlines the general approach to propagating uncertainties in a measurement.

Uncertainties in the final color values can be calculated using two different approaches described in Ref. [44]. In the first, the uncertainties in $R(\lambda)$ are propagated through uncertainties in the tristimulus values, which are then used to calculate the uncertainties in the chromaticity coordinates or color space values. In the second approach, the tristimulus definitions are substituted into the definitions of the final color values to obtain equations in terms of the $R(\lambda)$. The uncertainties in the final color values given by these equations are calculated directly from the uncertainties in the $R(\lambda)$. In both approaches, the correlations between the uncertainties in $R(\lambda)$ at different wavelengths must be included in the calculations.

The common sources of uncertainty in color measurements are signal noise, offset in the signal, and uncertainties in the standard and the wavelength. Signal noise is the result of random processes in both the optical source and detector. Offsets in the signal can arise from several sources including an improper zero, misalignment of the specimen, and stray light. An improper zero adds a constant value to all signals, while misalignment of the specimen is the result of a difference in location, with respect to the reference plane, between the specimen and the standard. For example, if the specimen and standard are physically different, it may not be possible to align one of the front surfaces with the reference plane. This can introduce sample recess errors. Heterochromatic stray light is radiant flux that reaches the detector at wavelengths outside the nominal spectral bandwidth of the illumination, which adds a value to each signal. This value depends upon the spectral shape of the illumination, the spectral reflectance of both the specimen and standard, and the responsivity of the receiver. The standard has an uncertainty from its calibration, and there is an uncertainty associated with the wavelength setting of the measuring instrument. Uncertainties in the normalized tristimulus weighting factors $W_{x,i}$, $W_{y,i}$, and $W_{z,i}$ can be an additional source of uncertainty for reflectance colorimetry. For example, the weighting factors chosen for the calculation may not be appropriate for the spectral bandwidth and interval of the measuring instrument. For complete uncertainty analysis, the correlations between signals at the same wavelength and between reflectance factors at different wavelengths for each of these sources should be considered in the analysis [45,46].

In a more simplistic approach, spectrophotometric errors are categorized as arising from *systematic errors* such as those generated by detector nonlinearity, wavelength registration, and bandpass variations. *Random errors* due to noise, drift, and sample presentation can contribute to the inaccuracy of a measurement; however, the systematic errors are of most concern to the manufacturers who strive to achieve *reproducibility* in their instruments. Results are said to be reproducible when the measurement of color is the same among different instruments of the same design. This is extremely important when the instruments are used over several years for quality and process control, especially in production facilities that might be separated by hundreds or thousands of miles!

Most manufacturers rely on "profile-based correction" (PBC) to model the systematic errors and to tighten reproducibility (sometimes referred to as "interinstrument agreement"). The approach was first described by Robertson [47] and further investigated by Berns and Petersen [39] who employed multivariate regression. A suite of stable colored standards which span the gamut of perceived color, typically the BCRA tile set series II from Ceram (now Lucideon) [48], are used to diagnose the systematic errors. The spectral reflectance (which may be SCI, SCE, 45°:0° or 0°:45°) of the colors comprising the tile set is first assigned by measuring the tiles with several instruments of the same type. These are normally the first or "pilot" instruments produced by the manufacturer. After vetting for obvious outliers, the average reflectance of each color as measured by the set of pilot instruments is computed, and these results become the assigned (but not traceable to any calibration) values of reflectance, $R_A(\lambda)$, for the PBC algorithm.

The errors are normally modeled as arising from five contributions: an offset error, such as that due to stray light or the use of a light trap with finite instead of zero reflectance; a photometric linear scale error, where the output at the top of scale is emphasized more than the bottom of the scale; a wavelength registration error, arising from an improper calibration of the wavelength scale which is proportional to the first derivative of the reflectance curve; a bandpass error, arising from a variation of the instrument bandpass with wavelength which is proportional to the second derivative of the reflectance curve; and a nonlinear photometric scale error, typically taken to be quadratic and arising from nonlinearity in the detector. With these in mind, the model equation that relates the difference between the assigned reflectance, R_A, and the reported reflectance, R,

$$R_A - R = c_1 + c_2 R + c_3 \frac{dR}{d\lambda} + c_4 \frac{d^2R}{d\lambda^2} + c_5(100 - R)R, \tag{10.19}$$

where the model terms are described above. By subjecting the reported reflectance to linear regression, the values for the model terms c_1–c_5 which minimize the difference over the suite of colors are found. The coefficients represent the sensitivity coefficients in this diagnostic model of the systematic errors. These coefficients are normally stored in the instrument memory and applied whenever a subsequent measurement is taken.

In the interest of some kind of traceability (but not in the rigorous sense as described earlier), the photometric top of scale of the instrument, that is, its white standard, is usually assigned by referencing its reflectance to that of a similar standard that was calibrated by an NMI such as NIST, or to a process such as that described by ASTM E259. Since the BCRA tiles are known to be thermochromic, it is important to observe constant environmental conditions when the PBC algorithm is implemented.

10.4 GONIOAPPARENT MATERIALS

Pigments are being manufactured with new and unique appearance attributes [49]. For example, gonioapparent materials such as metallic and pearlescent coatings exhibit differences in their perceived color with changes in the illumination or viewing angle, or both [50]. These coatings are very important to the generation of many unique effects used in the printing of currency, formulation of cosmetics, and application of paint to automobiles. For example, metallic coatings were developed in the 1930s, and today about 80% of all automobiles have metallic finishes [51]. These gonioapparent properties require exceptional processing conditions and characterization methods, which are different from the traditional single geometry method. The necessary set of measurement geometries is determined by the complexity of the scattering mechanisms present in these coatings. The combination of one illumination angle and three viewing angles is sufficient for the colorimetric characterization of metallic coatings but is not sufficient for the characterization of pearlescent interference coatings [52,53].

Table 10.1 shows a comparison of the optical characteristics of traditional color pigments and gonioapparent coatings. For all of these coatings, the front surface is a clear, smooth overcoat. The reflective properties of this overcoat are investigated at the specular geometry. In traditional color pigments, part of the incident light is absorbed, while the rest is diffusely scattered. Therefore, the perceived color is relatively independent of the measuring geometry, including the illumination and viewing angles, and the traditional single geometry of 0°:45° is sufficient to characterize the color of these coatings. In contrast, the primary interaction of light with metallic coatings is specular reflection from the flakes, and the perceived brightness depends on the viewing angle, but is independent of the illumination angle, while the perceived chroma and hue are independent of the measuring geometry. Metallic coatings appear brighter as they are rotated away from the viewer.

Pearl interference pigments or pearlescent coatings usually consist of thin metal oxide layers on transparent mica platelets, and their perceived chroma, hue, and brightness depend on both the illumination and viewing angles. The

TABLE 10.1 Comparison of the Optical Characteristics of Absorption and Gonioapparent (Metallic and Pearlescent) Coatings

Type of Coating	Absorption	Metallic	Pearlescent
Optical principle of pigments (first order)	 Diffuse scattering or absorption	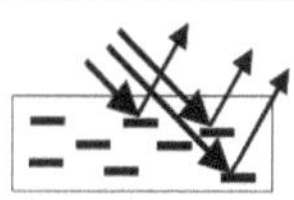 Specular reflection	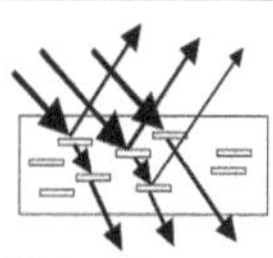 Thin-film pearl (mica) interference
Perceived chroma and hue	Independent of geometry	Independent of geometry	Dependent on geometry
Perceived brightness	Independent of geometry	Dependent on geometry	Dependent on geometry

unique pearlescent effect occurs because transparent shiny mica platelets allow a portion of the incident light to be transmitted. When the transmitted portion of the light meets further boundary surfaces with different refractive indices, a portion of the light is reflected. The total reflected light is then made up of portions that have traveled on different paths, producing optical interference.

Parallel alignment of the platelet pigments in a gonioapparent coating is essential for their unique effect. The wide angular dispersion of reflected light in gonioapparent coatings means that they can be fully characterized only if evaluated from many directions [54,55]. For example, the color of metallic coatings can be accurately characterized with an illumination angle of 45° and aspecular angles of 15°, 25°, and 110°. The aspecular angle is the viewing angle measured from the specular angle toward the sample normal. ASTM E2194-03 describes the measurement protocol including the recommended geometries and tolerances for the measuring instruments [56]. The number of measuring geometries needed to characterize pearlescent coatings is greater than that for metallic coatings. ASTM E2539-12 describes the measuring instrument requirements and parameters needed to make color measurements of thin-film interference pigmented coatings [57]. Instrument makers designed and built instruments that comply with the measurement geometries and specifications prescribed in these standards.

10.5 GEOMETRICAL ATTRIBUTES

Geometrical attributes are those attributes that vary the spatial or angular distribution of the light without significantly changing the spectral distribution [1]. Most objects have some degree of surface texture that determines the visual appearance. The smaller surface texture with lateral dimensions in the range of 0.01 mm results in a reduction in the gloss value and the first glimpse of reflection haze. As the surface texture increases to greater than 1 mm, the quality of the image forming starts to be affected, like distinctness of image (DOI).

There is no standard observer for the geometrical attributes, as there is for color. The geometrical scales are established by fabricating a set of samples with the attribute under investigation. Trained observers will visually rank these samples. The angular distribution of the reflected/transmitted light, or goniodistribution, is then measured under different measurement conditions, including incident and viewing angle, source and detector aperture, spectral power distribution of the illuminating source, and spectral responsivity of the detector. The scale is established when the correlation between the visual and experimental measurements is optimized. Visual assessment of these attributes is subject to error introduced by lighting condition and the size, color, and orientation of the surface being measured.

The total appearance of objects depends on the structure size and the viewing distance. Therefore, the measurement results depend on the type of measuring instrument, measurement geometry, calibration of the instrument, and the surface characteristics of the object to be measured. In order to be able to compare the measured values, the standardization of instruments has become increasingly important. This has resulted in the development of standard test method documents that give prescriptions on the instruments specifications including measurement geometry, source and detector apertures, spectral power distribution of the illumination source, and spectral power sensitivity of the detector.

10.5.1 Specular Gloss

Gloss is the perception by an observer of the mirror-like appearance of a surface and describes the ability of a surface to reflect light in the specular direction [58,59]. This perception changes whenever there is a change in the relative position or spectral distribution of the source, the sample, and the observer. Gloss is the second most utilized appearance attribute, after color, to evaluate the quality of products such as automotive coatings, textiles, and papers. High-gloss finishes are often used to make the products more visually appealing. Gloss is quantified by measuring the light reflected at the specular angle and is created in the uppermost surface layer of the object. Therefore, specular gloss strongly depends on the smoothness of the surface of the object. Different standard geometries are used to determine the specular gloss of materials for different applications. Table 10.2 lists examples of these geometries and their applications. These geometries were selected based on their

TABLE 10.2 Standard Geometries for Specular Gloss Measurements and Their Applications

Specular Angle (°)	Applications
20	High gloss of plastic films, appliances, and automotive finishes, typically greater than 70 gloss units
30	High gloss of image-reflecting surfaces
45	Porcelain enamels and plastics
60	Paints and plastics with gloss units between 10 and 70
75	Coated waxes and papers
85	Low gloss of flat matte paints and camouflage coatings, typically for less than 10 gloss units

ability to produce optimum discrimination between samples and to correlate with visual rankings.

The measurement of specular gloss consists of comparing the luminous reflectance from a test sample to that from a calibrated gloss standard, under the same experimental conditions. Therefore, gloss is a dimensionless and psychophysical quantity whose accurate determination depends on the characteristics of the measuring instrument and on the gloss standard. Several documentary standards describe the proper measurement conditions to determine specular gloss for specific surfaces. In particular, the ISO 2813 [60] and the ASTM D523 [61] describe the measurement procedure for specular gloss at 20°, 60°, and 85° geometries for nonmetallic paint samples from low- to high-gloss levels. These documentary standards specify the incident beam to be a collimated unpolarized beam, the specular angles, the source and receptor apertures, and the spectral flux distribution of the source and spectral response of the receptor. Problems with these documentary standards have been identified and are well documented [62,63]. The illumination beam is collimated with a spectral flux distribution approximating that of illuminant C. The luminous flux in the reflected beam is measured with a receiver having a spectral responsivity that approximates the CIE spectral luminous efficiency function V_λ. Figure 10.12 shows the spectral power distribution of the source and the spectral response of the detector. The dimensions of the source and receptor apertures are predetermined by the specular geometry, focal length of the collimator lens in the system, test material, and the correlation with visual ranking. The theoretical standard for specular gloss measurements is a highly polished planar black glass with a refractive index for the sodium D line ($\lambda_D = 589.3$ nm) of $n_D = 1.567$. To set the specular gloss scale, the specular gloss of this theoretical standard has an assigned value of 100 for each of the three standard specular geometries of 20°, 60°, and 85°. For primary gloss

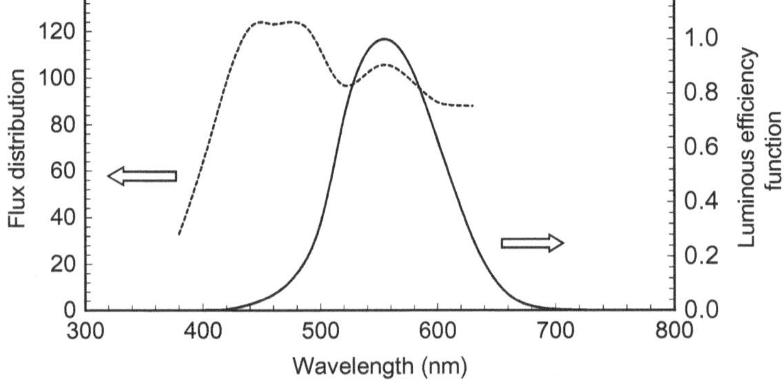

FIGURE 10.12 Spectral distributions of illuminant C (dashed line) and the CIE spectral luminous efficiency function (solid line).

standards with a refractive index different from $n_D = 1.567$, their specular gloss is computed from the refractive index and the Fresnel equations (see Chapter 2).

Measuring the specular gloss of a test sample involves the ratio of the luminous reflectances of the test sample and primary standard [64]. The specular gloss of a test sample at a nominal angle θ_0 is given by

$$G_t(\theta_0) = G_s(\theta_0) \cdot \frac{\rho_{v,t}(\theta_0)}{\rho_{v,s}(\theta_0)}, \tag{10.20}$$

where $G_s(\theta_0)$ is the specular gloss of the primary standard and $\rho_{v,t}(\theta_0)$ and $\rho_{v,s}(\theta_0)$ are the specular luminous reflectances of the test sample and primary standard, respectively. Furthermore, the specular gloss of the primary standard is given by

$$G_s(\theta_0) = G_0(\theta_0) \cdot \frac{\rho_s(\theta_0, \lambda_D)}{\rho_0(\theta_0, \lambda_D)}, \tag{10.21}$$

where $G_0(\theta_0)$ is the specular gloss of the theoretical standard and $\rho_s(\theta_0, \lambda_D)$ and $\rho_0(\theta_0, \lambda_D)$ are the specular reflectances of the primary and theoretical standards, respectively, at a wavelength $\lambda_D = 589.3$ nm. Combining Eqs. (10.20) and (10.21), the specular gloss of the sample under test is given by

$$G_t(\theta_0) = G_0(\theta_0) \cdot \frac{\rho_s(\theta_0, \lambda_D)}{\rho_0(\theta_0, \lambda_D)} \cdot \frac{\rho_{v,t}(\theta_0)}{\rho_{v,s}(\theta_0)}. \tag{10.22}$$

For each of the three standard geometries (20°, 60°, and 85°), the specular gloss of the theoretical standard is defined as $G_0(\theta_0) = 100$. The specular reflectance ρ from the surface of a dielectric sample depends on the incident angle θ defined relative to the normal of the sample, the index of refraction at the wavelength λ, and the polarization σ (p or s) of the incident radiant flux. The specular reflectance as a function of these variables is given by the Fresnel equations, Eq. (2.142). The specular reflectance for unpolarized incident radiant flux is calculated from

$$\rho(\theta, \lambda) = \frac{1}{2}[\rho(\theta, \lambda, p) + \rho(\theta, \lambda, s)]. \tag{10.23}$$

The specular reflectances at the standard angles of illumination for unpolarized radiant flux for the theoretical standard, $\rho_0(\theta_0, \lambda_D)$, are calculated from Eq. (10.23) and are listed in Table 10.3.

The specular gloss of the primary standard depends on its reflectance ρ_s. However, the documentary standards are ambiguous concerning the definition of this reflectance, since the reflectance is slightly wavelength dependent. One approach is to use the calculated $\rho_s(\theta_0, \lambda_D)$ of the primary standard for its index of refraction at $\lambda_D = 589.3$ nm. Other approaches determine the specular gloss of the standard by an integrated average of the reflectance, weighted by

TABLE 10.3 Specular Reflectance $\rho_0(\theta, \lambda_D)$ of the Theoretical Gloss Standard for Each Standard Illumination Angle at Wavelength $\lambda_D = 589.3$ nm

Illumination Angle (°)	$\rho_0(\theta, \lambda_D)$
20	0.049078
60	0.100056
85	0.619148

the product of the illuminant C, the CIE spectral luminous efficiency function, and the angular spread of the probing beam.

The components of uncertainty associated with gloss measurements are divided into those arising from experimentally measured quantities and those arising from deviations from standard recommendations. The components of uncertainties arising from the illuminator are the stability of the light source, polarization and angular distribution of the illuminator, and deviations of the illuminator spectral power distribution to the illuminant C. The components of uncertainties arising from the sample compartment include the errors in the angular positioning of the incident and viewing angle, repeatability of the angular setting, and sample alignment and repeatability of sample positioning. The components of uncertainties arising from the detector include responsivity and linearity of the detector, signal noise, and deviations from the specified spectral responsivity of the CIE spectral luminous efficiency function. The ISO 2813 standard specifies that gloss should be reported to the nearest full gloss unit. Therefore, the calibration of the reference instrument and primary standard should have an uncertainty of less than one gloss unit.

In addition, there are documentary standards for other applications. TAPPI T480 describes the measurement protocol for specular gloss at 75° for low to high gloss of coated or uncoated paper [65]. In this case, the primary standard has a defined refractive index of 1.54 at the helium-D line, 587.6 nm. For very high gloss paper like waxed paper or high-gloss ink films, TAPPI T653 is recommended [66].

10.5.2 Haze

The haziness of an object is the result of the scattering of the incident light due to particles dispersed in the substance, transmission haze, or, by imperfection on the surface, reflection haze. The perceived haze is described as the cloudiness of a product and is an important appearance attribute which is used to assess the quality of objects such as plastics, glass, and liquids.

10.5.2.1 Transmission Haze

Transmission haze is defined as the forward scatter of light going through the specimen. The ASTM method D1003-95 is the standard test method for haze and luminous transmittance of transparent plastics [67]. This standard describes the specification for the geometrical and spectral requirements of the measuring instrument and the measurement sequence. Measurement problems with this standard have been identified and documented [68,69]. This standard test method only applies to materials with haze values less than 30%, and for greater haze values, diffuse transmittance should be measured. Two procedures are described in this document for the measurement of haze: Procedure A uses a hazemeter, while Procedure B uses a spectrophotometer. Haze measurements are most commonly performed using a hazemeter, which is a purpose-built instrument. A spectrophotometer can be used provided it meets the recommendations of the standard. A schematic of a traditional hazemeter is shown in Fig. 10.13 [70].

According to the standard recommendations, the spectral power distribution of the source shall simulate the illuminant C or alternatively CIE standard illuminant A and the spectral responsivity of the detector shall simulate the CIE 1964 standard observer. This instrument uses a single collimated beam of light and an integrating sphere coated with a highly reflecting material to collect the transmitted flux. The standard recommendations also include dimensional specifications and locations for the sample, light trap, reflectance standard, and detector for the integrating sphere. The incident light enters through the sample port and can either go straight to a light trap or reflectance white standard or the sphere is rotated so the incident light hits the sphere wall. The reflectance white standard and the sphere wall should have similar reflectivity properties to reduce the error associated with the reflectance standard. The pivotable-sphere type instrument does not have this error as long as the sphere wall is uniform over its surface. When measuring transmission haze, the percentage of light diffusely scattered more than 2.5° from the incident is compared to the total light transmitted. The percent haze is given by

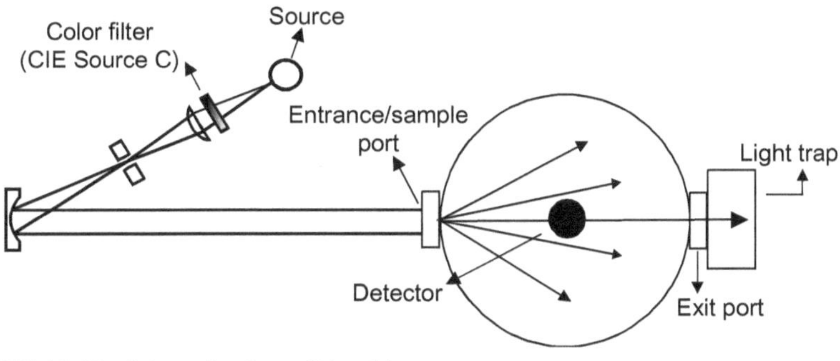

FIGURE 10.13 Schematic of a traditional hazemeter.

$$\%\,\text{Haze} = \frac{T_{\text{diffuse}}}{T_{\text{total}}} \times 100, \tag{10.24}$$

where $T = \%$ transmission. Percent haze is reported to the nearest 0.1 %. The measurement of the diffuse transmittance component, T_{diffuse}, is given by

$$T_{\text{diffuse}} = \frac{[T_4 - T_3\,(T_2/T_1)]}{T_1}, \tag{10.25}$$

where T_1 is the measurement of the incident light by rotating the sphere so the incident beam hits the internal wall of the sphere or the white reflectance standard, T_2 is the measurement of the total light transmitted by the sample, T_3 is the measurement of the scattered light by the instrument, and T_4 is the measurement of the light scattered by the instrument and sample. The measurement of the total diffuse component, T_{total}, is giving by

$$T_{\text{total}} = \frac{T_2}{T_1}, \tag{10.26}$$

where T_2 is the measurement of the total light transmittance by the sample and T_1 is the measurement of the incident light by rotating the sphere so the incident beam hits the internal wall of the sphere or the white reflectance standard. The percent haze measurement equation simplifies to

$$\%\text{Haze} = \left(\frac{T_4}{T_2} - \frac{T_3}{T_1}\right) \times 100. \tag{10.27}$$

Haze standards used to check the performance of these instruments consist of plastic chips with a range of percentage haze from 1% to 30%.

The components of uncertainty associated with transmission haze measurements are divided into those arising from experimentally measured quantities and those arising from deviations from the standard recommendations. The predominant components of uncertainties associated with the measured quantities are the stability of the source, unwanted scattered light, linearity of the detectors, reflectance of the integrating sphere coating, and reflectance and stability of the reference standard. The components of uncertainties arising from the standard recommendation include deviations of the illuminator spectral power distribution to the illuminant C, deviations from the spectral responsivity of the detector to the specified spectral responsivity, and deviations from the specified geometrical requirements for the integrating sphere.

10.5.2.2 Reflection Haze

Reflection haze is a measure of the spread of the specular component reflected from a surface and is caused by microscopic defects in the surface which scattered light adjacent to the direction of the specular reflection. Opaque objects with haze exhibit a milky finish or a halo effect around the reflection of a bright light source. Surface haze is a defect encountered in most coating applications including the automotive industry. Narrow-angle reflection haze is the measurement of the reflected light spread 2° from the specular reflection. The measurement of the reflected light spread 5° from the specular reflection is known as the wide-angle reflection haze. The ASTM standard D 4039 is the standard test method for measuring reflection haze of nonmetallic high-gloss surfaces [71]. This measurement method is based on the "Two-Parameter Gloss Methods" by Nimerrof [72]. The reflection haze index, H, is given by

$$H = G_{60} - G_{20}, \tag{10.28}$$

where G_{60} is the 60° specular gloss and G_{20} is the 20° specular gloss. The measurements for determining the specular gloss are specified in ASTM D523, "Standard Test Method for Specular Gloss." Because the specular gloss values of 20° and 60° depend on the refractive index of the samples, comparison of reflection haze index is limited to specimens with similar refractive index. The narrow-angle reflection haze is the measurement of the 2° specular component spread on either side of the specular component. This measurement is more sensitive than the traditional specular gloss measurement. In many cases, two samples can have the same specular gloss value, but their appearance is very different. The receptor apertures of the traditional glossmeter are too large to measure the difference, and it is only when you measure the reflection a few degrees away from the specular geometry that you can be sensitive to the differences between their appearances.

10.5.3 Distinctness of Image

A glossmeter is the most commonly used instrument to measure coating reflective appearance quality. However, it provides no information about the imaging forming qualities of a surface. This means that two samples with identical gloss values can have very different visual appearances. The receptor apertures used in the traditional glossmeter are too large, and the measurements are not sensitive to diffuse scattering due to surface defects [73].

Distinctness of Image (DOI) is a parameter which describes the visual distortion (blur) seen in the reflected image of an object from a high-gloss surface. The ASTM standard D5767-99 is the standard test method for the instrumental measurement of the spread of a beam of light after it has been reflected from a surface [74]. DOI is obtained by comparing the specular reflectance to that measured when the angle is changed slightly from the specular direction. The scale of DOI is a function of the sharpness of the reflected image from the object. DOI of 100 is defined for a surface which reflects an image perfectly without any distortion. A surface with no image clarity has a DOI of zero. The measurement of DOI is more sensitive than specular gloss for shiny surfaces and is used to characterize the visual appearance of high-gloss surfaces such as mirrors.

Orange peel is a surface defect that affects the image quality of the object and results in poor DOI. As orange peel reduces, the DOI improves. Orange peel is a subjective term used to describe a wavy pattern of light and dark areas on a high-gloss surface, like that found in many automotive finishes that resemble the skin of an orange. The surface defects in this observed orange peel are in the wavelength range of 0.1–30 mm. The visual scale for orange peel is based on ACT panels [75] or powder coating smoothness scale. This set consists of ten 101.6 mm × 152.4 mm panels which are painted black and ranked in a scale from 1, corresponding to high roughness and orange peel, to 10, corresponding to a very smooth coated surface. The operator judges which panel is closest visually to the test sample and assigns a numerical value.

REFERENCES

[1] R.S. Hunter, R.W. Harold, The Measurement of Appearance, second ed., John Wiley and Sons, New York, 1987.
[2] Colorimetry—Part 1: Standard Colorimetric Observers, ISO 11664-1:2007 (E)/CIE S014-1, CIE Central Bureau, Vienna, 2006.
[3] Colorimetry—Part 2: Standard Illuminants for Colorimetry, ISO 11664-2:2007 (E)/CIE S014-2, CIE Central Bureau, Vienna, 2006.
[4] M. Fairfield, Color Appearance Models, second ed., John Wiley and Sons, New York, 2005.
[5] A.H. Munsell, A Color Notation 2nd ed, Geo. H. Ellis Co., Boston, 1907.
[6] A.H. Munsell, Munsell Book of Color: Defining, Explaining, and Illustrating the Fundamental Characteristics of Color, vol. 1, Munsell Color Co., Baltimore MD, 1929.
[7] J. Schanda, Colorimetry: Understanding the CIE System, first ed., John Wiley and Sons, New York, 2007.
[8] J. Guild, The colorimetric properties of the spectrum, Philos. Trans. R. Soc. Lond. A 230 (1932) 149–187.
[9] D.B. Judd, Reduction of data on mixture of color stimuli, J. Res. Natl. Bur. Stand. 4 (1930) 515.
[10] D.B. Judd, The 1931 ICI standard observer and coordinate system for colorimetry, J. Opt. Soc. Am. 23 (1933) 359.
[11] D.B. Judd, G. Wyszecki, Color in Business, Science, and Industry, second ed., John Wiley and Sons, New York, 1963.
[12] D.B. Judd, 1964 CIE supplementary observer applied to the colorimetry of rutile and anatase forms of titanium dioxide, J. Opt. Soc. Am. 58 (1968) 1638–1649.
[13] I. Nimeroff, Field trial of the 1959 CIE supplementary standard observer proposal, J. Opt. Soc. Am. 54 (1964) 696–704.
[14] D.H. Brainard, Colorimetry, Handbook of Optics, McGraw-Hill Inc., New York, 26, vol. I, 1995.
[15] CIE No. 51 (TC-1.3), A Method for Assessing the Quality of Daylight Simulators for Colorimetry, CIE Central Bureau, Vienna, 1981.
[16] ASTM E308-13, Standard Practice for Computing the Colors of Objects by Using the CIE System, ASTM International, West Conshohocken, PA, 2013. http://dx.doi.org/10.1520/E0308. www.astm.org.
[17] ASTM D4086-92a, Standard Practice for Visual Evaluation of Metamerism. ASTM International, West Conshohocken, PA, 2011. http://dx.doi.org/10.1520/D2244-11. www.astm.org.

[18] Colorimetry—Part 3: CIE Tristimulus Values, ISO 11664-3:2012 (E)/CIE S014-3/E:2011, CIE Central Bureau, Vienna, 2011.

[19] Colorimetry—Part 4: CIE 1976 L^* a^* b^* Colour Space, ISO 11664-4:2008 (E)/CIE S014-4 E:2007, CIE Central Bureau, Vienna, 2007.

[20] Colorimetry—Part 5: L^* u^* V^* Colour Space and u', v' Uniform Chromaticity Scale Diagram, ISO 11664-5:2009 (E)/CIE S014-5 E:2009, CIE Central Bureau, Vienna.

[21] ASTM D2244-11, Standard Practice for Calculation of Color Tolerances and Color Differences from Instrumentally Measured Color Coordinates. ASTM International, West Conshohocken, PA, 2011. http://dx.doi.org/10.1520/D2244-11. www.astm.org.

[22] CIE Colorimetry – Part 6: CIE DE2000 Colour Difference Formula. CIE Central Bureau, Vienna, CIE S 014-6/E:2013.

[23] ASTM E259-06(2011), Standard Practice for Preparation of Pressed Powder White Reflectance Factor Transfer Standards for Hemispherical and Bidirectional Geometries. ASTM International, West Conshohocken, PA, 2013. http://dx.doi.org/10.1520/E0259-06R11. www.astm.org.

[24] M.E. Nadal, C.C. Miller, C. Zarobila, Reflectance factor measurement systematic errors due to near infrared fluorescence, Color. Res. Appl. 36 (2011) 96–100, http://dx.doi.org/10.1002/col.20614.

[25] ASTM E313-10, Standard Practice for Calculating Yellowness and Whiteness Indices from Instrumentally Measured Color Coordinates, ASTM International, West Conshohocken, PA, 2010. http://dx.doi.org/10.1520/E0313-10. www.astm.org.

[26] ASTM E1331-09, Standard Test Method for Reflectance Factor and Color by Spectrophotometry using Hemispherical Geometry, ASTM International, West Conshohocken, PA, 2009. http://dx.doi.org/10.1520/E1331-09. www.astm.org.

[27] ASTM E179-12, Standard Guide for Selection of Geometric Conditions for Measurement of Reflection and Transmission Properties of Materials, ASTM International, West Conshohocken, PA, 2012. http://dx.doi.org/10.1520/E0179-1. www.astm.org.

[28] ASTM E1767-11, Standard Practice for Specifying the Geometries of Observation and Measurement to Characterize the Appearance of Materials, ASTM International, West Conshohocken, PA, 2011. http://dx.doi.org/10.1520/E1767-11. www.astm.org.

[29] CIE 15.3:2004, Colorimetry, third ed., Commission Internationale de l'Eclairage, Vienna, 2004.

[30] ASTM E1349-06(2013), Standard Test Method for Reflectance Factor and Color by Spectrophotometry Using Bidirectional (45°:0° or 0°:45°) Geometry, ASTM International, West Conshohocken, PA, 2013. http://dx.doi.org/10.1520/E1349-06R13. www.astm.org.

[31] ASTM E1347-06(2011), Standard Test Method for Color and Color-Difference Measurement by Tristimulus Colorimetry, ASTM International, West Conshohocken, PA, 2013. http://dx.doi.org/10.1520/E1347-06R11. www.astm.org.

[32] F.J.J. Clarke, D.J. Parry, Helmholtz reciprocity: its validity and application to reflectometry, Light. Res. Technol. 17 (1985) 1–11.

[33] B.W. D'Andrade, S.R. Forrest, White organic light-emitting devices for solid-state lighting. Adv. Mater. 16 (18) (2004) 1585, http://dx.doi.org/10.1002/adma.200400684.

[34] I. Nimeroff, Propagation of errors in spectrophotometric colorimetry, J. Opt. Soc. Am. 43 (1953) 531–533.

[35] T. Shipley, G.L. Walker, Chromatic significance of spectrophotometric errors, J. Opt. Soc. Am. 46 (1956) 1052–1060.

[36] I. Nimeroff, Propagation of errors in tristimulus colorimetry, J. Opt. Soc. Am. 47 (1957) 697–702.

[37] A.R. Robertson, Colorimetric significance of spectrophotometric errors, J. Opt. Soc. Am. 57 (1967) 691–698.

[38] G. Wyszecki, W.S. Stiles, Color Science: Concepts and Methods, Quantitative Data and Formulae, second ed., John Wiley and Sons, New York, 1982, pp. 328–330.
[39] R.S. Berns, K.H. Petersen, Empirical modeling of systematic spectrophotometric errors, Color. Res. Appl. 13 (1988) 243–256.
[40] M.D. Fairchild, L. Reniff, Propagation of random errors in spectrophotometric colorimetry, Color. Res. Appl. 16 (1991) 360–367.
[41] P.D. Burns, R.S. Berns, Error propagation analysis in color measurement and imaging, Color. Res. Appl. 22 (1997) 280–289.
[42] J.C. Zwinkels, Errors in colorimetry caused by the measuring instrument, Text. Chem. Color. 21 (1989) 23–29.
[43] J. Taylor, European Commission. Good Practice Guide to Surface Colour Measurement: Final Report. Directorate-General Science, Research and Development, 2000.
[44] E.A. Early, M.E. Nadal, Uncertainty analysis for reflectance colorimetry, Color. Res. Appl. 29 (2004) 205–216.
[45] J.L. Gardner, R.B. Frenkel, Correlation coefficients for tristimulus response value uncertainties, Metrologia 36 (1999) 477–480.
[46] J.L. Gardner, Uncertainty estimation in colour measurement, Color. Res. Appl. 25 (2000) 349–355.
[47] A.R. Robertson, Diagnostic performance evaluation of spectrophotometers, in: C. Burgess, K.D. Mielenz (Eds.), Advances in Standards and Methodology in Spectrophotometry, Elsevier, Amsterdam, 1987, pp. 277–286.
[48] http://www.ceram.com/materials-development/colour-standards.
[49] R. Glausch, M. Kieser, R. Maisch, P. Gerhard, J. Weitzel, Special Effect Pigments, Curt, Germany, 1998.
[50] A.B. Rodrigues, Color and appearance measurements of metallic and pearlescent finishes, ASTM Standardization News (1995).
[51] A.B.J. Rodrigues, Measurement of metallic and pearlescent colors, Die Farbe 37 (1990) 65–78.
[52] C.S. McCamy, Observation and measurement of the appearance of metallic materials. Part I. Macro appearance, Color. Res. Appl. 21 (1996) 292–304.
[53] S. Teaney, G. Pfaff, K. Nitta, Pearlescent pigments: the next generation, Mod. Paint Coating (1999) 16–21.
[54] M.E. Nadal, E.A. Early, Color measurements for pearlescent coatings, Color. Res. Appl. 29 (2004) 38–42.
[55] W.R. Cramer, Examples of interference and color pigment mixtures green with red and red with green, Color. Res. Appl. 27 (2002) 276–281.
[56] ASTM E2194–03, Standard Practice for Multiangle Color Measurement of Metal Flake Pigmented Materials, ASTM International, West Conshohocken, PA, 2013. www.astm.org.
[57] ASTM E2539–12, Standard Practice for Multiangle Color Measurement of Interference Pigments. ASTM International, West Conshohocken, PA, 2013. www.astm.org.
[58] J.J. Hsia, The NBS 20-, 60-, and 85-degree specular gloss scales: NIST Tech. Note 594-10, 1975.
[59] F.E. Nicodemus, J.C. Richmond, J.J. Hsia, I.W. Ginsburg, T. Limperis, Geometrical considerations and nomenclature for reflectance: U.S. NBS Monograph 160, 1977.
[60] ISO 2813, Paint and Varnishes-Measurements of Specular Gloss of Nonmetallic Paint Films at 20°, 60°, and 85°, ISO, Geneva, 1978.
[61] ASTM D523, Standard test method for specular gloss, ASTM Standards on Color and Appearance Measurement, fifth ed., ASTM, Philadelphia, PA, 1996.

[62] R. Seve, Problems connected with the concept of gloss, Color. Res. Appl. 18 (1993) 241–252.
[63] F.B. Leloup, G. Obein, M.R. Pointer, P. Hanselaer, Toward the soft metrology of surface gloss: a review, Color Res. Appl. (2013) http://dx.doi.org/10.1002/col.21846.
[64] M.E. Nadal, E.A. Thompson, NIST reference goniophotometer for specular gloss measurements, J. Coat. Technol. 73 (2001) 73–80.
[65] TAPPI Test Method T480 om-09, Specular Gloss of Paper and Paperboard at 75 Degrees, www.tappi.org.
[66] Tappi Test Method T653 om-07, Specular Glass of Paper and Paperboard at 20 Degrees, www.tappi.org.
[67] ASTM D1003–00, Standard Test Method for Haze and Luminous Transmittance of Transparent Plastics, seventh ed., ASTM, Philadelphia, PA, 2004.
[68] F.W. Billmeyer Jr., Y. Chen, On the measurement of haze, Color. Res. Appl. 10 (1985) 219–224.
[69] H.L. Yu, C.C. Hsaio, Comparison of difference measurement methods for transmittance haze, Metrologia 46 (2009) S233–S237.
[70] V.R. Weidner, J.J. Hsia, NBS reference hazemeter: its development and testing, Appl. Opt. 18 (1979) 1619–1626.
[71] ASTM D 4039, Standard Test Method for Reflection Haze of Non-Metallic High-Gloss Surfaces. ASTM International, West Conshohocken, PA, 2013. www.astm.org.
[72] I. Nimerrof, Two-parameter gloss methods, J. Res. Natl. Bur. Stand. 58 (3) (1957) 127–135.
[73] M.R. Pointer, A framework for the measurement of visual appearance: CIE Technical Report, 2005.
[74] ASTM D5767–99, Standard Test Method for Instrumental Measurement of the Distinctness-of-Image Gloss of Coating Surfaces. ASTM International, West Conshohocken, PA, 2013. www.astm.org.
[75] ACT Test Panel Technologies, 273 Industrial Drive, Hillsdale Michigan 49242. www.acttestpanels.com.

Chapter 11

The Use of Spectrophotometry in the Pharmaceutical Industry

John P. Hammond
Starna Scientific Limited, Hainault, Essex, United Kingdom

Chapter Outline

11.1 Introduction **410**
11.2 Introduction to the Pharmaceutical Industry **410**
11.2.1 Process Analytical Technology 411
11.2.2 Quality by Design 411
11.2.3 Product Development Cycle 412
11.2.4 Discovery Research 412
11.3 Quality System for the Analytical Laboratory **414**
11.3.1 Consideration for Quality Systems in Data Quality Assurance/Control 414
11.4 UV and Visible Spectrophotometry **419**
11.4.1 Qualification of UV and Visible Spectrophotometers 421
11.4.2 Validation and Verification 430
11.4.3 Pharmaceutical Applications (UV–vis) 430
11.5 NIR Spectrometry **433**
11.5.1 Factors That Affect NIR Spectra 434
11.5.2 Instrumentation 435
11.5.3 Qualification of NIR Instruments 435
11.5.4 Method Validation 438
11.5.5 Pharmaceutical Applications (NIR) 439
11.6 Mid-IR Spectrometry **440**
11.6.1 Qualification of IR Spectrophotometers 441
11.6.2 Validation and Verification 445
11.6.3 Pharmaceutical Applications (Mid-IR) 445
11.7 Fluorescence Spectrometry **446**
11.7.1 Qualification of Fluorescence Instruments 447
11.7.2 Validation and Verification 453

Experimental Methods in the Physical Sciences, Vol. 46. http://dx.doi.org/10.1016/B978-0-12-386022-4.00011-X

11.7.3 Pharmaceutical Applications (Fluorescence) 454
11.8 Where Next? **454**
References **455**

11.1 INTRODUCTION

In this chapter, we cover the use of spectrophotometry in the pharmaceutical industry. It would not be unreasonable to state that these spectroscopic techniques constitute a large part of the analytical processes used in this industry, with the list expanding all the time. For example, modern products are taking advantage of new delivery methods, so that many are no longer simple solutions or powders, but complex colloids. In many cases, the particle size and distribution is critical to the functionality of the drugs, and a high degree of quality control (QC) of the dissolution process is required. In this specific application, ultraviolet (UV) and fluorescence spectrophotometers invariably provide the detectors for the instrumental-based systems. Coincidentally, they also perform the same role for the multitude of high-pressure liquid chromatography (HPLC) applications, which have their own unique control requirements, and are therefore not discussed in this chapter.

Much of the need for accurate optical property measurements stems from the need to control the purity and dose of the drugs, in the development, manufacturing, and QC of the product, and therefore at the beginning of the chapter, we discuss an overview of these processes.

Regulatory demands also place "evidence of control" requirements on the composition and stability of the pharmaceutical products, and thereby the proof data associated with this process. Spectrophotometric methods play an important role in the generation of this data, and the evaluation of the risks associated with this process, thereby assisting compliance. The validation of these processes is explained with respect to analytical control of the instrument systems and methodology used.

The practical implementation of the general quality concepts, to the control of ultraviolet–visible (UV–vis), near-infrared (NIR), mid-infrared (mid-IR), and fluorescence spectrophotometric systems is then discussed, together with examples of the specific applications areas in which each of these techniques are used.

Finally, the possible way regulation will influence the future use of spectrophotometry in the pharmaceutical industry is discussed.

11.2 INTRODUCTION TO THE PHARMACEUTICAL INDUSTRY

Pharmaceutical product development consists of a series of logical and systematic processes. When successful, the final outcome is a commercially available dosage form. However, this process can easily become a protracted and complicated process if there is a lack of control in any of the steps. The industry continues to change and evolve to match the regulatory demands placed on it, with the emphasis being increasingly placed on a risk-based approach. Two key documents introduced in recent years to support these concepts are the Process Analytical Technology (PAT) and the Quality by Design (QbD) guidance. Conversion of the validation and verification requirements detailed below will need to evolve to a risk-based protocol as this concept is generally adopted.

During the past 10 years, there have been global regulatory changes toward risk-based and lifecycle approaches to Current Good Manufacturing Practices (cGMPs) and regulatory issues relating to chemistry, manufacturing, and controls led particularly by the United States (US) Food and Drug Administration (FDA). In 2004, the FDA proposed that there should be science-based regulation of product quality [1].

11.2.1 Process Analytical Technology

As a consequence of the 2004 proposal, the FDA developed PAT guidance [1]. The FDA emphasized that one goal of the PAT guidance was to tailor the agency's usual regulatory scrutiny to meet the needs of PAT-based innovations that improve the scientific basis for establishing regulatory specifications.

Clearly, in order to implement this philosophy, there would have to be a structured quality system of document standards, guidance documents, etc. Given the complexity of the task, the FDA turned to an international standards body, namely, ASTM International, formally known as the American Society for Testing and Materials (ASTM), and in response to this request, Committee E55 on "Manufacture of Pharmaceutical Products" was formed in 2003.

11.2.2 Quality by Design

In November 2013, European regulators announced that they, along with the FDA, had released a second question-and-answer document intended to provide guidance to industry on the concept of QbD. The QbD concept is well known within most regulatory circles. Simply stated, it is the belief that quality should be designed, not tested, into the final product, including its manufacturing processes. In theory (and regulators say in practice as well), this results in fewer compliance problems because a manufacturer addresses problems before they exist, and more systematically when they occur. In the pharmaceutical sector, QbD concepts are broadly incorporated into the regulatory systems of any region that uses the International Conference on Harmonisation of Technical Requirements for Registration of Pharmaceuticals for Human Use (ICH) and its Q8 [2], Q9 [3], Q10 [4], and Q11 [5] documents. Both FDA and the European Medicines Agency (EMA) have been pushing QbD concepts heavily in recent years.

11.2.3 Product Development Cycle

The overall cycle of pharmaceutical product development is summarized in Fig. 11.1.

The clinical study of drug development is the most obvious and best known to laypersons and scientists. However, many associated behind-the-scene activities are also actively pursued in a parallel and timely manner to ensure the success of pharmaceutical product development. Clinical and commercial success cannot be achieved without successful completion of these other activities. It is important to note that the clinical phase boxes in Fig. 11.1 may not be aligned exactly chronologically with other development activities.

Historically, the time period for pharmaceutical drug product development has been in the order of 10–15 years. As in most commercial manufacturing organizations, with the ever-increasing demand for new products to add to the portfolio, there is constant pressure to reduce the time utilized to complete the development process, and make the product available to the market. However, the potential reduction in time achieved by this acceleration is often being offset by the increasing complexity and numbers required to achieve compliance in phase III clinical trials.

11.2.4 Discovery Research

In the discovery research phase of drug development, new compounds need to be created to meet targeted medical needs, hypotheses for model compounds

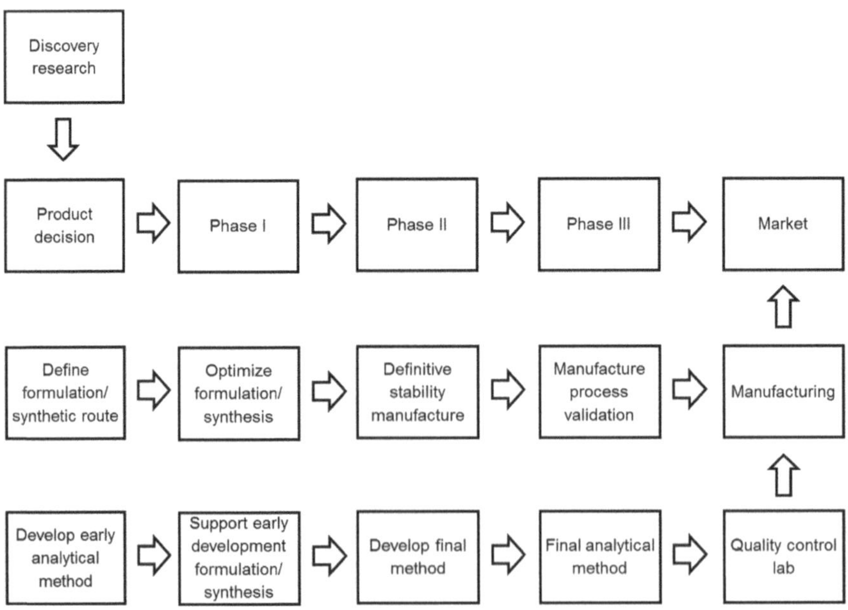

FIGURE 11.1 Overview of the drug development process.

are proposed, and various scientific leads are utilized to create and design new molecules. Thousands of molecules of similar structure are synthesized to develop a structure–activity relationship for the model. To reach this stage, large pharmaceutical companies rely on new technologies, such as combinatorial chemistry and high-throughput screening. These array technologies may have some form of associated spectroscopic measurement, for example, fluorescence correlation spectroscopy (FCS) and are the cornerstones in drug discovery. The new technologies increase the choice of compounds that can be synthesized and screened. Various *in vivo* and *in vitro* models are used to determine the value of these new candidate compounds. The sequencing of the complete human genome completed in 2000 through the Human Genome Project [6], is providing the basis for many possible targets for drug discovery through-genomics, proteonomics, and bioinformatics.

Analytical Development of Active Pharmaceutical Ingredient (API) and Drug Products. Early methods to support synthetic and formulation developments are often developed in the form of potency assay, impurities/related substance assay, dissolution, identity, chiral method, and content uniformity. These analytical methods are developed and validated in a fast and timely manner to support all phase II studies, and invariably are either spectroscopic or chromatographic by design.

Common Studies Performed on the API and Drug Product. At this stage of the development, it is important to gain preliminary information of the stability of the API and drug product. Therefore, open dish (i.e., nonprotected) stability studies are carried out to understand the chemical and physical stability of both the API and the drug product. Preliminary studies of packaging stability are conducted to obtain a preliminary assessment of the packaging materials that can be used, and photostability and thermal studies are conducted to determine the light and thermal stability of the API and drug product. Spectroscopic procedures, particular in the UV region, are critical in producing this data on stability and packaging suitability.

Impurities Level in New Drug Product. As the new drug product formulation progresses to this late stage of development, impurity profiles may differ from those of earlier formulations. The rationale for reporting and control of impurities in the new drug product and recommended storage conditions are often decided at this stage. Degradation products and those arising from excipient interaction and/or container closure systems will be isolated and identified. The impurity profile of the representative commercial process will be compared with the drug product used in development, and an investigation will be triggered if any difference is observed. Key spectroscopic areas used in these processes are the powerful qualitative techniques of NIR and mid-IR spectrophotometry, usually coupled with advanced chemometric processes. Identification of degradation products is required for those that are unusually potent and produce toxic effects at low levels.

Primary and developmental stability studies help scientists understand the degradation pathways. These studies are designed to get information on the stability of the drug product, expected expiry date, and recommended storage conditions. All specified degradation products, unspecified degradation products, and total degradation products are monitored in these studies.

Impurities in API. Treatment of the impurities in the API is similar to that for the new drug product. Impurities in the API include organic impurities (process and drug related), inorganic impurities, and residual solvents. QC analytical procedures are developed and validated to ensure appropriate detection and quantitation of the impurities. Specification limits for impurities are set based on data from stability studies and chemical development studies. A rationale for the inclusion or exclusion of impurities is set at this stage. The limits set should not be above the safety level or below the limit of the manufacturing process and analytical capability.

11.3 QUALITY SYSTEM FOR THE ANALYTICAL LABORATORY

Global regulatory requirements have become more similar as the result of deliberate harmonization under the auspices of the ICH, which brings together the regulatory authorities and pharmaceutical industry of Europe, Japan, and the United States to discuss scientific and technical aspects of drug registration, with the aim that analytical methods for global products must be able to meet global regulatory requirements. Using these protocols, it should be possible, for example, for a method to be developed and validated in the United States, and used without revalidation or additional support validation for use, in Japan or Europe. Within any given organization, the achievement of this objective is the responsibility of senior management and requires participation and commitment by personnel in many different functions at all levels within the establishment and by its suppliers. To achieve this objective reliably, there must be a comprehensively designed and correctly implemented system of quality standards incorporating Good Manufacturing Practices (GMP). It should be fully documented and effectively monitored. All parts of the quality systems should be adequately resourced with qualified personnel and suitable premises, equipment, and facilities.

11.3.1 Consideration for Quality Systems in Data Quality Assurance/Control

There are four critical components involved in the generation of reliable and consistent data (quality data) [7]. Figure 11.2 shows these components as layered activities within a quality triangle. Each layer adds to the overall quality. Analytical Instrument Qualification (AIQ) forms the base for generating quality data, and the other components essential for generating quality data are analytical method validation, system suitability tests, and QC check samples. Detailed below are the general considerations for these key areas. Later in the chapter, in the sections detailing the spectroscopic instrumentation used to assist compliance and control, the specific requirements relating to each technique will be discussed.

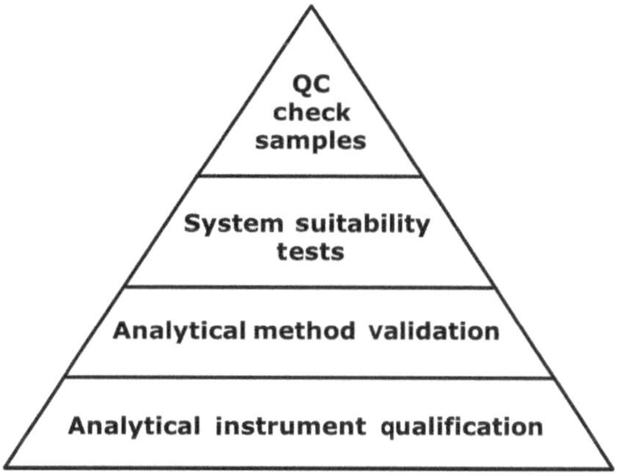

FIGURE 11.2 Components of data quality, redrawn and from United States Pharmacopeia (USP) <1058>.

Validation versus Qualification. In this chapter, the term validation is used for manufacturing processes, analytical procedures, and software procedures and the term qualification is used for the instrumental systems. Thus, the phrase "AIQ" is used for the process of ensuring that a spectrophotometer is suitable for its intended application.

11.3.1.1 Analytical Instrument Qualification

Spectrophotometers are just one group in the large variety of laboratory equipment, instruments, and computerized analytical systems, used in the pharmaceutical industry to acquire data to help ensure that products are suitable for their intended use. An analyst's primary objective is to consistently obtain reliable and valid data suitable for the intended purpose. Depending on the applications, users validate their procedures, calibrate their instruments, and perform additional instrument checks, such as system suitability tests and analysis of in-process QC check samples to help ensure that the acquired data are reliable. With the increasing sophistication and automation of analytical instruments, an increasing demand has been placed on users to qualify their instruments, both initially and on a continuing basis.

AIQ is the collection of documented evidence that an instrument performs suitably for its intended purpose, and thereby the use of a qualified instrument contributes to confidence in the validity of generated data. Instrument qualification is not a single continuous process, but instead results from several discrete activities. For convenience, these activities can be grouped into four phases: design qualification (DQ), installation qualification (IQ), operational qualification (OQ), and performance qualification (PQ). Some AIQ activities cover more than one qualification phase, and analysts potentially could perform them during more than one of the phases. However, in many instances, there is need for specific order to the AIQ activities; for example, IQ must occur first in order to initiate other qualification activities.

DQ—Design Qualification. Design Qualification is the documented collection of activities that define the functional and operational specifications of the instrument and criteria for selection of the vendor, based on the intended purpose of the instrument.

IQ—Installation Qualification. Installation Qualification is the documented collection of activities necessary to establish that an instrument is delivered as designed and specified, and is properly installed in the selected environment, deemed suitable for the instrument.

In all of the spectroscopic techniques described later in this chapter, evidence in compliance with the above definition is always required, but there are no specific additional requirements for these techniques, and therefore IQ is not discussed in the appropriate sections. In the OQ and PQ sections, additional technique-specific requirements are highlighted.

OQ—Operational Qualification. After a successful IQ, the instrument is ready for OQ testing. OQ is the documented collection of activities necessary to demonstrate that an instrument will function according to its operational specification testing in the selected environment.

PQ—Performance Qualification. PQ is the documented collection of activities necessary to demonstrate that an instrument consistently performs according to the specifications defined by the user and is appropriate for the intended use. After IQ and OQ have been performed, the instrument's continued suitability for its intended use is demonstrated through PQ.

Clearly, in a routine situation, only PQ may need to be performed on a regular basis, and OQ less frequently, for example, after maintenance or repair operations have been carried out on the instrument.

It should be remembered that AIQ does not prove that the analytical results generated by the spectrophotometer are correct. AIQ demonstrates that the instrument itself is working appropriately. However, many other factors contribute to the final result like sampling procedures, reagent quality, and sample handling.

Qualification of the instrument will assist the instrument user in making several important judgments, namely:

- Is the instrument performing to its manufacturer's specifications?
- Does it meet the requirements of the quality environment of the laboratory, for example, a pharmacopoeia?
- Is it performing suitably for the determination to be carried out?

While the pharmacopoeias and other standards may give generic recommendations for instrument performance, for a specific test, the required performance characteristics may be different and are usually given in the monograph for the test method. Furthermore, they can change over time, so claims by instrument manufacturers that a spectrophotometer is, for example, "pharmacopoeia compliant" should be treated with caution.

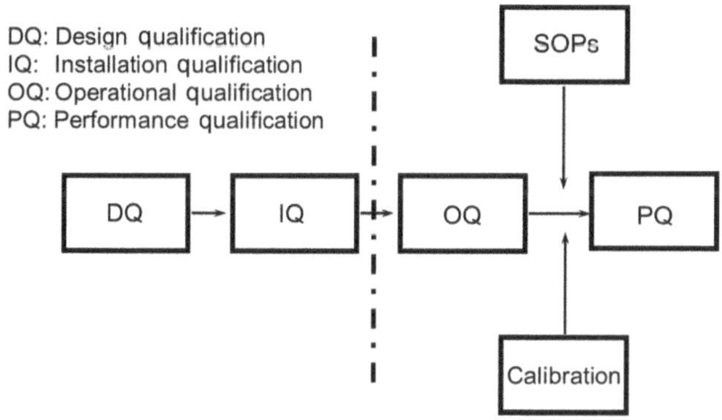

FIGURE 11.3 Timing, applicability, and activities for each phase of Analytical Instrument Qualification (AIQ).

In a regulated environment, the onus of proof is on the user to justify and prove, by the qualification of the system, that the instrument is "fit for purpose" and capable of providing the required accuracy and precision of data (Fig. 11.3).

11.3.1.2 Analytical Method Validation

Analytical method validation is the collection of documented evidence that an analytical procedure is suitable for its intended use. Use of a validated procedure with qualified analytical instrument systems provides confidence that the procedure will generate test data of acceptable quality.

Validation. Validation of an analytical procedure is the process by which it is established, by laboratory studies, that the performance characteristics of the procedure meet the requirements for the intended analytical applications. Typical analytical performance characteristics that should be considered in the validation of the types of procedures described in this document are shown below:

***Typical Analytical Performance Characteristics Used in Method Validation**

- Accuracy
- Precision
- Specificity
- Detection limit (DL)
- Quantitation limit (QL)
- Linearity
- Range
- Robustness

The effects of processing conditions and potential for segregation of materials should be considered when obtaining a representative sample to be used for validation of procedures. The ICH documents give guidance on the necessity for revalidation in the following circumstances: changes in the synthesis of the drug substance, changes in the composition of the drug product, and changes in the analytical procedure. The validity of an analytical procedure can be verified only by laboratory studies. Therefore, documentation of the successful completion of such studies is a basic requirement for determining whether a procedure is suitable for its intended application(s).

11.3.1.3 Method Verification

Verification consists of assessing selected analytical performance characteristics to generate appropriate, relevant data rather than repeating the validation process. In the US, this requirement is established in 21 CFR 211.194(a)(2) of the current cGMP regulations [8], which states that the "suitability of all testing methods used shall be verified under actual conditions of use."

The verification process is the assessment of whether the procedure can be used for its intended purpose, under the actual conditions of use for a specified drug substance and/or drug product matrix. Users should have the appropriate experience, knowledge, and training to understand and be able to perform the procedures as written. Verification should be conducted by the user that the results will provide confidence that the procedure will perform suitably as intended.

Verification Requirements. Verification requirements should be based on an assessment of the complexity of both the procedure and the particular material to which the procedure is applied. Although complete revalidation of a method is not required to verify the suitability of a procedure under actual conditions of use, some of the analytical performance characteristics listed under Section 11.5.4 may be used for the verification process. Only those characteristics that are considered to be appropriate for the verification of the particular procedure need to be evaluated. The process of assessing the suitability of an analytical test procedure under the conditions of actual use may or may not require actual laboratory verification of each analytical performance characteristic. The degree and extent of the verification process may depend on the level of training and experience of the user, on the type of procedure and its associated equipment or instrumentation, on the specific procedural steps, and on which pharmaceutical product is being tested.

Verification should assess whether the procedure is suitable for the drug substance and/or the drug product matrix, taking into account the drug substance's synthetic route, the method of manufacture for the drug product, or both, if applicable. Verification should include an assessment of elements such as the effect of the matrix on the recovery of impurities and drug substances from the drug product matrix, as well as the suitability of chromatographic conditions and column, the appropriateness of detector signal response, etc.

As an example, an assessment of specificity is a key parameter in verifying that a procedure is suitable for use in assaying drug substances and drug products. For instance, acceptable specificity for a chromatographic method may be verified by conformance with system suitability resolution requirements (if specified in the procedure). However, drug substances from different suppliers may have different impurity profiles that are not addressed by the compendial test procedure. Similarly, the nonactive ingredient (excipients) in a drug product can vary widely among manufacturers and may have the potential to directly interfere with the procedure or cause the formation of impurities that are not addressed by the compendial procedure. In addition, drug products containing different excipients, antioxidants, buffers, or container extractives may affect the recovery of the drug substance from the matrix. In these cases, a more thorough assessment of the matrix effects may be required to demonstrate suitability of the procedure for the particular drug substance or product. Other analytical performance characteristics such as an assessment of the DL or QL and precision for impurities procedures may be useful to demonstrate the suitability of the compendial procedure under actual conditions of use.

11.3.1.4 System Suitability Tests

System suitability tests verify that the system will perform in accordance with the criteria set forth in the procedure. These tests are performed along with the sample analyses to ensure that the system's performance is acceptable at the time of the test.

11.3.1.5 QC Check Samples

Many analysts carry out their tests on instruments standardized using reference materials and/or calibration standards. Some analyses also require the inclusion of QC check samples to provide an in-process or ongoing assurance of the test's suitable performance. In this manner, AIQ and analytical method validation contribute to the quality of analysis before analysts conduct the tests. System suitability tests and QC checks help ensure the quality of analytical results immediately before or during sample analysis.

11.4 UV AND VISIBLE SPECTROPHOTOMETRY

UV–vis spectrophotometry is one of the most widely used techniques in analytical chemistry, capable of producing accurate and precise results. For these reasons, procedures using this technique are found in analytical, clinical, and research laboratories, and find extensive use in quality assurance. As discussed earlier, in the pharmaceutical environment, validation is key to the generation of quality data. In these situations, it is essential to check instrument performance on a regular basis against reference materials having known characteristics that will provide that evidence.

Certified Reference Materials (CRMs) have been developed to give analytical chemists a convenient and dependable way of ensuring that their UV–vis instruments are performing to the requirements of their analytical methods. These materials are manufactured and calibrated under strictly controlled conditions that conform to all the relevant international standards.

Why Do We Need Reference Materials? To confirm the performance of an analytical instrument, it will be necessary to calibrate or test its performance against standards having known properties, and the pharmaceutical industry will require that any materials used to qualify an instrument can themselves be shown to be traceable to nationally or internationally recognized standards. Such standards are referred to as CRMs and must themselves be manufactured and calibrated under closely controlled conditions.

What is a CRM? As defined by ISO/REMCO (the International Standards Organisation Committee on Reference Materials) [9], a CRM is a

> Reference material, characterised by a metrologically valid procedure for one or more specified properties, accompanied by a certificate that provides the value of the specified property, its associated uncertainty, and a statement of metrological traceability.

Traceability. Traceability is a key concept in calibration and validation processes and is defined in ISO/IEC (International Organization for Standardization/International Electrotechnical Commission) Guide 99:2007 International Vocabulary of Metrology (VIM) [10]—as the

> property of a measurement result whereby the result can be related to a reference through a documented unbroken chain of calibrations, each contributing to the measurement uncertainty

International agreements between National Metrology Institutes (NMIs) at a high level by the Bureau International des Poids et Mesures (BIPM) mean that measurements made against traceable references in one country will be accepted in any other country that is a signatory within the agreement.

CRMs—The Manufacturing Environment. Historically, producers of CRMs have used either ISO/IEC 17025 [11] or ISO Guide 34 [12] to control their production processes but in 2004 the International Laboratory Accreditation Co-operation (ILAC), ruled that to be a Reference Material Producer, manufacturers should be accredited to both these standards simultaneously. There is considerable overlap between the two standards, but essentially ISO/IEC 17025 is mainly concerned with the technical and management competencies required to operate a testing and calibration laboratory, whereas ISO Guide 34 is more concerned with the supplier's organizational and production control systems. It also requires suppliers to operate a customer support and technical advice service. Any company accredited to these standards will be subject to regular inspections by the national accrediting body.

On October 26, 2012 at the 16th ILAC General Assembly in Rio De Janeirio, Brazil, the following resolution was adopted:

As ISO Guide 34:2009 includes normative references to ISO/IEC 17025 and ISO 15189, the General Assembly resolves that accreditation of reference material producers is conducted in accordance with ISO Guide 34:2009 alone.

In practice, the normative reference to ISO/IEC 17025 still requires the use of all appropriate and relevant sections of this standard, for the manufacture and production of a CRM. By definition, measured value assignments can only be effectively produced in an ISO/IEC 17025 controlled environment.

11.4.1 Qualification of UV and Visible Spectrophotometers

Instrument parameters checked as part of OQ and PQ of a UV–vis spectrophotometer include:

- Wavelength calibration
- Photometric accuracy
- Photometric linearity
- Photometric precision
- Resolution/bandwidth
- Stray light

These can all be checked using CRMs. The following notes briefly describe the various reference materials available for the OQ and PQ. Table 11.1 shows those reference materials accepted by various standardization bodies for compliance.

11.4.1.1 Operational Qualification

Acceptance criteria for critical instrument parameters that establish "fitness for purpose" are verified during IQ and OQ. Specifications for particular instruments and applications can vary depending on the analytical procedure used and the desired accuracy of the final result.

Wherever possible in the following procedures, analysts should employ CRMs in preference to laboratory-prepared solutions. CRMs obtained from a recognized accredited source include independently verified traceable value assignments with associated calculated uncertainty. These CRMs should be obtained from a recognized accredited source and include independently verified traceable value assignments with associated calculated uncertainty. CRMs should be kept clean and free from dust. Recertification should be performed periodically to maintain the validity of the qualification.

11.4.1.1.1 Wavelength Calibration

Accurate wavelength calibration is fundamental to good analytical results by UV–vis spectrophotometry. Most methods involve measurements at absorption maxima in the spectrum of the substance being measured, so clearly if the wavelength calibration of the instrument is incorrect, errors can be introduced. The sharper the absorption band being measured, the greater is this

TABLE 11.1 Compliance References

Parameter Tested	Wavelength Region	EP	DAB	USP	ASTM	TGA	BP	Material
Photometric accuracy	UV	•	•	•	•	•	•	Potassium dichromate solution
Photometric accuracy	Visible	•						Potassium dichromate solution
Photometric accuracy	Visible			•	•	•	•	ND glass filters
Wavelength calibration	UV–visible	•	•	•	•	•	•	Holmium oxide solution
Wavelength calibration	UV–visible			•	•	•	•	Holmium oxide glass filter
Wavelength calibration	Visible			•	•	•	•	Didymium glass filter
Wavelength calibration	Visible			•				Didymium oxide solution
Wavelength calibration	Far UV	•	•	•	•	•	•	Rare earth oxide solution
Wavelength calibration	UV–visible			•	•	•	•	Samarium oxide solution
Resolution/ bandwidth	UV–visible	•	•	•	•	•	•	Toluene in hexane
Resolution/ bandwidth	UV	•						Benzene vapor
Stray light	UV–visible	•	•	•	•	•	•	Stray light cut-off filters

Key: EP (Ph. Eur.), European Pharmacopoeia; DAB, Deutsches Arzneibuch (German Pharmacopoeia); USP, United States Pharmacopeia; ASTM, ASTM International; TGA, Therapeutic Goods Administration (Australia); BP, British Pharmacopoeia.

error. The absorption bands encountered in visible spectrophotometry are often quite broad, so minor wavelength errors may not seem too important at first sight, particularly if analytical standards are being used to calibrate the analysis. It is likely, however, that when wavelength errors exist, problems will be encountered when comparing results between laboratories and certainly when extinction coefficients are being determined or used to derive concentration values.

Calibration at the exact wavelength required for measurement may not be possible as no suitable wavelength standard may be available, but wavelength calibration should be checked at more than one wavelength, ideally bracketing the wavelength to be used for analysis. In some UV instruments, the facility exists to use an emission line (or sometimes two) from the installed Deuterium lamp to check the wavelength calibration. In principle, these are the best possible standards as they are fundamental atomic lines, but only two usable lines are available, both in the visible region at 486.0 and 656.1 nm. Even if both are used, they could not be said to qualify the UV range of the instrument.

A variety of wavelength CRMs are available to cover wavelengths from the far UV to the NIR. Solutions and solid filters containing rare earth elements are particularly useful, as these elements have sharp absorption bands in the UV, visible, and NIR regions of the spectrum. The most popular is holmium oxide, either incorporated into a solid filter (filter 11) or as holmium oxide solution in perchloric acid and sealed by heat fusion into a quartz cell. The solution offers 14 certified peaks from 240 to 650 nm and the filter 11, from 270 to 640 nm. "Didymium," a mixture of neodymium and praseodymium, is also popular and is available as either a filter (11 peaks from 430 to 890 nm) or as a solution reference (14 peaks from 290 to 870 nm). Samarium perchlorate is particularly useful for wavelengths from 230 to 560 nm, as it has 14 sharp peaks in this widely used region of the spectrum. For the far UV, a special rare earth reference cell has been developed having five certified peaks between 200 and 300 nm [13].

11.4.1.1.2 Photometric Accuracy, Linearity, and Precision

These parameters, also referred to as absorbance accuracy etc., describe the quantitative accuracy of the instrument, and as such are probably the most frequently checked. They test not only the optical performance of the instrument but also the quantitative accuracy of its electronic measuring system. It should be noted that deficiencies in any of the foregoing optical parameters could lead to errors in photometric accuracy.

International and national regulatory bodies recognize two reference materials for determining the accuracy and linearity of the absorbance scale of a UV absorption spectrophotometer. Potassium dichromate dissolved in dilute perchloric acid is recommended for use in the UV region. When prepared in 0.001 M perchloric acid, potassium dichromate gives a spectral scan containing characteristic peaks at 257 and 350 nm, and troughs at 235 and 313 nm. Potassium dichromate references are available in a range of concentrations: 20, 40, 60, 80, 100, 140, 160, 180, 200, 240, and 600 mg/L and 0.001 M perchloric acid blank, in UV quartz cells that have been permanently sealed by heat fusion. Absorbance values are certified at 235, 257, 313, 350 nm (20–240 mg/L), and 430 nm (600 mg/L) (Table 11.2).

TABLE 11.2 Typical Absorbance Values for Potassium Dichromate Solutions in 0.001 M Perchloric Acid in a 1-cm Cuvette

	235 nm	257 nm	313 nm	350 nm	430 nm
20 mg/l	0.246	0.286	0.096	0.214	
40 mg/l	0.493	0.574	0.193	0.428	
60 mg/l	0.743	0.865	0.290	0.643	
80 mg/l	0.994	1.157	0.387	0.858	
100 mg/l	1.247	1.452	0.484	1.074	
120 mg/l	1.501	1.749	0.582	1.290	
140 mg/l	1.758	2.050	0.679	1.506	
160 mg/l	2.016	2.351	0.777	1.723	
180 mg/l	2.275	2.655	0.875	1.940	
200 mg/l	2.536	2.960	0.974	2.157	
240 mg/l	3.043	3.552	1.168	2.589	
600 mg/l					0.954

Neutral density (ND) glass filters are recommended for the visible region (440–635 nm). Schott NG-type glasses have been used for over 30 years for the validation of the absorbance scale and linearity of spectrophotometers in the visible region. When manufactured to a specific thickness, filters with a range of transmittance and absorbance values can be produced. The spectral scan of these materials is essentially flat. ND filters are usually supplied in sets of three filters of varying transmission together with an empty aluminum holder, to be used as a blank, absorbance (A) and transmittance (T) values are certified at 440, 465, 546.1, 590, and 635 nm and the full range of filters covers nominal values from 0.03 to 3.5 A.

For those working outside these wavelength limits, two further materials are available. While not prescribed for pharmacopoeial compliance, they are widely recognized as absorbance references. These are nicotinic acid solutions for the far UV (210–260 nm) [13] and metal-on-quartz filters calibrated over the wavelength range of 250–635 nm.

11.4.1.1.3 Resolution

Resolution is the ability of the instrument to discriminate between two adjacent absorbance bands. This is usually described in terms of the spectral bandwidth (SBW) of the instrument, as defined in Chapter 3 which describes the range of wavelengths coming from the exit slit of the instrument's monochromator. As SBW increases, the measured peak absorption decreases and spectral detail is reduced and eventually obscured completely (Fig. 11.4).

Clearly, the smaller the SBW, the more accurately the instrument will measure the true absorbance of the absorption band. Sensitivity to the analyte will also improve. For the instrument, the SBW is defined as the width, at half the maximum intensity, of the band of light leaving the monochromator, see Fig. 11.5.

The accuracy of the measured absorbance will depend on the ratio of the SBW to the natural bandwidth (NBW) of the absorbing substance. The NBW is the width of the sample absorption band at half the absorption maximum. As a general rule, for satisfactory measurement, the SBW of the instrument should be 10% or less of the NBW of the absorption peak being measured.

This might suggest that a narrow SBW should always be used, but this invariably means the use of narrow slit widths. Narrow slits will also reduce the energy throughput and hence may adversely affect the signal-to-noise ratio (SNR) and hence the precision of the measurement. As SBW can have a significant effect on analytical performance, it is usually specified in the analytical method. The SBW of an instrument is usually quoted in the manufacturer's specification, but may need to be verified by testing against a suitable reference material, particularly if repairs or maintenance have been performed on the instrument's optical system.

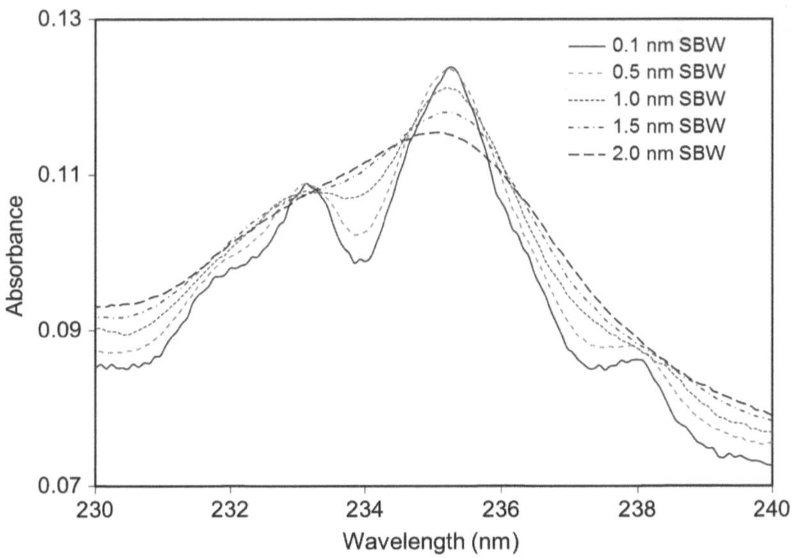

FIGURE 11.4 Effect of SBW on the measured absorbance of a samarium perchlorate wavelength standard.

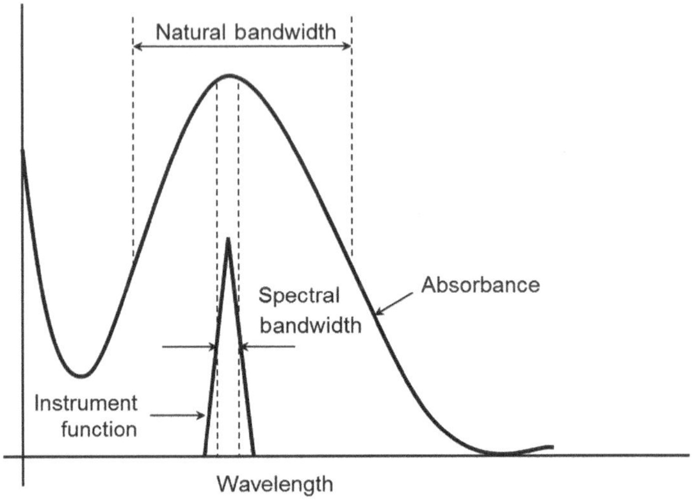

FIGURE 11.5 SBW: definition.

The most widely used reference for qualifying the SBW of a spectrophotometer is a solution of 0.020% toluene in hexane. This is supplied in a permanently heat-sealed quartz cell, with a hexane blank. Another approach is a benzene vapor reference. This contains 0.01 mL of benzene, which at room temperature is in the vapor state, sealed into a UV quartz cell. At SBWs less than 1 nm, the benzene vapor spectrum has characteristic features that may or may not be displayed, dependent on the SBW of the spectrophotometer. Note that benzene vapor will not work well with most photodiode array spectrophotometers as these instruments provide abridged spectral data where the peaks may not be resolved well enough to be useable. A toluene in methanol reference is also available for instrument resolution validation when using derivative spectroscopy.

11.4.1.1.4 Stray Light

Stray light, also called Stray Radiant Energy or Power, has previously been discussed in Chapter 3. As the detector cannot discriminate between the analytical wavelength and the stray light, the stray light contributes to the detector signal and introduces an error in the measured absorption. The stray light is not absorbed even at high concentrations of the absorbing species, so its effect is a negative deviation from the linear relationship between concentration and absorbance (the Beer–Lambert Law) on which most quantitative determinations are based (Fig. 11.6).

Stray light is wavelength and instrument dependent. It can be present at any wavelength but is most noticeable when the energy throughput of the system at the analytical wavelength is relatively low, for example, in the far UV region.

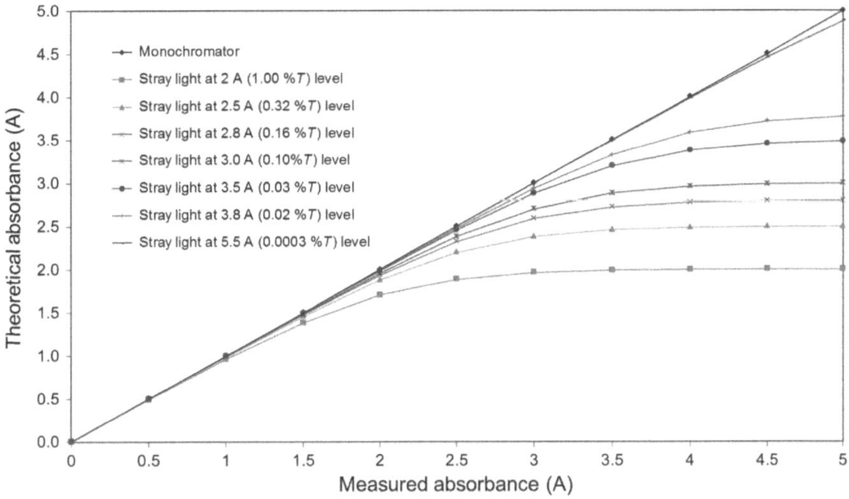

FIGURE 11.6 Illustration of the effect of stray light on linearity of absorbance measurements.

The usual way of assessing stray light is to measure, at the desired analytical wavelength, a sample that totally absorbs the radiation at that wavelength, but transmits at all other wavelengths. Any light detected by the instrument is then stray light. Practically, the usual method is to use cut-off filters or solutions that cut-off all light near the analytical wavelength. Figure 11.7 shows the spectra of some solutions specified in different standards for the evaluation of stray light.

Alkali halide salt solutions have very sharp transitional (cut-off) spectra, giving excellent filtering characteristics. Hence, below the specified "cut-off" wavelength, any indication of light transmittance must be stray light. The test for stray light is important even if the spectrophotometer is not used below 260 nm, because it is an excellent indication of the overall performance of the instrument optics, including grating and deuterium lamp (Table 11.3).

11.4.1.2 Performance Qualification

The purpose of PQ is to determine that the instrument is capable of meeting the user's requirements for all the parameters that may affect the quality of the measurement and to ensure that it will function properly over extended periods of time.

With few exceptions, pharmacopoeial spectrophotometric tests and assays call for comparison against an appropriate reference standard. This helps ensure measurement under identical conditions for the test specimen and the reference substance. These conditions could include wavelength setting, SBW selection, cell placement and correction, and transmittance levels. Cells that exhibit identical transmittance at a given wavelength may differ

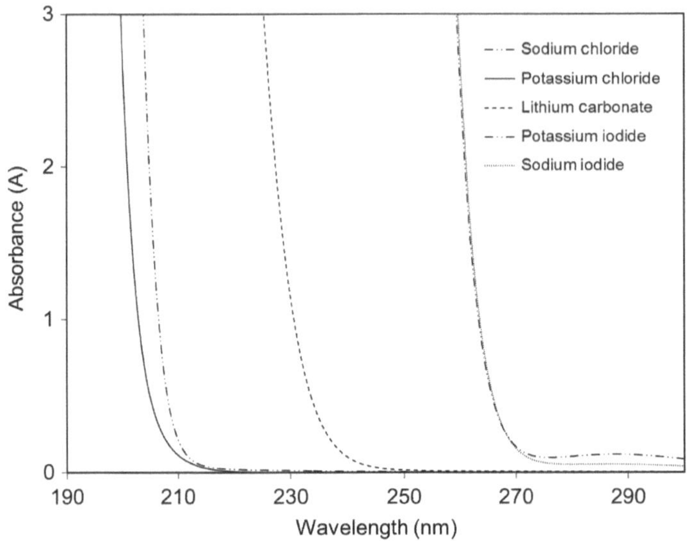

FIGURE 11.7 Stray light test solution spectra.

TABLE 11.3 Stray Light Solutions

Material	Cut-off (nm)	Concentration
Sodium nitrite	390	50 mg/L aqueous
Acetone	326	Spectroscopic grade
Potassium iodide	260	10 mg/L aqueous
Sodium iodide	260	10 mg/L aqueous
Lithium carbonate	227	Saturated aqueous
Sodium chloride	205	10 mg/L aqueous
Potassium chloride	200	12 mg/L aqueous

considerably in transmittance at other wavelengths. Appropriate cell corrections should be established and used where required.

Comparisons of a test specimen with a reference standard are best made at a peak of spectral absorption for the compound concerned. Assays that prescribe spectrophotometry give the commonly accepted wavelength for peak spectral absorption of the substance in question. Different spectrophotometers may show minor variation in the apparent wavelength of this peak. Good practice demands that comparisons be made at the wavelength at which peak absorption occurs.

The expressions "similar preparation" and "similar solution" as used in tests and assays involving spectrophotometry indicate that the reference comparator, generally a reference standard, should be prepared and observed in an identical manner for all practical purposes to that used for the test specimen. Usually, when analysts make up the solution of the specified reference standard, they prepare a solution of about (i.e., within 10%) the desired concentration, and they calculate the absorptivity on the basis of the exact amount weighed out. If a previously dried specimen of the reference standard has not been used, the absorptivity is calculated on the anhydrous basis. The expressions "concomitantly determine" and "concomitantly measure" as used in tests and assays involving spectrophotometry indicate that the absorbances of both the solution-containing the test specimen and the solution-containing the reference specimen, relative to the specified test blank, must be measured in immediate succession.

Sample Solution Preparation. For determinations using UV or visible spectrophotometry, the specimen generally is dissolved in a solvent. Unless otherwise directed in the monograph, analysts make determinations at room temperature using a path length of 1 cm. Many solvents are suitable for these ranges, including water, alcohols, chloroform, lower hydrocarbons, ethers, and dilute solutions of strong acids and alkalis. Precautions should be taken to use solvents that are free from contaminants that absorb in the spectral region under examination. For the solvent, analysts typically should use water-free methanol or alcohol or alcohol denatured by the addition of methanol but without benzene or other interfering impurities. Solvents of special spectrophotometric quality, guaranteed to be free from contaminants, are available commercially from several sources. Some other analytical reagent-grade organic solvents may contain traces of impurities that absorb strongly in the UV region. New lots of these solvents should be checked for their transparency, and analysts should take care to use the same lot of solvent for preparation of the test solution, the standard solution, and the blank. Best practice is to use solvents that transmit more than 40% ($39.9\%T = 0.399$ A) at the wavelength of interest.

Assays in the visible region usually call for concomitantly comparing the absorbance produced by the assay preparation with that produced by a standard preparation containing approximately an equal quantity of a reference standard. In some situations, analysts can omit the use of a reference standard (e.g., when spectrophotometric assays are made with routine frequency) when a suitable standard curve is available and is prepared with the appropriate reference standard, and when the substance assayed conforms to Beer's Law within the range of about 75–125% of the final concentration used in the assay. Under these circumstances, the absorbance found in the assay may be interpolated on the standard curve, and the assay result can be calculated. Such standard curves should be confirmed frequently and always when a new spectrophotometer or new lots of reagents are put into use.

11.4.2 Validation and Verification

Validation. The objective of UV–vis method validation is to demonstrate that the measurement is suitable for its intended purpose, including quantitative determination of the main component in a drug substance or a drug product (Category I assays), quantitative determination of impurities or limit tests (Category II), and identification tests (Category IV). Depending on the category of the test, the analytical method validation process for UV–vis requires testing for linearity, range, accuracy, specificity, precision, DL, QL, and robustness. These analytical performance characteristics apply to externally standardized procedures and those that use standard additions.

Verification. The objective of an UV–vis procedure verification is to demonstrate the suitability of a test procedure under actual conditions of use. Performance characteristics that verify the suitability of an UV–vis procedure are similar to those required for any analytical procedure. Verification should be performed using a reference material and a well-defined matrix. Verification of compendial UV–vis procedures should at minimum include the execution of the validation parameters for specificity, accuracy, precision, and QL, when appropriate, as indicated under "Validation."

11.4.3 Pharmaceutical Applications (UV–vis)

Of the four spectrophotometric techniques described in this chapter, currently the number of UV–vis procedures in use in the pharmaceutical industry probably accounts for more than all of the other three, that is, NIR, mid-IR, and fluorescence, combined.

The reasons for this preference are many, and varied, and owe as much to the history of the technique, as to the physical and chemical aspects. UV–vis spectrophotometry has been available as an instrumental technique for almost 70 years, since April 1947, when Applied Physics Corporation delivered the first commercial recording UV–vis spectrophotometer, a Cary 11, to Mellon Institute in Pittsburgh, USA. Given the availability of the instrumentation, and the fundamental linear relationship, thanks to Beer's Law; in the second half of the twentieth century, a vast array of chromogenic reagents and procedures were developed for the analytical quantitation of most elements in the periodic table. While many of these procedures have been superseded by elemental spectroscopic techniques such as inductively coupled plasma and related techniques; it has been often said that UV–vis is the "screwdriver of spectroscopy," that is, the tool found in everybody's toolbox, and as such will be found in most analytical laboratories across the world.

In the United States Pharmacopeia (USP), UV–vis still provides the majority of the spectrophotometric procedures, and quantifying this statement, there are still over two hundred specific monographs containing

UV–vis measurements in the current version (USP36-NF31), not counting any related HPLC methods using UV–vis detection. Not surprisingly, 99% of these procedures are quantitative in nature, with very few qualitative methods used for identification. These methods in use can be categorized as shown in Fig. 11.8.

As shown in Fig. 11.8, a significant percentage of UV–vis is used for dissolution testing of tablets and products in the pharmaceutical industry (Category III). Dissolution is a characterization test commonly used by the pharmaceutical industry to guide formulation design and control product quality. It is also the only test that measures the rate of *in vitro* drug release as a function of time, which can reflect either reproducibility of the product manufacturing process or, in limited cases, *in vivo* drug release.

In this technique, the UV–vis spectrophotometer is used as the detector, linked to a dissolution bath, and can be configured in a variety of options using various pumping systems, switching values, etc.

The three leading pharmacopoeia's, namely, the European Pharmacopoeia (EP), Japanese Pharmacopoeia (JP), and USP, all have their own mandatory control chapters for defining the physical characteristics of these systems, methodology, etc., which have been (Fig. 11.9) harmonized by the ICH in their document "Evaluation and Recommendation of Pharmacopoeial texts for use in the ICH Regions on Dissolution Test General Chapter, Q4B Annex 7(R2)."

A recently published paper [15] clearly demonstrates the important information that can be obtained from these dissolution studies. The variation in the drug release profile from four different sources is shown in Fig. 11.10.

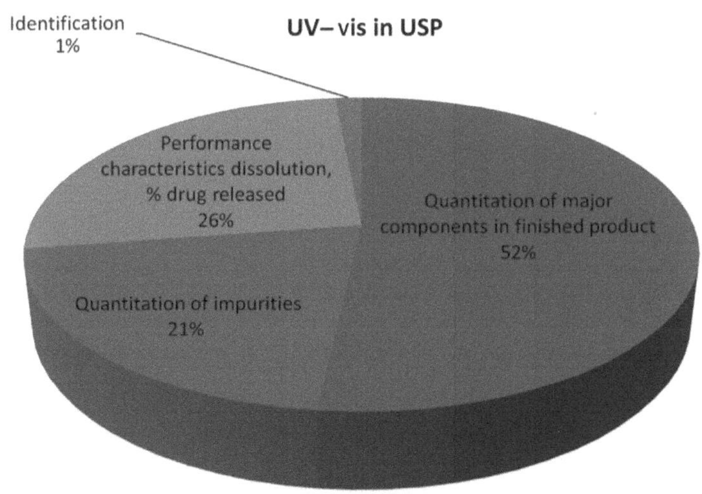

FIGURE 11.8 Application mix of UV–vis procedures found in the USP.

FIGURE 11.9 A typical dissolution bath, configured for paddle dissolution [14].

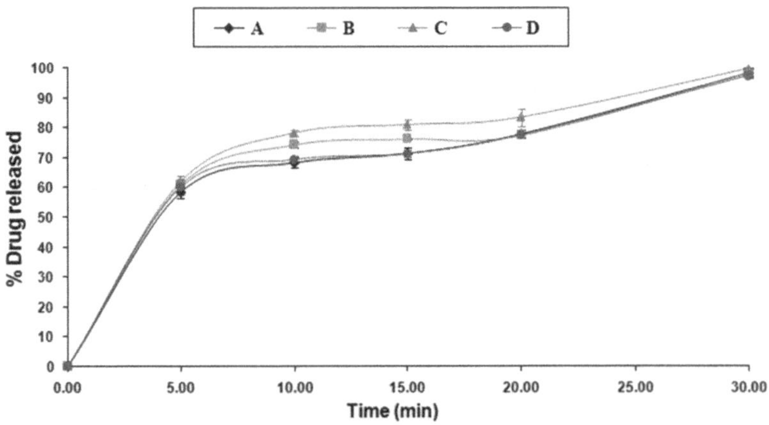

FIGURE 11.10 Comparative *in vitro* dissolution profiles of acetaminophen tablets [15].

11.5 NIR SPECTROMETRY

The NIR spectral region lies approximately between 780 and 2500 nm (4000–12,800 cm^{-1}) bridging the more well known and analytically used regions of the UV–vis (190–780 nm) and the infrared (4000–600 cm^{-1}). Until relatively recently, it has been called the "forgotten" region.

NIR spectrometry is a branch of vibrational spectroscopy that shares many of the principles that apply to other spectroscopic measurements. The NIR spectral region comprises two subranges associated with detectors used in the initial development of NIR instrumentation. The short wavelength (Herschel or silicon region) extends from approximately 780–1100 nm (12,821–9000 cm^{-1}); and longer wavelengths, between 1100 and 2500 nm (9000–4000 cm^{-1}) compose the traditional (lead sulfide (PbS)) NIR region.

Vibrational spectroscopy in the NIR region is dominated by overtones and combinations that are much weaker than the fundamental mid-IR vibrations from which they originate. Because molar absorptivities in the NIR range are low, radiation can penetrate several millimeters into materials, including solids. Many materials, such as glass, are relatively transparent in this region. Fiber-optic technology is readily implemented in the NIR range, which allows monitoring of processes in environments that might otherwise be inaccessible.

Given the range and diversity of type of NIR systems now available for use in the pharmaceutical industry, for example, the small handheld units being used for raw material checking; the instrument qualification tests and acceptance criteria described below may not be appropriate for all instrument configurations. In such cases, alternative instrument qualification and performance checks should be scientifically justified and documented. In addition, validation parameters discussed in this chapter may not be applicable for all applications of NIR spectrometry. Validation parameters characterized for a specific NIR application should demonstrate suitability of the NIR application for its intended use.

Transmission and Reflection. The most common measurements performed in the NIR spectral range are transmission and reflection spectrometry. Incident NIR radiation is absorbed or scattered by the sample and is measured as transmittance, T or reflectance, R, respectively. Transflection spectrometry is a hybrid of transmission and reflection wherein a reflector is placed behind the sample so that the optical path through the sample and back to the detector is doubled compared to a transmission measurement of a sample of the same thickness. Transflection is used to describe any double-pass transmission technique. The light may be reflected from a diffuse or specular (mirror) reflector placed behind the sample. This configuration can be adapted to share instrument geometry with certain reflection or fiber-optic probe systems in which the source and the detector are on the same side of the sample.

Most reflection measurements in the NIR are made of scattering samples such as powders and slurries. For such materials, NIR radiation can penetrate a substantial distance into the sample, where it can be absorbed when the wavelength of the radiation corresponds to a transition between the ground vibrational state of the analyte and either a harmonic of a given vibrational mode (an overtone) or the sum of two or more different modes (a combination band). Nonabsorbed radiation is scattered back from the sample to the detector. NIR reflection spectra are accessed by calculating and plotting $\log(1/R)$ versus wavelength. This logarithmic form is the pseudo-absorbance of the material and is commonly called absorbance.

11.5.1 Factors That Affect NIR Spectra

The following list is not exhaustive, but it includes many of the major factors that affect NIR spectra.

Sample Temperature. Sample temperature influences spectra obtained from aqueous solutions and other hydrogen-bonded liquids, and a difference of a few degrees may result in significant spectral changes. Temperature may also affect spectra obtained from less polar liquids, as well as solids that contain solvents and/or water.

Moisture and Solvent. Moisture and solvent present in the sample material and analytical system may change the spectrum of the sample. Both absorption by moisture and solvent and their influence on hydrogen bonding of the APIs and excipients can change the NIR spectrum.

Sample Thickness. Sample thickness is a known source of spectral variability and must be understood and/or controlled. The sample thickness in transmission mode is typically controlled by using a fixed optical path length for the sample. In diffuse reflection mode, the sample thickness is typically controlled by using samples that are "infinitely thick" relative to the detectable penetration depth of NIR light into a solid material. Here, "infinite thickness" implies that the reflection spectrum does not change if the thickness of the sample is increased.

Sample Optical Properties. In solids, both surface- and bulk-scattering properties of calibration standards and analytical samples must be taken into account. Surface morphology and refractive index properties affect the scattering properties of solid materials. For powder materials, particle size and bulk density influence scattering properties and the NIR spectrum.

Polymorphism. Variation in crystalline structure (polymorphism) from materials with the same chemical composition can influence NIR spectral response. Different polymorphs and amorphous forms of solid material may be distinguished from one another on the basis of their NIR spectral properties. Similarly, different crystalline hydration or solvation states of the same material can display different NIR spectral properties.

Age of Samples. Samples may exhibit changes in their chemical, physical, or optical properties over time. Care must be taken to ensure that both samples and standards used for NIR analysis are suitable for the intended application.

11.5.2 Instrumentation

All NIR measurements are based on exposing material to incident NIR light radiation and measuring the attenuation of the emerging (transmitted, scattered, or reflected) light. Several spectrophotometers are available; they are based on different operating principles—for example: filters, grating-based dispersive, acousto-optical tunable filter, Fourier-transform near-infrared (FT-NIR), and liquid crystal tunable filter. Silicon (Si), PbS, indium gallium arsenide, and deuterated triglycine sulfate (DTGS) are common detector materials. Conventional cuvette sample holders, fiber-optic probes, transmission dip cells, and spinning or traversing sample holders are common examples of sample interfaces for introducing the sample to the optical train of a spectrometer.

The selection of specific NIR instrumentation and sampling accessories should be based on the intended application, and particular attention should be paid to the suitability of the sampling interface for the type of sample that will be analyzed.

NIR Reference Spectra. NIR references, by providing known stable measurements to which other measurements can be compared, are used to minimize instrumental variations that would affect the measurement.

Transmittance. The measurement of transmittance requires a background reference spectrum for determining the absorption by the sample relative to the background. Suitable transmittance reference materials depend on the specific NIR application and include air, an empty cell, a solvent blank, or a reference sample.

Reflectance. The measurement of reflectance requires the measurement of a reference reflection spectrum to determine the attenuation of reflected light relative to the unattenuated incident beam. The reflectance spectrum is calculated as the ratio of the single-beam spectrum of the sample to that of the reference material. Suitable reflectance reference materials depend on the specific NIR application and include ceramic, perfluorinated polymers, gold, and other suitable materials.

11.5.3 Qualification of NIR Instruments

11.5.3.1 Operational Qualification

Because there are so many different approaches for measuring NIR spectra, OQ using standards with known spectral properties is recommended [16]. Using external traceable reference standard materials does not justify omitting the instrument's internal QC procedures. As is the case with any spectroscopic device, wavelength uncertainty, photometric linearity, and noise characteristics of NIR instruments should be qualified against target specifications for the intended application.

11.5.3.2 Performance Qualification

Performance qualification demonstrates that the NIR measurement consistently operates within target specifications defined by the user for a specific application; it is often referred to as system suitability. Performance qualification for NIR measurements can include comparing a sample or standard spectrum to previously recorded spectra. Comparisons of spectra taken over time from identical and stable samples or reference standard materials can form the basis for evaluating the long-term stability of an NIR measurement system. The objective is to demonstrate that no abnormal wavelength shift or change in detector sensitivity has occurred during ongoing analysis.

Characterizing Instrument Performance. Specific procedures, acceptance criteria, and time intervals for characterizing NIR instrument performance depend on the instrument and intended application. Many NIR applications use previously validated models that relate NIR spectral response to a physical or chemical property of interest. Demonstrating stable instrument performance over extended periods of time provides some assurance that reliable measurements can be taken from sample spectra using previously validated NIR models.

11.5.3.2.1 Wavelength Accuracy

In order for NIR spectrometry to yield accurate results for determinations performed by the application of multivariate statistical algorithms, the spectrum of the instrument that is used for the measurement, either on a wavenumber or a wavelength scale, must closely match the spectrum of the spectrometer used to acquire the calibration spectra. Similarly, the accuracy with which calibration data are transferred from instrument-to-instrument may be limited because of differences in the wavelength scale of the two instruments, thus precluding a broader use of NIR spectrometry. NIR spectra from sample and/or reference standard materials can be used to demonstrate an instrument's suitable wavelength dispersion performance against target specifications.

The USP NIR System Suitability Reference Standard or the National Institute of Standards and Technology (NIST) SRM 2036 for reflectance measurement and NIST SRM 2035 for transmittance measurement can be used for wavelength verification. Suitable materials for demonstrating wavelength dispersion performance include polystyrene, mixtures of rare earth oxides, and absorption by water vapor for instruments that use an interferometer for wavelength dispersion. With appropriate justification, alternative standards may be used. Wavelength uncertainty typically is characterized from a single spectrum (collected with the same spectral resolution to obtain the standard value) using a minimum of three peaks that cover a suitable spectral range of the instrument. Typical tolerances for agreement with standard values are ±1.0 nm below 2000 nm and ±1.5 nm from 2000 to 2500 nm. Alternative tolerances may be used when justified for specific applications.

11.5.3.2.2 Photometric Linearity and Response Stability

NIR spectra from samples and/or reference standard materials with known relative transmittance or reflectance can be used to demonstrate a suitable relationship between NIR light attenuation (due to absorption) and instrument response. For reflectance measurements, commercially available reflectance standards with known reflectance properties are often used. Spectra obtained from reflection standards are subject to variability as a result of the difference between the experimental conditions under which they were factory calibrated and those under which they are subsequently put to use. Hence, the reflectance values supplied with a set of calibration standards may not be useful in the attempt to establish an "absolute" calibration for a given instrument. Provided that: (1) the standards do not change chemically or physically, (2) the same reference background is also used to obtain the standard values, and (3) the instrument measures each standard under identical conditions (including precise sample positioning), the reproducibility of the photometric scale will be established over the range of standards. Subsequent measurements on the identical set of standards give information on long-term stability. Photometric linearity is typically characterized using a minimum of four reference standards in the range from 10% to 90% reflection (or transmission). NIR applications based on measuring an absorbance larger than 1.0 may require standards with reflectivity properties between 2% and 5% reflection (or transmission) for characterizing instrument performance at low reflectance. The purpose is to demonstrate a linear relationship between NIR reflectance and/or transmittance and instrument response over the scanning range of the instrument. Typical tolerances for a linear relationship are 1.00 ± 0.05 for the slope and 0.00 ± 0.05 for the intercept of a plot of the measured photometric response versus standard photometric response. Alternative tolerances may occur when justified for specific applications.

11.5.3.2.3 Spectroscopic Noise

NIR instrument software may include built-in procedures to automatically determine system noise and to provide a statistical report of noise or SNR over the instrument's operating range. In addition, it may be desirable to supplement such checks with measurements that do not rely directly on manufacturer-supplied procedures. Typical procedures involve measuring spectra of traceable reference materials with high and low reflectance. Tolerances for these procedures should demonstrate suitable SNR for the intended application.

High-Flux Noise. Instrument noise is evaluated at high-light flux by measuring reflectance or transmittance of the reference standard, with the reference material (e.g., 99% reflection standard) acting as both the sample and the background reference.

Low-Flux Noise. The same procedure may be used with a lower reflectivity reference material (e.g., 10% reflectance standard) to determine system noise at reduced light flux. The source, optics, detector, and electronics make significant contributions to the noise under these conditions.

11.5.4 Method Validation

The objective of NIR method validation, as is the case with the validation of any analytical procedure, is to demonstrate that the measurement is suitable for its intended purpose. NIR spectrometry is somewhat different from conventional analytical techniques because validation of the former generally is achieved by the assessment of chemometric parameters, but these parameters can still be related to the fundamental validation characteristics required for any analytical method.

Data pretreatment is often a vital step in the chemometric analysis of NIR spectral data. Data pretreatment can be defined as the mathematical transformation of NIR spectral data to enhance spectral features and/or remove or reduce unwanted sources of variation prior to using the spectrum. Calibration is the process of developing a mathematical relationship between NIR spectral response and properties of samples. Many suitable chemometric algorithms for data pretreatment and calibration exist; the selection should be based on sound scientific judgment and suitability for the intended application.

11.5.4.1 Ongoing Method Evaluation

Validated NIR methods should be subject to ongoing performance evaluation, which may include monitoring accuracy, precision, and other suitable method parameters. If performance is unacceptable, corrective action is necessary. It involves conducting an investigation to identify the cause of change in method performance and may indicate that the NIR method is not suitable for continued use. Improving the NIR method to meet measurement suitability criteria may require additional method development and documentation of validation experiments demonstrating that the improved method is suitable for the intended application. The extent of revalidation required depends on the cause of change in method performance and the nature of corrective action required in order to establish suitable method performance. Appropriate change controls should be implemented to document ongoing method improvement activities.

Revalidation of a qualitative model may be necessary as a result of the following:

- Addition of a new material to the spectral reference library
- Changes in the physical properties of the material
- Changes in the source of material supply
- Identification of previously unknown critical attribute(s) of material(s).

Revalidation of a quantitative model may be necessary as a result of the following:

- Changes in the composition of the test sample or finished product
- Changes in the manufacturing process
- Changes in the sources or grades of raw materials
- Changes in the reference analytical method
- Major changes in instrument hardware

Outliers. Sample spectra that produce an NIR response that differs from the qualitative or quantitative calibration model may produce an outlier. This does not necessarily indicate an out-of-specification result; but rather an outlier indicates that further testing of the sample may be required and is dependent on the particular NIR method. If subsequent testing of the sample by an appropriate method indicates that the property of interest is within specifications, then the sample meets its specifications. Outlier samples may be incorporated into an updated calibration model subsequent to execution and documentation of suitable validation studies.

11.5.4.2 Method Transfer

Controls and measures for demonstrating the suitability of NIR method performance following method transfer are similar to those required for any analytical procedure. Exceptions to general principles for conducting method transfer for NIR methods should be justified on a case-by-case basis. The transfer of an NIR method is often performed by using an NIR calibration model on a second instrument that is similar to the primary instrument used to develop and validate the method. When a calibration model is transferred to another instrument, procedures and criteria must be applied to demonstrate that the calibration model meets suitable measurement criteria on the second instrument. The selection of an appropriate calibration model transfer procedure should be based on sound scientific judgment.

11.5.5 Pharmaceutical Applications (NIR)

The NIR (900–2500 nm) region was largely neglected by analysts for many years because of the complex nature of the spectra produced by water, proteins, etc., in this region. However, thanks to the many and varied multivariate mathematical calibration approaches now available, broadly described by the term "chemometrics," it has recently found new and important applications, in the pharmaceutical industry for raw material and quality assurance procedures.

These applications often involve comparing an NIR spectrum from a sample to reference spectra and assessing similarities against acceptance criteria developed and validated for a specific application. In contrast, applications of quantitative analysis involve the development of a predictive relationship between NIR spectral attributes and sample properties. These applications typically use numerical models to quantitatively predict chemical and/or physical properties of the sample on the basis of NIR spectral attributes. Examples include; moisture content, API content uniformity, hardness, particle size, packing density, etc.

Initially, it was the arrival of FT-NIR instruments in the 1990s and the increasing awareness of the NIR's unique information content in the signal that can provide both physical and chemical information that catalyzed the deployment of NIR spectrometry systems for both qualitative and quantitative purposes within the industry.

It is the versatility of the available sample presentation modes in NIR, that is, fiber-optic probe-based systems, and latterly, small handheld units, combined with the availability of powerful signal process electronics that has revolutionized the use of the technique in two main pharmaceutical areas outside of the QC laboratory.

Firstly, in the goods-in receiving area of the raw materials warehouse, bulk materials may now be checked for identity and compliance to specification, immediately on receipt. As is the case with other spectroscopy measurements, interactions between NIR radiation and matter provide information that can be for both qualitative and quantitative assessment of the chemical composition of samples. In addition, qualitative and quantitative characterization of a sample's physical properties can be made because of the sample's influence on NIR spectra. Example application uses include the identification of fundamentally chemically different compounds, APIs, excipients, dyes, packaging material, polymorphs, isomers, and physical characteristics, such as crystalline versus amorphous.

Secondly, NIR spectrometry provides an invaluable tool in assisting compliance with the PAT initiative, because probes can easily be mounted within process streams, mixers, etc., to allow qualitative analysis of, for example, reaction pathways, drying of a product [17,18], granulation, blending and/or content uniformity, monitoring of blister packed final product, etc. In Ref. [17], NIR is used to track the drying process of an API, and in Ref. [18], the correlation between NIR spectral measurement and actual moisture content of a freeze-dried lyophilized preparation of folinic acid calcium salt (Leucovorin Ca) is described (Fig. 11.11).

11.6 MID-IR SPECTROMETRY

Mid-IR spectrometry is an instrumental method used in mostly qualitative identification procedures, and like its NIR counterpart can be used for quantitative analysis, often using a chemometric approach. The method involves

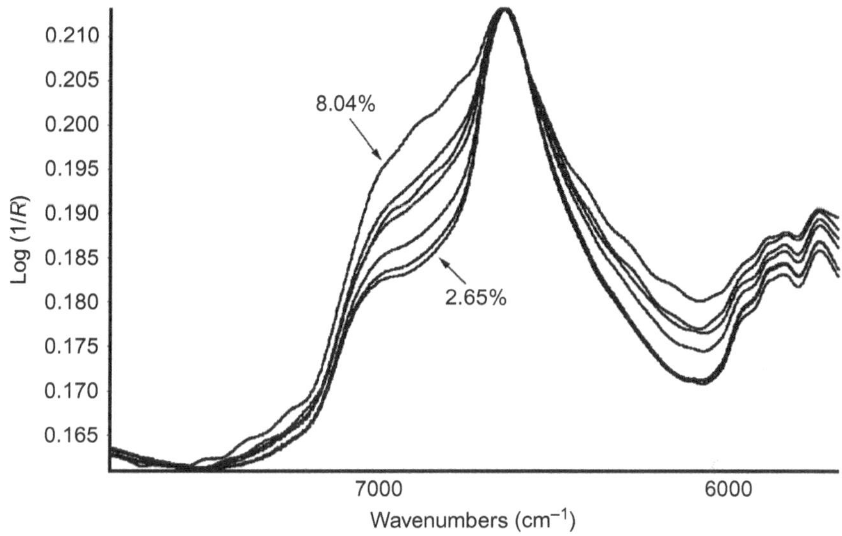

FIGURE 11.11 Effect of change of moisture content between 2.65% and 8.04% on the FT-NIR spectrum of lyophilized Leucovorin Ca [18].

measurement of the absorption of electromagnetic radiation over the wavenumber range between 4000 and 400 cm^{-1} (2.5 and 25 μm) caused by the promotion of molecules from the ground state of their vibrational modes to an excited vibrational state.

Vibrational modes involve the motion of all atoms of the molecule. When molecules contain a certain functional group, the transitions often occur in narrow spectral ranges. In this case, the wavenumbers at which these transitions occur are known as group frequencies. When a vibrational mode involves atomic motions of more than just a few atoms, the frequencies occur over wider spectral ranges and are not characteristic of a particular functional group but are more characteristic of the molecules as a whole. Such bands are known as fingerprint bands. All strong bands that absorb at wavenumbers above 1500 cm^{-1} are group frequencies. Strong bands that absorb below 1500 cm^{-1} can be either group frequencies or fingerprint bands.

11.6.1 Qualification of IR Spectrophotometers

11.6.1.1 Operational Qualification

While the science of mid-IR spectrometry may be performed on both dispersive and Fourier-transform IR (FT-IR) spectrometers, almost without exception those currently in use in the pharmaceutical industry are FT-IR [19].

11.6.1.1.1 Wavenumber Accuracy

The most commonly used wavenumber standard for infrared (IR) spectrometry is an approximately 35-μm thick, matte polystyrene film. The spectrum of such a film has several sharp bands at 3060.0, 2849.5, 1942.9, 1601.2, 1583.0, 1154.5, and 1028.3 cm^{-1}. The most frequently used is located at about 1601.2 cm^{-1}. The wavenumber of maximum response of the chosen band can be measured using the center-of-gravity, polynomial spline procedure, or other peak-picking algorithms.

11.6.1.1.2 Photometric Accuracy

There is no good way of measuring the absolute photometric accuracy of an FT-IR spectrometer. Photometric accuracy is partially covered in Chapter 4; however, a recognized procedure in the Pharmaceutical industry, for verifying the accuracy of the instrument's zero-energy level is to check the region of the single-beam spectrum below the detector cut-off between 200 and 300 cm^{-1} when a DTGS detector is used and between 400 and 300 cm^{-1} when a mercury cadmium telluride (MCT) detector is used. The average value of the single-beam spectrum in this region should be less than 1000 times the maximum value of the signal in the single-beam spectrum. This criterion is usually met when a DTGS detector is used but is rarely met with an MCT detector.

11.6.1.1.3 Sensitivity

The sensitivity of the instrument can be determined by measuring two single-beam spectra under exactly the same conditions and calculating their ratio to produce what is commonly known as a 100% line. The noise level in different spectral regions can be estimated either as the peak-to-peak noise, that is, the difference between the maximum and minimum values of the percent transmission in the selected spectral region(s), or the root-mean-square (RMS) noise, that is, the standard deviation of the spectrum in that region. The RMS noise level is the preferred metric because this calculation involves all the data in the selected region rather than just the two most deviant points. Typical measurement conditions to test the sensitivity of an FT-IR spectrometer equipped with a DTGS detector are 16 coadded scans, a resolution of 2 cm^{-1}, and Norton–Beer medium apodization. The most commonly used spectral region is 2200–2000 cm^{-1} because (a) this is where the performance of most mid-IR spectrometers is highest and (b) no common atmospheric interferent such as H_2O or CO_2 absorbs strongly in this region. However, other regions should be tested close to the ends of the spectrum, such as 650–450 cm^{-1} and 4000–3800 cm^{-1}. The SNR of the spectrometer operating with certain parameters in a given spectral region is estimated as 100/(RMS noise level in percent transmission).

11.6.1.1.4 Stability

The short- and long-term stability of the instrument also can be estimated from the deviation of the 100% line from 100% T at the short wavelength (high wavenumber) end of the spectrum. Short-term stability is estimated by measuring the two single-beam spectra a few minutes apart and calculating the 100% line. The long-term stability is measured by increasing the time between the two measurements to several hours.

11.6.1.1.5 Signal Averaging

The SNR should increase with the square root of the number of coadded scans. To test the signal-averaging capability, measure the SNR with the following numbers of scans: $N = 1$, 4, 16, 64, 256, 1024, and 4096. A plot of SNR versus $\sqrt{N}$ is linear if the instrument is correctly averaging signals. An alternative means of testing this is to plot log SNR versus log N. This plot should be linear with a slope of 2.00.

11.6.1.2 Performance Qualification

The purpose of PQ is to determine that the instrument is capable of meeting the user's requirements for all the parameters that may affect the quality of the measurement.

Procedure. Mid-IR spectra can be measured by transmission, external reflection, internal reflection (often called attenuated total reflection), diffuse reflection, and photoacoustic spectrometry (PAS). Different sample preparation techniques are available for these options. The most common sample preparation techniques are outlined below:

- *KBr Discs.* KBr will coalesce under high pressure and can be formed into self-supporting disks that are transparent to mid-IR radiation to about 400 cm^{-1}. Commercial presses and dies in a range of diameters are available for the preparation of alkali-halide and similar disks.
- *Mineral Oil Mulls.* 10–20 mg of the sample is placed into an agate or mullite mortar and then ground to a fine powder, and a small drop of the mulling agent is added. This is then mixed into a uniform paste, which is transferred to the center of a clean IR-transparent window (e.g., KBr, NaCl, AgBr, or CsI). A second matching window is placed on top of the mull, and the mull is squeezed to form a thin translucent film that is free from bubbles.

  The most widely use mulling agent for the mid-IR region is a saturated hydrocarbon mineral oil (liquid paraffin, Nujol).
- *Compression Cells.* A compression cell is useful when measuring a small or limited-quantity solid sample such as a single particle of an API or excipient, a contaminant such as a short length of fiber, or a small fragment from a packaging material. This is particularly the case for investigations

using an IR microscope system. When using a compression cell, the sample is placed between the windows of the cell, the cell is then tightened, and the sample thickness is reduced to an optimum for a transmission measurement. Because of the high strength of diamond, it is commonly used as the window material of compression cells.

- *Self-Supported Polymer Films.* The mid-IR transmission spectrum of many polymers used as packaging materials is recorded from samples prepared as thin self-supporting films using hot compression molding or microtoming.
- *Capillary Films.* Nonvolatile liquids can be examined neat in the form of a thin layer sandwiched between two matching windows that are transparent to mid-IR radiation. The liquid layer must be free of bubbles and must completely cover the diameter of the IR beam focused on the sample.
- *Liquids and Solutions in Transmission Cells.* For the examination of liquid and solution samples, transmission cell assemblies that comprise a window pair, spacer, filling ports, and a holder are available commercially in both macro- and microsample configurations. For laboratory applications, spacers typically are formed from lead, polytetrafluoroethylene, or polyethylene terephthalate and can be supplied, depending on spacer materials, in standard thickness path lengths from approximately 6 μm to 1 mm or larger.
- *Gases.* Mid-IR transmission cells for static or flow-through gas and vapor sampling are available in a wide range of materials to suit the application, from laboratory to process scale. In the laboratory, the traditional gas cell has been a 10-cm long cylinder made from borosilicate glass or stainless steel with an approximately 40-mm aperture at each end. Each open end is covered with an end cap that contains one of a pair of mid-IR—transparent windows constructed from, for example, KBr, ZnSe, or CaF_2.
- *Attenuated Total Reflection (ATR).* Attenuated total reflectance spectrometry relies on the optical phenomenon of radiation passing through a medium of high refractive index at a certain angle of incidence entirely reflected internally at a boundary in contact with a material of lower refractive index. The medium of high refractive index is also known as the internal reflection element (IRE). The sample under examination should be placed in close contact with the IRE such as diamond, germanium, ZnSe, or another suitable material of high refractive index. Ensure close and uniform contact between the substance and the whole crystal surface by applying pressure or by dissolving the substance in an appropriate solvent and then covering the IRE with the solution and evaporating to dryness.
- *Diffuse Reflection.* The most important and commonly used form of sample preparation for diffuse reflection is to dilute the sample by intimately mixing it with 90–99% of nonabsorbing diluents such as finely powdered KBr or KCl. The sample dilution has the added benefit of reducing absorption band intensities to an appropriate level.

- *Microscopic Sampling*. Coupling a light microscope with a mid-IR spectrometer allows spectra to be obtained from very small samples and is generally applied in transmittance or reflectance modes.

11.6.2 Validation and Verification

Validation. The objective of an IR method validation is to demonstrate that the measurement is suitable for its intended purpose, including: quantitative determination of the main component in a drug substance or a drug product (Category I assays), quantitative determination of impurities or limit tests (Category II), and identification tests (Category IV). Depending on the category of the test, the analytical procedure validation process for IR requires the testing of linearity, range, accuracy, specificity, precision, DL, quantitation limit, and robustness.

If the IR method results from a chemometrics model calculated against the response of another analytical technology (e.g., HPLC), then the same principles described in NIR spectrometry, specifically Section 11.5.4, can be applied.

11.6.3 Pharmaceutical Applications (Mid-IR)

Given that mid-IR is essentially a qualitative technique, applications of mid-IR spectrometry in the pharmaceutical industry, range widely from the basic identification and characterization of drug formulations to the study of kinetic pathways in new topic drug delivery systems.

Combining the various measurement modes and sample preparation techniques already described, provides a matrix of combinations, providing opportunities to exploit the advantages of mid-IR spectrometry. Again like NIR spectrometry, quantitation can be possible by the use of the appropriate chemometric technique, for example, the determination of methamphetamine using ATR which is illustrated in Fig. 11.12 for two different concentration levels [20].

ATR-based methodology can be used to study drug release in semisolid formulations, drug penetration, and influence of penetration modifiers and also provide in *in vivo* studies. It can also be used to obtain images of the surface of the skin and the spatial distribution of protein and lipid rich domains.

PAS can also be applied to measure drug content in semisolid and solid formulations, to determine drug penetration into artificial and biological membranes. The big advantage of this technique is the possibility of spectral depth profiling.

The small sample size made available by microscopic sampling can also be used, for example, to visualize the drug and excipient distribution in pharmaceutical formulations such as tablets and therapeutic transdermal system, contaminants in pharmaceutical samples, the mechanism of drug release, and study of penetrability of cosmetic ingredients in human hair.

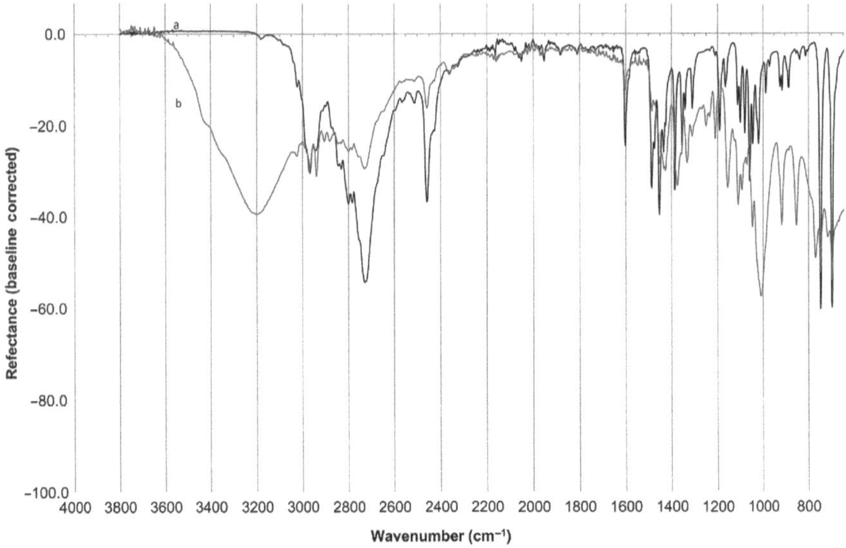

FIGURE 11.12 Typical ATR spectra of methamphetamine: (a) high concentration (78.6%) and (b) low concentration (10.3% cut with MSM (methylsulphonylmethane)) [20].

Mixtures of unknown materials may be found during the development and manufacturing of biopharmaceutical drug products. Their physiochemical properties must be identified to determine the root cause and to take corrective and preventive actions. Mid-IR spectrometry is widely employed as a powerful analytical tool to obtain the fingerprints of unknown materials, and by combining microspectroscopic techniques with spectral library searching, functional group analysis, and knowledge of the materials used in the processes, these incidents of unknown contaminants can be investigated.

11.7 FLUORESCENCE SPECTROMETRY

Fluorescence is a two-step process that requires absorption of light at a specific wavelength (excitation) followed by emission of light, usually at a higher wavelength. The emission of light is termed fluorescence.

The most common type of fluorescent sample is a submicromolar transparent solution that absorbs light following the Beer–Lambert–Bouguer Law and that fluoresces with an intensity that is directly proportional to the concentration, the absorptivity, and the fluorescence quantum yield of the fluorescent species or fluorophore.

Unlike absorption spectrometry, where deviation from linearity is the exception, fluorescence linearity can be affected by a number of sample-related effects; see Chapter 7.

A conventional fluorescence spectrometer has both excitation and emission wavelength selectors. The analyst collects a spectrum by fixing the wavelength of one of the selectors and scanning the other wavelength selector over a specified range.

Fluorescence methods also are termed background free because little excitation light reaches the detector. This characteristic makes fluorescence detection highly sensitive, down to single-molecule detection in some cases. Fluorescence detection also can be highly specific because a fluorophore emits a characteristic emission spectrum. Specificity and sensitivity are two of the more important strengths of fluorescence methods.

11.7.1 Qualification of Fluorescence Instruments

The purpose of this section is to provide test methods and acceptance criteria to ensure that the instrument is suitable for its intended use (OQ) and that it will continue to function properly over extended time periods (PQ) [21].

As with any spectrometric device, analysts must qualify a spectrofluorometer for accuracy and precision for both wavelength scale and relative intensity or detector signal scale. They also must establish sensitivity. OQ should span the operational ranges required within the laboratory for both intensity and wavelength scales.

11.7.1.1 Operational Qualification

The tolerances given in both the instrument OQ and PQ are applicable for general use. Specifications for particular instruments and applications can vary depending on the analytical procedure used and the desired accuracy of the final result.

Wherever possible, analysts should use CRMs for purposes of calibration in the steps detailed below in preference to laboratory-prepared solutions. When CRMs are obtained from a recognized accredited source, they have independently verified traceable value assignments with associated calculated uncertainties.

The criteria for acceptable instrument performance given in the following subsections are applicable for general use unless specified otherwise. Two general types of instrumental measurements are differentiated here: spectral, that is, those that measure intensity versus wavelength, and fixed, that is, those that measure intensity at a fixed wavelength and bandwidth.

11.7.1.1.1 Wavelength Accuracy

The level of confidence of measured peak positions is defined by wavelength accuracy for spectral measurements. Determining and calibrating the wavelength accuracy (see Section 7.6.1.4) at two points (one for emission and one for excitation) often are sufficient to qualify an instrument for this parameter because the relative wavelength accuracy typically does not change importantly, for example, on the order of 1 nm, across an instrument's effective wavelength range. However, determination of the accuracy of many wavelengths across the desired wavelength range demonstrates if further calibration, beyond a single point, is needed. Multipoint calibration involves measuring wavelength biases at multiple wavelengths and correcting for the wavelength dependence of the bias. A single-point calibration often can be applied to the wavelength axis in an instrument's software before data are collected, but a multipoint calibration may require that the correction be applied to spectra after they are collected. Additional information can be found in Chapter 7.

For fixed measurements, the wavelength position and bandwidth should be reproducible. For filter-based wavelength selection, this requires only that the same filter be used when analysts compare data over time. If a different filter must be used, for example, when data are compared across instruments and laboratories, then the transmission curves of the filters must be compared.

Atomic Line Spectra. This procedure is described as the fundamental method because the emission lines produced from a discharge lamp are characteristic of the source element, and, as a fundamental physical standard, these wavelengths have been measured with an uncertainty of no more than ±0.01 nm. In solution spectrofluorometry, the wavelength bias required rarely exceeds 1.0 nm. For these reasons, the atomic line standard values are cited without uncertainty. For this method, the lamp should be placed at the source position in the spectrofluorometer.

A commonly employed low-pressure mercury lamp has a number of intense lines that cover a large part of the UV and visible range, that is, 253.7, 296.7, 365.0, 404.7, 435.8, 546.1, 577.0, and 579.1 nm. Manufacturers often use two Xenon lines from the source at 260.5 and 541.9 nm as an internal calibration check because the accuracy of both the excitation and emission monochromators can be verified and can be used for diagnostic purposes.

Use of Rare Earth Oxide Solutions. This procedure uses a solution of a rare earth oxide prepared by dissolution in acid media. The most frequently used is holmium oxide in perchloric acid in combination with a diffuse reflector located at the sample position. Suitable CRMs are available commercially. The wavelength selector not being scanned should be removed; if removal is not practicable, it should be set to zero order (in this position, a grating behaves like a mirror reflecting all wavelengths). The diffuse reflector is scanned with and without the rare earth sample in place, and the ratio of the two intensities is calculated to obtain an effective transmittance spectrum. Minima in the intensity ratio correspond to absorption peaks of the sample. For a 40 mg/L solution of holmium oxide in perchloric acid, 1.0 nm SBW and a path length of 1 cm, 14 minima have established [22] wavelengths of 241.1, 249.9, 278.1, 287.2, 333.5, 345.4, 361.3, 385.6, 416.3, 451.3, 467.8, 485.2, 536.6, and 640.5 nm.

TABLE 11.4 Dopant, Excitation, and Emission Data for Selected Reference Materials

Dopant	Excitation λ (nm)	Emission λ (nm)
p-Terphenyl	295	338
Ovalene	342	482
Tetraphenylbutadiene	348	422
Anthracene	360	402

If the operational range of the spectrometer lies outside 240–650 nm, other certified rare earth oxides or other solutions are used. Didymium (a mixture of neodymium and praseodymium) is available as a traceable standard in both solution and glass presentations. Didymium is similar in preparation to the holmium materials and has useful peak characteristics in the 730–870 nm region.

Use of Polymethylmethacrylate (PMMA)-Doped References. This procedure uses solid reference materials manufactured by polymerization of a variety of fluorescent active aromatic ring compounds into an inert PMMA matrix. These materials are supplied as polished blocks for use in a standard cuvette holder (see Table 11.4).

Wavelength precision should be determined over the operational range using at least six replicate measurements. The standard deviation should not exceed ± 1 nm.

11.7.1.1.2 Performance Verification

Results from day-to-day testing of photostable intensity standards are used to verify the performance of an instrument. If the measured intensity does not change from that observed when the instrument was qualified, then instrument performance has not changed and remains qualified. Using such standards to determine an artifact-based or quasi-absolute intensity scale potentially enables measured intensities and instrument sensitivity to be compared over time or between instruments. Intensity measurements should be within the linear range of the instrument's detection system before analysts attempt intensity comparisons.

For instruments with filter-based wavelength selection, analysts should use fluorescence standards for spectral correction to determine expected intensity differences caused by filters with different transmission profiles. By compensating for these intensity differences due to spectral mismatch, the analyst can determine a quasi-absolute intensity scale for these instruments. Analysts should approach instrument-to-instrument comparisons with particular caution because of the relatively large and difficult-to-quantify uncertainties involved.

Use of Low-Conductivity (18 MΩ) water. The Raman band of water is used to measure SNRs in fluorescent instruments. The Raman band of water is inherently reproducible and does not degrade with time. Water is convenient to obtain in a pure state and allows interlaboratory comparisons to be made with a high level of confidence. No preparation or dilution is required. The Raman band is a low-level signal that provides a good test for both the optics and the electronics of an instrumental system.

The Raman band of water is not caused by fluorescence but is a result of Raman scattering. For water, the Raman band is always red-shifted 3382 cm^{-1} relative to the excitation. This band usually is measured by excitation at 350 nm, resulting in a Raman peak at 397 nm, but radiation up to 500 nm also can be used as the excitation wavelength, and the corresponding emission peak is 602 nm.

Use of Intensity Standards. Several solid-doped fluorescent materials are available. These polymers or glasses enable the relative spectral correction and day-to-day PQ of fluorescence instruments across the UV, visible, and NIR regions from 320 to 830 nm. The high photostability of the materials makes them particularly useful as day-to-day intensity standards, even when spectral correction is not needed or when the excitation wavelength differs from that used for certification.

Available from commercial vendors and from NIST as SRMs 2940 (orange emission), 2941 (green emission), 2942 (UV emission), 2943 (blue emission), and 2944 (red/NIR emission), these references are available in the form of a solid glass, standard-sized cuvette (12.5 mm × 12.5 mm × 45 mm) with three polished long faces for 90° detection and one frosted long face for front-face or epifluorescence detection, and come with a certified, steady-state emission spectrum, along with the estimated total uncertainties.

Alternatively, analysts can use fluorophores in solution that have been shown to be stable. Commercial vendors provide a 1 mg/L solution of quinine sulfate dihydrate in 0.105 M perchloric acid that has been fully characterized by NIST as SRM 936a, and a series of day-to-day intensity standards is available from the German Federal Institute for Materials Research and Testing (BAM).

As previously described, the pharmaceutical industry relies on validated methodology, produced on qualified instrumentation. The essential transfer of these fluorescence-based protocols, between for example, development and manufacturing, may involve instrumentation from different vendors, different physical locations, etc., and therefore must be constructed using corrected spectra.

In a recent paper [23], the NIST reference spectrum for quinine sulfate was compared with uncorrected and corrected spectra of quinine sulfate and analyzed on three commercially available spectrofluorometers. The

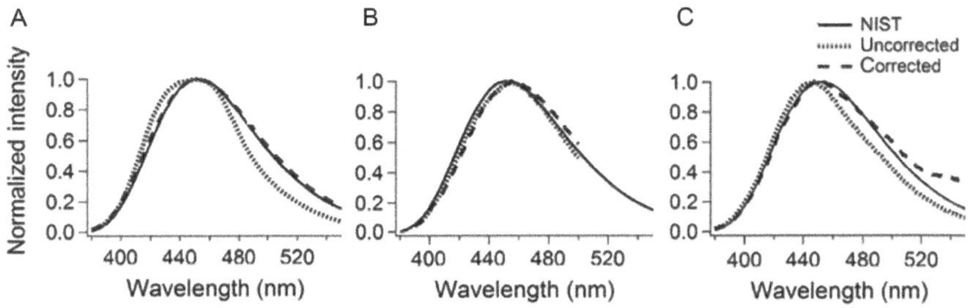

FIGURE 11.13 NIST reference spectrum of quinine sulfate compared to uncorrected quinine sulfate and corrected quinine sulfate emission collected on three spectrofluorometers, "a," "b," and "c." *Redrawn from Ref. [23].*

differences reported highlight the importance of always using corrected spectra. As shown in Fig. 11.13, before correction, none of the quinine sulfate spectra collected on the three instruments exhibited good agreement with the NIST reference spectrum across the whole wavelength range (370–550 nm). Specifically, none of the uncorrected spectra had an emission peak position centered at 450 nm as does the NIST spectrum. Spectrofluorometers "a" and "c" showed slightly blue-shifted peak positions, whereas "b" shows an emission peak slightly to the red of 450 nm. Furthermore, the overall shapes of the spectra did not agree with the NIST reference spectrum. For example, "a" had an emission curve that is broader than the NIST spectrum at wavelengths shorter than the peak position (i.e., <450 nm) and shallower than the NIST spectrum at longer wavelengths (>450 nm).

11.7.1.2 Qualitative and Quantitative Fluorescence Measurements

Two general classes of procedural measurements, namely qualitative and quantitative measurements are commonly performed by fluorescence spectrometry:

- *Qualitative Fluorescence Measurements.* Qualitative fluorescence measurements are used to detect the presence of particular analytes and yield a positive or negative answer. The excitation and emission wavelengths often are selected at the peak maximum of the fluorophore to be detected. The observation of fluorescence at the peak position above the limit of detection (usually three times the noise level) indicates a positive result.
- *Quantitative Fluorescence Measurements.* Quantitative fluorescence measurements are used to determine amounts or concentrations of analytes in unknown samples. These quantities may be determined in absolute units such as moles or moles per liter or in relative units such as the ratio of the concentrations of two fluorescent analytes contained in a single unknown solution. These determinations use the following proportionality

relating fluorescence signal (S) at a given pair of excitation and emission wavelengths (μ, λ) to fluorescent analyte concentration (c):

$$S \propto I_0 \Omega R_d \alpha \Phi c$$

I_0, intensity of the excitation beam
Ω, fraction of the fluorescence collected by the detection system
R_d, responsivity of the detection system
α, absorption coefficient
Φ, fluorescence quantum yield
c, concentration of the fluorescent analyte.

This linear proportionality with concentration applies to optically dilute samples, for example, solutions with an absorbance of less than 0.05 at a path length of 1 cm.

11.7.1.3 Good Spectroscopic Practice

Comparisons of a test specimen with a reference standard are best made at a peak of spectral emission for the compound of interest. Assays based on spectrofluorometry give the commonly accepted wavelengths for excitation and peak spectral emission of the substance in question. Different spectrofluorometers may show minor variation in the apparent wavelength of this peak. Comparisons should be made at the wavelength at which peak emission occurs. If this differs from the wavelength specified in the monograph by more than ±1 nm in the range of 200–400 nm or by more than ±2 nm in the range of 400–800 nm, recalibration of the instrument may be indicated.

Use of Reference Standards. With few exceptions, pharmacopoeial spectrofluorometric procedures provide results by comparison against a reference standard. This ensures measurement under identical conditions for the test specimen and the reference standard. These conditions could include wavelength setting, SBW selection, cell placement and correction, and intensity levels. Cells that exhibit identical optical fluorescence characteristics at a specific wavelength may differ considerably at other wavelengths. Analysts should establish and use appropriate cell corrections where required.

The terms *similar preparation* and *similar solution* in tests and assays that involve spectrofluorometry indicate that the reference standard should be prepared and observed in a manner that is identical to that used for the sample under test. Usually, when a solution of the specified reference standard is prepared at (i.e., within 10% of) the desired concentration, the fluorescence intensity is calculated on the basis of the exact amount weighed out. If they have not used a previously dried specimen of the reference standard, analysts should correct this intensity on the anhydrous basis. The expressions *concomitantly determine* and *concomitantly measured* as used in procedures that involve spectrofluorometry indicate that the intensities of both the sample solution and the standard solution, relative to the specified test blank, should be measured in immediate succession.

Sample Solution Preparation. For determinations using UV or visible spectrofluorometry, the specimen generally is dissolved in a solvent. Unless otherwise directed in the monograph, determinations are made at room temperature using a path length of 1 cm. Many solvents are suitable for these spectral ranges, including water, alcohols, chloroform, lower hydrocarbons, ethers, and dilute solutions of strong acids and alkalis. Solvents should be free from contaminants that fluoresce in the spectral region under examination. For the solvent, water-free methanol or alcohol or alcohol that has been denatured by the addition of methanol but does not contain benzene or other interfering impurities should be used. Spectrophotometric-quality solvents that are guaranteed to be free from contaminants are available commercially from several sources, but some analytical reagent-grade organic solvents may contain traces of impurities that fluoresce strongly in the UV region. New lots of these solvents should be checked for their transparency, and analysts should take care to use the same lot of solvent for preparation of the test solution, the standard solution, and the blank. Solvents that do not have an interfering fluorescence signature at the wavelength(s) of interest are optimal. In normal usage, the fluorescence baseline intensity should not be more than 2% of the expected measurement signal unless a larger value previously has been justified.

Assays in the visible region usually call for comparing concomitantly the fluorescence intensities produced by the sample solution with that produced by a standard solution that contains approximately an equal quantity of a reference standard. In some situations, it may be permissible to omit the use of a reference standard. This is true when spectrofluorometric assays are made with routine frequency, when a suitable standard curve is available and is prepared with the appropriate reference standard, and when the substance assayed conforms to Beer's Law within the range of about 75–125% of the final concentration used in the assay. Under these circumstances, the fluorescence intensity found in the assay can be interpolated on the standard curve, and the assay result can be calculated. Such standard curves should be confirmed frequently and always when a new spectrofluorometer or new lots of reagents are put into use.

11.7.2 Validation and Verification

Validation. The process for analytical procedure validation for fluorescence requires testing for linearity, range, accuracy, specificity, precision, DL, quantitation limit, and robustness. These analytical performance characteristics apply to externally standardized procedures and those that use standard additions.

Indirect Measurement Requirements. Some fluorescence procedures employ chromogenic derivatization, where a chemical reaction is used to produce the fluorophore. Generally, the requirements for the analytical performance characteristics should be used. In some instances, the required accuracy and precision for the direct measurements may not be achievable. Under these circumstances, the accuracy and precision requirements may be widened by as much as 50%. However, any such widening must be justified on scientific grounds and with documented evidence. Under these circumstances, the amount of replication required to produce a scientifically sound reportable value may be increased.

11.7.3 Pharmaceutical Applications (Fluorescence)

Fluorescence spectroscopy is an important investigational tool in many areas of analytical science, due to its extremely high sensitivity and selectivity, and the pharmaceutical industry is no exception. With many uses across a broad range of chemical, biochemical and medical research, it has become an essential technique allowing detailed, real-time investigation of the structure and dynamics of intact biological systems with extremely high resolution. It is particularly heavily used in pharmaceutical development where FCS has almost completely replaced radiochemical labeling in combinatorial chemistry and high-throughput screening [24].

Additionally, fluorescence methods in use can be categorized into the types already described.

Category I: Assay procedure for the API. Fluorescence can be used either as the detector for HPLC procedures, or directly (by comparison with a reference material), for example, Riboflavin, where the excitation wavelength is 444 nm, and the emission wavelength is 530 nm [25].

Category II: Impurities. Fluorescence is used for the determination of both organic and inorganic impurities, by reaction with the appropriate fluorogenic reagents. For example, aminobutanol in ethambutanol hydrochloride may be determined by derivatization with fluorescamine, where the excitation wavelength is 385 nm, and the emission wavelength is 485 nm [26]; and aluminum in sodium chloride may be quantified by the formation of the complex with 8-hydroxyquinoline, with the excitation wavelength at 392 nm and the emission wavelength at 518 nm [27].

Category III: Dissolution. Fluorescence detection is used for dissolution testing of tablets and products in the pharmaceutical industry when the use of UV absorption is not appropriate. Such a case is in the characterization of bromocriptine mesylate tablets, where the excitation wavelength is 315 nm, and the emission wavelength is 445 nm (Fig. 11.14) [28].

11.8 WHERE NEXT?

As we have seen, in the highly regulated environment of the pharmaceutical industry; fundamental, simple, qualitative and quantitative spectrophotometric applications, and the associated instrument systems can be fully specified

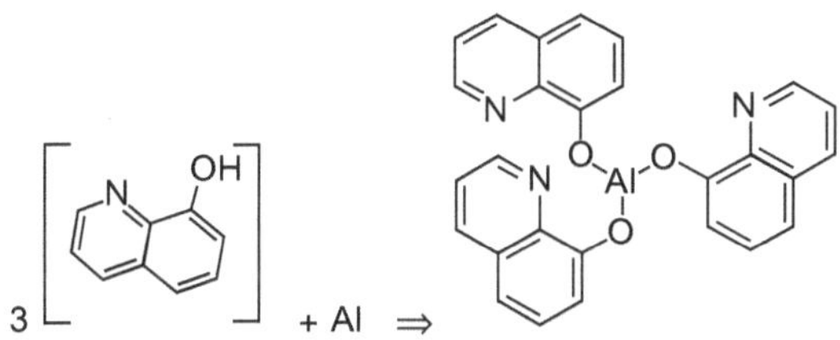

FIGURE 11.14 Reaction of 8-hydroxyquinoline with aluminium.

against defined control parameters, using established methodology, CRMs, etc., and the "evidence of control" established.

However, also, as discussed, various regulatory initiatives are driving the industry toward a risk-based approach for the complete process, which is moving the testing requirements out of the laboratory and on to the manufacturing floor, into the warehouse, etc. To solve the problems associated with producing spectrophotometric data within these regulatory environments, instrument systems have inevitably become more complex in their physical design, use of software algorithms, etc., even though they may appear simple in their operation.

This complexity poses its own set of challenges to the personnel tasked with the validation and verification of the measurement process, but there is "light at the end of the tunnel," as key personnel within the industry debate the way forward, using a lifecycle approach for analytical method validation, involving new key concepts, such as the analytical target profile [29].

REFERENCES

[1] US Food and Drug Administration (FDA), Guidance for Industry-PAT—A Framework for Innovative Pharmaceutical Development, Manufacturing, and Quality Assurance, Food and Drug Administration, Rockville, MD, USA, 2004.

[2] The International Conference on Harmonisation of Technical Requirements for Registration of Pharmaceuticals for Human Use (ICH), "Pharmaceutical Development, Q8(R2)", 2009.

[3] The International Conference on Harmonisation of Technical Requirements for Registration of Pharmaceuticals for Human Use (ICH), "Quality Risk Management, Q9", 2005.

[4] The International Conference on Harmonisation of Technical Requirements for Registration of Pharmaceuticals for Human Use (ICH), "Pharmaceutical Quality System, Q10", 2008.

[5] The International Conference on Harmonisation of Technical Requirements for Registration of Pharmaceuticals for Human Use (ICH), "Development and Manufacture of Drug Substances (Chemical Entities and Biotechnological/Biological Entities), Q11", 2012.

[6] US Department of Health and Human Services, "National Human Genome Research Institute", www.genome.gov, 2014.

[7] United States Pharmacopeia (USP), General Chapters, Analytical Instrument Qualification, 2013, USP 36–NF31 S2, <1058> p. 650.

[8] United States, Code of Federal Regulations, Current Good Manufacturing Practice for Finished Pharmaceuticals, Laboratory Records, 2013, Title 21, Volume 4, 21 CFR 211.194.

[9] International Organization for Standardization (ISO), Terms and Definitions used in Connection with Reference Materials, 2008, ISO Guide 30:1992/Amd. 1:2008.
[10] International Organization for Standardization (ISO), International Vocabulary of Metrology—Basic and General Concepts and Associated Terms (VIM), 2007, ISO Guide 99:2007.
[11] International Organization for Standardization (ISO), General Requirements for the Competence of Testing and Calibration Laboratories, 2005, ISO/IEC 17025:2005.
[12] International Organization for Standardization (ISO), General Requirements for the Competence of Reference Material Producers, 2009, ISO Guide 34:2009.
[13] J.P. Hammond, New References Assist Validation in Far Ultra Violet Spectroscopy, 2002. www.scientistlive.com/content/9427.
[14] Agilent Technologies Inc., Agilent 708-DS and 709-DS Dissolution Apparatus, 2011.
[15] A.K. Nayak, Comparative in vitro dissolution assessment of some commercially available paracetamol tablets, Int. J. Pharm. Sci. Rev. Res. 2 (1) (2010) Article 008 pp. 29–30.
[16] United States Pharmacopeia (USP), General Chapters, Near-Infrared Spectroscopy, 2013, USP 36–NF31 S2, <1119> p. 801.
[17] J. Parris, C. Airiau, R. Escott, J. Rydzak, R. Crocombe, Monitoring API drying operations with NIR, Spectroscopy 20 (2) (2005) 34.
[18] Thermo Scientific, Application Note 50780: Determination of Moisture Content in Freeze-Dried Materials by FT-NIR Spectroscopy, 2008.
[19] United States Pharmacopeia (USP), <854> Mid-Infrared Spectroscopy, Pharmacopeial Forum 39 (2) 2013 (March–April 2013).
[20] J. Hughes, G. Ayoko, S. Collett, G. Golding, Rapid quantification of methamphetamine: using attenuated total reflectance Fourier transform infrared spectroscopy (ATR-FTIR) and chemometrics, PLoS One 8 (7) (2013) e69609. www.plosone.org.
[21] United States Pharmacopeia (USP), <853> Fluorescence Spectroscopy, Pharmacopeial Forum 39 (1) 2013 (January–February 2013).
[22] J.C. Travis, et al., Intrinsic wavelength standard absorption bands in holmium oxide solution for UV/visible molecular absorption spectrophotometry, J. Phys. Chem. Ref. Data 34 (1) (2005) 41–56.
[23] R.M. Cory, M.P. Miller, D.M. McKnight, J.J. Guerard, P.L. Miller, Effect of instrument-specific response on the analysis of fulvic acid fluorescence spectra, Limnol. Oceanogr. Methods 8 (2010) 67–78.
[24] K.J. Moore, S. Turconi, S. Ashman, M. Ruediger, U. Haupts, V. Emerick, A.J. Pope, Single molecule detection technologies in miniaturized high throughput screening: fluorescence correlation spectroscopy, J. Biomol. Screen. 4 (6) (1999) 335.
[25] United States Pharmacopeia (USP), "Monograph: Riboflavin", 36-NF31 S2, p. 5035, 2013.
[26] United States Pharmacopeia (USP), "Monograph: Ethambutanol Hydrochloride", 36-NF31 S2, p. 3498, 2013.
[27] United States Pharmacopeia (USP), "Monograph: Sodium Chloride", 36-NF31 S2, p. 5155, 2013.
[28] United States Pharmacopeia (USP), "Monograph: Bromocriptine Mesylate tablets", 36-NF31 S2, p. 2692, 2013.
[29] United States Pharmacopeia (USP), Lifecycle management of analytical procedures: method development, procedure performance qualification, and procedure performance verification, Pharmacopeial Forum 39 (5) 2013 (September–October 2013).

Chapter 12

Spectrophotometry Applications: Remote Sensing

Carol J. Bruegge*, Roger Davies†, Florian M. Schwandner*
and Felix C. Seidel*
*Jet Propulsion Laboratory, California Institute of Technology, Pasadena, California, USA
†Department of Physics, University of Auckland, Auckland, New Zealand

Chapter Outline

12.1 Introduction 458
12.2 Measurement of Atmospheric Carbon Dioxide 458
12.3 The Remote Sensing of Clouds in the Earth's Atmosphere 464
12.3.1 Why Clouds? 465
12.3.2 The Spectrophotometry of Clouds: Forward Problem 466
12.3.3 The Spectrophotometry of Clouds: Inverse Problems 468
12.3.4 The Nakajima and King Approach for Plane-Parallel Clouds 468
12.3.5 The Optical Depth of Thin Clouds in General 469
12.3.6 The Optical Depth of Thick Clouds 469
12.3.7 Cloud Altitude 469
12.3.8 Cloud Motion Vectors 470
12.4 The Retrieval of Snow Properties 470
12.4.1 Snow-Covered Area 471
12.4.2 Snow-Grain Size 471
12.4.3 Snow Albedo 472
12.4.4 Radiative Forcing of Light-Absorbing Impurities in Snow 473
12.4.5 Benefits of Imaging Spectroscopy in Combination with Other Observations for Snow Remote Sensing 474
12.5 Volcanic Unrest 475
12.5.1 Early Remote Sensing Methods 475
12.5.2 Applications and Relevance 475
12.5.3 Imaging of Volcanic

Experimental Methods in the Physical Sciences, Vol. 46. http://dx.doi.org/10.1016/B978-0-12-386022-4.00012-1

Landforms, Eruption Detection, and Degassing 477
12.6 Calibration 479
12.7 Summary 481
Acknowledgments 482
References 483

12.1 INTRODUCTION

Remote sensing is the study of imagery and measurements made from balloon, aircraft, or in-orbit sensors. These sensors can provide routine and continuous sampling over large areas. In the case of space-borne sensors, global observations are made possible. The usage of satellite-borne instruments for scientific exploration began in the latter half of the twentieth century. The longest running program for monitoring the Earth is the Landsat enterprise. The first Landsat sensor was launched in 1972, and the most recent was that of Landsat 8, in 2013. Images from Landsat sensors have been used in agriculture, cartography, geology, forestry, regional planning, surveillance, and education applications. By the 1980s, spectra that spanned the entire solar-reflected range, from 400 to 2500 nm, were made available due to the Airborne Visible/Infrared Imaging Spectrometer (AVIRIS) [1]. These data have been used in investigations of atmospheric correction, ecology and vegetation, geology and soils, inland and coastal waters, the atmosphere, snow and ice hydrology, biomass burning, environmental hazards, human infrastructure, satellite data simulation and calibration, commercial applications, and spectral algorithms and modeling. This chapter discusses a few of these applications, specifically the retrieval of greenhouse gas abundance, and the study of cloud radiative effects, snow and ice, and volcano monitoring. It is concluded with a brief discussion on calibration.

12.2 MEASUREMENT OF ATMOSPHERIC CARBON DIOXIDE

It is well known that carbon dioxide (CO_2) concentration is rising in the atmosphere as a result of anthropogenic activities, such as fossil fuel burning and forest clearing. As CO_2 is a greenhouse gas and implicated as a contributor to global warming, the detection and measurement of atmospheric CO_2 is an important science. Carbon cycle researchers strive to better understand the sources and sinks of CO_2 and the processes that control greenhouse gas exchanges between the Earth's surface and its atmosphere. We know that more than half of the CO_2 emissions into the atmosphere are taken up by the land and ocean systems, but we do not know if or when this exchange will decrease or saturate, reducing uptake from the atmosphere.

Beyond the scientific goal of monitoring carbon fluxes, there is also a political need to establish accurate observing systems for quantifying sources and sinks of greenhouse gases on national levels in the context of the Kyoto

[2] protocol and its likely successors. The accuracy requirements for this goal are substantially higher than for the scientific objectives, because small changes in fluxes need to be measured over relatively small geographical areas and for long time periods.

Various methods to measure atmospheric greenhouse gases from the ground exist [3]. One technique makes use of direct measurement from towers. There are over 300 towers coordinated, through the FluxNet project alone, that use the eddy correlation technique. These towers measure wind speeds and direction, gas amounts, temperature, pressure, and humidity, at each tower. By doing so, they can determine gas fluxes. Another ground network is TCCON (Total Carbon Column Observing Network). Measurement stations exist at some two dozen sites around the world. These are ground-based Fourier Transform Spectrometers (FTSs) that record direct solar spectra at near-infrared (NIR) wavelengths. From these spectra, accurate and precise column-averaged abundance of CO_2, CH_4, N_2O, HF, CO, H_2O, and HDO are retrieved. Ground networks, such as these, provide datasets of high precision and sensitivity to CO_2 near the surface, where sources and sinks are located. They have also contributed to our understanding of atmospheric CO_2 growth rate.

In addition to ground-based measurements, space-based determinations are possible. These have the advantages of global coverage and on-going temporal measurements. They collect data over regions where ground stations cannot be deployed, such as over oceans, large forest regions, and remote areas of the globe. The quantity that we want measure on local to regional scales is the flux of CO_2 into or out of the atmosphere. We cannot measure these fluxes directly using space-based remote sensing observations, but we can acquire spatially and temporally resolved observations of the dry air mole fraction. This is a volume mixing ratio measured relative to dry air. By assimilating these spatially and temporally resolved concentration measurements into an atmospheric transport model, we can infer the flux distribution of interest. Although natural processes and human emissions can change the atmospheric CO_2 mixing ratio by as much as 8% near the surface (>30 ppm out of the ~385 ppm background), the amplitude of these variations decreases rapidly with altitude, such that CO_2 variations rarely exceed 2% (8 ppm) on regional to global scales. East-west variations are typically no larger than 0.3–0.5%. Because of this, modeling studies show that space-based measurements of CO_2 will dramatically improve our understanding of surface fluxes only if they have the accuracy, precision, coverage, spatial resolution, and temporal sampling needed to describe CO_2 variations with amplitudes no larger than 0.3–0.5 % (1–2 ppm) on scales ranging from <100 km over continents to ~1000 km over the ocean.

Greenhouse gas retrieval from space is an emerging technology, and therefore reliant upon the ground datasets for validation. Measurements of CO_2 from space have been reported by several sensors, including SCIAMACHY [4],

AIRS (Atmospheric Infrared Sounder) [5], GOSAT (Greenhouse gases Observing SATellite) [6,7], and OCO-2 (Orbiting Carbon Observatory) [8]. The latter two have been built specifically for greenhouse gas measurements from space. GOSAT was successfully launched in 2009. GOSAT contains several instruments, including the Thermal And Near infrared Sensor for carbon Observations-Fourier Transform Spectrometer (TANSO-FTS), where Tanso means "carbon" in Japanese. An attempt to launch the first OCO was also made that year; however the OCO sensor failed to achieve orbit upon launch. The replacement, OCO-2 is scheduled for launch in 2014. Both OCO-2 and the TANSO-FTS instrument aboard GOSAT make measurements in three NIR bands: one at 0.76 μm containing significant O_2 absorption (the so-called O_2 A-band), one at 1.6 μm containing weak CO_2 absorption, and one near 2.1 μm containing strong CO_2 absorption. High-resolution detection is required within each of these bands. GOSAT meets this need by use of an interferometer design. Conversely, the OCO design incorporates three, co-boresighted, imaging grating spectrometers.

The GOSAT and OCO-2 sensors measure optical radiation emitted by the sun and reflected either at the surface or through atmospheric scattering. One feature of these sensors is that they have nearly a uniform sensitivity to CO_2 from the surface up through the middle troposphere. This is highly desirable, and not true for sensors that measure in the thermal infrared (TIR). The reflected radiation they measure shows spectral variations (changes with wavelength) as a result of contributions from the sun, surface, and atmospheric absorption and scattering processes. The spectral nature of the sun can be described as a collection of absorption lines, called Fraunhofer lines, superimposed on a continuum. The variation in surface reflectance is relatively smooth. Across the OCO-2 bands, this reflectance is approximated as a linear function of wavelength. At some wavelengths, called the "atmospheric window" channels, the atmosphere is nearly transparent and the radiance at the top of the atmosphere is dominated by the sun and surface contributions. At other wavelengths, the atmosphere is absorbing and the measurement is also a function of the amount of absorbing material in the atmosphere. Figure 12.1 shows these absorption features, intermixed between the continuum, as would be observed at the spectral resolution of AVIRIS. In actuality there are rapid spectral variations, the result of many individual absorption lines. Such spectra are shown in Fig. 12.2, as measured by the OCO sensor. The depth of these lines is directly related to the amount of absorbing material along the atmospheric path, although the temperature and composition vertical profiles also impact the absorption. Therefore, with a measurement with a very high spectral resolution, it is possible to identify the absorption lines of various gases and to retrieve their concentration from the relative depth. Because the line absorption is a function of the atmospheric pressure, the technique may be used to retrieve a vertical profile with a very low vertical resolution.

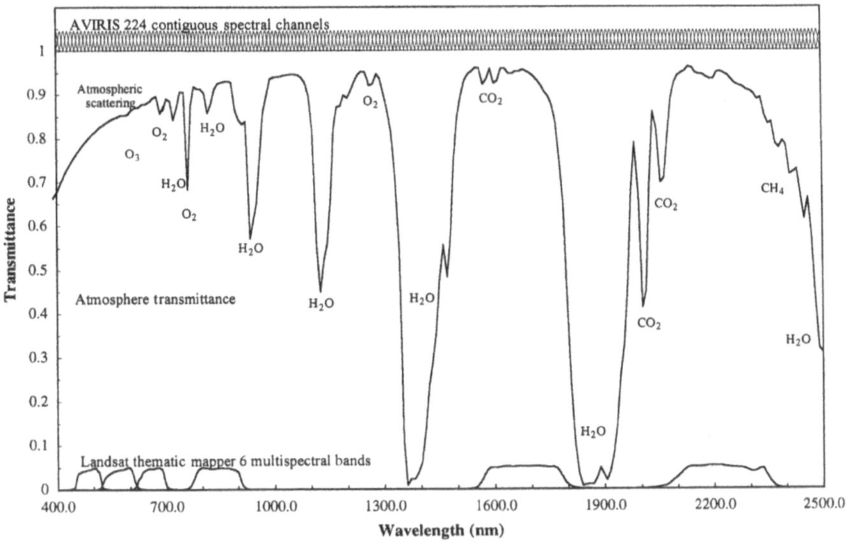

FIGURE 12.1 Atmospheric transmittance and absorption features, as compared to AVIRIS 10 nm bands and Landsat spectral bands [1].

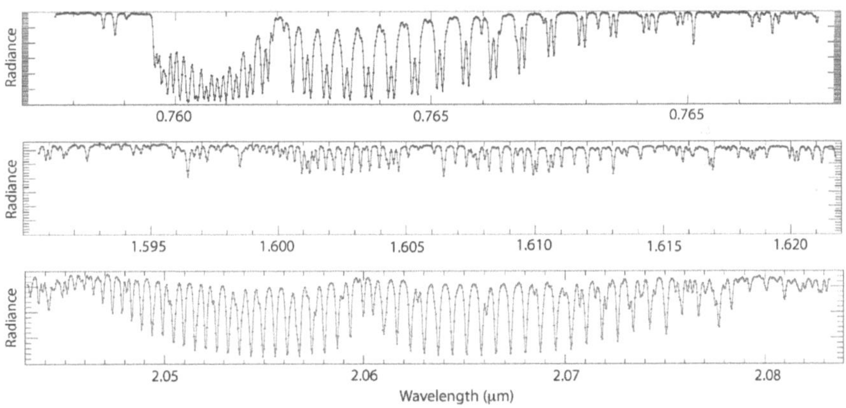

FIGURE 12.2 Spectra from the O_2, weak CO_2, and strong CO_2 OCO-2 bands. Data were taken using sunlight reflected from a diffuse panel during thermal vacuum instrument testing.

For CO_2 retrieval, the regions of interest are at wavelengths around 1.6 and 2 μm. Here, the absorption lines are well distinct from those of other gases. The differential absorption technique requires clear sky along the line of sight. It also requires the sun to be sufficiently high above the horizon so as to limit scattering in the atmosphere. These requirements limit the time-space sampling capabilities. In particular, no night measurement is possible, and high latitudes are not sampled during the winter season. This is a limitation in the attempt to quantify processes such as winter ecosystem respiration. Another constraint is that the surface reflectance must be large enough as it has a direct impact on the signal-to-noise ratio (SNR). Over land, natural surfaces are relatively dark at the wavelengths of interest, in particular when they are snow-covered. However, the low SNR problem exists mostly over water which is extremely dark in the infrared, and therefore not suitable for the measuring technique, except in the direction of the glint spot, where the light yield is greatly amplified.

The OCO science team is currently using their CO_2 retrieval algorithm on GOSAT data, under a collaboration agreement between NASA (National Aerospace and Space Administration) and JAXA (Japanese Aerospace eXploration Agency). The discussion to follow summarizes the OCO retrieval approach, as currently implemented on GOSAT data [9] and depicted in Fig. 12.3. It is implied that although results to date have been obtained through the processing of GOSAT data, these same algorithms will be used on OCO-2 datasets, following launch. Whereas GOSAT measures top-of-atmosphere (TOA) radiances in two polarized states, the OCO algorithm combines these and processes the total observed radiance. In the analysis, a prescreening step first removes cloudy data and then other bad data including those with a solar zenith angle greater than 85°. A best guess as to the geophysical parameters at each of the viewed locations is next made. This collection of parameters is called the state vector and includes vertical profiles of temperature, humidity, aerosol, and carbon dioxide, as well as surface pressure, and surface reflectance. The state

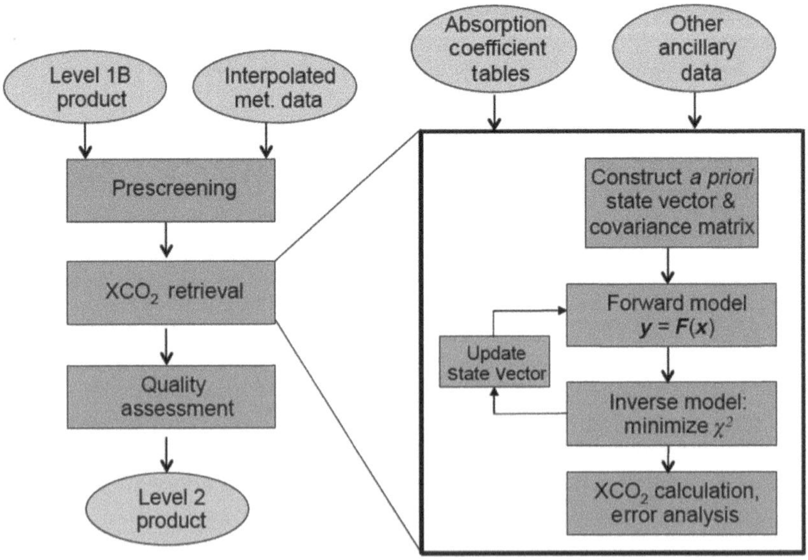

FIGURE 12.3 Flowchart showing the basic steps for the end-to-end CO_2 retrieval [9].

vector will be modified during the retrieval process, until the TOA predicted from it matches that measured by GOSAT.

The first guess to the state vector is constructed as follows. Surface pressure and profiles of temperature and water vapor are taken from 3-hourly European Centre for Medium-Range Weather Forecasting (ECMWF) model forecast fields, interpolated linearly in space and time to the GOSAT field-of-view. The surface is assumed to be Lambertian, reflecting optical radiation equally in all directions. This is a valid assumption, in that there is very little atmospheric scattering, under clear conditions and at GOSAT wavelengths. Without atmospheric scattering, knowledge of surface reflectance with angle is not needed. The surface reflectance itself is estimated from the level of the continuum in the observed spectrum of each band, assuming a clear sky, absorption-free atmosphere. For atmospheric scattering, four types are included in the model: water cloud, ice cloud, and two different types of aerosol. By combining them appropriately, the retrieval can reproduce the TOA in all three spectral bands. The first aerosol type is a mixture of course and fine-mode dust, while the second is a carbonaceous mixture, as taken from the aerosol climatology of Ref. [10]. The nature of the retrieval is such that values that start at zero *a priori* will stay at zero. Thus, the retrieval is incapable of putting aerosols at pressures lower than about 300 hPa and cannot put scatterers of any type at pressures lower than 100 hPa, both of which are physically sensible restrictions.

The solar model, also required in the simulation of TOA, is the product of two parts. The first is a high-resolution, empirical line list for the full solar disk, with over 20,000 lines. The line list is based on a series of balloon and telescope observations of the disk center. The solar continuum model is a ninth-order polynomial fit to the NIR part of the solar spectrum, measured by the SOLSPEC (Solar Spectral irradiance measurements) instrument [11]. Other than the variation in the sun–earth distance that modulates the overall intensity, the solar continuum is assumed to be time invariant; sunspots and other solar activity are currently ignored. These two spectra are multiplied to obtain the solar spectrum at high resolution.

Once the *a priori* state vector is assembled for a given observation, a radiative transfer code is used to simulate TOA from the state vector. The process of computing TOA from a state vector is called the forward model. In actuality, the inverse is required. That is, the retrieval must take the space-borne radiance measurements and infer the state vector. This is done by using the *a priori* state vector, running the forward model, then from these determining what changes to this state vector are needed to minimize the residuals between the measured TOA and modeled spectra.

Input to the forward model includes the viewing geometry and instrument line-shape (ILS) (spectral response terms). As stated, the model simulates TOA spectra in the three NIR bands, as would be produced by GOSAT. To do the model breaks the atmosphere into 20 layers, at equally spaced pressure

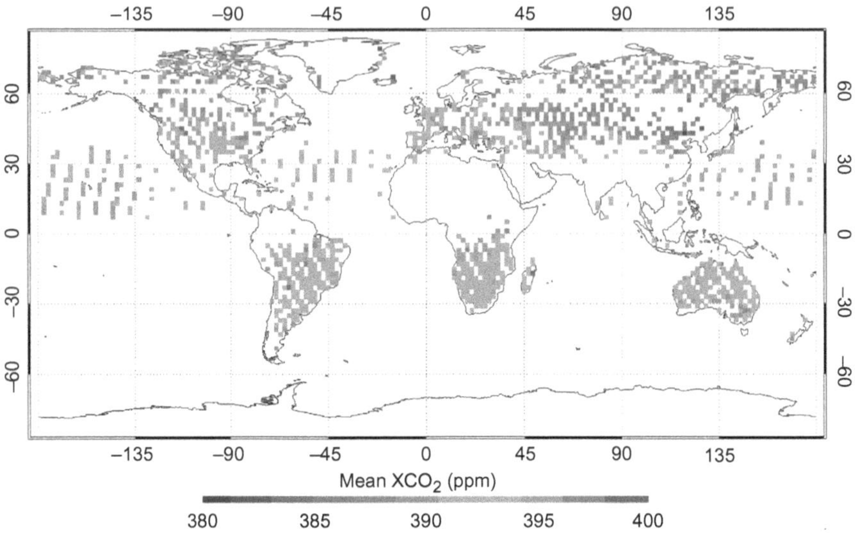

FIGURE 12.4 Globally retrieved CO_2, July 2012, as processed by version B3.4 of the OCO-2 code.

levels. As the state vector is retrieved at these 20 layers, vertical estimates of carbon dioxide are made. The code computes absorption and scattering as radiation propagates through each layer. It also produces radiance Jacobians, which are derivatives of the radiances with respect to all the state vector parameters. These derivatives are used by the inverse model to adjust the state vector to its retrieved state.

Since the column amount of measured CO_2 would naturally vary depending on surface elevation, it is desirable to report CO_2 scaled by surface pressure. For the OCO community, they elect to use the retrieval of O_2 as a proxy for surface pressure. This can be done because this gas is evenly mixed in the atmosphere. Figure 12.4 shows a global CO_2 map, for July 2012. These retrievals were made using GOSAT data, as processed by the OCO-2 processing algorithm.

On-going research into this retrieval algorithm includes improved processing algorithms that account for atmospheric scattering affects. Several of these algorithms have shown relatively good agreement as compared with simultaneous and collocated TCCON observations [9].

12.3 THE REMOTE SENSING OF CLOUDS IN THE EARTH'S ATMOSPHERE

This section explores a few of the modern applications of spectrophotometry to the remote sensing of clouds. The examples chosen are by no means exhaustive and have been chosen subjectively to be illustrative of the range of possibilities.

12.3.1 Why Clouds?

Modern geophysics is actively pursuing a quantitative understanding of the properties of clouds, primarily because of their importance to the Earth's climate system, and motivated by rapid technological advances in global remote sensing. There are two key areas of interest. One is the broadband radiometric effect of clouds on the global distribution of radiant energy, and any long-term changes that can be detected. The other is the connection between clouds and the hydrological cycle, especially the net release of latent heat, and its vertical distribution. In order to properly understand or model either of these areas, detailed knowledge of cloud morphology, cloud phase (and if glaciated, ice crystal habit), particle density, and particle size-distribution is required, at least to the point where reliable parameterizations or reproducible statistical summaries can be developed or used.

An important example is the dominant contribution of clouds to the planetary albedo. While the TOA broadband result can be measured fairly well, explaining its origin requires consideration of spectral reflection across the entire solar spectrum, 90% of which lies in the wavelength range of 0.33–2.1 μm. The spectral albedo is relatively independent of wavelength in the visible portion of this range, but becomes progressively more complex in the solar infrared due to absorption by water vapor, which takes place mainly in the column above the cloud, and due to absorption by condensed water within the cloud in the spectral regions between the vapor absorption bands.

Of similar importance is the role of upper tropospheric clouds to the greenhouse effect, with transmission and absorption of longwave radiation by thin cirrus clouds being of special interest. Ninety percent of the longwave radiant energy lies in the wavelength range of 6.3–43 μm. The spectral nature of gaseous absorption, particularly not only by water vapor but also by carbon dioxide, and to a lesser extent by ozone and other minor gases, dominates this spectral region, with cloud boundaries acting like grey bodies, and being most effective in the window regions between the gaseous absorption bands. Key variables here tend to be the altitudes and temperatures of the cloud boundaries.

Turning to the hydrological importance of clouds, the relationship between hydrology and radiation is crucial to understanding the role of clouds in the climate system, but is often imprecisely measured. Here, the main variable of interest is the cloud water content. On a global scale, most clouds are low-level (i.e., with altitudes at approximately 1 km above the surface) boundary layer clouds that are relatively dry, with column water amounts on the order of 10 g m^{-2}. As such, they play a smaller role in global hydrology compared with the much thicker and wetter cumulonimbus and nimbostratus clouds that make up about 10% of global clouds. These provide most of the global precipitation and have water contents on the order of 1000 g m^{-2}.

To first order, there is a linear relationship between cloud water content and cloud optical depth, so that in the context of passive remote sensing, once the optical depth has been determined, together with a measure of effective radius, r_e, the column water content, W, can be estimated from the relationship

$$W = \frac{2\rho\tau}{3r_e}, \tag{12.1}$$

where ρ is the liquid water density and τ is the cloud optical thickness.[1]

12.3.2 The Spectrophotometry of Clouds: Forward Problem

The standard radiative transfer, described previously in this chapter, can of course be applied to clouds, and a variety of solution techniques [12] were developed for this in the latter half of twentieth century. These became progressively more sophisticated as the available computers became more powerful, but as a generalization suffered from two major limitations: they applied only to monochromatic radiative transfer, and only to plane-parallel (i.e., one-dimensional) media. For cloud remote sensing, the monochromatic limitation was not a major limitation, except for the treatment of broadband energetics where the combination of droplet scattering and gaseous absorption makes the problem more challenging. But the one-dimensional approach ultimately proved fatal to serious cloud remote sensing because real-world clouds show very strong horizontal variability. At near-visible wavelengths, the radiation field within a cloud is only weakly absorbed, and the resultant multiple scattering means that the entire field is strongly affected by the full morphology of the cloud, so that clouds can only occasionally be approximated as being locally plane parallel (the Independent Pixel Approximation [13]). This is especially true of cumulus cloud fields and deep convective clouds, but is also true of stratiform cloud decks that often show variations in local shortwave radiances by two orders of magnitude over horizontal distances of 10 km, and almost always present a rough top surface that affects the anisotropy of reflected radiation. Even the Helmholtz Principle of reciprocity does not apply to clouds when taken in its usual directional form [14]. Instead, it must be replaced by its more general form, and this is not of much use in a geophysical context. This is because the directional form assumes no horizontal variations in either the illumination from the sun or the reflected radiance, in order to reverse their directions. While the solar illumination is reasonably invariant, reflection from heterogeneous clouds may be highly variable, making the directional problem non reciprocal (as then the solar illumination would have to vary spatially too).

1. The symbol τ is commonly used for optical thickness by the remote sensing community. The CIE International Lighting Vocabulary (ILV), which forms the basis for the nomenclature used for most of this book, assigns the symbol δ for optical thickness.

Because of this greater challenge, solutions to the radiative transfer problem in three-dimensions were developed specifically for use in cloud fields. The two most useful of these currently are the Monte Carlo approach to radiative transfer (which is extremely versatile and can in principle handle all types of problems), and the spherical-harmonics discrete-ordinates approach (SHDOM) in three-dimensions. These and other techniques are well described by Ref. [15] and tips on their practical implementation are available through the Intercomparison of 3D Radiation Codes (*I3RC*) [16]. Suffice it to say that with modern computational power, the forward monochromatic radiative transfer problem for fields of heterogeneous water clouds is now essentially solved. The main practical problem is one of specifying the three-dimensional distribution of cloud microphysical quantities that act as the input field to the model. Radiative transfer in ice clouds has not yet been completely solved, as the crystals are decidedly nonspherical, and in general may not be randomly oriented. However, there are quite reasonable approximations that can be applied [17].

The problem of non-monochromatic radiative transfer in clouds could be treated by direct integration over wavelength, but such a line-by-line approach coupled with multiple scattering is not practical. Instead, there are two reasonably efficient solutions for narrowband or broadband radiation. For simpler problems, the spectrum is divided into a finite sum of equivalent monochromatic regions with coefficients chosen using the *correlated-k* approach [18]. More generally, the Monte Carlo output can be used to generate monochromatic photon pathlength distributions for conservative scattering, and these may then be convoluted with a narrowband transmission function, applying van de Hulst's equivalence theorem [19] to obtain the narrowband radiance. That is, the narrowband radiance

$$L(\lambda,\Delta\lambda)=\frac{1}{\Delta\lambda}\int_{\lambda}^{\lambda+\Delta\lambda}L_{\lambda}\mathrm{d}\lambda \tag{12.2}$$

can be obtained from the Monte Carlo solution, following [20] as

$$L(\lambda,\Delta\lambda)=E_{\lambda}\int_{0}^{\infty}p(l)\{T[\lambda,\Delta\lambda,y_0]-T[\lambda,\Delta\lambda,y_0+y(l)]\}\exp(-k_{\lambda}l)\mathrm{d}l, \tag{12.3}$$

where $p(l)\mathrm{d}l$ is the fraction of photons emerging with pathlengths between l and $l+\mathrm{d}l$ from an equivalent conservative cloud with the same scattering coefficients, found from the Monte Carlo model; E_{λ} is the spectral irradiance illuminating the scene before any absorption takes place, assumed to be slowly varying with wavelength; $T[\lambda,\Delta\lambda,y]$ is the narrowband transmission through an absorbing gas of amount y, obtained separately from a narrowband transmission model. Note that T decreases monotonically with y, but need not do so exponentially; y_0 is the absorber amount encountered before the cloud top is reached, and is essential for example when there is water vapor above the cloud top; and k_{λ} is the volume absorption coefficient of the cloud droplets, assumed to be slowly varying with wavelength. The total broadband radiance is found similarly by summing over its narrowband components,

$$L_{\mathrm{bb}} = \sum_{i=1}^{n} L_{\lambda_i} \Delta\lambda_i, \tag{12.4}$$

where n is on the order of 10.

12.3.3 The Spectrophotometry of Clouds: Inverse Problems

Clouds present a significant remote sensing challenge. Being essentially white, they provide little information when viewed spectrally at visible wavelengths. The single-scatter albedo, a, of pure water cloud droplets is so close to unity ($1-a \approx 10^{-4}$), that light may be scattered over a thousand times before emerging from a very thick cloud. Consequently, the scattered radiance that is ultimately measured by a distant sensor is usually more affected by the geometric shape (morphology) of the cloud than by its intrinsic properties. As the wavelength is lengthened into the solar infrared, however, the imaginary part of the refractive index of water rises, the single-scatter albedo drops below unity, and the scattered light becomes dependent on low order scatter. Now, it is far more sensitive to droplet properties, especially the effective radius of the droplet size distribution. Unfortunately, by the same token, the lack of multiple scattering means that the scattered radiance comes only from the illuminated portion of the cloud boundary, and interior properties are unavailable at these wavelengths.

As the wavelength used increases further, typically starting at 3.7 μm, and continuing through the TIR window of the atmosphere, clouds make their presence felt through their emission, and remotely measured radiances now give information mainly about the temperature of the cloud boundary. The exception is thin cirrus, which combines aspects of both emission and transmission that can be a blessing or a curse to the remote sensor, depending on the context. Here, there is just enough difference in the scattering properties of ice crystals at wavelengths between 8 and 12 μm that information about thin cirrus cloud microphysics can sometimes be obtained [21].

Here, we consider a few examples of the cloud properties that can be measured from space, starting first with a plane-parallel case, and proceeding to examples that take into account the horizontal heterogeneity of clouds.

12.3.4 The Nakajima and King Approach for Plane-Parallel Clouds

For plane-parallel clouds and a given viewing geometry, the measured spectral radiance is a function of mainly the cloud optical thickness, τ, the asymmetry parameter, g, and the single-scatter albedo, a. To first order, these depend only on cloud thickness and effective radius, so that a forward model of radiative transfer can be used to yield maps of spectral radiance as a function of τ and r_e [22]. The inverse problem is then reduced to some version of matching two or more measured spectral radiances to the optimal solution for τ and r_e using a look-up procedure.

12.3.5 The Optical Depth of Thin Clouds in General

Horváth and Davies [23] used the multiangle capability of the Multiangle Imaging SpectroRadiometer (MISR) instrument on the Terra satellite to examine optical depth retrievals using the one-dimensional assumption. They showed that the inversion of radiances measured simultaneously from nine different viewing directions generally yielded nine different estimates of cloud optical depth if the cloud was assumed to be plane parallel. While this highlights the problem of cloud heterogeneity, multiple measurements can be used to reduce the uncertainty in optical depth, and at least provide a measure of uncertainty in the retrieval [24].

12.3.6 The Optical Depth of Thick Clouds

The spectral radiance reflected by a cloud of given type must increase monotonically as the cloud thickness increases. However, this increase tends to saturate for optical depths ($1/e$) greater than ≈ 50. In other words, once the cloud becomes bright enough, the effect of adding more thickness becomes hard to measure. Yet, it is the thick clouds that are of most interest in terms of their hydrological contribution. The saturation of reflected radiances affects both plane parallel and heterogeneous clouds, but for the latter there is at times a solution to the thick problem. Reference [25] showed that when the side of a deep convective cloud is visible, the multiangle approach can yield both a measure of the cloud geometry as well radiances at multiple angles. Instead of performing an inversion directly for τ, they used an iterative Monte Carlo model of three-dimensional radiative transfer to find a best fit to the extinction coefficient. Once this was found, the total optical depth could be estimated from the product of extinction coefficient and geometric cloud depth.

12.3.7 Cloud Altitude

While cloud altitude is often estimated by using measured brightness temperature or by CO_2 slicing [26], this is an indirect approach that also needs knowledge about the atmospheric temperature structure to relate temperature or pressure to height. Direct measurement can of course be made using space-borne lidar [27], but cloud heights can also be measured directly using measurements of reflected solar radiance—provided these are made at reasonably high spatial resolution simultaneously from at least two directions.

This approach takes advantage of the fact that most clouds are not plane parallel giving rise to spatial contrast patterns of reflectivity at cloud top that can be matched from two directions. The resulting parallax provides a stereo estimate of cloud-top height that is independent of the atmospheric temperature structure, and that is insensitive to radiometric calibration. In the case of MISR, for example, the effective, global, annual cloud-top height can be measured to within a sampling error of about 8 m [28].

12.3.8 Cloud Motion Vectors

Given more angles, and a short but finite time between measurements from different angles, the stereo approach of the previous section can be extended to also yield height-resolved cloud motion vectors (or cloud-tracked winds). The nine cameras on MISR, for example, span a time interval of about 7 min from the most oblique forward view to the most oblique aft view, allowing two sets of camera triplets to estimate the cloud motion vector with a good degree of quality control [29]. As for the cloud altitudes, this technique relies on the presence of cloud heterogeneity to provide sufficient contrast in the reflectivity patterns for the stereo matchers to work.

12.4 THE RETRIEVAL OF SNOW PROPERTIES

Snow and ice are important natural, economic, and political resources. They store freshwater, reducing flood risk; and runoff from snow and ice melt produces hydropower and contributes to the water supply for drinking, agriculture, recreation, and wildlife. In many parts of the world, large populations depend heavily on snowpack for their water supply.

Snow and ice also play a significant role in the Earth's climate with strong leverage on the energy balance. Similar to clouds, fresh snow reflects most of its incident solar light, while other natural surface types absorb most of the light and convert it to thermal energy. Thus, it is believed that small changes in the cryosphere can have large effects on Earth's climate and life, for example, due to sea level rise.

To understand changes in the cryosphere, it is essential to understand and predict snow properties over time using measurements from microphysical to global scales. Significant progress in this area has been made over the past decades thanks to emerging remote sensing tools, which enable retrieval of snow properties in a quantitative and effective manner. Comprehensive overviews of common remote sensing techniques and instruments for the mapping of snow properties are given by Dozier and Painter [30], Nolin [31], and Dietz *et al.* [32].

In this section, we provide an overview on the use of spectroscopy and methods to determine the snow cover, grain size, albedo, and radiative forcing of light-absorbing impurities. More details on the related methods and products are provided in Refs. [33,34].

12.4.1 Snow-Covered Area

Snow cover is an important observable. It has been estimated from space remote sensing techniques since the launch of TIROS-6 in 1962. The strong intensity contrast between the signals of snow and other surfaces allows simple binary retrievals. However, more than half a century later, we still struggle to distinguish snow from clouds with passive optical remote sensing techniques.

The Normalized Difference Snow Index was developed to detect fully snow-covered pixels [35,36]. Linear spectral mixture analysis methods were introduced by Rosenthal and Dozier [37,38] to derive snow-covered areas including partially snow-covered pixels from multispectral data. The linear approximation requires that surface types (e.g., snow, vegetation, rocks, etc.) not exchange photons. An exchange of photons would lead to nonlinear spectral mixtures. The vastly improved spectral information from imaging spectrometer instruments, such AVIRIS, adds important constraints for the inversion of the mixed spectral signatures. Painter [39] and Sommerfield [40] showed the improvements of the fractional snow-covered area retrievals based on this approach.

12.4.2 Snow-Grain Size

The properties of a snowpack are influenced by snow accumulation, densification, and thermodynamically driven metamorphism. The size and shape of snow grains can provide information about the occurrence of these processes. Snow crystals can grow to complex shapes (e.g., depth-hoar) or become more round and clustered due to melt–freeze processes [41,42]. Snow-grain size can be used as a first-order approximation for snow density, which is a prerequisite together with the snow volume to determine how much water is contained in the snowpack (snow water equivalent) [42]. Larger grain sizes are associated with an increased snow density and a reduction in broadband albedo, an effect which is often used in snow remote sensing techniques [43,44].

The scientific observation of snow grains has a long tradition dating back to Kepler in 1611 [45,46] and Descartes in 1635 [47], to name only a few. New measurement methods, such as contact spectroscopy, X-ray tomography, and mathematical models, have emerged and improve our understanding of snow properties. However, the complexity and variability of the snow matrix requires us to complement snow-grain size estimates with additional size parameters [48,49]. The snow-grain size parameters retrieved by different methods, including remote sensing approaches, are not directly intercomparable, which complicates validation efforts.

Spectroscopy applications in the visual, NIR, and shortwave infrared region make use of the fact that the absorption of light is related to snow-grain size. Light absorption by water ice crystals is generally weak in the ultraviolet through visual spectral region (0.2–0.7 μm) and increasingly effective toward longer wavelengths up to the microwave region (ca. 500 μm). Local maxima in ice absorption suitable for spectroscopy are found at 1.03, 1.26, 1.49, and 2.0 μm [50]. The first and second maxima distinguish absorption due to ice versus liquid water. This distinction is especially important in melting snow conditions.

Painter [38] suggested fitting absolute intensity measurements at 1.03 μm to a look-up based on radiative transfer model results of snow directional reflectance with respect to the optical effective snow-grain size. This method was refined by Nolin and Dozier [51] using the integral of the absorption feature scaled by its continuum. The new approach is more robust to errors in the absolute signal due to terrain and radiometric calibration uncertainties. However, the old method remains superior in the case of spectral calibration uncertainties. Painter *et al.* [52] combined this new method with a linear spectral mixture analysis approach to retrieve the snow-grain size simultaneously with the snow-covered area. This combined approach allows a first-order approximation of the water contained in the snow based on assumptions of the snow depth and of the relationship between snow-grain size and snow density.

A limitation of optical snow-grain size remote sensing methods is their dependence on corrections for atmospheric water vapor, which absorbs light in a very similar way as water ice crystals. Liquid water- and light-absorbing impurities, such as dust particles, in the snow pack can also change the retrieval results [53,54].

12.4.3 Snow Albedo

Snow and clouds reflect most of the solar radiation and are therefore an important, yet not fully understood factor of the Earth's energy balance [55]. Their high variability in time and space can alter the total broadband albedo of the surface significantly. Snow albedo is a physical surface property, which directly influences the energy fluxes (direct and diffuse solar irradiance, thermal and latent heat) together with terrain features, the atmosphere, and the intensity of solar radiation. Furthermore, the albedo of snow determines how much of the solar radiation is being absorbed and driving snowmelt processes directly. Snow albedo is therefore an important variable to predict water runoff from snow in hydrological catchments.

Calibrated optical remote sensing instruments measure the radiances reflected by a surface and scattered by the atmosphere. For snow observations, effects of atmospheric extinction have to be removed first and normalized by the irradiance at the surface to obtain the directional reflectance of the surface. This quantity still depends on the combination of solar, viewing, and terrain geometry, as well as surface properties. Models are normally used to account for these effects in the atmosphere and at the surface, which are often referred to as atmospheric and bidirectional reflectance distribution function correction, respectively. One possible approach to convert the observed directional reflectance R^{obs} to the observed albedo A^{obs} is to apply the anisotropy factor, which describes the ratio of a modeled albedo A^{mdl} to the modeled directional reflectance function R^{mdl} at a given solar and observation geometry, such that:

$$A^{obs} = R^{obs} \frac{A^{mdl}}{R^{mdl}}. \tag{12.5}$$

Spectroscopy instruments provide continuous spectral data, which are required to retrieve the broadband, or spectrally integrated albedo weighted by the spectral (indicated by λ subscript) solar irradiance at the surface, E_λ^{mdl}:

$$A^{obs} = \frac{\int A_\lambda^{obs} E_\lambda^{mdl} \Delta\lambda}{\int E_\lambda^{mdl} \Delta\lambda}. \tag{12.6}$$

The directional snow reflectance and the snow albedo are both inversely related to the snow-grain size, where growing snow grains tend to reduce the albedo particularly at wavelengths beyond 0.85 μm [56,57].

12.4.4 Radiative Forcing of Light-Absorbing Impurities in Snow

Another important observable, based on spectroscopy, is the radiative forcing of light-absorbing impurities in snow. These impurities consist of dust, soot particles, and other absorbing particles brought into the snowpack via dry and wet deposition.

As the snow melts, light-absorbing impurities accumulate at the snow surface, accelerating the snowmelt. They also tend to increase snow-grain size, thus reducing the albedo further. Finally, when snow thins to a few centimeters, solar radiance is absorbed by the underlying surface and enhances snowmelt. This self-reinforcing mechanism (positive feedback) impacts the energy balance on all scales [58].

The additional absorption is derived from the difference between a modeled clean snow albedo and the observed snow albedo, scaled by the available instantaneous irradiance at the snow surface:

$$\mathrm{RF} = \int E_\lambda^{mdl} \left(A_\lambda^{mdl} - A_\lambda^{obs} \right) \Delta\lambda. \tag{12.7}$$

The influence of light-absorbing impurities on the snow albedo is most prominent in the visible part of the spectra where the snow is strongly reflecting. It is therefore sufficient to derive the spectral integration in this spectral range. In addition, the irradiance term can be integrated over time and adjusted for cloudy conditions in order to provide a total radiative forcing for the corresponding timespan to improve snowmelt prediction models. An example of snow-grain size, albedo, and radiative forcing retrieval products as described in this section is shown in Fig. 12.5.

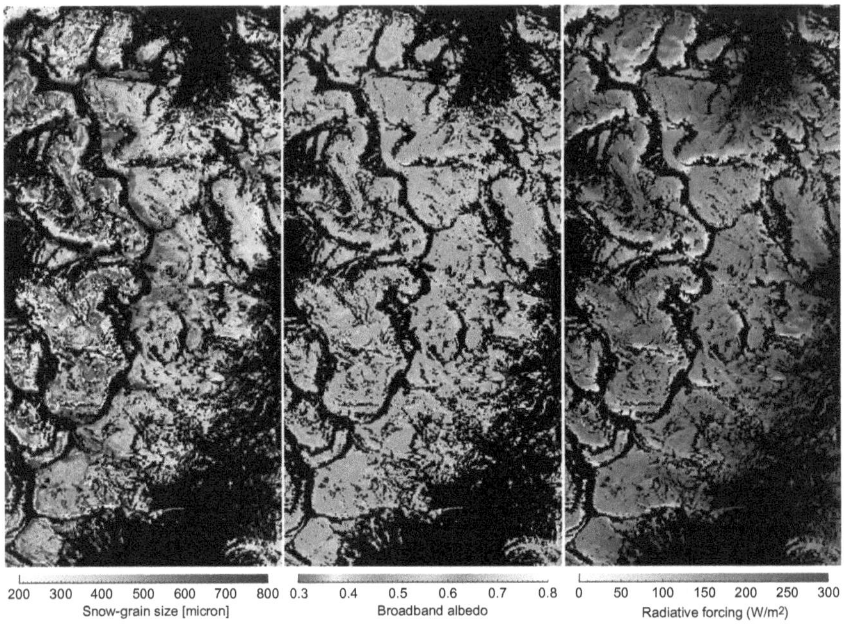

FIGURE 12.5 Snow-grain size (left), albedo (middle), and radiative forcing (right) retrieval products. The underlying data were acquired with AVIRIS on the ER-2 platform over the Juan Mountains, Colorado, on June 15, 2011 [34].

12.4.5 Benefits of Imaging Spectroscopy in Combination with Other Observations for Snow Remote Sensing

Recent studies, based on *in situ* field spectroscopy and energy flux tower measurements, have shown how strong the effects of light-absorbing impurities are on snowmelt in mountains [59,60]. Complete coverage of the solar spectral range is an important characteristic of spectroscopy instruments allowing quantitative retrievals of broadband snow albedo as well as the radiative forcing of light-absorbing impurities in snow. As a complement to the *in situ* measurements, airborne imaging spectroscopy observations allow us to retrieve snow-grain size, albedo, and radiative forcing of snow impurities on large scales, which in turn allow us to extend our understanding of spatial and temporal aspects of snowmelt processes to even larger scales [33,34].

Retrievals based on remote sensing data, however, are sensitive to sensor calibration, atmospheric correction, and digital surface model uncertainties. Distinguishing the impacts due to light-absorbing impurities in snow from those due to absorbing aerosols in the atmosphere remains a challenge. Seidel and Popp [61] showed that the optical depth of absorbing aerosols over snow can be retrieved with sufficient sensitivity, given that the snow reflectance is known at least in one location. Moreover, a combination of imaging spectroscopy with multiangle polarization data, for example AVIRIS and AirMSPI

(AIRborne Multiangle SpectroPolarimetric Imager) [62], allows simultaneous retrievals of atmospheric and snow surface properties. The Airborne Snow Observatory, at NASA's Jet Propulsion Laboratory, has shown how a combination of imaging spectroscopy and lidar data has improved the retrievals of snow properties, snow depth, and snow water equivalent.

12.5 VOLCANIC UNREST

In this section, we summarize remote spectroscopic methods used to characterize and detect unrest on dormant and active volcanoes, a natural hazards application of some of the above-described remote sensing techniques. For the past three decades, remote sensing of volcanoes has been routinely performed by volcano observatories and research groups, using ground-based, airborne, and space-borne (satellite-based) sensors. Applications include baseline characterization of volcanic landforms, long-term characterization of pre-eruptive unrest, short-term thermal, and gas compositional monitoring for prediction of eruptions during unrest, as well as posteruptive ash dispersion studies.

12.5.1 Early Remote Sensing Methods

Earliest instrumental applications of remote spectroscopy to volcanoes were upward-looking optical extinction measurements performed after the 1902 Pelée (Martinique) [63] and Katmai (Alaska) 1912 eruptions [64]. After the 1963 Agung (Bali, Indonesia) eruption, aerosol optical depth variations were already observed by several observatories and research groups [65–67].

Since the early 1970s, Correlation Spectroscopy in the Ultraviolet (CoSPEC), originally developed for quantification of stack emissions, quickly became standard instrumentation to measure sulfur dioxide (SO_2) plume variations before eruptions in order to assess, monitor, and predict volcanic unrest [68,69]. Airborne spectroscopic methods in the TIR first confirmed the thermally active fissures on top of the 1980 pre-eruptive lateral bulge on Mount St. Helens (USA), but image processing and interpretation were not yet fast enough to provide near-real-time information for observational volcano monitoring operations [70]. First satellite observations of SO_2 from volcanoes using the Total Ozone Mapping Spectrometer (TOMS) instrument on the Nimbus-7 satellite were successfully done for the 1982 El Chichón eruption [71]. TOMS could only detect major eruptions but a vibrant scientific field emerged from there.

12.5.2 Applications and Relevance

Today, remote spectroscopic methods constitute one of the major backbone methods of operational volcano observatories, together with seismic monitoring, and other ground-based geophysical and geochemical methods. Information derived from all these methods permeate through the entire observatory task chain. Volcano observatories are nonresearch institutions that operationally monitor the several hundred potentially active and unrestful volcanoes worldwide. Their tasks include baseline characterization and hazard zonation, long-term geophysical and geochemical monitoring, short-term unrest monitoring, and eruption prediction, issuing alert level changes to authorities, providing Volcanic Ash Advisory Centers (VAACs) with source term information during eruptions, outreach, education, and mitigation planning, as well as communication with decision authorities, civil defense, nongovernment organizations, the public, and the press. Figure 12.6 shows VAAC region boundaries, along with active volcanoes and major air routes.

To understand the magnitude of this application, the Volcano Observatories' umbrella organization, the World Organization of Volcano Observatories (WOVO) has over 70 active member agencies, some of which operate multiple individual observatories. For example, one of these member organizations, Indonesia's government agency tasked with monitoring it's over 130 active volcanoes (Center for Volcanology and Geological Hazard Mitigation), currently staffs 76 instrumented observatories [71]. More typically, one of the five volcano observatories currently (i.e., in late 2013) operated by the U.S. Geological Survey, the Alaska Volcano Observatory,

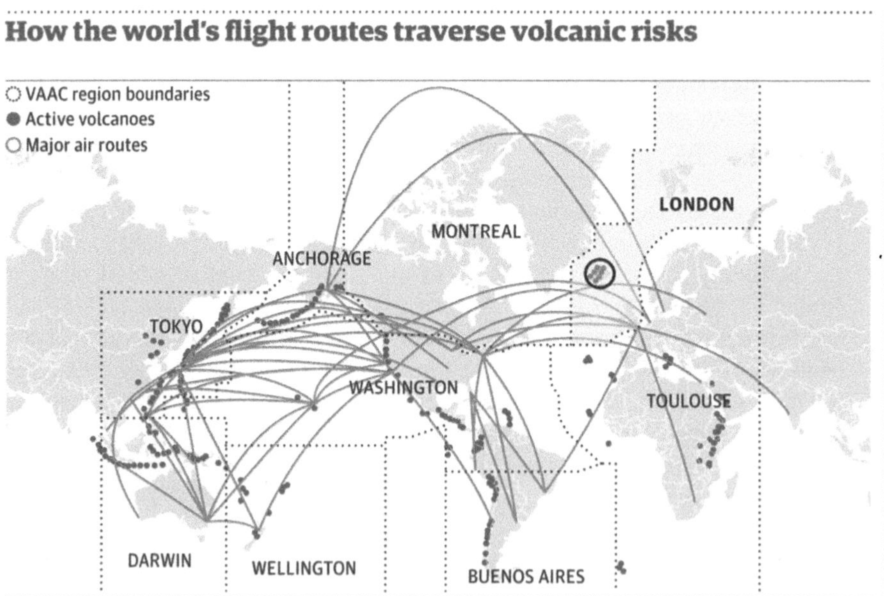

FIGURE 12.6 Global active volcanic regions and Volcanic Ash Advisory Centers (VAACs). The London VAAC (highlighted) responded to the 2010 Eyjafjallajökull eruption [72].

monitors 20 volcanoes in Alaska and the Aleutians. Together, all observatories actively monitor hundreds of active and potentially active volcanoes.

WOVO principally operates within two projects, both of which routinely and centrally use remote sensing spectroscopy data: (a) WOVO's global database of volcanic unrest (WOVOdat), which is primarily used by and for operational observatories, and (b) the VAAC, which are cosponsored between WOVO (relaying source term information like injection heights, erupted ash and gas mass flux, volcano alert levels, and eruption forecasts) and regional meteorological offices (dispersion modeling and plume forecasting).

12.5.3 Imaging of Volcanic Landforms, Eruption Detection, and Degassing

Volcano observatories use space-borne and occasionally airborne remote spectroscopic imaging methods in their daily operations to assess ground deformation, degassing behavior, vegetation changes, to map out hazard zones and eruption product areas, volumes, and directions. For example, space-borne imagery and data are obtained from several NASA and ESA (European Space Agency) satellite instruments, including ASTER (Advanced Spaceborne Thermal Emission and Reflection Radiometer, aboard the Terra satellite), generations of Landsat, MODIS (Moderate Resolution Imaging Spectroradiometer, aboard the Terra and Aqua satellites), various RADAR (Radio Detection And Ranging) missions, as well as TOMS, GOME (Global Ozone Monitoring Experiment, aboard the ESA's ESR-2 satellite), and OMI (Ozone Monitoring Instrument, aboard NASA's Aura satellite). These (among others) instruments are used in their individual capabilities to detect thermal anomalies, swiftly detect eruptions otherwise unnoticed (frequently used in the Aleutians, Kamchatka, and in East Africa, where no dense ground instrumentation networks exist), derive long-term inflation–deflation episodes using Synthetic Aperture RADAR Interferometry (InSAR), and map out SO_2 plumes.

12.5.3.1 Gas Monitoring of Continuous, Quiescent Volcanic Emissions

Dormant and active volcanoes continuously emit gases through craters, flanks, active vents, and fissures. Episodes of eruptive activity are often preceded by an increase in degassing up to months before cataclysmic events. Such increases are thought to be caused by changes in the magmatic reservoirs kilometers underground such as deep recharge, mixing, and crystallization, and by conduit processes. However, detection of early deep signals can be masked by strong hydrothermal systems. While early signals may not be diagnostic, they do permit early warning and increased monitoring activity. Gases emitted are usually dominated by varying degrees of water vapor, but some meteoric water can have been entrained after gas exsolution from the magma. The principal and strongly dominating component of anhydrous volcanic gas is CO_2, accompanied by minor components such as SO_2, CH_4, H_2S, H_2, CO, halides, and a large variety of inorganic and organic trace species [73,74]. Methods used to detect and monitor these gases comprise ground-based closed- and open-path spectroscopic methods, but airborne and space-borne methods are showing promising advances recently as well.

Open-path ground-based methods include solar occultation spectroscopy through the plume (mainly for CO_2 using FTS instruments), and differential optical absorption spectroscopy (DOAS) in the UV, which is used to monitor plume gas fluxes of SO_2 in campaign (mobile) or stationary (monitoring) mode. Recently, the latter have seen worldwide application in standardized slant-column scanning miniaturized automated DOAS systems in the 50+ instruments global NOVAC network [75,76]. Most recent developments employ UV camera systems to monitor degassing laterally, with capabilities to perform DOAS in image processing in an automated fashion, deriving most accurate real-time series of SO_2 emissions [77,78]. Closed-path methods, using either an FTS with across-crater mirror systems or magma as a radiation source, can measure multiple chemical gas species for monitoring compositional changes of CO_2, CH_4, H_2, HCL, and others [79]. Diffuse surface flux systems employ nondispersive NIR sensors for monitoring flank CO_2 emissions indicative of deeper magmatic processes [80]—however, these are not remote sensing methods *sensu stricto*, but rather remotely operated *in situ* spectroscopic methods.

Airborne methods have been successfully applied for multiple gas species in Japan [81]. Satellite platforms and sensors including TOMS, GOME, and OMI are used to assess SO_2 emissions routinely [82,83] and volcanic CO_2 detection might become a reality soon, with advances from JAXA's GOSAT satellite and the 2014 launch of NASA's OCO-2 satellite. Furthermore, infrared spectroscopy has successfully been used to detect eruptions even as far as extraterrestrial volcanism on Jupiter's moon Io [83].

12.5.3.2 Volcanic Eruption Plumes

Once an explosive volcanic eruption occurs, the resulting particle plume (*Note*: "volcanic ash" is not an incineration product but finely blasted pieces of quenched magma: microscopic shards of glassy rock) is of concern for any air-assisted combustion engine on the ground and in the air. Volcanic ash entrained in jet engines can melt onto parts of the engine and form a rock coating which may quickly result in engine failure. Several such deleterious ash aircraft encounters have occurred in the past, and consequently, the aviation industry, and its regulating agencies are concerned with early warning and prediction of areas where potential ash aircraft encounters may occur. Globally, VAACs disseminate such information.

Famously, the 2010 Eyjafjallajökull eruption in Iceland, a small eruption in terms of amount of energy, gases, and magma produced, also produced a vast amount of ash in a runaway effect of magmatic heat-induced glacial ice melt which kept fracturing the magmatic surface, exposing fresh magma continuously. The European airspace was closed down for weeks. The validation of ash dispersal models by the London VAAC was conducted by several European nations, using specialized airborne remote sensing instruments like spectral lidar, as well as using various satellite remote sensing products sensitive in the spectral regions where volcanic ash absorbs. Limb sounding aircraft mounted active infrared spectroscopic instruments (AVOID system, Airborne Volcanic Object Imaging Detector) have recently been implemented by a variety of air carriers [84].

Sulfur dioxide detection in eruption plumes has to take into account the observationally and thermodynamically well-known gas–particle conversion from SO_2 to sulfate in cooling plumes, facilitated by available condensation nuclei in the form of ash particles. Therefore, a variety of particle shapes from concave surfaced glass fragments to microlites (microscopic crystals) and spherical droplets are known to occur, which ideally should be taken into account in models of particle–light interactions and in radiative transfer models.

Detection of thermal anomalies and surface characterization in the TIR is a now well-established technique and applied to volcanoes using Landsat TIR bands, MODIS, and ASTER scenes [85]. A global volcanic hot spot tracking software application called MODVOLC (MODIS Volcano Detection, using MODIS bands 21, 22, and 32) and its resulting data, are freely available via the World Wide Web [86].

Newer and available sophisticated volcanic gas and ash plume identification and tracking routines promise a greater degree of national autonomy for observatories and VAACs alike: for example, the Plume Tracker software [87] enables users to select spectral information from a variety of available sensors and products themselves, and to identify, track, and quantify different aerosol retrievals, thermal signals, and SO_2 burdens.

12.6 CALIBRATION

Most remote sensing systems measure incident light[2] using several channels, where each channel is designed with a specified color sensitivity. That is, the instrument's perchannel output is designed to be a function of both the amount of incident light as well as its wavelength. The color selection may be achieved by using spectral filters within the instrument, or may be accomplished by use of a grating that disperses light according to the wavelength. These systems, if used for scientific studies, are calibrated for their radiometric and spectral response. The radiometric calibration relates the sensor output digital count to the sensor-incident radiance. For space-borne systems, this is first done prior to launch. For visible-sensing instruments, a large flat-field

2. "Light" is used loosely in this section to mean "optical radiation."

source, such as that from an integrating sphere, is characterized by a reference detector or by comparison to a reference source. This determines the output in absolute units. Traceability is made to SI (*Système International d'Unités*) units, a standard agreed upon by the international science community. SI traceability is typically facilitated by a national standards laboratory, which in turn disseminates this traceability to commercial vendors. These vendors, in turn, transfer this traceability through secondary detector and source standards that they make available to the calibration laboratories. For thermal sensing instrument or channels, a blackbody source is used. Radiometric response determinations are typically made to a relative uncertainty of 3% to meet the science requirements.

Once on-orbit, sensors begin to degrade. This is especially true in the initial months following launch. In order to maintain radiometric response knowledge, experiments must continue on throughout the sensor's mission life. These experiments might utilize on-board sources, such as lamps or solar-reflecting diffuser plates. Sensors that utilize on-board calibrators, and those that do not have this capability, both make use of Earth-views to determine radiometric calibration. Vicarious calibration experiments are those for which the surface and atmospheric state are measured at uniform areas, such as desert playa. These measurements can be made by field teams, or by using automated facilities [88]. With these data, the sensor-incident radiance can be computed for the specific fly-by time. The comparison of this radiance to the sensor output while viewing the test site provides an update to the radiometric response.

Equally important to the radiometric response is the determination of the spectral response of a system. It is rare that these measurements are made on-orbit; hence, the laboratory preflight determination is assumed to be valid over time. For the determination, a broad-band source, such as a tungsten–halogen lamp, is typically used. This is fed into a monochromator, which in turn uses internal gratings to disperse the light, and direct monochromatic light onto the sensor under test. The spectral response is the output response of the sensor versus wavelength of illumination. The characterization should measure both the in-band response, near the region of peak sensitivity, and also the far-wing response. As this distribution can only be determined in a relative sense, it is normalized such that the area under this curve is unity.

It is common practice to summarize a sensor's spectral properties by tabulating the center wavelength and spectral width. This has been done by quoting the wavelength-of-peak response, and full-width half-maximum (FWHM). The latter is given by wavelengths at which the response falls to half of its peak value. These parameters can be misleading in cases where the SRF may be double-peaked, asymmetric, or have a large out-of-band response. A better representation can be obtained from an equivalent square band response analysis [89].

For instruments with very narrow spectral response functions, a tunable diode laser can be used, or an FTS. For OCO preflight testing, both these techniques were used [90], as summarized here. The spectral calibration includes two characteristics: the spectral dispersion and the ILS function of each detector element. The spectral dispersion describes the wavelength at which each detector element had its centroid response. The ILS, also commonly referred to as the spectral response function, represents the response of a single detector element to any wavelength of light incident upon that element. The ILS and dispersion parameters were determined using laser-diode sources, and validated using solar spectra collected by the instrument and compared against simultaneously collected spectra from a collocated high-resolution FTS observing the same spectral regions.

The challenge for OCO ILS determination was that it needed to be analyzed for an enormous amount of data: three bands, eight footprints per band, and 1016 spectral pixels per footprint yield in theory 24,384 individual ILS functions. Luckily, the physics of the OCO instrument design indicates that the ILS and centroid wavelength response (dispersion) of OCO should vary smoothly in the spectral dimension across each band, mitigating the challenge. The ILS functions and spectral dispersion experiments were conducted with three separate tunable diode lasers having instantaneous line widths <1 MHz, one for each of the OCO bands. Each tunable diode-laser test consisted of setting the lasers to a specific wavelength, waiting for frequency, and intensity stabilization, averaging several seconds of data, and then scanning the laser wavelength by a small fraction of the FWHM to repeat the process. The lasers typically required tens of seconds to properly stabilize at the desired wavelength. Scanning the entire spectral range of each band would have taken many days, and thus, was impractical. Therefore, detailed ILS characterizations were performed over a limited set of wavelengths that spanned each spectral range and could be interpreted to parameterize all of the ILS functions.

Once these scans were completed, the wavelength of peak response was made known for each detector element (pixel), as shown in Fig. 12.7. A Gaussian fit to the centroid wavelength response enabled a simple determination of the spectral dispersion for each of eight spatial footprints and three spectral bands of the OCO instrument. It was found that the centroid wavelengths of the pixels of a given footprint had a spacing that varied smoothly across the band in question and could be described by a fourth order polynomial.

12.7 SUMMARY

Remote sensing measurements, and the spectral analysis of these data, provide scientists with a rich information source to better understand Earth processes. Sensors such as GOSAT and OCO-2 have been built to measure CO_2 sources and sinks. These data will allow improved modeling of carbon fluxes and yield improved predictive tools for understanding greenhouse warming. A quantitative understanding of clouds, snow, and ice properties are also important to understanding Earth's climate system. They are important contributors to the global fluxes of radiant energy, as well as the hydrological cycle. Remote sensing measurements allow researchers to determine cloud and snow

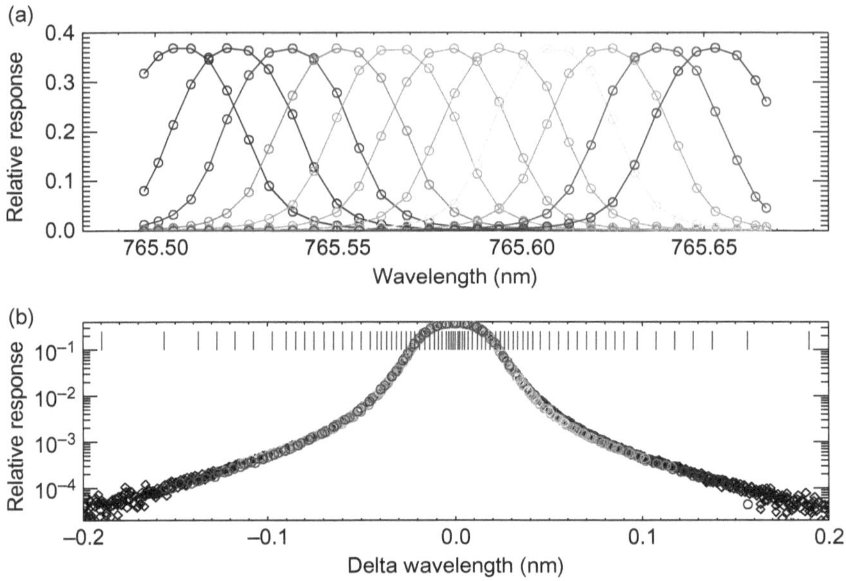

FIGURE 12.7 Data recorded from the laser scans as seen by each of several neighboring pixels in the O_2 A-band. (a) The signal from a set of 11 pixels is displayed in different colors. (b) These channels are combined to give a single set of data with a much higher resolution, as shown by the colored circles. Additional pixels with smaller responses are shown as black diamonds; their wavelength centroids were approximated. These pixels contained information on the far wings. The red vertical lines (dark gray in print) near the top of panel (b) represent the wavelength positions used to represent the ILS on a fixed wavelength grid [90].

cover, grain size, albedo, and the radiative forcing of light. For snow, the study of absorbing impurities is an important research area in the study of the cryosphere. Along with ever evolving models and the combination with other instruments, such as lidar and AirMSPI, we are able to improve our observations of cloud and snow properties to understand better their role in our climate. Remote sensing observations are a major source of information for those who study volcano hazards, perform long-term geophysical and geochemical monitoring, short-term unrest monitoring, and eruption prediction, frequently resulting in the issuing of alert level changes to authorities. These are but a few examples of an emerging and ever-expanding science.

ACKNOWLEDGMENTS

Much of the research and programs described in this chapter were carried out at the Jet Propulsion Laboratory, California Institute of Technology, Pasadena, California, under contract with the National Aeronautics and Space Administration (NASA). The authors also wish to acknowledge the help of David Crisp, Brendan Fisher, Rob Green, Cate Heneghan, Chris O'Dell, and Rob Rosenberg and for references and figures used to develop this chapter.

REFERENCES

[1] R.O. Green, M.L. Eastwood, C.M. Sarture, T.G. Chrien, M. Aronsson, B.J. Chippendale, J.A. Faust, B.E. Pavri, C.J. Chovit, M. Solis, M.R. Olah, O. Williams, Imaging spectroscopy and the airborne visible/infrared imaging spectrometer (AVIRIS), Remote Sens. Environ. 65 (1998) 227–248.

[2] Kyoto Protocol to the United Nations Framework Convention on Climate Change (UNFCCC). Signed December 11, 1997. http://unfccc.int/kyoto_protocol/items/3145.php.

[3] F.-M. Breon, P. Ciais, Spaceborne remote sensing of greenhouse gas concentrations, Compt. Rendus Geosci. 342 (2010) 412–424.

[4] H. Bovensmann, J.P. Burrows, M. Buchwitz, J. Frerick, S. Noel, V.V. Rozanov, K. V. Chance, A.P.H. Goede, SCIAMACHY: mission objectives and measurement modes, J. Atmos. Sci. 56 (1999) 127–150.

[5] T.S. Pagano, H.H. Aumann, D.E. Hagan, K. Overoye, Prelaunch and in-flight radiometric calibration of the atmospheric infrared sounder (AIRS), IEEE Trans. Geosci. Remote Sens. 41 (2) (2003) 265–273.

[6] A. Kuze, H. Suto, M. Nakajima, T. Hamazaki, Thermal and near infrared sensor for carbon observation Fourier-transform spectrometer on the Greenhouse Gases Observing Satellite for greenhouse gases monitoring, Appl. Opt. 48 (2009) 6716–6733.

[7] T. Yokota, Y. Yoshida, N. Eguchi, Y. Ota, T. Tanaka, H. Watanabe, S. Maksyutov, Global concentrations of CO_2 and CH_4 retrieved from GOSAT: first preliminary results, SOLA 5 (2009) 160–163.

[8] D. Crisp, R.M. Atlas, F.M. Breon, L.R. Brown, J.P. Burrows, P. Ciais, B.J. Connor, S.C. Doney, I.Y. Fung, D.J. Jacob, C.E. Miller, D. O'Brien, S. Pawson, J.T. Randerson, P. Rayner, R.J. Salawitch, S.P. Sander, B. Sen, G.L. Stephens, P.P. Tans, G.C. Toon, P.O. Wennberg, S.C. Wofsy, Y.L. Yung, Z. Kuang, B. Chudasama, G. Sprague, B. Weiss, R. Pollock, D. Kenyon, S. Schroll, The orbiting carbon observatory (OCO) mission, Adv. Space Res. 34 (2004) 700–709.

[9] C.W. O'Dell, B. Connor, H. Boesch, D. O'Brien, C. Frankenberg, R. Castano, M. Christi, D. Eldering, B. Fisher, M. Gunson, J. McDuffie, C.E. Miller, V. Natraj, F. Oyafuso, I. Polonsky, M. Smyth, T. Taylor, G.C. Toon, P.O. Wennberg, D. Wunch, The ACOS CO2 retrieval algorithm—part 1: description and validation against synthetic observations, Atmos. Meas. Tech. 5 (2012) 99–121.

[10] R. Kahn, P. Banerjee, D. McDonald, Sensitivity of multiangle imaging to natural mixtures of aerosols over ocean, J. Geophys. Res. 106 (2001) 18219–18238.

[11] G. Thuillier, M. Hersé, D. Labs, T. Foujols, W. Peetermans, D. Gillotay, P.C. Simon, H. Mandel, The solar spectral irradiance from 200 to 2400 nm as measured by the SOLSPEC spectrometer from the Atlas and Eureca Missions, Solar Phys. 214 (2003) 1–22.

[12] J. Lenoble, Radiative Transfer in Scattering and Absorbing Atmospheres: Standard Computational Procedures, A Deepak Publishing, Hampton, Virginia, 1985.

[13] R.F. Cahalan, W. Ridgway, W. Wiscombe, T.L. Bell, J.B. Snider, The albedo of fractal stratocumulus clouds, J. Atmos. Sci. 51 (1994) 2434–2455.

[14] R. Davies, Spatial autocorrelation of radiation measured by the Earth Radiation Budget Experiment: scene inhomogeneity and reciprocity violation, J. Geophys. Res. 99 (1994) 20879–20887.

[15] A. Marshak, A. Davis (Eds.), 3D Radiative Transfer in Cloudy Atmospheres, Springer, Berlin, 2005.

[16] R.F. Cahalan, et al., The I3RC: bringing together the most advanced radiative transfer tools for cloudy atmospheres. Bull. Am. Meteorol. Soc. 86 (2005) 1275–1293, http://dx.doi.org/10.1175/BAMS-86-9-1275.

[17] P. Yang, K.-N. Liou, Geometric-optics–integral-equation method for light scattering by non-spherical ice crystals, Appl. Opt. 35 (1996) 6568–6584.

[18] A.A. Lacis, V. Oinas, A description of the correlated *k* distribution method for modeling nongray gaseous absorption, thermal emission, and multiple scattering in vertically inhomogeneous atmospheres, J. Geophys. Res. 96 (1991) 9027–9063.

[19] H.C. van de Hulst, Multiple Light Scattering, Academic Press, New York, 1980.

[20] R. Davies, W.L. Ridgway, K.-E. Kim, Spectral absorption of solar radiation in cloudy atmospheres: a 20 cm^{-1} model, J. Atmos. Sci. 41 (1984) 2126–2137.

[21] W.L. Smith, X.L. Ma, S.A. Ackerman, H.E. Revercomb, R.O. Knuteson, Remote sensing cloud properties from high spectral resolution infrared observations, J. Atmos. Sci. 50 (1993) 1708–1720.

[22] T. Nakajima, M.D. King, Determination of the optical thickness and effective particle radius of clouds from reflected solar radiation measurements. Part I: theory, J. Atmos. Sci. 47 (1990) 1878–1893.

[23] Á. Horváth, R. Davies, Anisotropy of water cloud reflectance: a comparison of measurements and 1D theory. Geophys. Res. Lett. 31 (2004), http://dx.doi.org/10.1029/2003GL018386, L01102.

[24] L. Liang, L. Di Girolamo, S. Platnick, View-angle consistency in reflectance, optical thickness and spherical albedo of marine water-clouds over the northeastern Pacific through MISR-MODIS fusion, Geophys. Res. Lett. 36 (2009), L09811.

[25] C. Cornet, R. Davies, Use of MISR measurements to study the radiative transfer of an isolated convective cloud: implications for cloud optical thickness retrieval. J. Geophys. Res. 113 (2008), http://dx.doi.org/10.1029/2007JD008921, D04202.

[26] W.P. Menzel, W.L. Smith, T.R. Stewart, Improved cloud motion wind vector and altitude assignment using VAS, J. Clim. Appl. Meteorol. 22 (1983) 377–384.

[27] D.M. Winker, M.A. Vaughan, A. Omar, Y. Hu, K.A. Powell, Z. Liu, W.H. Hunt, S. A. Young, Overview of the CALIPSO Mission and CALIOP Data Processing Algorithms, J. Atmos. Ocean. Technol. 26 (2009) 2310–2323.

[28] R. Davies, M. Molloy, Global cloud height fluctuations measured by MISR on Terra from 2000 to 2010, Geophys. Res. Lett. 39 (2012).

[29] R. Davies, Á. Horváth, C. Moroney, B. Zhang, Y. Zhu, Cloud motion vectors from MISR using sub-pixel enhancements. Remote Sens. Environ. 107 (2007) 194–199, http://dx.doi.org/10.1016/j.rse.2006.09.023.

[30] J. Dozier, T.H. Painter, Multispectral and hyperspectral remote sensing of alpine snow properties, Annu. Rev. Earth Planet. Sci. 32 (2004) 465–494.

[31] A.W. Nolin, Recent advances in remote sensing of seasonal snow, J. Glaciol. 56 (200) (2010) 1141–1150.

[32] A.J. Dietz, C. Kuenzer, U. Gessner, S. Dech, Remote sensing of snow—a review of available methods, Int. J. Remote Sens. 33 (2011), 2014/02/13.

[33] J. Dozier, R.O. Green, A.W. Nolin, T.H. Painter, Interpretation of snow properties from imaging spectrometry, Remote Sens. Environ. 113 (Suppl. 1) (2009) S25–S37, 9.

[34] T.H. Painter, F.C. Seidel, A.C. Bryant, S.M. Skiles, K. Rittger, Imaging spectroscopy of albedo and radiative forcing by light-absorbing impurities in mountain snow, J. Geophys. Res. Atmos. 118 (2013) 9511–9523.

[35] J. Dozier, Spectral signature of alpine snow cover from the Landsat thematic mapper, Remote Sens. Environ. 28 (1989) 9–22.

[36] D.K. Hall, G.A. Riggs, V.V. Salomonson, Development of methods for mapping global snow cover using moderate resolution imaging spectroradiometer data, Remote Sens. Environ. 54 (1995) 127–140, 11.

[37] W. Rosenthal, J. Dozier, Automated mapping of montane snow cover at subpixel resolution from the Landsat thematic mapper, Water Resour. Res. 32 (1) (1996) 115–130.

[38] A.W. Nolin, J. Dozier, L.A.K. Mertes, Mapping alpine snow using a spectral mixture modelling technique, in: K. Steffen (Ed.), Proceedings of the Symposium on Remote Sensing of Snow and Ice, Ann. Glaciol. 17 (1993) 121–124.

[39] T.H. Painter, D.A. Roberts, R.O. Green, J. Dozier, The effect of grain size on spectral mixture analysis of snow-covered area from aviris data, Remote Sens. Environ. 65 (1998) 320–332, 9.

[40] R. Sommerfeld, E.R. LaChapelle, The classification of snow metamorphism, J. Glaciology 9 (1970) 3–17.

[41] S.C. Colbeck, An overview of seasonal snow metamorphism, Rev. Geophys. 20 (1) (1982) 45–61.

[42] R.L. Armstrong, R. Brown, R.E. Jordan, M.R.A. Brun, E. Brun, J.C. King, J.W. Pomeroy, D.M. Gray, C. Fierz, R.J. Harding, C. Plss, P.M.B. Fhn, E. Martin, Z.-L. Yang, R. Essery, J. Cohen, Snow and Climate: Physical Processes, Surface Energy Exchange and Modeling, Cambridge University Press, Cambridge, 2008.

[43] S.G. Warren, Optical properties of snow, Rev. Geophys. 20 (1) (1982) 67–89.

[44] T.C. Grenfell, S.G. Warren, P.C. Mullen, Reflection of solar radiation by the antarctic snow surface at ultraviolet, visible, and near-infrared wavelengths, J. Geophys. Res. Atmos. 99 (D9) (1994) 18669–18684.

[45] J. Kepler, The Six-Cornered Snowflake, 1611 translated by L.L. Whyte 1966 (Oxford Univ. Press), 1611.

[46] P. Ball, In retrospect: on the six-cornered snowflake, Nature 480 (2011) 455, 12.

[47] F.C. Frank, Descartes' observations on the amsterdam snowfalls of 4, 5, 6 and 9 february 1634, J. Glaciol. 13 (1974) 535.

[48] C. Mätzler, Relation between grain-size and correlation length of snow, J. Glaciol. 48 (162) (2002) 461–466.

[49] S. Colbeck, E. Akitaya, R. Armostrong, H. Gubler, J. Lafeuille, K. Lied, D. McClung, E. Morris, "The international classification for seasonal snow on the ground", tech. rep., ICSSG, 2009.

[50] S.G. Warren, R.E. Brandt, Optical constants of ice from the ultraviolet to the microwave: a revised compilation, J. Geophys. Res. Atmos. 113 (D14) (2008) D14220.

[51] A.W. Nolin, J. Dozier, A hyperspectral method for remotely sensing the grain size of snow, Remote Sens. Environ. 74 (2000) 207–216.

[52] T.H. Painter, J. Dozier, D.A. Roberts, R.E. Davis, R.O. Green, Retrieval of subpixel snow-covered area and grain size from imaging spectrometer data, Remote Sens. Environ. 85 (2003) 64–77.

[53] J. Dozier, S.R. Schneider, D.F. McGinnis, Effect of grain size and snowpack water equivalence on visible and near-infrared satellite observations of snow, Water Resour. Res. 17 (4) (1981) 1213–1221.

[54] R.O. Green, J. Dozier, D. Roberts, T. Painter, Spectral snow-reflectance models for grain-size and liquid-water fraction in melting snow for the solar-reflected spectrum, Ann. Glaciol. 34 (1) (2002) 71–73.

[55] G.L. Stephens, J. Li, M. Wild, C.A. Clayson, N. Loeb, S. Kato, T. L'Ecuyer, P.W. Stackhouse, P.W. Stackhouse, M. Leb-Sock, T. Andrews, An update on earth's energy balance in light of the latest global observations, Nat. Geosci. 5 (2012) 691–696.

[56] W.J. Wiscombe, S.G. Warren, A model for the spectral albedo of snow. i: pure snow, J. Atmos. Sci. 37 (1980) 2712–2733.

[57] S.G. Warren, W.J. Wiscombe, A model for the spectral albedo of snow. ii: snow containing atmospheric aerosols, J. Atmos. Sci. 37 (1980) 2734–2745.

[58] J. Hansen, L. Nazarenko, Soot climate forcing via snow and ice albedos, Proc. Natl. Acad. Sci. U. S. A. 101 (2) (2004) 423–428.

[59] T.H. Painter, S.M. Skiles, J.S. Deems, A.C. Bryant, C.C. Landry, Dust radiative forcing in snow of the upper colorado river basin: 1. A 6 year record of energy balance, radiation, and dust concentrations, Water Resour. Res. 48 (7) (2012).

[60] S.M. Skiles, T.H. Painter, J.S. Deems, A.C. Bryant, C.C. Landry, Dust radiative forcing in snow of the upper colorado river basin: 2. Interannual variability in radiative forcing and snowmelt rates, Water Resour. Res. 48 (7) (2012) W07522.

[61] F.C. Seidel, C. Popp, Critical surface albedo and its implications to aerosol remote sensing, Atmos. Meas. Tech. 5 (7) (2012) 1653–1665.

[62] D.J. Diner, F. Xu, M.J. Garay, J.V. Martonchik, B.E. Rheingans, S. Geier, A. Davis, B.R. Hancock, V.M. Jovanovic, M.A. Bull, K. Capraro, R.A. Chipman, S.C. McClain, The airborne multiangle spectropolarimetric imager (airmspi): a new tool for aerosol and cloud remote sensing, Atmos. Meas. Tech. Discuss. 6 (2013) 1717–1769.

[63] H.H. Kimball, Solar radiation, atmospheric absorption, and sky polarization, at Washington, D.C., Bull. Mt. Weather Obs. 3 (1910) 69–126.

[64] C. Dorno, Himmelshelligkeit, Himmelspolarisation und Sonnenintensität in Davos 1911–1918, Abh. Preuss. Met. Inst. 6 (303) (1919) 1–290.

[65] E.C. Flowers, H.J. Viebrock, Solar radiation: an anomalous decrease of direct solar radiation, Science 148 (1965) 493–494.

[66] F.E. Volz, Note on the global variation of stratospheric turbidity since the eruption of Agung Volcano, Tellus 17 (4) (1965) 513–515.

[67] A.J. Dyer, B.B. Hicks, Global spread of volcanic dust from the Bali eruption of 1963, Quart. J. Roy. Meteorol. Soc. 94 (1968) 545–554.

[68] R.E. Stoiber, A. Jepsen, Sulfur dioxide contributions to the atmosphere by volcanoes, Science 182 (4112) (1973) 577–578.

[69] G. Williams-Jones, J. Stix, C. Hickson, The COSPEC Cookbook: Making SO_2 Measurements at Active Volcanoes, Methods in Volcanology, vol. 1, IAVCEI, Barcelona, Spain, 2008.

[70] H.H. Kieffer, D. Frank, J.D. Friedman, Thermal Infrared Surveys: Observations Prior to the Eruption of May 18, in: The 1980 eruptions of Mount St. Helens, vol. 1250, 1982, pp. 257–278, USGS Professional Paper.

[71] A.J. Krueger, Sighting of El Chichón sulphur dioxide clouds with the Nimbus-7 Total Ozone Mapping Spectrometer, Science 220 (1983) 1377–1379.

[72] S. Rogers, "Volcanic ash: how do you spot the next volcano to disrupt flights? Every one listed," *The Guardian* (United Kingdom), Datablog http://www.theguardian.com/news/datablog/2010/apr/20/volcanic-ash-smithsonian-icao, 2010.

[73] R.B. Symonds, W.I. Rose, G.J.S. Bluth, T.M. Gerlach, Volcanic gas studies—methods, results, and applications, Rev. Mineral. Geochem. 30 (1994) 1–66.

[74] F.M. Schwandner, T.M. Seward, A.P. Gize, K. Hall, V.J. Dietrich, Halocarbons and other trace heteroatomic organic compounds in volcanic gases from Vulcano (Aeolian Islands, Italy), Geochim. Cosmochim. Acta 101 (2013) 191–221.

[75] B. Galle, M. Johansson, C. Rivera, Y. Zhang, M. Kihlman, C. Kern, T. Lehmann, U. Platt, S. Arellano, S. Hidalgo, Network for observation of volcanic and atmospheric change

(NOVAC)—a global network for volcanic gas monitoring: network layout and instrument description, J. Geophys. Res. D 115 (2010), 19p.
[76] C. Kern, T. Deutschmann, C. Werner, A.J. Sutton, T. Elias, P.J. Kelly, Improving the accuracy of SO_2 column densities and emission rates obtained from upward-looking UV-spectroscopic measurements of volcanic plumes by taking realistic radiative transfer into account, J. Geophys. Res. D 117 (2012) D20302-1-23.
[77] C. Kern, F. Kick, P. Lübcke, L. Vogel, M. Wöhrbach, U. Platt, Theoretical description of functionality, applications, and limitations of SO_2 cameras for the remote sensing of volcanic plumes, Atmos. Meas. Tech. 3 (2010) 733–749.
[78] A.J. Prata, Measuring SO_2 ship emissions with an ultra-violet imaging camera, Atmos. Meas. Tech. Discuss. 6 (2013) 9467–9511.
[79] M. Burton, P. Allard, F. Mure, C. Oppenheimer, FTIR remote sensing of fractional magma degassing at Mount Etna, Sicily, Geol. Soc. London. Spec. Publ. 213 (2003) 281–293.
[80] M. Liuzzo, S. Gurrieri, G. Giudice, G. Giuffrida, Ten years of soil CO2 continuous monitoring on Mt. Etna: exploring the relationship between processes of soil degassing and volcanic activity, Geochem. Geophys. Geosyst. 14 (2013) 2886–2899.
[81] H. Shinohara, Composition of volcanic gases emitted during repeating Vulcanian eruption stage of Shinmoedake, Kirishima volcano, Japan, Earth Planets Space 65 (2013) 667–675.
[82] S.A. Carn, N.A. Krotkov, K. Yang, A.J. Krueger, Measuring global volcanic degassing with the Ozone Monitoring Instrument (OMI), in: D.M. Pyle, T.A. Mather, J. Biggs (Eds.), Remote Sensing of Volcanoes and Volcanic Processes: Integrating Observation and Modeling, Geol. Soc. London, Spec. Pub., vol. 380, Geological Society, London, 2013.
[83] R.J. Andres, A.D. Kasgnoc, A time-averaged inventory of subaerial volcanic sulfur emissions, J. Geophys. Res. D 103 (1998) 25251–25261.
[84] F. Prata, A. Durant, A. Kylling, The airborne volcanic object imaging detector (AVOID): a new tool for airborne atmospheric remote sensing of clouds, Geophys. Res. Abstr. 14 (2012) 8977.
[85] M.S. Ramsey, A.J.L. Harris, Volcanology 2020: how will thermal remote sensing of volcanic surface activity evolve over the next decade? J. Volcanol. Geotherm. Res. 249 (2012) 217–233.
[86] R. Wright, L.P. Flynn, H. Garbeil, A.J.L. Harris, E. Pilger, MODVOLC: near-real-time thermal monitoring of global volcanism, J. Volcanol. Geotherm. Res. 135 (2004) 29–49.
[87] V.J. Realmuto, S. Baxter, P.W. Webley, Plume tracker: a new toolkit for the mapping of volcanic plumes with multispectral thermal infrared remote sensing, in: AGU Fall Meeting, 2011, abstract.
[88] D.X. Kerola, C.J. Bruegge, H.N. Gross, M.C. Helmlinger, Vicarious calibration of visible-near infrared earth remote sensors using LED spectrometer (LSpec) facility measurements in conjunction with radiative transfer models, IEEE Trans. Geosci. Remote Sens. 47 (4) (2008) 1244–1255.
[89] J.M. Palmer, Effective bandwidths for landsat-4 and landsat-D' multispectral scanner and thematic mapper subsystems, IEEE Trans. Geosci. Remote Sens. GE-22 (3) (1984) 336–338.
[90] J.O. Day, C.W. O'Dell, R. Pollock, C.J. Bruegge, R. David, D. Crisp, C.E. Miller, Preflight spectral calibration of the orbiting carbon observatory, IEEE Trans. Geosci. Remote Sens. 49 (7) (2011) 2793–2801.

Chapter 13

Microspectrophotometry

Paul C. Martin* and Michael B. Eyring[†,‡]
*CRAIC Technologies, San Dimas, California, USA
[†]Micro Forensics, Ltd., Phoenix, Arizona, USA
[‡]Arizona Department of Public Safety Crime Laboratory, Phoenix, Arizona, USA

Chapter Outline

13.1 Introduction 489
13.2 Microspectrophotometer Instrument Design and Construction 490
13.2.1 The Microscope 491
13.2.2 The Spectrophotometer 493
13.2.3 Computer Interface and Software Control 494
13.2.4 MSP System Capabilities 495
13.3 Using the MSP System 496
13.4 Current Applications of MSP in Industry and Research 498
13.4.1 Forensic Applications of MSP 498
13.4.2 Microspot Thin-Film Thickness 502
13.4.3 Graphene and Carbon Nanotubes 504
13.4.4 Heavy Element Chemistry 505
13.4.5 Lignins 506
13.4.6 Cellulosic Nanomaterials 507
13.4.7 The MSP Analysis of Gemstones and Glass Fragments 508
13.4.8 Identifying Protein Crystals 512
13.4.9 Gold and Silver Nanomaterials 514
13.5 Conclusion 515
References 516

13.1 INTRODUCTION

Instrumental spectrophotometry became commercially available about 1940 with the introduction of Arnold Beckman's DU spectrophotometer (Model D Ultraviolet). That instrument and its early competitors' machines were designed for manual selection of analytical wavelength and visual reading and logging of the instrument's analog output values. As time progressed,

Experimental Methods in the Physical Sciences, Vol. 46. http://dx.doi.org/10.1016/B978-0-12-386022-4.00013-3

wavelength scanning versus direct graphing of the photometer's output became the norm. Instrument complexity continued to increase to include fluorescence analysis and the simultaneous scanning of both illumination input and output wavelengths, but direct graphical outputs were still the norm, and instruments grew to the size of large desks.

About 1950, a microscope was interfaced with an existing, direct reading spectrophotometer, creating the first microspectrophotometer (MSP). In 1959, another MSP system was invented the allowed the use of dual (sample vs. reference) beams. Microscopy had the potential to greatly expand the application of spectrophotometry as a scientific tool, but commercial development and wide use of the technique grew slowly. Narosi [1] provides some early history of MSP.

The advent of semiconductors, integrated circuits, and the personal computer led to a rapid reduction in spectrophotometer instrument size and complexity and added the ability to store and mathematically evaluate spectrophotometric data. These developments also removed the need for dual beam optics and facilitated the application of spectroscopy to microscope systems. Interestingly, the semiconductor industry itself became an early driving force in the use of MSP, with its need for accurate measurements of microscopic thin films and features. This synergy continues today.

Commercial MSPs appeared as new microscopes were designed to interface with monochromators, and multiple types of illumination and new, broad wavelength range optics were adapted to existing microscopes. At the same time, scientists in such diverse fields as physics, biology, chemistry, and engineering became more interested in using the MSP to analyze ever smaller and thinner samples in addition to new sample types. Ultraviolet (UV), visible (vis), and near-infrared (NIR) illumination was a relatively standard part of the existing microscopy instrument design package. This capability added to the interest in the MSP in large part due to UV–vis–NIR spectrophotometry filling an energy gap in analytical analysis, between that of infrared and electron or X-ray excitation.

The UV–vis–NIR energy range of roughly 1–5 eV encompasses the bond energies between the atoms of a majority of compounds and crystal structures, and of the interfacial bonds between atomic and molecular structures. The excitation of samples within this energy range and the resultant absorption, fluorescence, or phosphorescence produced, as well as the possible time or angular dependence of energy absorption, offer insights into a host of chemical and physical properties within the samples.

13.2 MICROSPECTROPHOTOMETER INSTRUMENT DESIGN AND CONSTRUCTION

The MSP is a purpose-designed instrument intended to measure the electronic spectra of either microscopic samples or microscopic specimen areas on much larger samples. As with spectrometers for macroscale samples, there are many types of MSPs. MSPs have been built for infrared, Raman, cathodoluminescence, and even nuclear magnetic resonance spectroscopy. The topic of this section is the MSP designed to operate from the deep ultraviolet, through the visible and into the NIR region. The modern MSP can be configured to collect electronic spectra by transmission or absorbance through the sample, emission from the sample, or reflectance from the sample. Additionally, it can collect all of these spectra with the same system. A typical system will also allow the user to acquire images of the sample being measured and many systems offer additional features such as the ability to map the spectral response from a surface with micron-scale spatial resolution. With the addition of specialized algorithms, the MSP can also be used to measure the thickness of thin films or act as a colorimeter for microscopic samples.

The MSP is a hyphenated instrument that integrates a light microscope with a UV–vis–NIR spectrophotometer (see Fig. 13.1). The spectrophotometer measures the intensity of various wavelengths of light from the ultraviolet through the visible and NIR regions. The microscope is an imaging tool designed to enlarge an image of microscopic objects so that they may be studied easily. Integrated as the UV–vis–NIR MSP, such a device can acquire absorbance, reflectance, and emission spectra with sampling areas on the micron scale.

13.2.1 The Microscope

The microscope component of an MSP system can be represented by a wide array of microscope platforms, from discrete component optical bench systems to phase contrast, polarized light, UV–vis, or fluorescence systems. They can be designed to provide a single analytical measurement type such as spectral reflectance or multiple types of measurement. The former has its obvious limitations, while the multiple measurement instruments might sacrifice performance in some measurement modes in order to provide application versatility. This wide selection o f instruments and options demands that the users be well informed in their microscopy knowledge and performance needs in order to select the best instrument for their application(s).

The wavelength region(s) of interest as well as the type of samples to be measured will also have a profound effect on the selection of illumination type and optical path. The most versatile instruments will feature multiple illumination sources, ranging from tungsten filament to a variety of arc lamps, and will also probably employ Schwarzschild-type objectives and metalized front surfaced mirrors to avoid illuminant energy attenuation, chromatic aberrations, and polarization of the illumination and sample light beams. More specialized or limited systems might perform well with quartz optics, normal optical glass, or fluoride beam path components.

The illumination system will typically need to utilize stabilized power supplies: the short-term stability and long-term drift specifications of lamp

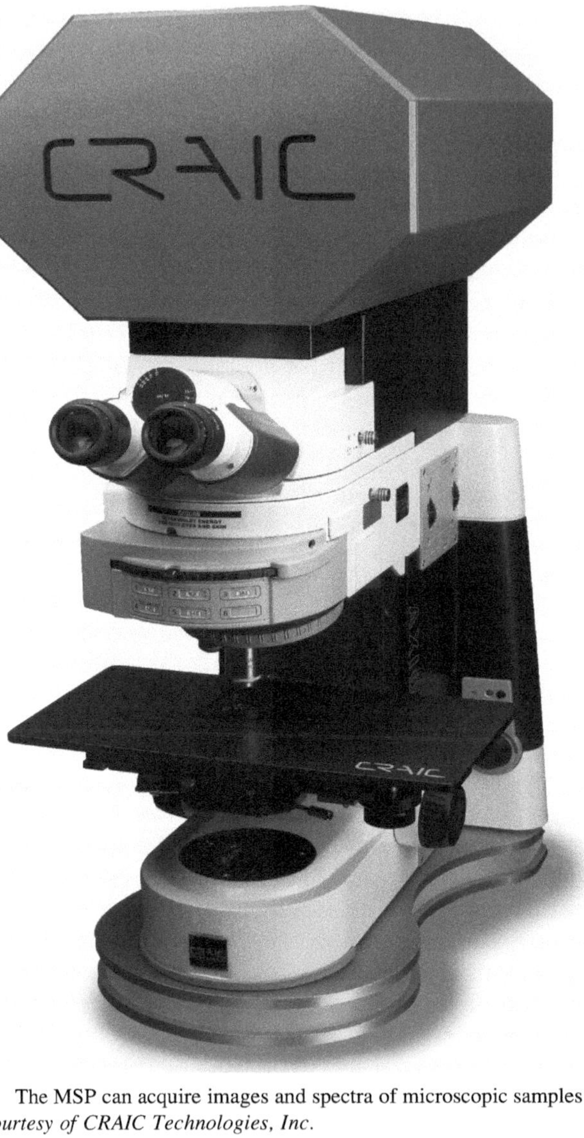

FIGURE 13.1 The MSP can acquire images and spectra of microscopic samples in the UV–vis–NIR. *Image courtesy of CRAIC Technologies, Inc.*

voltage will be a significant point of system selection. Although the eye can readily adapt to small voltage-based shifts or drifts in intensity and changes in color balance, modern detectors require very stable illumination to function accurately and precisely. A typical, high-quality microscope for normal optical or video monitor viewing will generally not be adequate for this task. Simply adding a spectrophotometer module to an existing optical microscope will generally not provide the optimum spectral performance.

The highest performance MSPs utilize custom-built microscopes that are optimized for the application. In order to cover such a broad spectral range while maintaining good image and spectral quality, the microscope must be integrated with the spectrophotometer. Standard optical microscopes, if used, limit the usable spectral range of the MSP as they only cover a small portion of the visible region due to the materials used for the optics as well as the light sources themselves.

13.2.2 The Spectrophotometer

The primary challenge in the original design of the MSP revolved around reducing photometer noise and increasing photometer sensitivity, or more specifically, increasing signal-to-noise (S/N) ratio. This was a profound challenge with the first spectral scanning, direct-recording systems. The most sensitive, broad-spectrum detector available at the time was the photomultiplier tube (PMT). PMT driver/amplifier electronics were well developed with stable drivers and low noise, analog amplification designs. The major drawback of the PMT was in its relatively slow dynamic response and its need for a scanning monochromator in the MSP system.

The second challenge was to provide a compact, precise, and accurate, broad-spectrum monochromator. This challenge and its solution revolved around the dispersive element of the system. Prismatic dispersive elements limited the spectral range and energy throughput of the monochromator. Ruled gratings, and later holographic gratings, resolved these shortcomings but still required relatively bulky and slow moving servo-mechanical drive systems and associated power supplies. These moving parts also increased maintenance, precision, and reliability challenges, although well-designed systems operated well; had extended dynamic range; and were precise, accurate, and robust. Their primary limitations were that they were physically large, had no NIR capability, and had slow acquisition rates that needed several minutes to acquire a single absorbance spectrum over a limited spectral range. Despite these limitations, PMT-based systems with scanning monochromators are suitable for many applications.

The advent of large area, multiple element semiconductor detectors, such as photodiode arrays and later charge-coupled devices (CCDs), allowed MSP designers to eliminate the mechanical scanning component of the system, to extend the detection range to longer wavelengths, and to reduce acquisition times to a fraction of a second. These systems still required a dispersive grating and an associated entrance slit but had no exit slit and recorded the entire spectrum as a unit divided across the number of detector elements. The resolution of the semiconductor detector-based systems is controlled by a combination of physical detector length, the number of detector elements across that length, and the distance of the detector from the entrance slit and grating. As distance increases, detector length and associated detector elements can be increased to improve resolution but light intensity decreases and the S/N ratio can degrade. Currently, CCDs have become the norm for MSP detectors in the UV–vis region, while InGaAs arrays cover the NIR region.

Once again, detector noise is a design challenge due to the inherent thermal-electron noise in the detector elements. With the current power of computer hardware and software, the ability to collect and average multiple spectra in subsecond times and to also decrease the detector element sampling rate can go a long way toward increasing S/N. But noise can still be a problem at the extremes of the illuminating optical system's spectrum, where intensity approaches zero. The S/N can then be further improved by applying thermoelectric (Peltier) cooling to the detector element. Detector cooling is especially critical when using InGaAs arrays for NIR microspectroscopy.

Today, the useful short wavelength limit for MSPs is approximately 200 nm ($\sim$50,000 cm^{-1}) and the long wavelength limit is approximately 2100 nm ($\sim$4800 cm^{-1}), with an associated dynamic range of about three absorbance units (A) or a minimum transmittance (T) of about 0.1%T. This dynamic range can be increased at the expense of narrowing the scanned spectral region to better optimize the MSP for that region.

For optimum MSP performance, the spectrophotometer should also be purpose-built for microspectroscopy. It requires a high dynamic range since users commonly switch between different types of measurements requiring differing levels of sensitivity: changing from transmittance or reflectance microspectroscopy to fluorescence microspectroscopy in order to measure different types of spectra of the same sample for example. This allows the user to obtain different types of spectra from exactly the same location on the sample with the same instrument.

It should be noted that detectors, illumination systems, microscopes, and optical systems will continue to evolve and change as new technology and new user demands develop. As an example, where sample area selection and microscope viewing were once done through the microscope oculars, they are now generally selected and viewed on the computer screen. Though convenient, this does not remove the need to use the microscope's optical viewing system to align and focus a number of microscope components, from the illuminating lamp or arc, to apertures and to the various objective elements.

13.2.3 Computer Interface and Software Control

The most dramatic changes have occurred in the controlling hardware and software. The very first commercial systems utilized purpose-built computers and had limited data analysis and storage capabilities. With the advent of the personal computer and graphical operating systems, the MSP has changed in terms of both operation and capabilities. The modern MSP uses a personal computer running the latest operating systems. Sophisticated software packages are designed to provide integrated control of the instrument, including running all facets of the spectrophotometer, imaging systems, and the microscope. Modern software can even run a fully automated MSP in which all aspects of the measurements are computer controlled. This includes mapping features in which spectral measurements can be acquired of many predefined areas over a large sample surface. A four-dimensional data cube can then be created including the *X* and *Y* spatial axes of the sample surface in addition to the axes for wavelength range and intensity (or absorbance or reflectance depending upon the type of measurement). This degree of computer control can also have its drawbacks, in that a system can produce apparently useful data, even when it is not running properly or is malfunctioning. There is no substitute for a user having a deep knowledge of microscopy, their instrument, and their samples.

Modern software also includes many sophisticated data analysis features that can add to the capabilities of the MSP. This includes basic arithmetic to sophisticated statistical analysis of the data to the creation and manipulation of databases. Software can also add capabilities such as the ability to perform colorimetric measurements and to determine the thickness of multilayer films of microscopic sampling areas. This latter capability is especially useful in the semiconductor industry in which film thickness is measured at many locations across a wafer and then a map of the thickness is generated for the wafer with high spatial resolution.

13.2.4 MSP System Capabilities

Macro (optically unaided, or "naked eye") spectrophotometry systems have a long history of use in virtually all areas of the color control industry, from paint and pigments to fabrics and dyes, stains, inks, color television, and video monitors; plastics; and illuminating lamps to paper manufacturing. Some of these are diffuse reflectance spectrophotometry systems that utilize an integrating sphere and use sample sizes up to 25–50 mm in diameter.

Some macro- and microsystem designs employ selectable or continuously variable goniometer-based mounting assemblies for their illuminators and detector assemblies, to evaluate the angular dependence of spectrophotometric data. (The addition of a single condensing lens to an optical imaging beam path can constitute a "micro"assembly.) As an example, the angular dependence of automotive paint surface color has become an important measurement, as pigment companies produce new products, such as holographic pigments, that produce color "flop" or change with the viewing angle. Measuring and controlling this flop effect and the application of these "effect" pigmented paint films are vital to an industry that might assemble a vehicle from prepainted parts produced in different factories. The automotive repair industry also relies on such measurements as repair shops shift to buying prepainted parts in order to reduce their use of paint booths. This practice is especially true in shops that customize new cars in order to satisfy buyers at new car dealerships. Pigment manufacturers also need to measure the interference colors of pigments made by layering one or more thin dielectric films on an otherwise colorless substrate such as glass, mica, and aluminum.

Architectural glazing materials have also seen the need for thin-film analysis as new products utilize a wide range of interference effects to produce different window colors or to provide environmental protection from heat or ultraviolet light. The advent of large area, continuous process, sputter coaters have helped make these products economical to produce.

Each of these aforementioned applications utilizes either diffuse or spectral reflectance measurements, but MSP systems also offer an array of transmittance and emission-based measuring setups. These formats allow the analysis of gemstones, glass, thin films, light-emitting diodes, liquid crystal displays, biotechnological devices, and advanced materials for nanotechnology and much more.

13.3 USING THE MSP SYSTEM

Integration of the major components of the MSP (the spectrophotometer, microscope, and software) is critically important. While the spectrophotometer and microscope are optimized for microspectroscopy, integrating the two seamlessly enables them to work together and yield the highest quality spectra. The microscope–spectrophotometer interface must meet several design challenges. Most importantly, the interface must direct the electromagnetic energy collected by the microscope from the sample into the spectrophotometer. The user also needs to visualize the sample measurement area as part of a larger field of view. This is done by having the entrance aperture of the spectrophotometer in the same focal plane as the sample image. The sample stage is then moved until the image of the entrance aperture is over the sampling area. In Fig. 13.2, the black square in the center of the image is the entrance aperture of the spectrophotometer. The imaging and the spectroscopy are carried out simultaneously to make the MSP easier to use.

Figure 13.3 is a simple optical diagram of an MSP configured for absorbance microspectroscopy. In the figure, the microscope optics focus light onto the sample. Some photons will be absorbed by the molecules in the sample and the remaining light will be collected by the microscope objective after being transmitted through the sample. From the objective, the light is focused onto the entrance aperture of the spectrophotometer. As this aperture has a mirrored surface, most of the light is reflected from the entrance aperture surface onto the camera. Thus, the spectrophotometer entrance aperture is imaged so that it appears as a black square on the sample (see Fig. 13.2). This allows for easy usage of the MSP. The light that passes through the entrance aperture then passes into the spectrophotometer where a spectrum is measured.

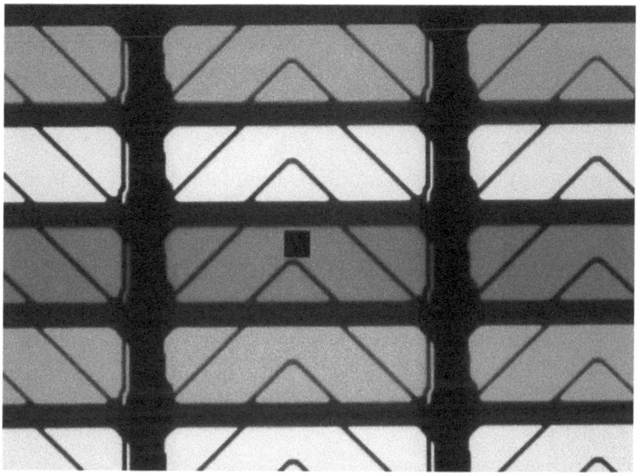

FIGURE 13.2 Image of a liquid crystal display with an MSP. The black square in the center of the image is the entrance aperture for the spectrophotometer and is on the same focal plane as the sample. The user simply moves the sample stage, so the black square is superimposed on the area to be measured.

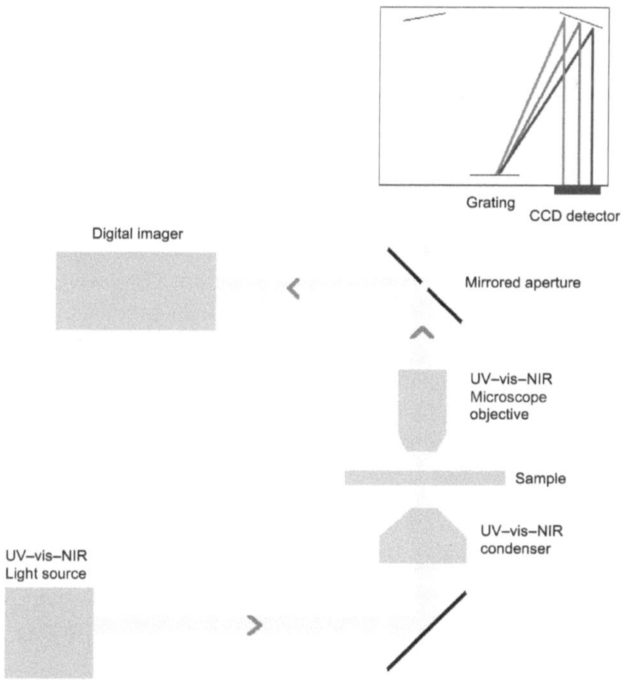

FIGURE 13.3 Schematic showing the optical path for an MSP configured for absorbance microspectroscopy and imaging. *Image courtesy of CRAIC Technologies, Inc.*

The microscope can be configured with different illumination schemes depending upon the type of experiment to be performed. Figure 13.3 shows a diagram of an MSP configured to measure the transmission spectrum of a microscopic sample. Similarly, incident white light illumination allows for reflectance microspectroscopy from the deep UV to the near IR. Incident illumination at wavelengths in the excitation range of photoluminescent samples can also be used for fluorescence or phosphorescence microspectroscopy. Additional features can also be incorporated in the MSP to add more capabilities. The most common one is the ability to measure the thickness of microspot areas of thin films for semiconductor metrology.

13.4 CURRENT APPLICATIONS OF MSP IN INDUSTRY AND RESEARCH

13.4.1 Forensic Applications of MSP

The prosaic assumption is that forensic science is only applied to criminal cases and investigations. In a more inclusive definition, forensic science is the application of science to the resolution of legal questions or disputes. This definition yields a wide array of forensic science applications from patent challenges, to environmental regulation violations, paternity disputes, tort law, or claims of civil damages, and of course criminal cases of assault, burglary, homicide, forgery, poisoning, explosives device evaluations, and hit and run accident investigations. MSP can be applied to any of these cases that involve the presence or transfer of microscopic colored materials or materials with UV absorbers.

Paint transfers occur in a variety of circumstances, including accidental over spraying during commercial painting, accidental or intentional vehicular impacts with other vehicles or structures, ingestion of paints by children, criminal damage or assault using painted weapons, bullets passing through or glancing off painted surfaces, and pigmented inks used in producing forged documents or currency.

Below is an example of paint fragment analysis that might be used to produce distinguishing MSP spectra, either by diffuse reflectance or by transmission analysis. Figure 13.4 shows opposite sides of an automotive paint film, mounted on a scanning electron microscope sample stub. Both the red color or "base" coat of the paint film (on the right) and the bottom primer coat (on the left) can provide spectra for use in comparative analysis. Primer spectra can be useful in that they are often more variable than the overlaying base coat and can provide discriminating data between different samples that are otherwise intended to appear to be the same color. But, whether two colors are the "same" is a question of visual perception.

Sometimes, MSP information can also be derived from thin sections of paint films, such as the red paint film shown in Fig. 13.5, which is a section

FIGURE 13.4 Typical paint film sample (scale: 1 mm per division).

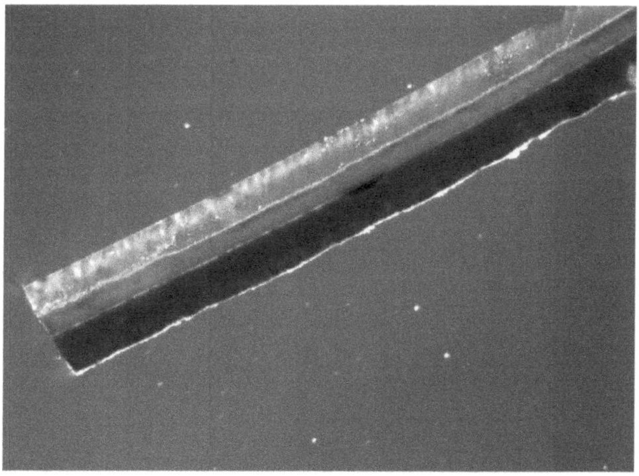

FIGURE 13.5 Cross section, ~100×.

of the paint film shown in Fig. 13.4. The section shows that the film includes a clear coat, red color coat, and at least one primer layer that appears opaque. In this case, the diffuse reflectance spectrum of the primer surface would be more useful for comparison than the transmission spectrum of the cross-sectioned primer spectrum.

A similar situation is shown in Fig. 13.6 which is a section of a white paint fragment. Clear coat, color coat, primer, and electro-coat layers are shown in this section, but it is not obvious that the color coat is white.

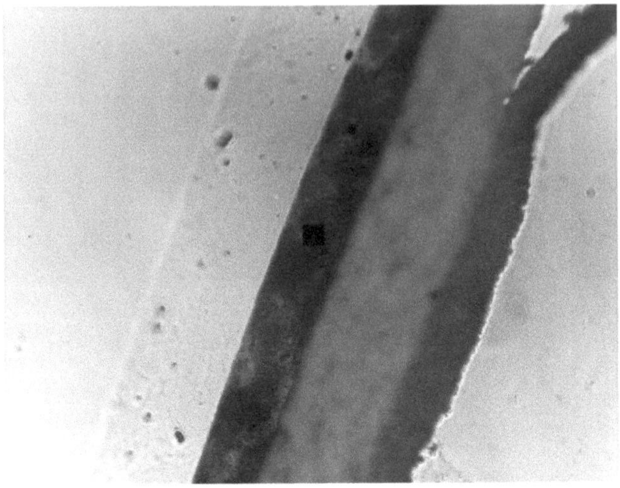

FIGURE 13.6 Cross section, ~200×.

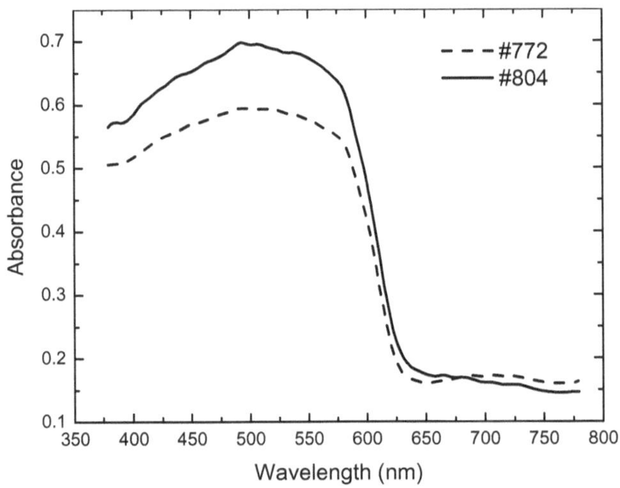

FIGURE 13.7 Comparison of averaged transmission MSP spectra of PR-4's #772 and #804.

Figure 13.7 exhibits a typical example of the importance of careful MSP validation and operation, in that they show the transmission (plotted in absorbance units) curves of two red (Chrysler Primrose Red, color code PR-4) paint films from two Dodge Durango vehicles, painted a year apart, at the same factory. The vehicles were intended to be the same color, and indeed, they cannot be distinguished by the naked eye under daylight illumination.

Samples of the two paint films were mounted, side by side, in the same polymer sectioning block and microtomed together, to insure that their comparative thicknesses were the same. Multiple spectra were run at different points along each color layer and each sample set was averaged. A comparison of the two averaged sets clearly shows that one sample has lower overall absorption in the blue-to-yellow region and higher absorption in the red region of the spectrum with a crossover at about 680 nm.

These samples are metameric and that point would be essential information in any forensic comparison of the paints. The manufacturer was obviously happy with both paint films. This is what makes MSP so essential in forensic color analysis, in order to take advantage of situations where it would be economically disadvantageous for a manufacturer to control paint colors more closely than the eye can perceive under normal viewing conditions. A general guideline to the use of MSP in forensic paint examinations is given by ASTM E2808-11 [2].

Figure 13.8 gives an example of distinguishing the deep UV absorbance of two different textile fibers that appear to be the same color. Although the visible region of the spectra will not distinguish the two fibers, the deep UV region shows a distinct difference between them. That difference is characteristic of different UV absorbers or fluorescent optical brighteners that are added to fiber polymer systems to either protect the fabrics from sunlight deterioration or enhance their appearance after washing.

The MSP evaluation of gemstones and colored glass samples is shown in Section 13.4.7, and a general discussion of forensic MSP applications, including dyes, inks, paints, and fibers, is given by Eyring [3]. Some references relating to the application of MSP to fiber analysis are given by other authors [4–7]. It is obvious that MSP has been recognized as a useful forensic tool for quite some time.

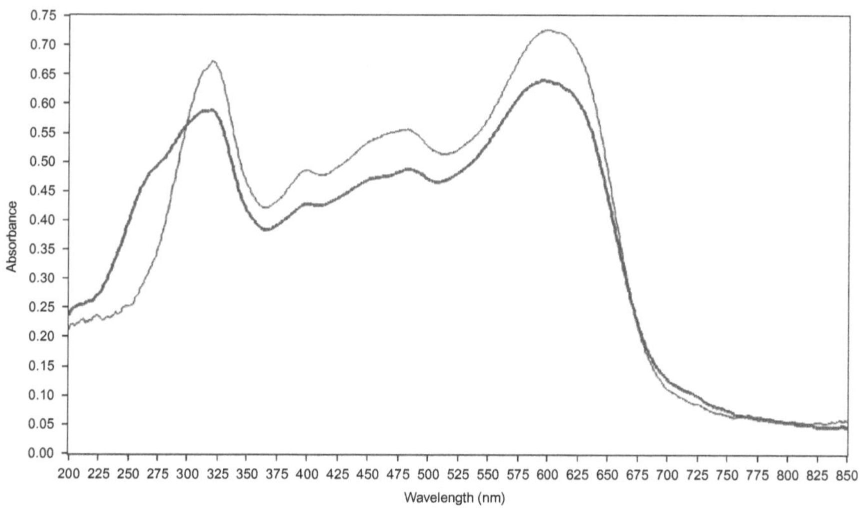

FIGURE 13.8 Fiber "B" (blue; dark gray in print) versus fiber "C" (red; light gray in print) UV–vis–NIR absorbance spectra.

13.4.2 Microspot Thin-Film Thickness

Thin films range from atomic or molecular monolayers to microns thick. They can form naturally but are also manufactured for use in many different applications. An example of a naturally forming film is the oxide layer that forms on aluminum and lends it resistance to further oxidation. Aluminum components are therefore resistant to the degradation caused by rust seen in components made out of iron. Interestingly, one of the early applications of thin-film coatings was the deposition of zinc on iron and steel components so that they were rust resistant.

Two common industrial applications of thin films are their usage in building semiconductors and as optical coatings. With semiconductors, materials are deposited layer by layer and circuits are etched in extensive processes needed for the construction of integrated circuits [8]. By using materials with different properties in each layer, such as conductors, semiconductors, and insulators, complex integrated circuits can be built. However, the thickness of each of these layers must be known accurately in order for the devices to yield consistent results. Optic devices are also created using thin films. In optics, thin films are used as protective layers and antireflective and reflective coatings [9]. Again, multiple layers of different materials are combined to yield various optical affects. As such, the thickness of such coatings must be closely monitored in order to yield the desired results from the optical device.

Optical interference patterns are a commonly used method to determine the thickness of thin films [10]. See Section 2.6.3 for more detail on the theory of these optical phenomena. MSPs can measure the optical interference pattern by illuminating the device with a broadband beam. The incident light is reflected from both the upper and lower boundaries of the film (see Fig. 13.9). When both rays are reflected, they interfere with one another in a manner that can be characterized by the Fresnel equations. This degree of interference depends upon the difference in phase of the two rays, a factor determined by the refractive index of the film material, the wavelength of the light, the angle of incidence, and the thickness of the film. With white light, the interference patterns appear as multicolored bands. The MSP measures the spectrum of light from the thin film (see Fig. 13.10) and, using variations of the Fresnel equations and knowledge of the film material, determines the thickness of the film by fitting the spectral data to a model of that film.

MSPs are commonly used for thin-film thickness measurements as they possess a number of advantages. First, they do not contact the device and therefore are less likely to contaminate it. Second, the measurements are performed rapidly and can even be done inline of a manufacturing process. Third, they can also be used to measure a wide variety of films with a large range of thicknesses. Fourth, these instruments can be configured to measure either the

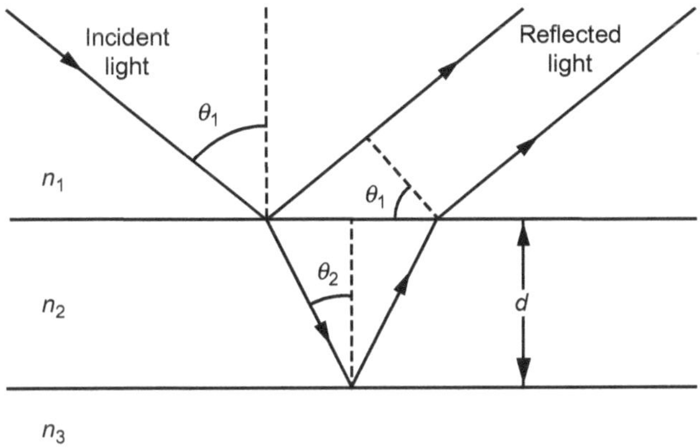

FIGURE 13.9 Image of light rays reflecting off top and bottom boundary layers of a thin film. The substrate is opaque in this diagram.

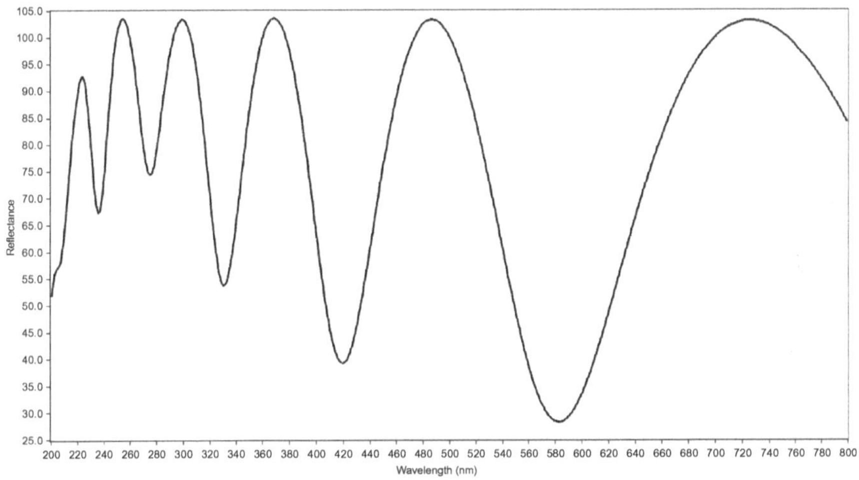

FIGURE 13.10 Interference spectrum from a thin film measured with an MSP. The film is 502-nm thick SiO_2 on a silicon substrate. This spectrum is not normalized but is relative to silicon which was used as the reference material.

reflected or transmitted spectral interference so that film thickness measurements can be made on both opaque and transparent substrates. Fifth, they can measure the thickness of stacks of multiple films. Finally, MSPs can measure the thickness of microscopic sample areas. This allows them to be used as metrology devices for microscopic circuits as well as to create film thickness maps with high spatial resolution to determine the film thickness variation across a device.

13.4.3 Graphene and Carbon Nanotubes

Carbon has many forms. Graphene is a lattice of carbon atoms with each unit cell in a hexagonal pattern. It is essentially a two-dimensional sheet of carbon atoms that is one atom thick. Carbon nanotubes have a three-dimensional structure that can be described as cylindrical carbon molecules with a small diameter relative to the length of the tube. Both materials exhibit unique physical, optical, and electrical features, though their characteristics differ from one another and those of other forms of carbon. These materials can also be engineered to have specific properties and thus represent another type of metamaterial.

MSPs are used to study the optical characteristics of graphene. Characterizing this material is still ongoing [11] as is development of devices and applications that use it [12]. Graphene structures are also being developed that have variable optical characteristics [13]. By studying the optical spectra of these materials under different conditions, such as temperature, the electronic and physical structures of these materials may be better understood. Such work is also leading to the engineering of new graphene materials because of an enhanced understanding of their physics and chemistry (Fig. 13.11).

Carbon nanotubes have dramatically different properties [14,15] to that of graphene and other allotropes of carbon. This cylindrical form of carbon is finding use in many different applications, and quality control testing of it can be done by optical spectroscopy. Minor changes to the structure of carbon nanotubes yield detectable changes in the optical spectrum, hence the utility of microspectroscopy in production of these materials [16]. The optical properties of the carbon nanotubes themselves are also of interest [17]. For example, MSPs are used to study the optical switching performance of carbon nanotubes and characterize their novel nonlinear optical properties (Fig. 13.12).

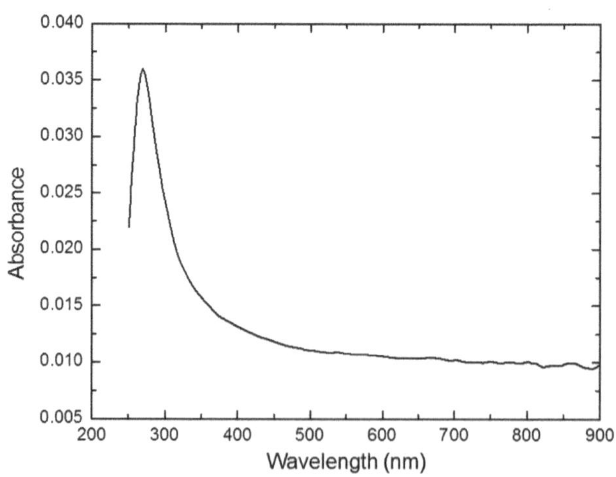

FIGURE 13.11 Absorbance spectrum of a single layer graphene film on a quartz substrate.

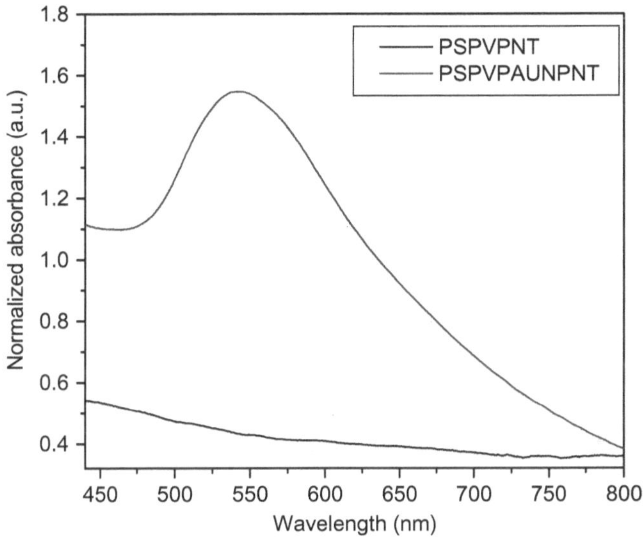

FIGURE 13.12 Absorbance microscopic spectra of nanotubes formed both with and without gold nanoparticles. *Image courtesy of Vladimir V. Tsukruk, Georgia Institute of Technology.*

13.4.4 Heavy Element Chemistry

The chemistry and physical properties of heavy elements are quite unique and are receiving more attention as potential applications are developed. One fruitful field is in the research into actinide compounds. Since only thorium, uranium, and to a lesser extent plutonium are found in nature, much about these elements is still to be understood. As such, complexes with actinides are constantly being developed and utilized for different fields. For example, actinides are radioactive and have been used in everything from nuclear weapons to radioisotopic power supplies. Understanding the chemistry of solid-state actinide compounds is critical, and this is done by measuring the UV–vis–NIR absorption spectra and fluorescent emission spectra of single crystals. This yields information about the electronic states of these materials in various complexes. It also enhances understanding of the oxidation states of the actinides and on how actinides [18] form bonds with other atoms and differ from lanthanides in similar complexes [19].

Lanthanide complexes are also a topic of study, alongside the actinides, in that while their electronic structures are similar, their chemistry and physical properties can be quite different.

MSPs are an important tool for this research. This is because many of these compounds are synthesized as microscopic crystals and in very small amounts [20]. Any study of their solid-state electronic structure is therefore done with single crystals by absorption microspectroscopy with supplementary data coming from fluorescent emission microspectroscopy. Interestingly,

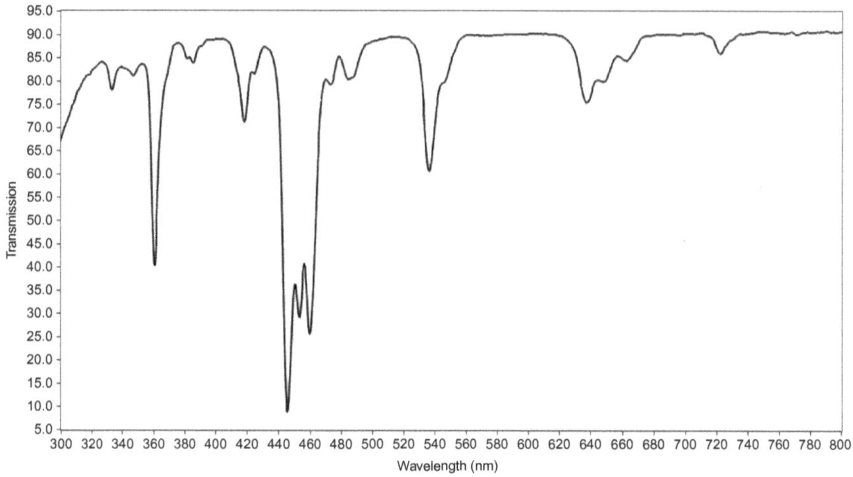

FIGURE 13.13 UV–vis spectrum of a holmium oxide glass standard taken with an MSP.

the electronic spectra of these materials yield complex spectra that can be termed as "fingerprint" spectra in that they can be used to readily identify these compounds. In fact, materials such as holmium oxide (see Fig. 13.13) are used to calibrate the wavelength axes of spectrophotometers and MSPs due to their complex electronic spectra that feature numerous peaks [21].

13.4.5 Lignins

Lignin is a biopolymer commonly found in the cell walls of wood [22]. Cellulosic fibers are embedded in a matrix of lignin to form the majority of the structure of wood [23]. Different types of wood are differentiated by the percentage volume of lignin as well as the types and orientations of this biopolymer. Also, the distribution of lignin within the cell walls aids in the determination of the type of wood as well as determining the physical properties of the material [24].

MSPs are used to study the properties of wood, lignin, and cellulose in a number of different manners. Both absorbance and fluorescent emission microspectroscopy are used to measure the distribution of lignin within the cellular walls. Because of the high spatial resolution of the instrument, cross sections of the woody material may be mounted on quartz slides and the distribution mapped within the section. This is mainly done with ultraviolet absorbance microspectroscopy as these materials absorb strongly in the UV region.

Fluorescence emission microspectroscopy has also been utilized to measure the orientation of cellulosic microfibrils in wood that has been subjected to rot [25]. One major advantage to using ultraviolet fluorescence in this study

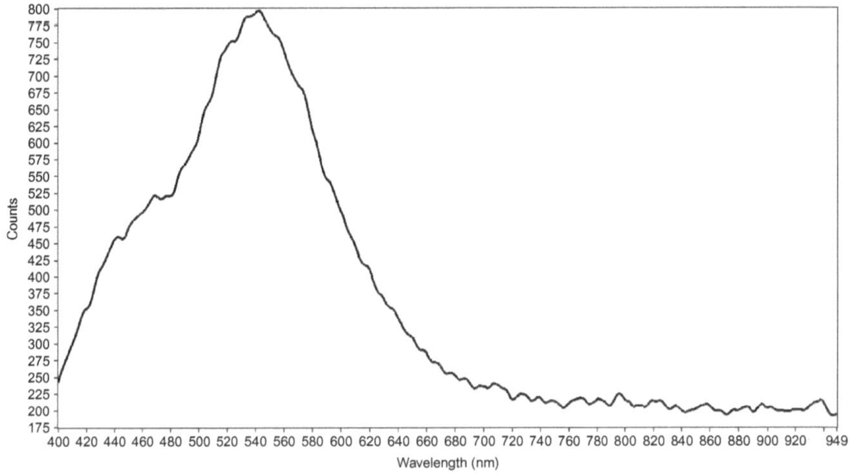

FIGURE 13.14 Fluorescence emission spectra of Tracheid Radial wall of Red Maple.

is that none of the samples needed to be stained and images of the materials could be simultaneously captured with the spectra. The variations in both the intensity and shape of the emission spectra can show differences in the lignin compositions of different types of fibers. These differences in the spectra of the different types of fibers can be supported by other histochemical tests (Fig. 13.14).

13.4.6 Cellulosic Nanomaterials

The demand for high-performance, engineered materials is steadily growing. So it is the demand for recyclable materials and green manufacturing processes. Substances that may potentially fit these requirements are cellulosic nanomaterials. Crystalline cellulose has mechanical properties similar to that of other reinforcement materials while having a reactive surface on which functional groups may be added to yield different surface properties [26]. Cellulosic nanomaterials can be made from readily available materials such as wood, plants, bacteria, and algae. However, different source materials yield different types and shapes of cellulosic nanomaterials all of which can be optimized for different properties by bioengineering. From these, cellulosic nanocrystals in the form of whiskers are obtained. Again, the actual shape of the whiskers is determined by the source material. Other structures are also found and research is being conducted into their physical properties.

Cellulosic nanomaterials in suspension act as liquid crystals (see Fig. 13.15). Much of the optical and physical characteristics of a liquid crystal are due to the asymmetric rod structure, and this is seen with cellulosic nanocrystals. Some of the unique structural features of these materials, such as the helical twist normal to the rod axis, can be the cause of features such as optical bandgaps which can be studied by MSPs. These materials also exhibit other optical phenomena such as birefringence which is not a property seen in common cellulose. The addition of functional groups can also change the optical properties of this material.

MSPs can be used in this field as with other nanomaterials. Absorption and emission microspectroscopy [27,28] can aid in studying their electronic structure (see Fig. 13.16). And as these materials can be optimized for various optical effects, the engineering processes can be monitored by microspectroscopy. The major advantages of microspectroscopy are that these compounds can be studied in both solid form and a suspension. Additionally, an MSP can analyze a single fiber in order to better characterize and tune its properties or can be used to understand the spectral variation across a sample with micron-scale spatial resolution.

13.4.7 The MSP Analysis of Gemstones and Glass Fragments

13.4.7.1 *Synthetic Versus Natural Gems*

Synthetic gem manufacturing has a long history of successes and failures as chemists have tried to replicate nature's rare beauty. As Kunz [29] stated in 1886, "and there is little reason to fear that the true ruby will ever lose the place it has occupied for so many centuries. These stones [man-made Geneva rubies] show the triumph of modern science in chemistry, it is true; and although some may be willing to have the easily attainable, there are others

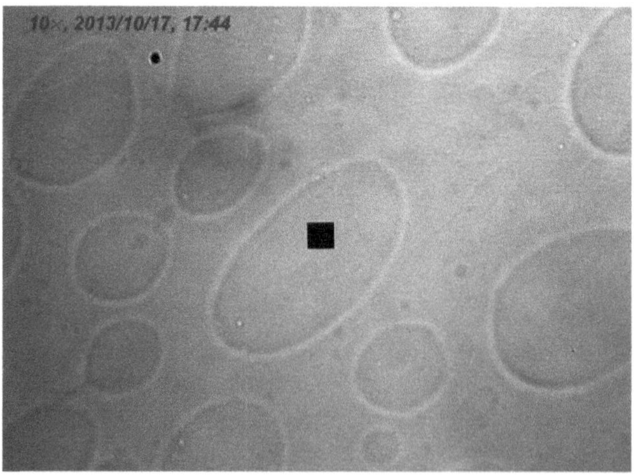

FIGURE 13.15 Image of cellulosic nanomaterials in an aqueous solution on a quartz slide under a quartz cover slip. The black square in the center is the entrance aperture of the MSP. The aperture is 12 μm × 12 μm.

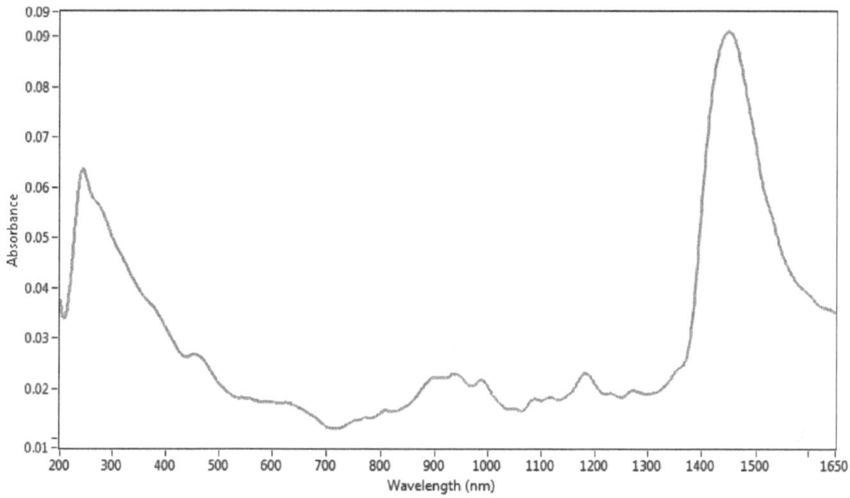

FIGURE 13.16 UV–vis–NIR microscopic spectra of nanocrystalline cellulose. Note the absorbance peaks both at 244 nm and at 1451 nm.

who will almost want, what the true ruby is becoming today, the unattainable. One will be nature's gem and the other the *gem made by man*." The rubies Kunz was describing were not of gem quality, and his concern was expressed long before rubies were the necessary heart of early lasers.

Since that time, synthetic gem manufacturing has been an ongoing research and commercial enterprise and is described in depth by Nassau [30], where he notes the first gem quality ruby being produced in 1902 by Verneuil [31]. It is important to note that these gems are described as "synthetic" or sometimes "artificial," and they are not described as "imitation" or "fake" until they are in the hands of a fraudulent gem dealer who claims they are nature's own.

Given that fraud is often a human occupation, the detection of fake gems has been a constant effort within the jewelry world. This leads to the development of the Chelsea Colour Filter for emeralds in 1934 that transmitted light in the 570 and 690 nm regions where true emeralds get their color and helped single out synthetic stones with different transmittance regions. Coincidentally, the Chelsea Filter offers insight into what several other minerals such as ruby and sapphire "are not" but they cannot confirm what a mineral "is." As synthetic gem compositions changed with new formulas and manufacturing processes, the Chelsea filter became less useful and additional filter types were developed, such as the Hanneman Gem Filter Set [32]. A splendid discussion of all of these filters is given by John Harris and the GemLab.UK [33], with examples of the filters' transmission spectra.

The advent of microspectrophotometry offered the obvious advantage of more accurate and precise gem color analysis but was hampered by the severe refraction effects at the air–gem interface. This problem can be solved by submerging a gem in a colorless liquid of relatively high refractive index such as microscope objective immersion oil of $n_d = 1.515$. The oil does not have to exactly match the n_d of the sample stone to be effective in removing detrimental refractions. A 3-mm deep immersion well is shown in Fig. 13.17 and wells of various depths are easily made. Parallel surfaces of the slide and the top edge of the well are essential. Normal bowl shaped well slides are not well suited for this application.

The resulting complete transmission spectra of immersed gems offer a rapid, nondestructive, and accurate comparison of natural and synthetic materials. A discussion of this gemstone color measurement application is given by Martin and Eyring [34]. Future work in this area will include the use of a combination of the MSP with Chelsea and other filter types in the beam path to more accurately determine and differentiate the filter induced color shifts on gems in both plane and polarized light, which are now only visually evaluated.

13.4.7.2 Discrimination of Similar Glass Fragment Colors

The possible origin of glass fragments can be an issue in many different scenarios, from accidental contaminants in food or drink to intentional malicious assaults, a hit and run accident or a fight in a bar. Unfortunately, in any of these scenarios, each individual glass fragment is a separate analytical item. One cannot assume that two pieces of colored glass lying on a barroom floor came from the same bottle. Proximity does not prove similarity or common origin.

A spectroscopist working on a glass fragment comparison is faced with first sorting out a possible pile of glass particles and color that can be difficult to visually judge due to the variety of particle shapes and thicknesses at hand. Microspectrophotometry can be the first step in sorting and analyzing a collection of fragments. Figure 13.18 shows the effect of viewing immersed fragments versus viewing them in air.

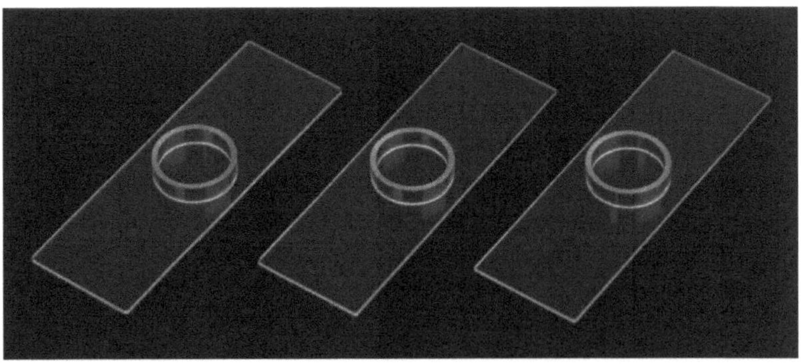

FIGURE 13.17 Quartz well slides with 3-mm well. *Image courtesy of CRAIC Technologies, Inc.*

FIGURE 13.18 Immersion effect.

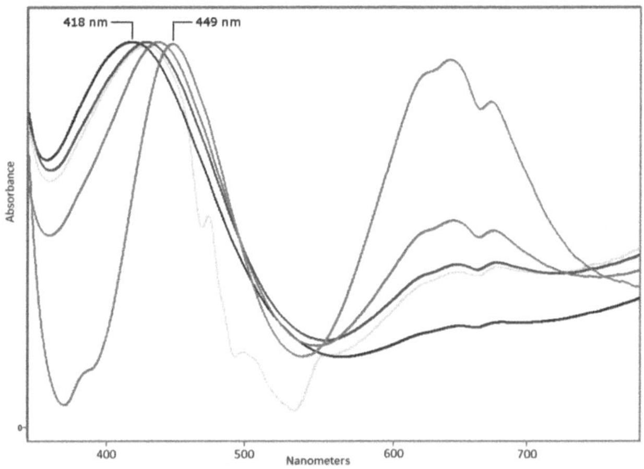

FIGURE 13.19 Average maxima of four absorbance groups of green glass samples.

The spectra in Fig. 13.19 are those of 25 different green glass bottle fragments. Microspectrophotometric analysis sorted them into four distinct groups for further analysis. In addition, although one of the spectra (green curve) falls in the second group from the left, its spectra are quite different from the average of the group. This type of variation offers even more differentiation of fragments.

A list of the samples and their maximum absorbance values is shown in Fig. 13.20, which illustrates the roughly 10 nm difference in absorbance maxima shown in Fig. 13.19.

A detailed discussion of this work and the associated refractive indices and elemental compositions of the green glass samples will appear in the journal *The Microscope* [35] where it is in prepublication review as of this writing.

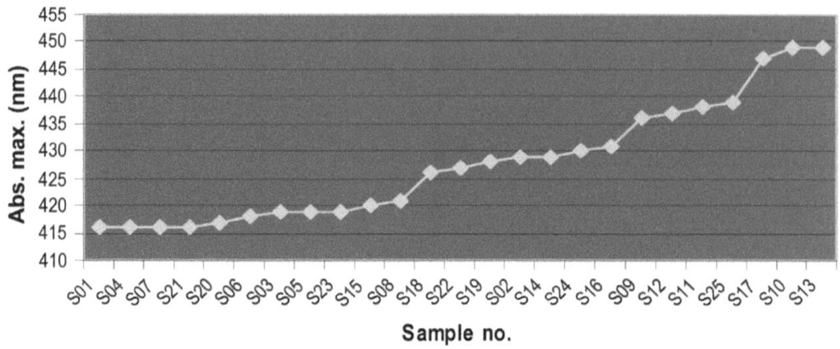

FIGURE 13.20 Wavelength maximum MSP absorbance values of green glass samples.

13.4.8 Identifying Protein Crystals

Crystallization of biological macromolecules has helped further our understanding of their three-dimensional structures, the way they interact with one another, and the mechanism behind their function [36]. The analysis of protein crystals is a critical aspect of this research. However, while the techniques to grow and analyze the crystalline structure are mature, rapid means to locate and identify the crystals themselves are problematic [37]. Two types of microspectroscopy have been used to locate, identify, and qualify protein crystals [38] prior to X-ray crystallographic analysis. This ability to isolate and qualify the protein crystals prior to X-ray analysis has the potential to greatly speed up research in this field.

Protein can contain tryptophan residues. Tryptophan strongly absorbs light at 280 nm while emitting at approximately 353 nm by what is known as "intrinsic protein fluorescence." Other residues in protein also absorb in this region but not nearly as strongly. A common method of protein crystal growth utilizes evaporative diffusion in which a salt-protein solution is the source material for any crystals that form. However, a drawback to this method is that salt crystals also form and they cannot be readily differentiated from any protein crystals. To date, most qualifying techniques have yielded ambiguous results and are often destructive. However, by using an MSP, one can readily distinguish and qualify the protein crystal prior to preparing it for X-ray crystallography.

Fluorescence microspectroscopy was the first technique to be used to qualify protein crystals rapidly and accurately. Crystalline material can be analyzed either in solution or after having been recovered. The MSP excites the crystal with 280 nm light and then observes the emission at 353 nm. If there is no emission, then the crystal is salt. This test is accurate but care must be taken not to denature the crystal with extended exposure to ultraviolet light which can occur since the emission is quite weak. Similarly, there are many mechanisms which can cause quenching of the fluorescent emission.

Another technique is to directly observe the absorbance of the protein crystal at 280 nm. As salt crystals do not contain tryptophan, they do not yield an absorbance spectrum when analyzed by this method (see Figs. 13.21 and 13.22). The advantages of this method are that it requires a much shorter exposure time to ultraviolet light and it is much faster. It is also quite effective in locating high-purity protein crystals. However, it does require the use of UV transparent cells and mounting materials in order to transmit UV light for microspectroscopy.

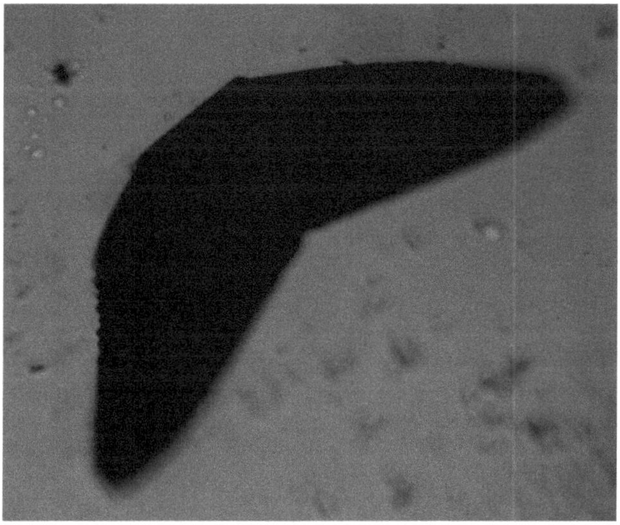

FIGURE 13.21 Ultraviolet absorbance image of a protein crystal.

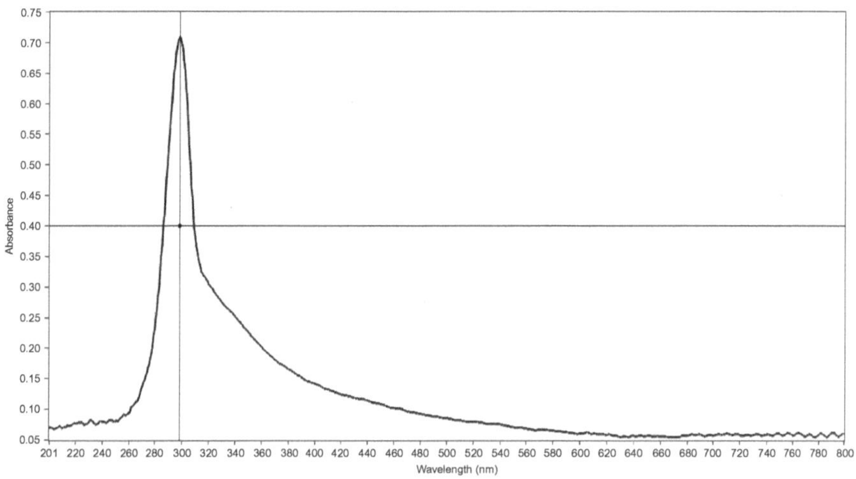

FIGURE 13.22 Ultraviolet–visible absorbance spectrum of a single protein crystal.

13.4.9 Gold and Silver Nanomaterials

In the last decade, there has been a large amount of work done to develop and study gold and silver nanostructures that exhibit plasmonic characteristics [39,40]. These metamaterials can be tuned to exhibit specific optical characteristics under well-defined conditions. As such, they are finding applications in highly sensitive detectors used in biomedicine, biology, physics, and chemistry. Further advances and fine-tuning of these engineered materials have the potential to enhance them to even the single-molecule detection level. They could then be used in the treatment of many diseases, detection of biological and chemical weapons, as well as advancing our understanding of the physics of metamaterials and plasmonics.

Gold and silver nanoparticles have been developed that exhibit surface plasmon resonance (see Fig. 13.23). Surface plasmon resonance occurs when light is reflected off these metallic nanoparticles. The result is a novel spectral response that is characteristic of the size, shape, and composition of the metallic nanoparticles. As surface plasmon resonance is a boundary effect, changes in the surface of the gold or silver nanomaterials, such as the adsorption of a target molecule, cause a change in the spectral characteristics of these nanostructures. Thus, surface plasmon resonance-based sensors can be very sensitive to minute changes in their environment.

Many types and shapes of gold and silver nanoparticles [41] and films have been developed in order to better understand the physics and chemistry of these unique materials. Devices have also been built that either exhibit surface plasmon resonance [42], such as gold nanowells, or contain nanoparticles that do, such as lab-on-a-chip sensors (see Fig. 13.24). MSPs are used for much of this research and development for their ability to study devices with

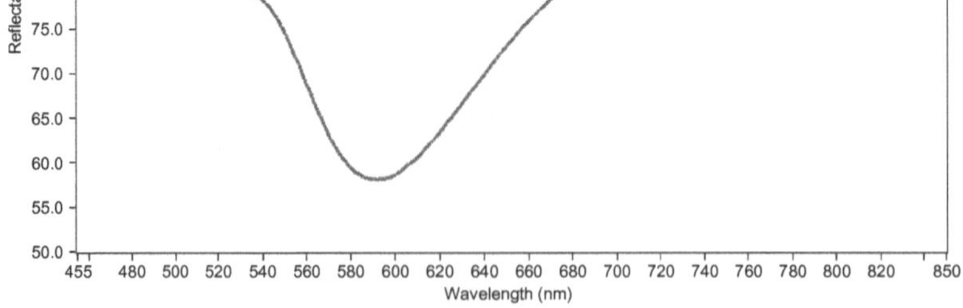

FIGURE 13.23 Reflectance spectrum of gold nanoparticles deposited on glass surface. This spectrum is not normalized but is relative to glass which was used as the reference material.

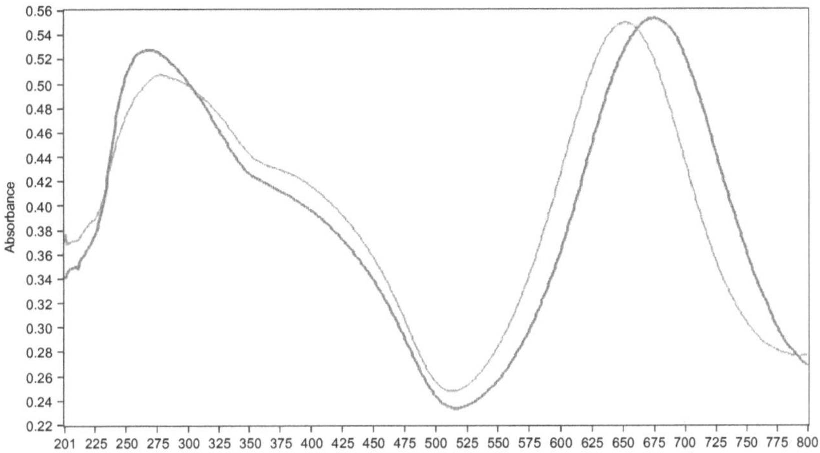

FIGURE 13.24 Changes in optical characteristics of surface plasmon resonance-based liquid sensor film in wet (purple; dark gray in print) and dry (green; light gray in print) conditions.

a high spatial resolution. An MSP can measure the reflectance or transmission spectrum of individual sensors or it can be used to spectrally characterize the plasmonic variations across a surface containing nanowells. MSPs are also used in the creation of nanomaterials and devices in order to ensure that the device patterns and nanoparticles shapes and sizes conform to predicted designs and optical characteristics [43].

13.5 CONCLUSION

The MSP is a device that combines the magnification capabilities of the optical microscope with the abilities of a spectrophotometer. Such instruments are capable of measuring absorbance and reflectance spectra from the deep UV through the visible and into the NIR regions. They also have the capabilities to measure emission and polarization spectra in addition to acquiring the corresponding UV, visible, and NIR images.

These combined capabilities mean that MSPs are used in both laboratory and production environments for many different fields. These include the analysis of trace evidence in a forensic laboratory, grading the energy content of coal to measuring the thickness of thin films on a semiconductor production line. In research, the modern MSP is used to study metamaterials, to devise methods of storing radioactive waste, and to develop the next generation of microsensors. A hybrid instrument, the UV–vis–NIR MSP, serves to open the micro-world to spectroscopic analysis.

REFERENCES

[1] F.I. Harosi, Microspectrophotometry: the technique and some of its pitfalls, in: Vision in Fishes, New Approaches in Research, NATO Advanced Study Institutes Series, vol. 1, Springer-Verlag, USA, 1975, pp. 43–54.

[2] E2808-11, Standard guide for microspectrophotometry and color measurement in forensic paint analysis, ASTM Book of Standards Volume: 14.02, 2013.

[3] M.B. Eyring, Visible microscopical spectrophotometry in the forensic sciences, in: R. Saferstein (Ed.), second ed., Forensic Science Handbook, vol. 1, Prentice Hall, New Jersey, 2002, pp. 321–387.

[4] J. Dunlop, Colour analysis by microspectrophotometry, in: J. Robertson (Ed.), Forensic Examination of Textile Fibres, Ellis Horwood, Chichester, England, 1992, pp. 127–140.

[5] M.B. Eyring, Spectromicrography and colorimetry: sample and instrumental effects, Anal. Chim. Acta 288 (1994) 25–34.

[6] M.C. Grieve, J. Dunlop, P.S. Haddock, An assessment of the value of blue, red, and black cotton fibers as target fibers in forensic science investigations, J. Forensic Sci. 33 (1988) 1332–1344.

[7] R. Macrae, R.J. Dudley, K.W. Smalldon, The characterization of dyestuffs on wool fibers with special reference to microspectrophotometry, J. Forensic Sci. 24 (1976) 117–129.

[8] Peter Van Zant, Microchip Fabrication, fourth ed., McGraw-Hill, London, UK, 2000.

[9] Oliver S. Heavens, Optical Properties of Thin Solid films, Dover Publications, New York, 1965.

[10] W.A. Pliskin, Nondestructive optical techniques for thin-film thickness measurements, in: E.M. Murt, W.G. Gulder (Eds.), Physical Measurements and Analysis of Thin Films, Plenum, New York, 1969 (Chapter 1).

[11] Zhe Fei, et al., High-energy optical conductivity of graphene determined by reflection contrast spectroscopy, Phys. Rev. B 78 (20) (2008) 1–4, 201402.

[12] Jun Liu, et al., Graphene-based materials for energy applications, MRS Bull. 37 (12) (2012) 1265–1272.

[13] Jun-Yu. Ou, et al., Reconfigurable photonic metamaterials, Nano Lett. 11 (5) (2011) 2142–2144.

[14] H. Kataura, et al., Optical properties of single-wall carbon nanotubes, Synth. Met. 103 (1–3) (1999) 2555–2558.

[15] K. Iakoubovskii, et al., Optical characterization of double-wall carbon nanotubes: evidence for inner tube shielding, J. Phys. Chem. C 112 (30) (2008) 11194–11198.

[16] Sehoon Chang, et al., Responsive hybrid nanotubes composed of block copolymer and gold nanoparticles, Macromolecules 42 (15) (2009) 5781–5785.

[17] Andrey E. Nikolaenko, et al., THz bandwidth optical switching with carbon nanotube metamaterial, Opt. Express 20 (6) (2012) 6068–6079.

[18] S.A. Wang, E.V. Alekseev, D.W. Juan, H.M. Miller, A.G. Oliver, G.K. Liu, W. Depmeier, T.E. Albrecht-Schmitt, Functionalization of borate networks by the incorporation of fluoride: syntheses, crystal structures, and nonlinear optical properties of novel actinide fluoroborates, Chem. Mater. 23 (11) (2011) 2931–2939.

[19] Matthew J. Polinski, et al., Differentiating between trivalent lanthanides and actinides, J. Am. Chem. Soc. 134 (25) (2012) 10682–10692.

[20] Jian Lin, Justin N. Cross, Juan Diwu, Nathan A. Meredith, Thomas E. Albrecht-Schmitt, Comparisons of plutonium, thorium, and cerium tellurite sulfates, Inorg. Chem. 52 (8) (2013) 4277–4281.

[21] D. Allen, Holmium oxide glass wavelength standards, J. Res. Natl. Inst. Stand. Technol. 112 (2007) 303–306.

[22] Stuart E. Lebo Jr., Jerry D. Gargulak, Timothy J. McNally, "Lignin". Kirk-Othmer Encyclopedia of Chemical Technology, John Wiley & Sons Inc., Hoboken, New Jersey, 2001
[23] A.B. Wardrop, The structure of the cell wall in lignified collenchyma of Eryngium sp., Aust. J. Bot. 17 (1969) 229–240.
[24] Yan Yu, Distribution of lignin in Chinese fir branches determined by ultraviolet microspectrometer, Spectrosc. Spectr. Anal. 32 (6) (2012) 1685–1688.
[25] Iris Vazquez-Cooz, Robert W. Meyer, Fundamental differences between two fiber types in Acer, IAWA J. 29 (2) (2008) 129.
[26] Robert J. Moon, et al., Cellulose nanomaterials review: structure, properties and nanocomposites, Chem. Soc. Rev. 40 (7) (2011) 3941–3994.
[27] L. Zhang, Q. Li, J. Zhou, L. Zhang, Synthesis and photophysical behavior of pyrene-bearing cellulose nanocrystals for Fe^{3+} sensing, Macromol. Chem. Phys. 213 (2012) 1612–1617.
[28] A. Mangalam, J. Simonsen, A.S. Benight, DNA/cellulose hybrid nanomaterials, in: NSTI-Nanotech, Vol. 2, 2006, pp. 144–147.
[29] G.F. Kunz, On the new artificial rubies, Trans. N. Y. Acad. Sci. October 4 (1886) 3–10.
[30] K. Nassau, Gems Made By Man, Chilton Book Co., Pennsylvania, 1980.
[31] A. Verneuil, Memoire sur la reproduction du rubis par fusion, Ann. Chim. Phys. Ser. 83 (1904) 20–48.
[32] W.W. Hanneman, Gem Filter Set Model 4000, Hanneman Gemological Instruments, hanneman@netzero.net.
[33] John and Kathleen Harris, Jewellers and Gemologists, 4 The Crescent Carlisle, Cumbria, UK, CA1 1QN, http://www.gemlab.jkharris.co.uk/.
[34] P. Martin, M. Eyring, Microspectral Analysis of Gemstones and the Applications to Forensic Science. Presented at the American Academy of Forensic Sciences, San Antonio, TX (2007)
[35] Microscope Publications, 2820 S. Michigan Ave., Chicago, IL, 60616, Gary J. Laughlin, PhD, editor. themicroscope@mcri.org.
[36] I. Russo Krauss, A. Merlino, A. Vergara, F. Sica, An overview of biological macromolecule crystallization, Int. J. Mol. Sci. 14 (2013) 11643–11691.
[37] S. Desbois, S.A. Seabrook, J. Newman, Some practical guidelines for UV imaging in the protein crystallization laboratory, Acta Crystallogr. F Struct. Biol. Cryst. Commun. 69 (2013) 201–208.
[38] H.S. Gill, Evaluating the efficacy of tryptophan fluorescence and absorbance as a selection tool for identifying protein crystals, Acta Crystallogr. F 66 (2010) 364–372.
[39] N.I. Zheludev, Y.S. Kivshar, From metamaterials to metadevices, Nat. Mater. 11 (2012) 917.
[40] Y. Jin, Engineering plasmonic gold nanostructures and metamaterials for biosensing and nanomedicine, Adv. Mater. 24 (2012) 5153–5165.
[41] Chen Ling, Liu Fan-Xin, Zhan Peng, Pan Jian, Wang Zhen-Lin, Ordered gold nanobowl arrays as substrates for surface-enhanced Raman spectroscopy, Chin. Phys. Lett. 28 (2011) 57801, National Laboratory of Solid State Microstructures and Department of Physics.
[42] Kebin Li, Liviu Clime, Lilin Tay, Bo Cui, Matthias Geissler, Teodor Veres, Multiple surface plasmon resonances and near-infrared field enhancement of gold nanowells, Anal. Chem. 80 (2008) 4945–4950, Industrial Materials Institute; Institute of Microstructural Sciences, Canada.
[43] Xinlong Xu, et al., Flexible visible–infrared metamaterials and their applications in highly sensitive chemical and biological sensing, Nano Lett. 11 (8) (2011) 3232–3238.

Index

Note: Page numbers followed by "*f*" indicate figures, and "*t*" indicate tables.

A

Absolute measurements, 4–5, 136, 383–384
  angle-resolved, 292, 294, 315
    over-illumination method, 295–296
    under-illumination method, 297–301
  diffuse reflectance, 192, 193–212, 248, 318–321
  fluorescence, 228, 231, 235, 256, 259–260, 278–286
  regular reflectance and transmittance, 151–153, 152*f*, 156–158, 158*f*, 169–170, 176*f*
Absorbance, 34–35, 36, 87*f*, 92, 134, 137, 226*t*, 274–275, 423, 424, 425, 426, 427, 429, 437
  effect on fluorescence measurements, 231, 257, 258–259, 259*f*, 265, 274–275
  in microspectrophotometry, 491, 493, 496, 497*f*, 500, 500*f*, 501, 501*f*, 504*f*, 505*f*, 506, 509*f*, 511, 511*f*, 512*f*, 513, 513*f*, 515, 515*f*
  in pharmaceutical industry, 423, 424, 425, 425*f*, 427*f*, 429, 433–434, 437, 452
  specific, 36
Absorptance, 1–2, 34–36, 37–39, 42, 184, 230, 231, 234–235, 345
Absorption coefficient, 34–35, 36, 50–51, 225*t*, 226*t*, 452
Accuracy, 3, 4–5
  angular, 161–162, 166
  FTS measurements, 137–138
  numerical, 318–319
  photometric, 147, 251, 422*t*, 423–424, 442
  slit-width, 262–264
  wavelength, 88–89, 113–114, 262–264, 436–437, 442, 447–449
Active pharmaceutical ingredient (API), 413, 414, 440, 443, 454
Adaptation, 19–20, 368–369, 370
Aging
  of instrument, 262, 384
  of sample, 435
Air bearing, 117–118, 119
Airborne instruments
  airborne multiangle spectropolarimetric imager (AirMSPI), 474–475, 481–482
  airborne visible/infrared imaging spectrometer (AVIRIS), 458, 460, 461*f*, 471, 474–475, 474*f*
Albedo. *See also* Reflectance, diffuse
  clouds, 465, 468
  snow, 470–475, 474*f*, 481–482
Alignment, 118, 119, 123, 124–125, 132–133, 145, 146–147, 306
Analog to digital converter (ADC), 106–107, 120, 128
Analytical instrument qualification (AIQ), 414, 415–417, 419
Angle-resolved scatter (ARS), 293–294
Annular geometry, 250, 267, 388–389
Antireflection coatings, 43–44, 151–153, 298
Aperture
  entrance, 115–116, 129*f*, 132, 140, 392, 496, 497*f*, 508*f*
  slit, 69–71, 69*f*, 72, 73–74, 82, 83, 83*f*, 84–85, 84*f*, 88, 89–90, 149–150, 251, 262–264, 270–271, 425, 493–494
Apodization, 103–104, 103*f*, 105–106, 108–109, 139, 442
Appearance
  color (*see* Color)
  geometry
    distinctness of image (DOI), 403–404
    gloss, 213, 368, 386–387, 388, 397–400, 397*t*, 400*t*, 403
    haze, 400–403
    orange peel, 404
Applied Physics Corporation, 430
Artifacts, spectral, 120, 148–150
Assay, 245–246, 413, 419, 427–428, 429, 430, 445, 452–453, 454
Atmospheric absorption
  in measurements, 105–106, 113, 121, 124–125, 134–135, 139, 338–339
  in remote sensing, 460, 461*f*
Atomic force microscopy (AFM), 322–325, 324*f*
Atomic line spectra, 85–86, 99–101, 105–106, 113, 423, 448. *See also* Calibration, wavelength
Attenuated total reflection (ATR), 443–445, 446*f*

Attenuation
to ensure linearity, 136–137, 139
linearity testing, 92–93, 126, 266*f*
of a wave, 20–21
Auto-zero mode, 170, 171*f*
Averaging, signal, 110–112, 443

B

Background, 90–91, 99–101, 120, 121–122, 126, 128, 131, 132*f*, 134, 139, 146, 231, 259–260, 270, 275, 336–337, 340, 341–342, 344, 358–359, 435, 437, 438
Bandpass function, 70–74, 85–86, 149–150, 422*t*, 424–425, 448
accuracy, 262–264
circular slit, 72–73
color measurement, 375–376, 383, 392, 393–394
double monochromator, 73
fluorescence measurements, 225, 227, 238, 245–246, 257, 263, 264, 265, 271, 272–273
FTS, 106–107
measuring, 73–74
rectangular slit, 71–72
Bandwidth. *See also* Bandpass function
coherence, 49–50
Barium sulfate ($BaSO_4$), 191–192, 206*f*, 208, 211, 248, 258–259
BBRDF. *See* Bispectral BRDF
Beam profile, 303
Beamsplitter
in FTS, 98, 98*f*, 104, 107–108, 113, 114*f*, 116–118, 122–123, 122*f*, 126, 129, 134, 140, 338–339
in IV method, 175
in reference signal, 266–267, 278–279, 299
Beckman, A., 489–490
Beer-Lambert law, 36, 54, 92, 231, 274–275, 426, 446
Benford, F.A., 196
Biconical reflectance, 185*t*
Bidirectional geometry, 225*t*, 226*t*, 232–233, 250, 268–269
Bidirectional reflectance. *See* Bidirectional reflectance distribution function (BRDF)
Bidirectional reflectance distribution function (BRDF), 28–30, 32–34, 291–331
cosine space, 319–321
extrapolation, 326–328
facet model, 57–58
infrared, 316
in-plane, 304–305
integration, 31–32, 203, 318–321
Mueller matrix, 41–42, 316–317
noise-equivalent, 303, 309–311
out-of-plane, 305–309, 319–322, 325, 326–328, 327*f*
polarization, 41, 42–43, 316–317
Rayleigh–Rice theory, 282
reference standards, 192, 318–321, 318*f*
relationship to reflectance, 30–32, 185–186, 185*t*
remote sensing, 472–473
ultraviolet, 316
Bidirectional scattering-surface reflectance distribution function (BSSRDF), 28–30
ideal, 32–34
Bidirectional transmittance distribution function (BTDF), 28–30
measurement, 317
Rayleigh–Rice theory, 64–65
Bihemispherical reflectance, 185*t*, 195–196, 199
Birefringence, 507–508
Bispectral BRDF (BBRDF), 36–38, 227, 235
Bispectral fluorescence colorimeters, 250–251
Blackbody
color temperature, 373, 373*f*
heat transfer, 334–335
ideal, 18–19, 334
radiation, 18–19, 334–335
reference, 336–337, 339, 342–343, 343*f*, 344, 348*t*, 358, 359, 360*f*, 479–480
source, 89–90, 89*f*, 114–115
BRDF. *See* Bidirectional reflectance distribution function (BRDF)
Brewster angle, 46, 47
Brightness
fluorescent, 214, 224, 231, 232, 252, 253
perceived, 395, 395*t*
BSSRDF. *See* Bidirectional scattering-surface reflectance distribution function (BSSRDF)
BTDF. *See* Bidirectional transmittance distribution function (BTDF)

C

Calibrated detector (CD), 251, 278–279, 281
Calibrated reflector (CR), 251, 278–279, 279*f*
Calibrated source (CS), 251, 265, 279, 279*f*, 281
Calibration
atomic line lamps, 262, 423, 448

rare earth standards, 85–86, 87*f*, 139, 422*t*, 423, 436–437, 448, 449
water Raman band, 263, 283, 450
in color measurements, 375–376, 380, 393
in emissivity measurements, 340–342, 348*t*, 359
fluorescence, 278–286
mirror, 146–147, 151, 153–154, 156–158, 167–168, 171–172
remote sensing, 479–481
sources (CS), 251
vicarious, 480
wavelength, 84–87, 113–114, 263–264, 392, 421–423, 422*t*, 436–437, 447–448
Candela, 20
Carbon dioxide (CO_2), 121, 135
atmospheric, 458–464
in laboratory measurements, 121, 135, 135*f*, 359
Carbon nanotubes, 504, 505*f*
Cat's eye reflector, 117–118. *See also* Corner cube reflector
Cauchy relationship, 78–79, 87
Cavity ringdown, 171–172
CCD. *See* Detector, charge coupled device (CCD)
Cellulose, 506, 507–508, 509*f*
Center burst. *See* Zero path difference
Center mount, 213, 213*f*
Certified reference materials (CRM), 208–209, 282, 420, 423, 447, 448, 454–455
Changeover points, 149
Chirp, 107
Chopper. *See* Lock-in detection
Chroma compression, 381
Chromaticity coordinates, 234, 285, 377–380, 379*f*
CIELAB, 285, 378, 380–383, 382*f*
CIELUV, 378–380
CIEXYZ, 378
CIE 1924 spectral luminous efficiency, 20*f*, 371–372
Clouds, 52, 53–54, 464–470
Coated glass, 148, 153, 159
Coatings
Kubelka–Munk analysis, 52–54, 223
metallic, 77–78, 81, 394–395, 396
pearlescent, 394–395, 395*t*, 396
thick films, 49–51
Coblentz sphere, 202–203, 202*f*
Coblentz, W.W., 202
Coherency matrix, 40–41
Color, 367–404
CIE 1964 color matching functions, 372
CIE 1931 standard colorimetric observer, 376, 377–378
difference, 380–383
geometry, 367, 370, 387–388, 395, 395*t*, 397
hue, 368, 378, 380, 381, 382*f*, 395, 395*t*
lightness, 368, 378, 380, 381
matching functions, 371–372
metamerism, 376–377, 384
saturation, 368
temperature, 284, 373, 373*f*
Colorimeter, 375, 384, 388, 390, 391, 490–491
Combinatorial chemistry, 412–413, 454
Commission Internationale de l'Eclairage (CIE), 13, 186–187, 368
Comparison methods, 212–214
Compression cells, 443
Computer control, 139, 300, 345–347, 494–495
Concentration, analyte, 223–224, 230, 231, 245–246, 274–275, 283, 426, 445, 451
Confidence level, 6–7, 7*t*, 147, 154–156
Conical-directional reflectance, 185–186, 185*t*
Conical-hemispherical reflectance, 185*t*
Constitutive constants. *See* Optical constants
Constitutive relations, 22
Corner cube reflector, 117–118, 121–122, 339*f*
Correlated color temperature (CCT), 284, 373
Cosine space, 319–321
Critical angle, 47
Cuvette, 85–86, 213, 231, 244–245, 270, 274–275, 286, 424*t*, 435, 449, 450

D

Daylight illuminant. *See* Illuminant, CIE: D50, D55, D65 and D75
Degradation products, 413–414
Delta function, 33, 103
Depolarizer, 159–
Detection gate, 246, 277
Detection limit (DL), 417, 419, 430, 445, 453
Detector, 90–93
charge coupled device (CCD), 87–88, 240, 242–243, 269, 493–494
in color measurement, 375–376, 383–394
complementary metal oxide semiconductor (CMOS), 91
deuterated triglycine sulfate (DTGS), 120, 435, 442
dynamic range, 84, 91, 120, 136–137, 139, 242–243, 264–265, 300–301, 494
in FTS, 119, 120, 121
germanium (Ge), 90–91, 120
indium antimonide (InSb), 120, 126, 132–133

Detector (*Continued*)
indium gallium arsenide (InGaAs), 90–91, 111–112, 120, 149, 158, 494
lead sulfide (PbS), 90–91, 149, 433, 435
mercury cadmium telluride (MCT, HgCdTe), 109–110, 120, 126, 132–133, 136, 339*f*, 442
multichannel, 84, 240, 242–243
nonequivalence, 128–129
photodiode array, 87–88, 240, 242, 493–494
photomultiplier tube (PMT), 87–88, 90–91, 92, 149, 158, 241–243, 262, 263, 264, 266*f*, 279, 280*f*, 300–301, 304, 493
photopic, 239–240
pyroelectrics, 90–91, 120, 128–129, 131, 133, 139
silicon (Si), 87–88, 91, 242–243, 300–301, 390, 433, 435
spectral sensitivity, 90–91
standard, 278–286
thermopile, 90–91, 119–120, 260, 359
triglycine sulfate (TGS), 338–339
uniformity, 90–91, 92, 133, 136–137, 139, 300–301, 313
Diamond turned optics, 292, 325, 326*f*
Differential optical absorption spectroscopy (DOAS), 478
Diffraction equation, 74, 323, 326–328
Diffuser, 37, 92, 184, 205, 262, 267, 293, 383–384, 479–480
Directional-conical reflectance, 185*t*
Directional-hemispherical reflectance, 32, 34, 185*t*, 198–200, 202, 203, 213
Discovery research, 412–414
Dispersion
FTS elements, 107–108
in gases, 113
prism, 2–3, 78–81, 80*f*, 87
Dispersive element, 68–71, 82–83, 84, 493
Dispersive spectrometer, 67–93
Dissolution, 410, 413, 431, 432*f*, 448, 454
Dissolution bath, 432*f*
Distinctness of image (DOI), 396, 403–404
Donaldson radiance factor matrix, 210, 226*t*, 235–237, 273, 274
Double beam, 163*f*, 170, 213*f*, 240
Drift, 110, 113, 121, 124–125, 135–136, 154, 163–164, 250, 281, 339, 393, 491–492
Drude model, 336

E

Earth observing systems. *See* Satellite-borne instruments
Electrical pickup, 124
Electromagnetics, 12–13, 20–26, 59–65
Ellipsometer, 338, 352–354
Emissivity, 38–39, 334–362
blackbody comparison method, 338–340
ceramics, 357–362
cold crucible technique, 354–355
containerless techniques, 351–357
Kirchhoff's law, 38–39, 184, 334, 338, 345
long wavelengths, 336
molten metals, 349–357
near ambient temperature, 338–345
oxidized metals, 345–349
volumetric, 357
X-point, 350–351
Emittance. *See* Emissivity
Energy, radiant, 13–14
photon, 37–38
Stokes vector, 41
Erb, W., 200
Error analysis. *See* Uncertainty, analysis
Eulerian cradle, 305, 306*f*, 307
Evidence of control, 410, 454–455
Excitation-emission matrix (EEM), 226*t*, 232. *See also* Donaldson radiance factor matrix
Extinction coefficient, 34–36, 226*t*, 231, 338, 421–422
in clouds, 469
in paint coatings, 52, 54
Extinction cross section, 35–36
Extreme ultraviolet (EUV), 305, 323
Extrinsic molecular property, 229

F

Fabry–Perot fringes, 135–136, 137
Facet scattering model, 54–59
Fellgett's advantage, 110–111
Field of view (FOV), 190, 195–196, 296, 297, 300, 301–302, 303, 304, 326*f*, 344, 370, 371, 372, 496
Field stop, 115–116, 130, 132, 297*f*, 300–301
Films, 48–49, 107, 117, 156–158, 213, 395, 396, 397–398, 490, 495–496, 498, 503*f*
capillary, 444
self-supported polymer, 444
standard observer, 417
thick, 49–51
thickness measurement, 348–349, 502–503
thin, 48–49, 213, 490–491, 495–496, 498, 502–503, 515
Filter
attenuating, 254, 265, 278
balancing, 239–240

bandpass, 88, 239–240, 245–246, 247, 270, 277–278
cutoff, 252, 254
fluorescence weakening, 254
neutral density, 36, 126, 131, 132*f*, 137, 232, 254
Fluorescence resonance energy transfer (FRET), 276–277
Fluorescence
abridged measurements, 239–240
analytical applications, 223–225, 226*t*, 228, 230–232, 244–247, 256, 259–260, 263, 264–265, 270, 281
anisotropy, 225*t*, 268, 282
bispectral measurements, 238–239, 250–251, 271–272
colorimetric applications, 223, 224–230, 226*t*, 232–237, 247–256, 264–265, 268–269, 270, 272–273, 278, 284
in color measurements, 385
diagonal fluorescence, 272–273
in diffuse reflectance measurements, 205, 210–211
emission range, 222, 231–232, 234, 235
emission spectrum, 225*t*, 227–228, 238, 239, 240, 241, 251, 264–265, 281, 286, 447, 450
epifluorescence, 244–245, 244*f*, 450
excitation-emission matrix, 226*t*, 232
excitation range, 222, 254, 264, 269, 270, 281, 498
excitation spectrum, 225*t*, 227–229, 238, 239, 241, 281–282, 286, 385
front face geometry, 244*f*, 245, 282, 450
general considerations, 36–38
0°/90° geometry, 244–245, 260, 274–275, 282, 286
instrumentation, 240–243, 244–260
lifetime, 222–223, 241–242, 246, 277
microsecond, 246
in microspectrophotometry, 489–490, 491, 494, 498, 501, 505–507, 507*f*, 512
one-monochromator method, 234, 240, 245–246, 248–249, 248*f*, 252–256, 255*f*, 260–261, 261*f*, 265, 270–271, 284
qualitative *vs.* quantitative measurements, 451–452
quantum efficiency, 225*t*, 226*t*, 229, 230, 234–235, 258–259, 260
quantum yield, 225*t*, 226*t*, 229, 230, 231–232, 234–235, 256–260, 277, 280, 281–282, 286
quenching, 229, 243, 276–277, 286, 512
reflection overspill, 272–273
sampling effects, 228, 271–274
spectrofluorimeter, 241, 249–250, 278
reference, 251, 265, 267, 267*f*
spectrometry, 231, 446–454
spectrophotometer, 241, 262–263, 264, 269, 410
standards, 224, 238–239, 248, 251, 252, 253, 256–257, 258–259, 260, 265, 271, 278–286, 449–450, 449*t*, 452–453
steady state, 223, 241–242, 246, 450
terminology, 224–237
two-monochromator method, 238, 240, 250–251, 260–261, 264–265, 270–271, 284
yield, 226*t*, 230, 234–235 (*see also* Fluorescence, quantum yield)
Fluorescence weakening method, 254–255, 255*f*, 256
Fluorescent lamp illuminant. *See* Illuminant, CIE: F2, F7 and F12
Fluorescent whitening agents (FWA), 224, 228, 229*f*, 252, 385
Fluorometer, 223, 231–232, 244–246, 247, 260, 263, 281, 447, 448, 450–451, 451*f*, 452–453. *See also* Fluorescence, spectrofluorimeter
Flux, radiant. *See* Power radiant
Focusing, 69–70, 69*f*, 73–74, 78, 82, 83–84, 133, 247, 298, 300
Forensics, 498–501
Fourier expansion, 20–21, 25–26, 60–61
Fourier transform infrared spectrometer (FTIR). *See* Fourier transform spectrometer (FTS)
Fourier transform interferometer, 104
Fourier transform spectrometer (FTS), 98–140
in emissivity measurements, 338–339, 339*f*, 340, 341–342, 343, 343*f*, 344, 345, 358
Fresnel–Huygens principle, 180–181
Genzel interferometer, 122–123, 122*f*
Michelson interferometer, 98–101, 98*f*, 100*f*, 114*f*, 117–118, 121–123
in regular reflectance and transmittance measurements, 112–113, 149–151, 156–158, 176, 176*f*
in remote sensing, 459–460, 478, 480–481
Fresnel reflection, 44–47
FTS. *See* Fourier transform spectrometer (FTS)
FWA. *See* Fluorescent whitening agents (FWA)

G

Ganz–Griesser method, 253
Gases
  in emissivity measurements, 352–354, 359
  in fluorescence measurements, 277
  in FTS, 113, 114–115, 121
  in remote sensing, 458–460, 461–462, 465, 466, 467–468, 469–470, 475, 477–479
  in UV scattering measurements, 316
Gems, 496, 501, 508–510
General Electric Corporation, 3
Geometric extent, 17*f*, 43–44
Glass fragments, 479, 508–511, 511*f*
Glazing, 154, 496
Gloss
  appearance, 249, 368, 369–370, 386–387, 388, 396, 397–399, 397*t*, 400, 400*t*, 402–403, 404
  in diffuse reflectance measurements, 204, 205, 208–209, 213
  standard geometries, 397*t*
Gloss trap. *See* Specular excluded
GMP. *See* Good manufacturing practices (GMP)
Gonioappearance, 367, 394–396, 395*t*
Goniometer
  in angle-resolved measurements, 301, 303, 316
    in-plane, 304–305, 305*f*
    out-of-plane, 305–309, 306*f*, 308*f*
    robotic arm, 307–308, 307*f*
  in FTS, 136–137
  in microspectrophotometry, 495–496
  in specular reflectance measurements, 170, 171*f*
Good manufacturing practices (GMP), 414
Graphene, 504
Grating, 69, 74–78, 74*f*, 81–83, 83*f*, 84–85, 84*f*, 87, 88
  blazing, 76*f*, 77–78
  changeover, 149, 150
  comparison with FTS, 112–114, 134
  efficiency, 75–76, 76*f*, 87, 267, 269
  equation, 74, 277–278, 323, 326–328
  groove density, 74, 149–150
  holographic, 77–78, 77*f*, 493, 495–496
Ground-based measurements
  Total Carbon Column Observing Network (TCCON), 459, 464
Grum, F., 191–192

H

Hagen-Rubens relationship, 336
Hardy, A.C., 3
Harmonics, in FTS, 126, 129
Haze, 368, 369–370, 400–403
  reflection, 402–403
  transmission, 401–402
Heating
  ceramic, 89–90, 114–115, 351, 352, 360*f*, 362
  gas burner, 358–361, 360*f*
  induction, 353, 354–355
  laser, 356–357, 358
  resistive, 351
Heavy element, 505–506
Helium-neon (HeNe) laser, 106, 118–119, 343, 343*f*
Helmholtz equation, 22
Helmholtz reciprocity. *See* Reciprocity, Helmholtz
Hemispherical-conical reflectance, 185*t*
Hemispherical-directional reflectance, 185*t*, 186–187, 198–199, 207*f*, 213–214
HeNe laser. *See* Helium-neon (HeNe) laser
High throughput screening, 412–413, 454
HSRE. *See* Stray radiant energy, heterochromatic
Hue. *See* Color, Hue
Human visual system (HVS), 239–240, 368, 369*f*, 370, 371, 376–377
Humidity. *See* Water vapor
HVS. *See* Human visual system (HVS)

I

Ice
  atmospheric, 463, 465, 467, 468, 481–482
  ground, 459, 470, 471–472, 478–479, 481–482
IFE. *See* Inner filter effects (IFE)
Illuminance, 19–20
Illuminant, CIE
  A, 373, 374*f*, 390–391, 401–402
  C, 253, 284, 373–374, 374*f*, 390–391, 398–399, 398*f*, 400, 401–402
  D65, 253, 253*f*, 255*f*, 256, 284, 374, 374*f*, 389
  D50, D55 and D75, 374
  F2, F7 and F12, 374
  white point, 375, 378–380
Illuminator, in angle-resolved measurements, 295–296, 298–299
ILS. *See* Instrument line shape (ILS)
ILV. *See* International lighting vocabulary (ILV)
Impedance of free space, 23, 98–99
Impurities
  in diffuse reflectance standards, 206–207, 210–211

in emissivity measurements, 352, 362
in fluorescence measurements, 243, 276–277, 286
in pharmaceuticals, 413, 414, 418, 420, 429, 430, 445, 453, 454
snow, 471–473, 481–482
Index matching, 509–510
Indicatrix, 180–181
Induction furnace, 353
Inner filter effects (IFE), 257, 274–275, 276
Instrument line shape (ILS), 103, 104, 111, 270–271, 280, 463–464, 480–481, 482*f*
Instrument signature, 301–303, 302*f*, 313, 316, 323, 324*f*
Integrating sphere
in angle-resolved measurements, 295, 296*f*, 297*f*, 300–301, 300*f*, 317
baffles, 187–188, 190–191, 193, 197–199, 200, 270, 296*f*
coating materials, 191–192
in color and appearance measurements, 384, 386–387, 388*f*, 401–402, 401*f*
in diffuse reflectance and transmittance, 179–219
recommended geometries, 186–187
in dispersive spectrometer, 73–74, 84–85, 92
in emissivity measurements, 345–349
in fluorescence measurements, 227, 233, 235, 249, 250, 251, 256, 258–259, 260, 261*f*, 278, 279*f*
in FTS, 132–133, 136–139, 138*f*
general principles, 187–193
homogeneity, 192–193
multiplier, 188–189
in regular reflectance and transmittance, 146, 147–148, 148*f*, 151, 152*f*, 153, 156–158, 160–161, 161*f*, 163*f*, 165, 170, 173–174
in remote sensing, 481–482
theory, 187–195
Intensity, radiant, 17, 18
relationship to fields, 24–25
spectral, 18
Interferometer. *See* Fourier transform interferometer
Interlaboratory comparison, 137–138, 153, 158, 159, 450, 467
International lighting vocabulary (ILV), 13, 37, 225*t*, 226*t*, 233, 234–235
Interreflections, 93, 121–122, 129–130, 129*f*, 130*f*, 139, 163, 223
Intrinsic molecular property, 36, 186–187, 231
Irradiance, 14–15
relationship to fields, 23–24
spectral, 18, 130, 233–234, 281, 467–468
stokes vector, 41

J

Jacquinot's advantage, 112
Jacquinot stop, 115–116

K

KBr. *See* Potassium bromide (KBr)
Kirchhoff relationship, 38–39, 184, 334, 338, 345
Korte, H., 200
Kubelka–Munk theory, 52–54, 223

L

Lambertian, 33–34, 37, 84–85, 114–115, 181, 182, 188–189, 190, 191–192, 193, 199–200, 250, 293, 304, 318–319, 321–322, 463
Laser, 19, 23
in angle-resolved scattering, 297*f*, 298, 299, 300, 301, 302*f*, 309, 313–315, 316
in color measurements, 392
crystals, 263
in dispersive methods, 73–74, 87–88, 90
in emissivity measurements, 343, 343*f*, 353–356, 357, 358–361, 360*f*
in fluorescence measurements, 239, 245, 247
in FTS, 98, 99–101, 106, 113, 114*f*, 115, 117, 118–119, 133, 135, 140
heating, 353, 356–357, 358, 359–361, 360*f*
mirrors, 144, 171–172, 173–174
remote sensing calibration, 312, 480–481
speckle, 298, 313–314
LED. *See* Light emitting diode (LED)
Levitation, 353–354
electromagnetic, 353–354, 354*f*
electrostatic, 355–356, 356*f*
Lifetime, luminescence, 222–223, 241–242, 246, 277
Light emitting diode (LED), 193, 230, 239–240, 245–246, 247, 389–390, 496
Lightness. *See* Color, lightness
Light trap, 213, 214–215, 215*f*, 250, 270, 386–388, 391–392, 394, 401–402, 401*f*
Lignins, 506–507

Linearity
in angle-resolved measurements, 304, 313, 315
in color and appearance measurements, 391–392, 393–394, 400, 402
in dispersive applications, 90–91, 92, 93, 93*f*
in emissivity measurements, 336–337, 340, 344
in fluorescence measurements, 241, 252, 264, 265, 266, 266*f*, 274–275, 286
in Fourier transform methods, 101–103, 113, 120, 126, 128–129, 134, 136–137, 139
in pharmaceutical applications, 417, 421, 423–424, 427*f*, 430, 445–446, 454
in regular reflectance and transmittance, 145, 149, 156–158
Linear shift invariance, 326–328
Lippert equation, 275
Liquid crystal, 435, 496, 497*f*, 507–508
Liquids, 229, 235, 244–245, 256, 257, 260, 277, 308–309, 349–350, 351, 352–355, 356–357, 400, 434, 444
Little, W.F., 198–199
Littrow geometry, 74–75, 76, 76*f*, 77, 82–83, 83*f*
Lock-in detection, 90–92, 110, 130–131, 297*f*, 298–299, 304, 353–354, 359
Luckey, G.W., 191–192
Luminance, 19–20
Luminous flux, 19–20
Luminous intensity, 19–20

M

Massachusetts Institute of Technology, 3
Maxwell equations, 21–22, 24–26
Measurement equation, 5, 6, 16, 145–146, 165, 311, 401–402
Metamerism. *See* Color, metamerism
Methamphetamine, 445, 446*f*
Method transfer, 439
Michelson interferometer. *See* Fourier transform spectrometer, Michelson interferometer
Microarray (MA). *See* Microwell plate
Microscope, 136, 247, 443, 490–493, 494–495, 496, 498, 515
Microscopic sampling, 445, 495
Microspectrophotometer (MSP), 489–515
Microwell plate (MP), 245, 247
Mid infrared (Mid-IR) spectrometry, 440–446
Mineral oil mulls, 443
Mirror, moving, 98–103, 98*f*, 104, 105*f*, 107, 114*f*, 117–119, 123, 339*f*, 343
Mirror, reference, 136, 144, 146–147, 151, 154–156, 155*t*, 166–168, 171–172
Misalignment, 113, 124–125, 132–133, 146–147, 151–153, 156–158, 165, 173, 393
Moisture, 434, 440, 441*f*
Molar absorption coefficient, 36, 231. *See also* Specific absorbance
Molar extinction coefficient, 36, 226*t*, 231. *See also* Specific absorbance
Molecular optical properties, 224–225, 231–232
Monitor detector, 240, 241*f*, 242, 262, 278–279, 280, 281, 299, 391, 391*f*
Monochromator
aberrations, 68–69, 73–74, 83–84
in angle resolved scattering, 296*f*, 297, 314–315
in color measurements, 384–385
comparison to FTS, 104, 119–120, 137–138
Czerny–Turner, 82–83, 83*f*, 84, 84*f*, 345–347
dispersive analyzer, 68–70, 71, 73–74, 75, 78–79, 81, 82–85, 87–90, 92
double, 73, 88
Ebert–Fastie, 82–83, 83*f*
in fluorescence measurements, 238, 239, 240, 241–242, 241*f*, 245, 251, 261, 262, 263, 264, 266–267, 270–271, 277, 280
Littrow, 82–83, 83*f*
in microspectrophotometry, 412
in pharmaceutical applications, 425, 448
prism premonochromator, 75, 81
in reflectance and transmittance measurements, 150
in remote sensing, 480, 490
stray light, 68–69, 73, 81, 83, 87
MP. *See* Microwell plate (MP)
MSP. *See* Microspectrophotometer (MSP)
Mueller matrix, 41–43, 292, 316–317
Multichannel spectrometer. *See* Polychromator
Multilayer coatings, 48–49, 133, 144, 171–172, 173–174, 292, 323
Multiple reflections
in emissivity measurements, 339–340
in a transparent optic, 148, 160–161
Multiplexing, 110–112, 113, 124, 137, 295, 297
Munsell system, 370–371, 378, 380, 381–383, 382*f*

N

Nanomaterials, 507–508, 508*f*, 514–515, 514*f*
Naperian absorption coefficient. *See* Absorption coefficient

National metrology institute (NMI), 137–138, 193, 204, 272–273, 282, 312, 383–384, 392, 394
Near infrared (NIR) spectrometry, 433–440
NEP. *See* Power radiant, noise equivalent
Newton, Isaac, 2–3
NIR spectrometry. *See* Near infrared (NIR) spectrometry
Nitrogen cooling, 120, 316, 338–339, 344
Nitrogen gas (N_2), 121, 211
NMI. *See* National metrology institute (NMI)
Noise, 87–88, 91–92
  in angle-resolved measurements, 299, 301–302, 303, 309–311, 316, 323
  in color and appearance measurements, 383, 393, 400
  in diffuse reflectance measurements, 202, 213–214
  in emissivity measurements, 337–338, 353–354
  in fluorescence measurements, 246
  in FTS, 104, 106–108, 110–112, 117–118, 120, 124, 125–126, 135–136
  high and low flux noise, 438
  in regular reflectance and transmittance measurements, 159–160
  spectroscopic, 437
Nonlinearity. *See* Linearity
Nonspherical optics
  ellipsoidal mirror, 115–116, 202
  paraboloidal mirror, 115–116
  toroidal mirror, 83–84
Normalization
  in BRDF measurments, 315–316
    relative normalization, 315
    specular normalization, 316
  fluorescence measurements, 228, 264, 270–271, 281
Normalized data (for fluorescence measurements), 251, 279–280
Nyquist frequency, 106, 118–119, 124

O

Opacity, 204, 205, 207
OPD. *See* Optical path difference (OPD)
Optical constants, 20–21, 144, 338, 348, 353–354. *See also* Refractive index
Optical density. *See* Absorbance
Optical depth, of clouds, 465–466, 469, 474–475
Optical head, 240
Optical path difference (OPD), 98–99, 100*f*, 101–103, 102*f*, 104, 105–107, 105*f*, 109*f*, 111, 113, 117–119, 123, 124, 126, 127*f*, 130–131, 341–342
Optical Society of America, 3
Optical theorem, 35–36
Orange peel, 368, 404
Order sorting, 75, 81, 83, 270, 277–278
Outliers, 393–394, 439

P

Paint, 52, 53*f*, 54, 394–395, 397*t*, 398–399, 404, 495–496
  black, 321–322, 322*f*
  fluorescent, 236*f*, 285*f*
  forensics, 498–501, 499*f*, 500*f*, 501*f*
PAT. *See* Process analytical technology (PAT)
Pen lamp, 262–263. *See also* Atomic line spectra
Perfectly reflecting diffuser (PRD), 182–183, 190, 201–202, 203–204, 205, 227, 232, 233–234, 235–236, 247, 248, 286, 383–384, 391–392. *See also* Lambertian
Permeability, 22
Permittivity, 22
Phase
  anomalous, 341–342
  correction, 107–110, 109*f*, 119, 125–126, 125*f*, 341–342
  error, 107–110, 119, 125–126, 135–136, 341–342
  function, 52
Phosphorescence, 27–28, 222–223, 246, 490, 498
Photobleaching, 246
Photodegradation, 246
Photodiode array (PDA). *See* Detector, photodiode array
Photoluminescence. *See* Fluorescence
Photoluminescent quantum yield, 226*t*, 234–235, 286
Photoluminescent radiant yield, 226*t*, 234–235, 259
Photomultiplier (PMT). *See* Detector, photomultiplier
Physical transfer standards (PTS), 251, 260–261, 265, 278–281
Planck's Law, 18–19, 38–39, 334–335, 343
Plasmon, 514–515, 515*f*
PMMA. *See* Polymethylmethacrylate (PMMA)

Polarization, 20–21, 30, 39–42, 44–45
in angle-resolved measurements, 292, 299, 301–302, 316–317, 321–322, 323
anisotropy, 268
degree of, 150, 266–269
in fluorescence measurements, 251, 266–267, 268–269, 278, 281, 286
in FTS, 123, 133–134
in gratings, 76, 76*f*, 78
in regular reflectance and transmittance, 144, 146–147, 149, 150–151, 150*f*, 156, 159–160, 159*f*, 163–164, 165, 167–168, 169–170, 173, 176*f*
remote sensing, 474–475
Polarizer, 123, 159–160, 163*f*, 175, 261–262, 265–266, 268–269, 296*f*, 297*f*, 299
extinction ratio, 160
Glan–Thompson, 160
wire grid, 160
Polychromator, 68–70, 75, 84, 84*f*, 87–88, 242
Polymethylmethacrylate (PMMA), 449
Polymorphism, 434
Polytetrafluoroethylene (PTFE)
fluorescence in, 211, 260
integrating sphere coating, 158, 191, 192, 260
packed/pressed powder, 209, 211, 248, 258–259, 315, 318, 318*f*, 319–322, 320*f*
sintered, 206*f*, 209–210, 211, 260, 286, 315, 317, 318–319
Potassium bromide (KBr), 116, 117, 139–140, 338–339, 339*f*, 443, 444
Power radiant, 13
noise equivalent (NEP), 91–92, 110–111, 120
spectral, 18–19, 36–37, 227
stokes vector radiant, 41
Power spectral density function (PSD), of a surface, 59, 65, 322–325, 324*f*
Poynting vector, 23–24, 25–26, 98–101
PRD. *See* Perfectly reflecting diffuser (PRD)
Precision, 3, 4–5
Precision aperture, 116, 300*f*, 312
Prism, 2–3, 69, 75, 78–83, 80*f*, 81*f*, 87, 112, 118–119, 123, 137
Process analytical technology (PAT), 411, 440
Product development cycle, 412
Profile based correction (PBC), 393–394
Proteins, 244–245, 268, 439, 445, 512–513, 513*f*
PTFE. *See* Polytetrafluoroethylene (PTFE)
PTS. *See* Physical transfer standards (PTS)
Purge, 114–115, 114*f*, 117, 121, 134, 135, 139–140, 316
Pyrometer, 352, 356–357

Q

QbD. *See* Quality by design (QbD)
QC. *See* Quality control (QC)
QL. *See* Quantitation limit (QL)
Qualification
analytical instrument qualification (AIQ), 415–417, 417*f*, 419
design (DQ), 415–416
installation (IQ), 415–416, 421
operational (OQ), 415–416, 421–427, 435–436, 447–451
performance (PQ), 415–416, 421, 427–429, 443–445, 447, 450
Quality by design (QbD), 411
Quality control (QC), 410, 414, 415, 419, 435–436, 440
Quality system, 411, 414–419
Quantitation limit (QL), 417, 419, 430, 445, 453
Quantum efficiency, fluorescence, 225*t*, 229, 230, 234–235, 260
external, 229, 235, 258–259
internal, 230
Quantum efficiency, spectral
of the fluorescence process, 226*t*, 234–235, 259
Quantum properties, 37–38
Quantum yield
absolute methods, 259–260
fluorescence, 225*t*, 229, 230, 231–232
photoluminescent, 226*t*, 234–235
relative methods, 256–259
Quartz tungsten halogen (QTH). *See* Source: optical radiation, quartz tungsten halogen (QTH)

R

Radiance, 16–17
correcting for background, 342
relationship to fields, 25
spectral, 18–19, 67
stokes vector, 41
top of atmosphere (TOA), 462–463, 465
Radiance coefficient, 292–293
Radiance factor, 183, 185–186, 201–202, 204, 236
bispectral, 232, 234, 258–259, 272–273

bispectral luminescent, 36–38, 226*t*, 227, 232, 233, 235, 236*f*, 237*t*, 251, 259, 268–269, 272, 281, 284, 285*f*
Donaldson matrix (*see* Donaldson radiance factor matrix)
luminescent, 233–234, 251, 255*f*, 269
reflected, 233–234, 236, 251, 254, 255–256, 255*f*, 269, 272, 280, 285*f*
spectral, 233–234, 254, 258–259, 279, 286
total spectral, 226*t*, 233–234, 235–236, 249, 252, 254, 285, 285*f*
Radiant energy. *See* Energy, radiant
Radiant intensity. *See* Intensity, radiant
Radiant power. *See* Power radiant
Radiant yield, photoluminescent, 226*t*, 234–235, 259
Radiation heat transfer, 334–335, 348–349
Radiative transfer, 52, 466, 467, 469, 472, 479
Raman band, of water, 263, 283, 450
Raw data (as for fluorescence measurements). *See* Uncorrected data
Rayleigh–Rice theory, 59–65
Ray tracing, 55–57, 326
Reabsorption, 257, 274–275
Receiver, 296*f*, 297–301, 297*f*, 300*f*, 305*f*, 308–309, 313–314, 315, 316, 317, 327*f*, 393, 398–399
Reciprocity, Helmholtz, 30, 33–34, 292, 337–338, 387, 388–389, 466
Mueller matrix, 42–43
Reference-free methods. *See* Absolute measurements
Reference materials (RMs). *See* Reference standards
Reference standards
cleaning, 205, 207, 211, 243
diffuse reflectance
barium sulfate ($BaSO_4$), 191–192, 208, 211, 248, 258–259
ceramics, 205, 206, 206*f*, 209, 211
magnesium oxide (MgO), 203, 206–208
opal glasses, 205, 206*f*, 208–209, 210–211, 210*f*
polytetrafluoroethylene (PTFE), 206*f*, 209–210, 211, 248, 259, 286, 315, 317, 318, 318*f*, 320*f*
Vitrolite, 206–208
fluorescence
color standards, 284–285
quantum yield standards, 286
handling, 211–212, 243
mirror, 136, 144, 146–147, 151, 154, 155*t*, 156, 166–168
primary, 182, 204, 247, 286, 399–400
requirements, 204–205
transfer, 204–205, 248, 251, 276–277, 278–281, 282–283
working, 204, 205
Reflectance, 19, 20–21, 30–32
angle-resolved, measurement, 292–317, 321–322, 325, 326–328
definition, 19, 30–32
diffuse, 30–32, 180–181, 182–216
Benford method, 195–197
Coblentz sphere method, 202–203
first Taylor method, 195–197
goniometric method, 203, 318–321
Korte–Schmidt method, 200–202
Kubelka–Munk, 52–54
models, 52–65
nomenclature, 184–187, 185*t*, 186*f*, 187*t*
recommended geometries, 187, 187*t*
relative methods, 212–214
Sharp-Little method, 198–199
third Taylor method, 197–198
van den Akker method, 199–200
intrinsic, 54
in microspectrophotometry, 491, 494–496, 503*f*, 514*f*
regular (*see* Reflectance, specular)
specular, 32–33, 44–47, 48–51, 48*f*, 50*f*, 144–170, 171–176
distinctness of image, 396, 403
films, 48–49, 153
in FTS, 136–137
gloss, 204, 205, 208–209, 212–213, 368, 369–370, 386–387, 388, 396, 397–400, 402–403, 404
high reflectance, 171–176
low reflectance, 153–154
normal incidence, 150–158
oblique incidence, 158–170
total internal, 46–47
Reflectance accessories
IV, 153, 173–174, 175–176, 176*f*
retro-mirror pair, 166, 167*f*
VN, 151–153, 152*f*
VW, 153, 172–174, 172*f*, 173*f*, 174*f*, 175–176, 175*f*
Reflectance factor, 182–183, 184, 185–186, 198–199, 228, 234, 292–293, 337–338, 347*f*, 383–384, 388, 391–392, 393
Reflection coefficients
Fresnel, 45–46
scattering, 64–65, 322–323
thin film, 49–50

Refraction, 2–3, 43–44, 78–79, 144, 180–181, 184, 275–276, 509–510
double, 88, 270
index of, 20–21, 22, 24, 30, 34–35, 49–50
air, 301–302
effect on fluorescence, 275–276, 283
effect on reflection and transmission, 44–45, 144–145, 147–148, 160–161, 338, 385–386, 434, 502, 509–510
gloss standards, 398–400, 402–403
measurement of, 113, 137
prism, 78–79, 79*f*, 87
Refractive index, 137, 275–276
Remote sensing, 458–482
Reproducibility, 8, 126, 154–156, 164, 206–207, 250, 264, 284, 352, 375, 383, 385, 393–394, 431, 437
Resistive heating, rapid, 352–353
Resistivity, of liquids *vs.* solids, 349–350
Resolution. *See also* Bandpass function
angular, 294–295, 296, 298, 300
spatial, 298
wavelength, 68–69, 78, 81–82, 84–85, 101, 104, 105–106, 111, 112–114, 115–116, 117–118, 123, 124, 132, 135, 135*f*, 137, 138–139, 149–150, 383, 419, 426
Resolving power, 81–82, 104, 105–106, 111, 112
Retroreflection, 180–181, 306, 309
Robot, 307–308, 307*f*
Roughness, 33, 184, 202, 292, 322–325, 326–328, 334, 385–386, 404
large scale, 54–59
during oxidation, 348–349, 349*f*, 351
slightly rough, 59–65, 322–325

S

Sample holder, 3–4, 200, 213, 240, 243, 244–245, 303, 307–308, 316, 319–321, 339–340, 339*f*, 340*f*, 344, 359–361, 360*f*, 435
Sampling interval, 106–107, 124
Satellite-borne instruments
advanced spaceborne thermal emission and reflection radiometer (ASTER), 477, 479
airborne visible/infrared imaging spectrometer (AVIRIS), 458, 460, 461*f*, 471, 474–475, 474*f*
atmospheric infrared sounder (AIRS), 459–460
global ozone monitoring experiment (GOME), 477, 478
greenhouse gases observing satellite (GOSAT), 459–460, 462–464, 478, 481–482
landsat, 458, 461*f*, 477, 479
moderate resolution imaging spectroradiometer (MODIS), 392, 477
orbiting carbon observatory (OCO), 459–460, 461*f*, 462–463, 464, 464*f*, 478, 480–481
ozone monitoring instrument (OMI), 477, 478
SCanning Imaging Absorption spectroMeter for Atmospheric CHartographY (SCIAMACHY), 459–460
total ozone mapping spectrometer (TOMS), 475, 477, 478
Saturation. *See* Color, saturation
Scan length, 101–106
Scatter
total integrated (TIS), 202
total scatter (TS), 202
Scattering, 1–2, 12–13, 20–21
in angle-resolved measurements (*see* Instrument signature)
anisotropic surfaces, 325
elastic, 28, 36–37, 239
generalized scattering function, 27–28
grain boundary, 358–359
inelastic, 26–27, 184, 213–214
Lambertian, 33–34
models, 52–59
facet, 54–59
Kubelka–Munk, 52–54
Rayleigh–Rice, 59–65
multilayer mirrors, 292, 323
Rayleigh scatter, 301–302, 303, 316
roughness, 33, 54–65, 184, 202, 292, 322–325, 326–328
volume, 52–54, 53*f*, 184, 211, 318, 322–323
Schmidt, M., 200
Schwarzschild objective, 491
Secondary beams, 148, 160–161, 163
Self-absorption, 228, 229
Sellmeier equation, 78–79
Sensitivity
alignment and geometry, 120, 130, 135–136, 156–158, 262–263, 298, 388–389
analyte, 425, 447, 454
carbon dioxide (CO_2), 459
color, 381–383, 479–480
detector, 87–88, 90–91, 137, 149, 181, 262, 269, 298, 299, 300–301, 304, 338–339, 390, 436
human visual system, 369*f*, 376–377

instrument, 114, 300, 303, 309, 316, 442, 449, 479–480, 494
measurement, 224–225, 250, 261–262, 270, 299, 349–350, 447, 453–454, 494
optical depth, 474–475
polarization, 266–268, 317
spectral, 251, 261–262, 334–335, 397
Sequence, measurement, 154–158, 155*t*, 163–164, 167–168, 401
Shadow and obscuration function, 59
Sharp, C.H., 198–199
Shutter, 270, 299, 347–348, 347*f*
Signal to noise ratio (SNR), 91–92, 106–107, 110–112, 119, 120, 124, 126, 132–133, 134, 139, 213–214, 246, 353–354, 383, 425, 437, 442, 443, 450, 461–462
Singlet–singlet transition, 222
Size of source effect, 344
Skull melting, 354–355, 356–357
Snell's law of refraction, 30, 43–44, 46–47, 144, 180–181, 184, 275–276, 385–386
applied to prism, 78–81
Snow
Albedo, 471, 472–473, 474, 474*f*, 481–482
covered area, 471, 472
grain size, 470–473, 474, 474*f*
impurities, 470–471, 472, 473, 474–475, 481–482
SNR. *See* Signal to noise ratio (SNR)
Software, 104, 105–106, 110, 139, 146, 307–308, 415, 437, 447–448, 455, 479, 494–495, 496
Solar energy, 2, 3–4, 144
Solar occultation spectroscopy, 478
Solar protection, 216
Solid angle, 13, 15–17, 15*f*, 27, 31, 36–37, 43, 55–57, 105, 110–111, 115–116, 144–145, 183, 184–186, 185*t*, 292, 293–295, 296, 297, 298, 299, 303, 315, 317
and *f*/number, 70–71, 105–106
projected, 15*f*, 29, 201–202, 233–234, 294–295, 299
Snell's law, 43
Solutions, liquid, 85–86, 92, 230, 231, 243, 244–245, 258, 258*f*, 259*f*, 260, 263, 265, 266*f*, 275, 278, 282, 286, 410, 421, 422*t*, 423, 424, 424*t*, 426, 427, 428*f*, 428*t*, 429, 434, 444, 446, 447, 448, 449, 450, 451, 452–453, 507–508, 512
Solvent
blank, 231, 257, 270, 435
cleaning, 211, 243
effects, 257, 275, 283
impurities, 276, 286, 429, 453
oxygen in, 277
residual, 414, 434
solution, 92, 257, 258, 429, 444, 453
Solvent-matrix interactions, 257, 275, 283
Source, optical radiation, 3–4, 88–90
aging, 262, 384
blackbody, 89–90, 338, 340, 373, 479–480
in color measurements, 368–369, 372–373, 374, 375, 376–377, 384–386, 387–391, 390*f*, 391*f*, 393, 396, 397–399, 400, 401–403, 401*f*
deuterium (D_2), 84–85, 89–90, 89*f*, 115, 423, 427
fluctuations, 112, 124, 241–242, 299
for FTS, 114–115
globar, 114–115
laser, 87–88, 90, 98, 115, 239, 245–246, 247, 297*f*, 298–299, 300, 301–302, 302*f*, 309, 313–315, 316, 480–481, 482*f*
mercury xenon (HgXe), 115
Nernst glower, 89–90
quartz tungsten halogen (QTH), 89–90, 89*f*, 115, 279, 280*f*, 345–347, 347*f*, 389
silicon nitride (Si_3N_4), 345–347, 347*f*
size effect (*see* Size of source effect)
spectral distribution, 89*f*
standard, 278
synchrotron, 115, 316
tungsten, 89–90, 261, 261*f*, 279, 281, 373, 389–390, 491
xenon, 89–90, 89*f*, 241, 255*f*, 256, 261, 261*f*, 262, 281, 389–390
Spatial filter, 297*f*, 298–299
Spatial frequency, 65, 322–325
Specific absorbance, 36, 226*t*, 231
Speckle, 298, 313–314
Spectral analyzer, 68–70, 68*f*, 69*f*, 75, 76, 77, 78, 79–84, 87
Spectral concentration, 18, 67, 227, 235, 264–265
Spectral efficiency factor, 226*t*, 232–233, 239–240, 251, 264
Spectral irradiance. *See* Irradiance, spectral
Spectral power. *See* Power radiant, spectral
Spectral radiance. *See* Radiance, spectral
Spectral radiance factor. *See* Radiance factor, spectral

Spectral radiant intensity. *See* Intensity, radiant, spectral
Spectrofluorimeter. *See* Fluorescence, spectrophotometer
Specular excluded (SCE), 202, 250, 386–388
Specular included (SCI), 202, 212–213, 214–215, 250, 386–387
SRE. *See* Stray radiant energy (SRE)
Stability
  drug product, 410, 413, 414
  instrument, 124–125, 126, 138–139, 140, 266, 436, 443, 491
  mechanical, 106–107, 121
  photo-, 282, 283, 286, 413, 450
  response, 437
  sample, 205, 207, 208, 282, 284, 356–357, 401–402
  source, 89–90, 114–115, 313, 359–361, 389–390, 400, 402
  thermal, 124–125, 284, 345, 358–359, 362
Standard deviation, 6–8, 146–147, 154–156, 284, 303, 313–314, 442
Standard illuminant. *See* Illuminant
Standard observer, 368, 372*f*, 375, 384, 390–391, 396, 401–402
Stefan-Boltzmann law, 334–335
Stokes shift, 222–223, 275, 286
Stokes vector, 40–43, 316–317
Stray light, 3, 5, 68–69, 73, 81, 83, 84, 87–88, 134, 149, 153, 163–164, 167–168, 245, 251, 260, 269–270, 277, 292, 296, 298, 300, 314–315, 317, 321–322, 326, 393, 394, 421, 422*t*, 426–427, 427*f*, 428*f*, 428*t*. *See also* Stray radiant energy
Stray radiant energy (SRE), 73, 87, 88, 93*f*, 269–270, 426
  heterochromatic (HSRE), 88, 270
  isochromatic, 88, 269–270
Student's *t*-distribution, 6, 147, 154–156
Substitution methods, 136, 212, 213, 250, 258–259
Sulfur dioxide (SO_2), 475, 477–478, 479
Sumpner, W.E., 189
Surround, viewing, 370
Synchronous scan, 239

T

Taylor, A.H., 195
Technically corrected data (for fluorescence measurements), 228, 229*f*, 251, 258, 260, 283
Textiles, 54, 216, 386–387, 388–389, 397–398
Texture, surface, 2, 205, 210–211, 386–387, 388, 396
Thermal conduction, 334–335, 358
Thermal radiation, 334–337, 344, 351, 353–354. *See also* Emissivity; Planck's law
Thermochromism, 68, 205, 209, 213–214, 394
Thermometer, 334–335, 336, 339–340, 339*f*, 343, 350–351, 358–359
Three point support, 151, 359–361
Throughput, 77, 82, 110–111, 112–113, 114, 116, 121–122, 136–137, 166–167, 168, 191, 193, 212, 262, 280, 425, 426, 493
Time gate, 246, 277
Traceability, 4, 85–86, 139–140, 204, 205, 278, 284, 292, 312–313, 342–343, 343*f*, 383–384, 391–392, 394, 420, 421, 435–436, 437, 447, 449, 479–480
Translucency, 192, 207, 208, 214, 388
Transmission spectrometry, 433, 443–445
Transmittance, 1–2, 4, 20–21, 30–32, 34–35, 36, 38–39, 93
  atmospheric, 461*f*
  bidirectional, 317
  color, 367, 370, 375, 383, 384, 390–391, 401–402
  diffuse, 32, 65, 181, 183–184, 186–187, 187*t*, 212, 213, 214–216, 215*f*
  in fluorescence measurements, 229, 230, 238, 239, 254, 255*f*, 256, 261–262, 261*f*, 266*f*, 270, 276
  fresnel, 45–46, 49–51
  in FTS, 110, 126, 128–129, 128*f*, 129*f*, 130, 130*f*, 131, 132–133, 132*f*, 134, 136–138, 138*f*, 153–154, 156–158
  Kubelka–Munk, 53–54
  in microspectrophotometry, 494, 496, 509
  Mueller matrix, 42
  normal incidence, 147–148, 150–158
  oblique incidence, 158–170
  in pharmaceutical measurements, 424, 427–428, 433, 435, 436–437, 438, 445, 448
  recommended geometries, 187*t*
  regular, 144, 145–146, 150–170, 171–176
  standards, 85–86, 87*f*
Transmittance factor, 184
Triplet–singlet transition, 222–223
Tristimulus values, CIE, 234, 253, 285, 371–372, 372*f*, 375–376, 377, 378, 383–384, 392–393

True spectra (for fluorescent measurements), 228, 276, 279, 280*f*, 282–283
Tungsten lamp. *See* Source, optical radiation, quartz tungsten halogen (QTH)
Turbidity, 277, 388. *See also* Scattering
Two-monochromator method, 238, 240, 250–251, 260–261, 264–265, 270–271, 284

U

Ulbricht, F.R., 190
Ulbricht sphere. *See* Integrating sphere
Uncertainty
- analysis, 4, 5–6, 311–315, 393
- BRDF measurements, 300–301, 303, 311–315
- color measurements, 392–393
- combined, 5
- covariance, 5–6
- diffuse reflectance, 192, 203–204, 214
- emissivity measurements, 342–343, 344–345
- expanded, 6
- fluorescence measurements, 262–263, 265–266, 278
- FTS, 123, 124–125, 125*f*, 126, 130, 132–134, 135–139, 135*f*, 138*f*
- general discussion, 4–8
- gloss measurements, 400
- haze measurements, 402
- pharmaceutical measurements, 420, 421, 435–437, 448
- regular reflectance and transmittance, 145, 146–147, 151, 154–158, 159, 160, 164, 165, 166*f*, 168, 169*f*, 171–172, 173, 176
- remote sensing, 469, 479–480
- standard, 5–6
- types A and B, 6–8

Uncorrected data (for fluorescence measurements), 228, 236*f*, 238–240, 284–285, 285*f*
Uniformity
- detector, 90–91, 92, 132–133, 135–137, 139, 313, 314–315
- material, 204, 205, 207, 260, 413, 440
- source, 90–91, 92, 132–133
- temperature, 345, 358–359

V

Validation, 410–411, 414, 415, 417–418, 419, 420, 424, 426, 430, 433, 438–439, 445, 453–454, 455, 459–460, 471, 478–479, 500
van den Akker, J.A., 199
Vector perturbation theory, 59, 322–323
Verification, 264, 283, 284, 410–411, 418, 430, 436–437, 445, 449, 453–454, 455
Vibrations
- lattice, 116, 137, 357
- mechanical, 112, 119, 122–123, 124, 323, 325
- molecular, 222–223, 433–434, 440–441

Volcanic ash, 475–479, 476*f*
Volcanos, 458, 475–478, 479

W

Water vapor, 121, 135, 436–437, 463, 465, 467–468, 472, 477–478
Wavelength scale. *See* Calibration, wavelength
Wavenumber, definition, 98–99
Wavenumber scale. *See* Calibration, wavelength
Waves
- plane, 23–24
- spherical, 24–25

Welch–Satterthwaite formula, 6–7
White light interferometry (WLI), 323–325, 324*f*, 326*f*
Whiteness, 224, 232, 253, 271, 285, 385
Wien's displacement law, 335–336
Wood's anomalies, 76, 76*f*, 238, 267, 268–269, 281

Z

Zeiss Elrepho®, 208, 214
Zero path difference (ZPD), 98–99, 101–103, 107, 110, 117–118, 119, 120, 126, 128*f*, 140, 341–342

FSC
www.fsc.org
MIX
Papier | Fördert gute Waldnutzung
FSC® C083411

Kolibri 360 GmbH
Ferdinand-Jühlke-Straße 7
99095 Erfurt, Deutschland
produktsicherheit@kolibri360.de